Nuclear Magnetic Resonance Spectroscopy of Liquid Crystals

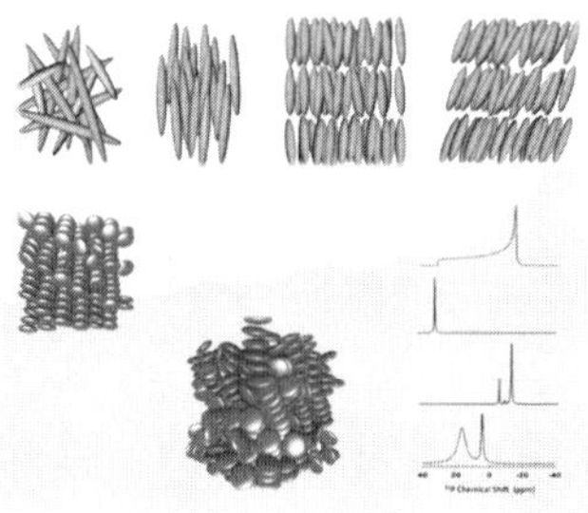

Nuclear Magnetic Resonance Spectroscopy of Liquid Crystals

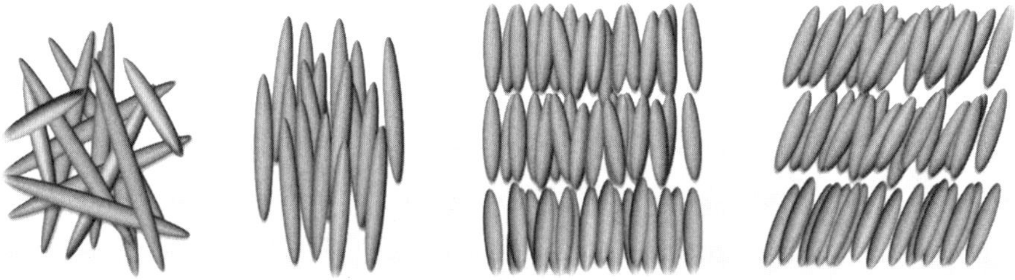

Editor

Ronald Y. Dong

University of British Columbia, Canada

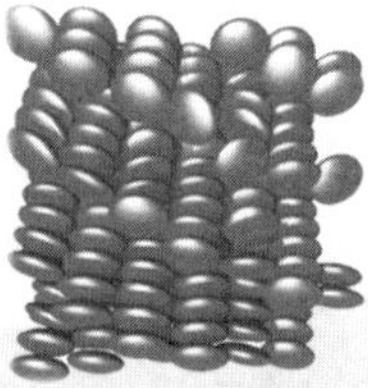

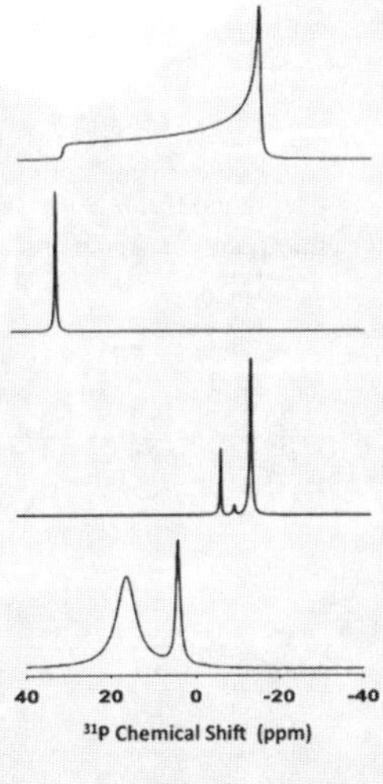

World Scientific

NEW JERSEY • LONDON • SINGAPORE • BEIJING • SHANGHAI • HONG KONG • TAIPEI • CHENNAI

Published by

World Scientific Publishing Co. Pte. Ltd.
5 Toh Tuck Link, Singapore 596224
USA office: 27 Warren Street, Suite 401-402, Hackensack, NJ 07601
UK office: 57 Shelton Street, Covent Garden, London WC2H 9HE

British Library Cataloguing-in-Publication Data
A catalogue record for this book is available from the British Library.

NUCLEAR MAGNETIC RESONANCE SPECTROSCOPY OF LIQUID CRYSTALS

ISBN-13 978-981-4273-66-4
ISBN-10 981-4273-66-X

Printed in Singapore by Mainland Press Pte Ltd

Preface

The diversity of Liquid Crystals is evident in everyday life in incarnations ranging from liquid-crystal displays to biological membranes. It is therefore important to have a good understanding of these materials to further the progress in this field of research and development. To achieve this, no single technique, whether it is theory or experiment, can provide a clear understanding of the fundamental physics and chemistry of liquid crystals. There is no doubt that the liquid-crystal field is interdisciplinary in nature and requires the expertise of biologists, chemists, physicists, engineers and materials scientists. Indeed it also provides a ground for cross-fertilization of different research areas to achieve new potential applications in the fields of telecommunication, medicine, memory devices and beyond. Nuclear magnetic resonance is one of the experimental techniques that were used early on (in the early nineteen sixties) to unravel and harness the intrinsic anisotropic potential that exists in these materials. Given the number of NMR active nuclei, the wealth of modern NMR pulse techniques and the technical advances in NMR instrumentation, the happy marriage between NMR and liquid crystals is clearly demonstrated in this volume. The Editor has written a text that bears the same name as this volume some sixteen years ago; he mentioned then the limitation of a single-author book is necessarily hampered by the limited research interests of the writer. However, the main aim was to lay down the theoretical NMR and liquid-crystal physics background for the benefit of newcomers to the fields of NMR and Liquid Crystals. An indication of the success of that aim is that the text is being used by many graduate students and other researchers. The present edited volume certainly covers more diverse applications of NMR in liquid crystals, and it is hoped that this can serve as a compendium to the earlier text.

The volume contains the latest state-of-the-art developments in using NMR to study liquid crystals. Small solutes (as well as not so small proteins) in thermotropic and lyotropic liquid-crystal phases are discussed in

Chapters 1 to 4. These chapters are mainly focused on observed spectroscopic (equilibrium) properties. These are followed by chapters on the liquid-crystal molecules using deuterium, proton or carbon-13 nuclei in these molecules as probes to sense dynamical properties and/or static properties like the orientational ordering and phase structure. The question of quenched disorder of a liquid crystal confined in silica aerosol gel networks is studied in Chapter 7 by using deuterium probes located within the liquid crystal via chemical deuteriation. Chapters 5 and 8 deal with proton and deuterium spin relaxation, respectively, of numerous different liquid-crystalline phases formed by molecules of increasingly complex anisometric shapes. Chapter 6 deals with deuterium NMR of chiral tilted phases and the effect of the NMR field. Chapter 9 describes a proton study of self-diffusion of liquid-crystal molecules by utilizing the large fringe field gradient in a superconducting magnet. Chapter 10 describes a means of combining the external electric field and the NMR field to probe the director reorientational dynamics in a deuterated nematic liquid crystal. The director dynamics and individual molecular motions (translation or reorientation) are closely tied to the elastic and viscous properties of liquid crystals. Chapter 11 provides a survey of theoretical approaches to treat viscoelastic properties of liquid crystals and to give possible relations between theoretical predictions and NMR observables. The remaining two chapters have to do with carbon-13 NMR of calamitic liquid crystals. The former one deals with the study of orientational order parameters, and molecular conformation in rod-like and bent-core mesogens. The latter one provides a combined density functional theory and NMR study of a specific bent-core liquid-crystal molecule that is known to exhibit in bulk a uniaxial and a biaxial nematic phase upon decreasing temperature.

The scope of this volume is necessarily limited by its size and the expertise of the contributors. However, it is our hope that it can provide a glimpse of the latest news in the field of Liquid Crystals, in particular what NMR can offer to advance the understanding of the underlying principles in this type of soft matter. The Editor is most grateful to all the contributors (and friends) to make his dream come true. He is also indebted to his wife Natalia for putting up with him days and nights as well as weekends on this book project.

Ronald Y. Dong
UBC, Vancouver, May 1, 2009

Contents

Chapter 1

Novel Strategies for Solving Highly Complex NMR Spectra of Solutes in Liquid Crystals

E. Elliott Burnell[1], Cornelis A. de Lange[2], and W. Leo Meerts[3]

[1]*Chemistry Department, University of British Columbia, 2036 Main Mall, Vancouver, BC, Canada V6T 1Z1*
email: elliott.burnell@ubc.ca

[2]*Laser Centre, Vrije Universiteit, De Boelelaan 1081, 1081 HV Amsterdam, The Netherlands*
email: cdelange@few.vu.nl

[3]*Molecular and Biophysics Group, Institute for Molecules and Materials, Radboud University Nijmegen, P.O. Box 9010, 6500 GL Nijmegen, The Netherlands*
and
Physical Chemistry Department, Vrije Universiteit, De Boelelaan 1083, 1081 HV Amsterdam, The Netherlands
email: leo.meerts@science.ru.nl

The application of liquid-crystal NMR to solutes with more than eight spins has for decades been hampered by the extremely high complexity of the spectra. We describe recent experimental and novel theoretical advances to address the central problem of spectral analysis. Experimentally the application of multiple-quantum NMR is shown to be a powerful tool. Moreover, theoretical algorithms from evolutionary biology are applied to NMR of solutes in anisotropic environments. These novel spectral fitting procedures are aided by estimates of the solute degree of orientational order, based on simple phenomenological size and shape models. The apparent success of these approaches has considerably extended the realm of applications of NMR in ordered liquids.

1.1. Introduction

The Hamiltonian that describes the high-resolution NMR spectra of solutes dissolved in anisotropic solvents is well known and has proved to be an excellent predictor for such detailed experimental information. The parameters that determine the NMR-spectra include the chemical shieldings and indirect spin-spin couplings that are well known from NMR in isotropic liquids. Additionally, direct dipole-dipole and quadrupolar couplings are not averaged to zero in anisotropic phases and usually dominate the spectra of solutes in such solvents. The complexity of these NMR-spectra increases rapidly with the number of interacting nuclear spins. In this chapter we shall limit ourselves to ^{1}H NMR spectra.

We shall limit our discussion to the use of nematic liquid-crystal solvents that are apolar and possess uniaxial symmetry. There exists a large literature on the study of a great variety of solutes in these anisotropic phases. For simple solutes that contain a small number of nuclear spins and a high degree of symmetry, the problem posed by the spin Hamiltonian can be solved analytically.[1,2] For larger solutes with less symmetry this is no longer possible and other means have been developed. Commonly computer programmes are employed that calculate a spectrum from estimated input values for chemical shieldings, and indirect and direct dipole-dipole couplings. The calculated spectrum is then compared to the experimental one to see whether some degree of correspondence between calculated and experimental spectra can be detected. If this is the case, least-squares fitting procedures are employed for adjusting the spectral parameters in such a way that the correspondence is improved. In an iterative fashion this process hopefully converges, and good agreement between calculated and experimental spectra is obtained. If an insufficient number of transitions in the calculated spectrum can be sensibly assigned to experimental ones, the procedure is doomed to failure. In practice this approach is often notoriously time-consuming and usually limited to solutes with fewer than 8 spins.

The problems are exacerbated for molecules with low symmetry that may require as many as five independent and *a priori* unknown orientational order parameters, or for molecules that exist in several conformers that undergo fast interconversion on the NMR time scale. For such cases even crude guesses of the dipolar couplings are difficult to obtain. In addition, the anisotropic contributions to the chemical shifts make it difficult to predict values for solutes dissolved in nematic phases.

In this chapter we shall discuss both experimental and theoretical methods of considerable sophistication to address the problem of analyzing very complex spectra of large and sometimes flexible solutes in liquid-crystal solutions, including those that undergo extensive conformational change. Experimentally we shall employ the method of multiple-quantum (MQ)

NMR to develop a strategy that leads in a step-wise fashion to an analysis of the normal one-quantum NMR spectrum. Application of MQ-NMR to the highly complex spectra of solutes in nematic phases that could not be solved otherwise has shown the inherent power of this method. Several successful examples of the MQ approach will be given.

A drawback of the MQ-method is its experimental sophistication whose implementation requires detailed technical knowledge. For many years attempts have been made to develop automated spectral fitting routines, but with limited success. Recently novel and extremely powerful methods of analyzing extremely complicated NMR spectra of solutes in nematic phases have been developed. These methods are based on Evolutionary Algorithms (EA's). We use in particular Genetic Algorithms (GA's) and Evolutionary Strategies (ES's) which can be implemented into automated spectral fitting procedures that are capable of reliably locating the global minimum in an extensive search space.

It is important to realize that the novel experimental and theoretical developments that will be discussed in this chapter have led to a significant extension of the scope of liquid-crystal methods to obtain detailed information about ordered solutes that was previously thought unattainable. In the archives of many NMR laboratories there are numerous examples of highly complex NMR spectra of partially ordered solutes that never yielded to serious, painstaking and time-consuming attempts to solve them. With the present techniques the study of all kinds of interesting and rather large solutes in anisotropic liquids has now become eminently feasible, as will be shown in the following. In that sense the NMR liquid-crystal method has obtained a new lease of life.

1.2. Theory

In order to calculate high-resolution single-quantum NMR spectra of solutes in anisotropic solvents the starting point, as always in quantum mechanics, is a Hamiltonian. As is common in molecular spectroscopy we define a basis set consisting of products of nuclear spin wave functions, set up the spin Hamiltonian matrix, and perform matrix diagonalization. Transitions between matrix eigenvalues represent spectral frequencies, while intensities are obtained by calculating the squares of matrix elements of the magnetic-dipole transition operator sandwiched between pairs of eigenvectors obtained in the diagonalization procedure.[1,2]

The spin Hamiltonian (in Hz) for a collection of spin $I = \frac{1}{2}$ nuclei in a solute orientationally ordered in a liquid-crystalline solvent is

$$\begin{aligned}\mathcal{H} = &-\frac{B_Z}{2\pi}\sum_{\mu}\left[\gamma_\mu(1-\sigma_\mu^{\text{iso}}-\sigma_\mu^{\text{aniso}})\right] I_{\mu,Z} \\ &+\sum_{\mu}\sum_{\nu>\mu} J_{\mu\nu}^{\text{iso}}\mathbf{I}_\mu\cdot\mathbf{I}_\nu \\ &+\sum_{\mu}\sum_{\nu>\mu}(2D_{\mu\nu}+J_{\mu\nu}^{\text{aniso}})\left[I_{\mu,Z}I_{\nu,Z}-\frac{1}{4}(I_{\mu,+}I_{\nu,-}+I_{\mu,-}I_{\nu,+}\right]\end{aligned} \tag{1.1}$$

where B_Z is the external magnetic field, γ_μ is the gyromagnetic ratio of nucleus μ, σ_μ is the chemical shielding (which has isotropic and anisotropic terms), and $J_{\mu\nu}$ is the indirect spin-spin coupling between nuclei μ and ν. For protons the anisotropic part J^{aniso} is to an excellent approximation zero.

The dipolar coupling between nuclei μ and ν is:

$$D_{\mu\nu} = -\frac{h\gamma_\mu\gamma_\nu}{4\pi^2}\left\langle r_{\mu\nu}^{-3}\left(\frac{3}{2}\cos^2\theta_{\mu\nu,Z}-\frac{1}{2}\right)\right\rangle \tag{1.2}$$

where $r_{\mu\nu}$ is the internuclear separation and $\theta_{\mu\nu,Z}$ is the angle between the $\mu\nu$ and magnetic field directions, and the angle brackets denote an average over solute internal and reorientational motions. With the assumption that the molecule rapidly interconverts among several 'rigid' conformers Eq. (1.2) can be written

$$D_{\mu\nu} = -\frac{h\gamma_\mu\gamma_\nu}{4\pi^2}\sum_i p_i r_{i,\mu\nu}^{-3}\sum_{k,l} S_{i,kl}\cos\theta_{\mu\nu,k}\cos\theta_{\mu\nu,l} \tag{1.3}$$

where i labels the conformer which occurs with probability p_i and $\theta_{\mu\nu,k}$ is the angle between the conformer-fixed k axis and the vector joining nuclei μ and ν which have internuclear separation $r_{i,\mu\nu}$. $S_{i,kl}$ is the kl element of the order parameter tensor for conformer i

$$S_{i,kl} = \left\langle\frac{3}{2}\cos\theta_{k,Z}\cos\theta_{l,Z}-\frac{1}{2}\delta_{kl}\right\rangle_i \tag{1.4}$$

where the angle brackets denote an average over conformer reorientational motions and $\theta_{k,Z}$ is the angle between the conformer-fixed k axis and the space-fixed Z axis which for the liquid crystals used here lies along both the director and magnetic field directions.[3] In this picture small-amplitude vibrational corrections and vibration-reorientation coupling are neglected.

It is useful to distinguish among three different categories of solutes: (i) 'Rigid' solutes with a single conformation that only undergo normal internal vibrational motion. The degree of orientational order of such solutes, depending on solute symmetry, is determined by at most five independent order parameters S_{kl}. Examples are azulene and biphenylene (Fig. 1.1); (ii) Solutes that undergo conformational change between conformers that

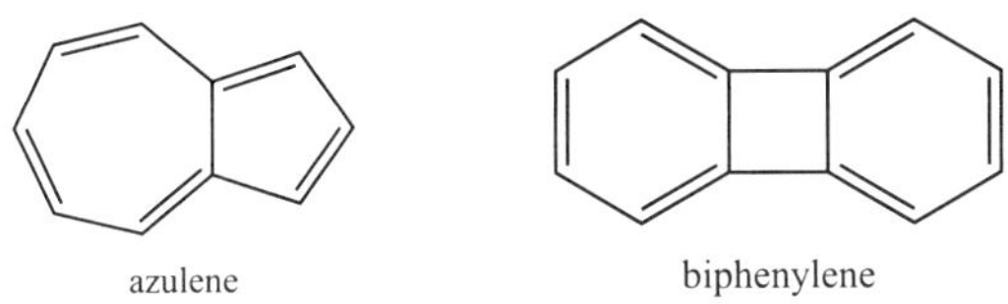

Fig. 1.1. The molecules azulene and biphenylene.

are related by symmetry. Although each conformer in principle requires up to five independent order parameters, symmetry confines this number. Examples are 2,6-dichloro-1-ethenylbenzene, biphenyl and *p*-bromo-biphenyl (Fig. 1.2); (iii) Solutes that interconvert rapidly among several symmetry-

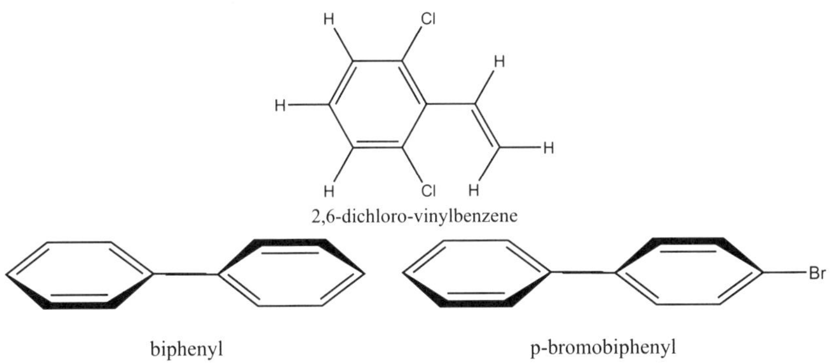

Fig. 1.2. 2,6-dichloro-1-ethenylbenzene, biphenyl and *p*-bromo-biphenyl.

unrelated conformers. Each conformer i requires its own set of independent order parameters $S_{i,kl}$. Examples are butane (C_4H_{10}, Fig. 1.3) and pentane (C_5H_{12}, Fig. 1.3).

Since conformational probabilities and order parameters for each conformer only occur as products in the expressions for the dipolar couplings, a separation can only be achieved if we have some source of independent information about either property. From a large body of research it is now accepted wisdom that a key contribution to the average degree of orientational order is the short-range interaction between the solute and the solvent

Fig. 1.3. The molecules butane (C_4H_{10}, left) and pentane (C_5H_{12}, right).

environment. An important breakthrough in understanding the physical processes underlying partial orientation in anisotropic phases was achieved by the introduction of so-called 'magic mixtures' in which the short-range contribution is deemed to be by far the dominant one. Moreover, simple phenomenological models based on the size and shape of the solute have been developed that can predict order parameters at approximately the 10% level. For an overview of this highly significant development we refer to the literature.[4,5] Clearly, by employing one of the simple size and shape models for estimating order parameters developed by several research groups, a very useful tool for restricting the possible values of the dipolar couplings is available to us. Employing such models has been a crucial factor in the successes of our 1Q high-resolution and multiple-quantum NMR investigations, as well as to the automated fitting procedures.

1.3. Multiple-quantum NMR

The normal single-quantum NMR spectrum consists of transitions between spin energy states that involve the flip of a single spin. For high-spin systems there are many such transitions. The idea of MQ-NMR is to use transitions that involve the mutual flip of many spins. When the number of spins flipped approaches the number of spins in the molecule, the number of possible transitions is greatly reduced. Weitekamp wrote an excellent review that covers a wide range of ideas and experiments.[6] Here we concentrate on the application to the analysis of the high-resolution NMR spectra of solutes orientationally ordered in liquid-crystal solvents, where the technique is used for spectral analysis. We use examples mainly from our own work to demonstrate the use and limitations.

The idea is exemplified by the example of benzene,[6] Fig. 1.4. The normal 1Q spectrum consists of 74 lines which show no regular pattern. The 6Q spectrum is a single line, involving the flip of all six spins. The 5Q spectrum is a doublet because all spins are equivalent: there is one transition involving $|\alpha\alpha\alpha\alpha\alpha\alpha\rangle \rightarrow |\alpha\beta\beta\beta\beta\beta\rangle$ and another involving $|\alpha\alpha\alpha\alpha\alpha\beta\rangle \rightarrow |\beta\beta\beta\beta\beta\beta\rangle$.

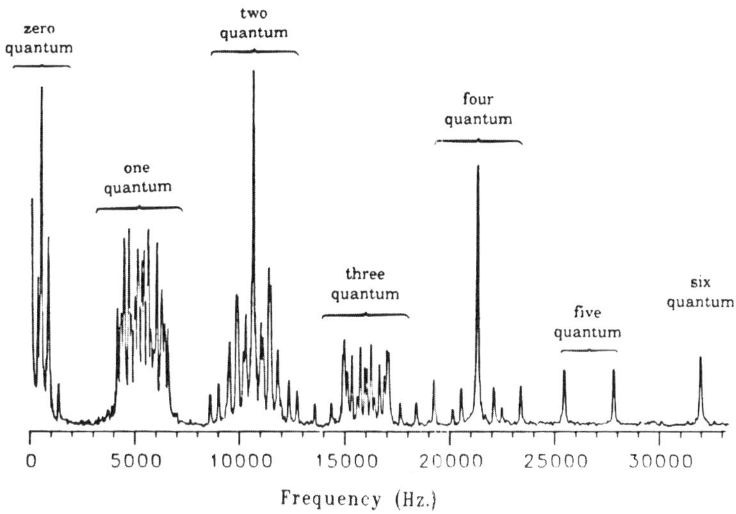

Fig. 1.4. The multiple-quantum spectrum of benzene dissolved in N-(4-ethoxybenzylidene)-4'-*n*-butylaniline (EBBA) obtained using pulse sequence A of Table 1.1. Reproduced with permission from Ref. [7].

The 4Q spectrum consists of three triplets because there are three ways of choosing the two spins that are not flipped, being those that are either *ortho*, *meta* or *para* to each other. The central line of this triplet involves 4Q coherences between spin states such as $|\beta\alpha\alpha\alpha\alpha\alpha\rangle$ and $|\alpha\beta\beta\beta\beta\beta\rangle$. It can be shown that all spectral parameters can be obtained by analysis of the three highest order and relatively simple MQ spectra.[6]

Of course in the case of benzene the 1Q spectrum is readily analyzable with the aid of a computer — in this case computer analysis is 'trivial' because of the high D_{6h} symmetry and hexagonal structure which means there is essentially a single unknown (the order parameter S) needed to predict the three independent dipolar couplings and a single chemical shift. As we shall discuss below, for more complicated spin systems, the MQ-NMR spectra are an invaluable tool to aid spectral analysis.

1.3.1. *The MQ experiment and some variants*

Since the selection rule for NMR spectroscopy is, in general, that only 1Q transitions are directly observable, the MQ spectra must be obtained indirectly. In general, any three-pulse sequence (see Fig. 1.5) can be used

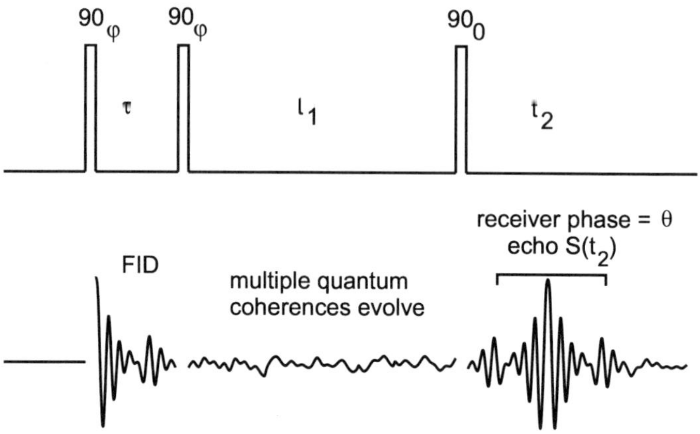

Fig. 1.5. The pulse sequences discussed are all variants of the sequence shown, with phase cycling as described in Table 1.1. Sequence A is a simple example of a generic three-pulse sequence, any of which can in general produce a MQ-NMR spectrum. In general, some of the MQ-NMR coherences that are generated by the second pulse, and that evolve during the variable time t_1, are transferred back into observable coherences by the third pulse. The MQ-NMR spectrum is obtained by Fourier transforming as a function of t_1 the echo signal that follows the third pulse. In practice this is accomplished via a 2D transform of the signals acquired as a function of t_1 and t_2, and projecting the absolute value of the resultant spectrum onto the F_1 axis to give the MQ-NMR spectrum, as shown in Fig. 1.4. (Reproduced from Ref. [8] with permission from Academic Press.)

to generate the MQ spectrum. The first pulse generates observable X and Y magnetizations from equilibrium Z magnetization. During a waiting time τ the X and Y magnetizations evolve under the spin Hamiltonian, and other unobservable 1Q coherences are generated by the dipolar and spin-spin parts of this Hamiltonian. The second pulse converts these other coherences into MQ coherences of all possible orders. Between the second and third pulses the MQ coherences evolve under the spin Hamiltonian. An important feature of this evolution is that the effect of chemical shift offset (in the rotating-frame) is enhanced by the MQ order such that NQ coherences evolve proportional to N times this offset frequency. The third pulse serves to rotate some of the MQ coherences back to unobservable 1Q coherences, which then further evolve under the Hamiltonian to form an observable echo at a time τ following the third pulse. A Fourier transform of the echo amplitude as a function of t_1, the time between pulses 2 and 3, gives in general a spectrum containing all MQ orders as shown in Fig. 1.4. In practice, a two-dimensional experiment is performed in which an echo

Table 1.1. Phase cycles for the three pulse sequences[a].

sequence	phase	phase list
A[b]	ϕ	0
	θ	0
B[c]	ϕ	0, $90/N$, $2 \times 90/N$, $3 \times 90/N$ .... $(4N-1) \times 90/N$
	θ	0, 90, 180, 270
C[d]	ϕ	0, $360/n$, $2 \times 360/n$, $3 \times 360/n$ .... $(n-1) \times 360/n$
	θ	0

[a] As defined in Fig. 1.5, ϕ refers to the phase of pulses 1 and 2, and θ refers to the receiver phase. (Reproduced from Ref. [8] with permission from Academic Press.)
[b] Sequence A gives all MQ orders in a single, wide spectrum (see *e.g.* Fig. 1.4). For this experiment the phases ϕ and θ can be any arbitrary constant angle. Zero degrees was chosen for convenience.
[c] Sequence B is used for the selective detection of the NQ (and $2NQ$, $3NQ$...) spectrum. N is the quantum order being selectively detected.
[d] Sequence C is used for the 3D experiment, where n is an arbitrary integer that should be at least $2N_{\text{max}}+1$. A separate signal is acquired and stored for each value of n, giving a three-dimensional interferogram that is a function of t_1, ϕ, and t_2.

signal is collected for each increment of t_1, and the absolute value of the double Fourier transform with respect to t_1 and t_2 is projected onto the F_1 axis to give the MQ spectrum. Collection of the entire echo in this manner both improves the signal-to-noise ratio and sometimes reveals more transitions in the MQ spectrum than are observable with simply using the echo amplitude.

Historically, as shown in Fig. 1.4, sequence A of Table 1.1 was used with a large frequency offset to spread out the various MQ orders which precess during time t_1 (between pulses 2 and 3) at N times this offset. The large spectral width required to separate the various orders leads to a very wide spectral width with poor digital resolution, and even to problems in exciting the entire spectrum with the pulse powers and widths normally used.[8]

Several schemes have been employed to improve the digital resolution by the selective detection of one order at a time. In one method, field gradient pulses are applied before and after the third pulse.[9] The relative integrated amplitude of these pulses is chosen such that before the third pulse, the effective Z gradient adds $N\nu$ to the offset, which is recovered after pulse 3 by a gradient N times as large. Only the NQ coherences are coherently added to the echo signal, the others being destroyed by the gradients which are set to refocus only the Nth quantum order.

An alternative method is to phase-cycle the first and second pulses and the receiver phase in the manner described by sequence B of Table 1.1. In this way only the desired quantum order (and multiples of it) survive the signal addition and contribute to the MQ spectrum.[10] In addition, only the $+N$Q or $-N$Q spectrum is obtained (see Fig. 1.6), and one can work on-resonance. The spectral width is then much narrower, and thus the digital resolution can be much improved. One problem with this and the previous gradient technique is that the experiment measures a single quantum order, and must be repeated many times in order to collect each MQ order independently; the collection of several orders requires a lot of time.

A solution to the problem of needing a separate experiment to collect each order is to collect all orders at the same time in a 3D experiment, as suggested by Weitekamp[6] and as demonstrated by Syvitski *et al.*[8] The idea is (for each time t_1) to collect an echo for each value of the phase ϕ of pulses 1 and 2 which are varied between 0 and 360° in n increments, sequence C of Table 1.1. Fourier transformation with respect to the phase ϕ gives planes of signal F_1, F_2 where the various MQ spectra are the projections onto the F_1 axis, as demonstrated in Fig. 1.7. As can be seen in Fig. 1.6, the 4Q and 6Q spectra are virtually identical to those collected using the selective detection method of sequence B in Table 1.1: the important difference is that it took the same time using the 3D sequence to collect spectra for *all* orders as it did to collect a single order in the selective experiment.

1.3.2. *Spectral analysis with MQ*

As pointed out in the introduction, with fewer than about 8 spins $I = \frac{1}{2}$, especially for solutes with high enough symmetry (for which there are only one or two independent order parameters) the spectrum can usually be solved using the high-resolution 1Q spectrum and line assignment techniques. More complicated systems are difficult, and the difficulty increases with increasing the number of spins, with lowering the symmetry, and with solute flexibility. Several examples where analysis of the MQ solute spectra proved invaluable, and for which it is not clear that the 1Q spectrum could have been analyzed, are the following.

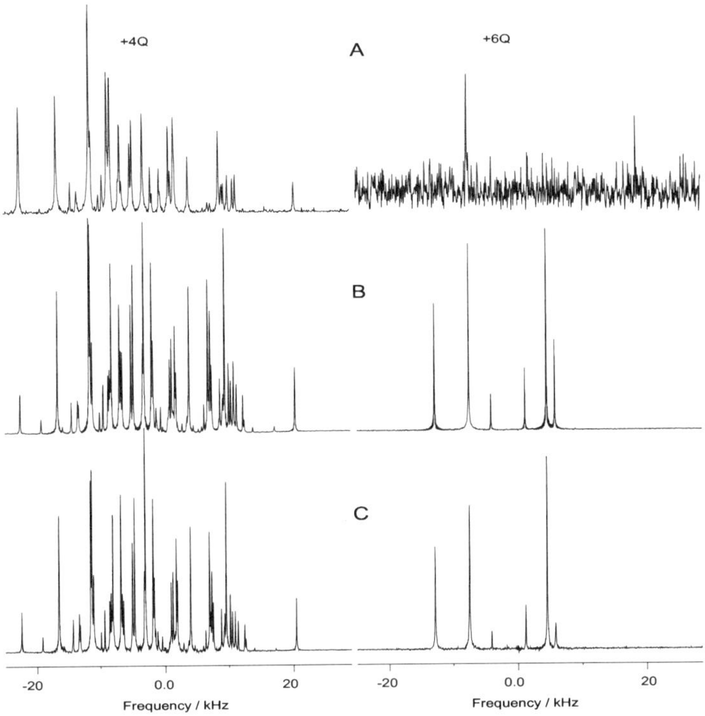

Fig. 1.6. Comparison of the $+4Q$ and $+6Q$ MQ-NMR spectra of *para*-chlorotoluene in ZLI-1132 obtained with different techniques. (A) Two sections of the spectrum collected in 14 h 5 m using sequence A of Table 1.1 with τ set to 20 ms, four scans/t_1 increment, 8192 t_1 increments, F_2 spectral width of 83333 Hz, and F_1 spectral width of 1MHz. The absolute value of the projection of the complex F_2 spectra onto the F_1 axis gives the F_1 spectrum, which is the entire MQ-NMR spectrum. The spectrometer frequency was offset by 60000 Hz from the center of the spectrum in order to separate the various orders. Under these conditions, the higher order coherences are poorly detected, presumably due to the combination of the large offset with the long duration (10 μs) of the 90° pulses. The line at ≈18 kHz in the 6 Q spectrum is an artifact. (B) The absolute value of the spectrum obtained from a selective detection of the +4 Q complex spectrum requiring 14 h 10 m (and the +6 Q complex spectrum requiring another 10 h 40 m) obtained using sequence B of Table 1.1 with N equal 4 (and 6), τ set to 20 ms, two (and one) scans for each of 16 (and 24) phase increments per t_1 increment, 1024 t_1 increments, F_3 spectral width of 33333 Hz, and F_1 spectral width of 70 kHz. (C) Two slices of the absolute value of the 3D experiment (displayed in Fig. 1.7) obtained using sequence C of Table 1.1. The intensity differences between the 6 Q spectra obtained using sequences B and C are most likely due to spectrometer fluctuations. For comparison, the ratios of vertical scales for the 6 Q to the 4 Q spectra are 17 × for (A), 1.35 × for (B), and 1.42 × for (C). The t_1 signals were zero filled once before the Fourier transform. The digital resolution is then 61 Hz per point in (A) and 34 Hz per point in (B) and (C). The peaks of very strong lines have been chopped off for clarity. (Reproduced from Ref. [8] with permission from Academic Press.)

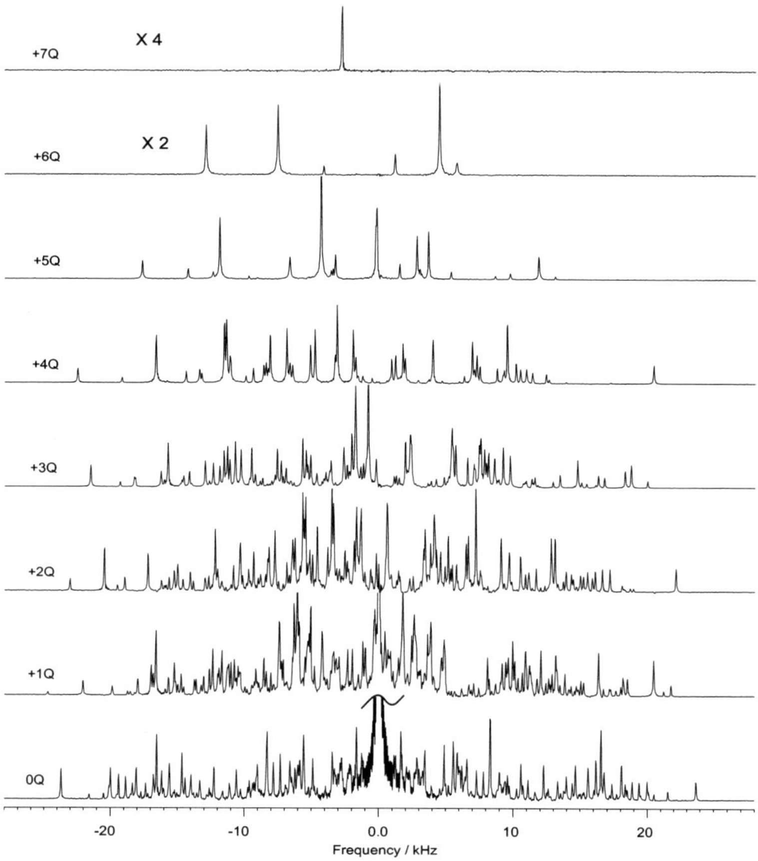

Fig. 1.7. Absolute values of multiple-quantum spectra of *para*-chlorotoluene in ZLI-1132 obtained in 14 h 25 m from the 3D experiment using sequence C of Table 1.1 with τ set to 20 ms, two scans for each of n=16 phase increments per t_1 increment, 1024 t_1 increments, F_3 spectral width of 33333 Hz, and F_1 spectral width of 70 kHz. The Fourier transforms F_3 of the echo signals $S(t_2)$ have been summed onto the F_1 dimension. The result is a two-dimensional spectrum where F_2 are the pseudo ϕ spectra, and F_1 the desired multiple quantum spectra. The positive MQ-NMR orders (F_1 slices 9 to 16) are shown, which correspond to the 0 to $+7Q$ spectra. The peaks of very strong lines have been chopped off for clarity. The Bruker software multiplies the first point of $S(t_2)$ by $\frac{1}{2}$ to correct the rolling baseline caused by the staggered acquisition of real and imaginary signals. While this is important for normal high-resolution NMR experiments, for the 3D MQ NMR experiment it results in small amounts of all MQ orders showing up in the spectrum for each MQ order: it is necessary to set FCOR = 1 in the Bruker software[11] in order to correct this problem and to obtain MQ spectra that contain only one order. In the spectra displayed, the extra signal from all MQ orders was removed by subtracting the signal obtained from an order higher than $N_{\max}$. (Reproduced from Ref. [8] with permission from Academic Press.)

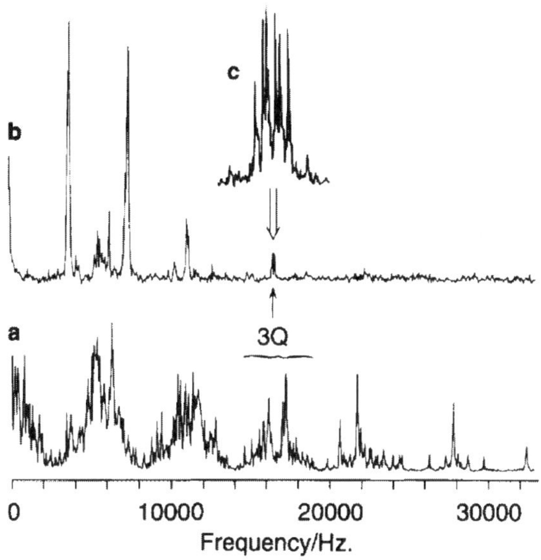

Fig. 1.8. (a) The 400 MHz nonselective MQ-NMR spectrum of 1,3-dichloro-2-ethenylbenzene dissolved in EBBA acquired using pulse sequence A of Table 1.1. (b) The 400 MHz selective excitation MQ-NMR spectrum of 1,3-dichloro-2-ethenylbenzene dissolved in EBBA using pulse sequence A of Table 1.1 with the first two pulses being frequency selective for the aromatic protons (see text). (c) An expanded view of the 3Q region noted in (b). (Reproduced from Ref. [12] with permission from Academic Press.)

1.3.2.1. *2,6-dichloro-1-ethenylbenzene*

This solute (see Fig. 1.2) has only six protons and hence it might be expected to produce an orientationally ordered six-spin NMR spectrum that is easy to analyze. That is not the case. One problem is that the ethenyl group makes an unknown dihedral angle with the benzene ring, and the molecule has no symmetry when this group is out of plane. Hence, there are five unknown order parameters. In addition, the anisotropies in chemical shifts of the ethenyl protons lead to problems in assigning initial values to these shifts. It turns out that assigning the wings of the spectrum to the aromatic protons is easy, and all the problems lie in the cluttered central clump of transitions. This central part from the ethenyl protons was only assigned once the MQ spectrum was obtained and solved.[12,13]

This molecule was also used to investigate the possibility of selectively exciting a subset of spins and so leading to a spectrum from which long-range intergroup couplings between aromatic and ethenyl protons could more easily be assigned. Here this is possible because there is only weak coupling between the distinct groups of spins. This weak coupling exists because the dipolar couplings between groups of spins are much weaker than those within a group, and there is negligible mixing of basis states between different spin groups. The same non-mixing of basis states occurs in high-resolution NMR when the chemical-shift difference between two spins is much greater than their J coupling. Hence, it is possible to irradiate transitions that involve only one spin group.

It turns out that the outer lines in the normal 1Q NMR spectrum involve only aromatic protons and do not depend on intra-ethenyl couplings. Thus it is possible to excite MQ-coherences among the three aromatic protons by making the first two pulses of sequence A of Table 1.1 frequency selective for outer lines of the spectrum, and hence for the aromatic protons only (a so-called DANTE[14] pulse is used for this purpose). The highest order spectrum then observed is the three-quantum spectrum (from the three aromatic protons) that then consists of eight transitions (see Fig. 1.8) which all depend on the long-range dipolar couplings between aromatic and ethenyl protons. In this way information about these long-range couplings is obtained without needing knowledge of the much stronger short-range couplings and the anisotropies in the chemical shifts. These small long-range couplings are difficult to obtain otherwise.

1.3.2.2. *Biphenyl*

Biphenyl (see Fig. 1.2) is interesting because it forms the core of many liquid crystals. It has 10 spins $I = \frac{1}{2}$ and its orientational order is described by two independent order parameters. Its interest lies in determining the dihedral angle which might be affected by the anisotropic environment. The 1Q spectrum is very complicated. The eight-quantum spectrum was analyzed and provided excellent starting parameters for a fit to the 1Q high-resolution spectrum.[15]

1.3.2.3. *Butane*

Butane (see Fig. 1.3) is arguably the most complicated solute whose spectrum has been analyzed using MQ-NMR. There are 10 spins $I = \frac{1}{2}$, three conformers (only two of which are related by symmetry), and three unknown independent order parameters for each of the two symmetry-unrelated conformers. The conformational problem further complicates matters because the relative populations of conformers is the subject of debate, and may well be affected by the interaction between the solute and the surrounding molecules in the condensed phase as well as the anisotropic part of the ordering potential. Of course the investigation of these issues is what makes the study of butane interesting. The horribly complicated 1Q spectrum which has few distinguishing features (Fig. 1.9 left) was analyzed by first fitting the selectively detected 7Q and 8Q spectra (Fig. 1.9 right) which gave sufficiently accurate NMR parameters that almost all transitions in the 1Q spectrum could be assigned immediately.[16]

1.3.2.4. *Some problems*

Although the MQ technique is quite successful in extending the number of spins that can be accommodated, there remain problems. It is still necessary to generate trial MQ spectra and to match transitions with the experimental spectrum. Without a lot of effort (*i.e.* following the evolution of the density matrix through the three-pulse sequence), the intensities of the calculated spectrum are meaningless, and one must deal with matching frequencies only and ignoring intensities. The advantage is that there are far fewer lines than in the 1Q spectrum. In general, all spectral parameters are determinable from analysis of a combination of the $N_{\max}$, $N_{\max} - 1$ and $N_{\max} - 2$ spectra.[6] For simple systems for which there is a reasonable estimate of the structure, a crude starting point can be an initial fit with the order parameters and chemical shifts — in this case, one may be able to get away with fitting the relatively sparse $N_{\max}$ and $N_{\max} - 1$ spectra. In the case of butane (see above), the $N_{\max} - 2$ and $N_{\max} - 3$ spectra were used for the analysis of the dipolar couplings and chemical shifts.

Studies in our laboratory demonstrate that analysis of relatively poor-resolution MQ spectra give NMR parameters that are quite accurate. This accuracy may be sufficient for many purposes, but is lower than that obtainable from the high-resolution 1Q spectrum. For more accurate determination of the spectral parameters, those obtained from analysis of the MQ-NMR spectra provide initial estimates for fitting using 1Q high-resolution line assignment techniques — in every case in our laboratory, the MQ parameters predict a spectrum in which one can immediately assign, with no ambiguity, almost all transitions, and hence obtain spectral parameters to very high accuracy.

While MQ-NMR techniques have proven to be quite successful in the analysis of complicated NMR spectra of orientationally ordered molecules,

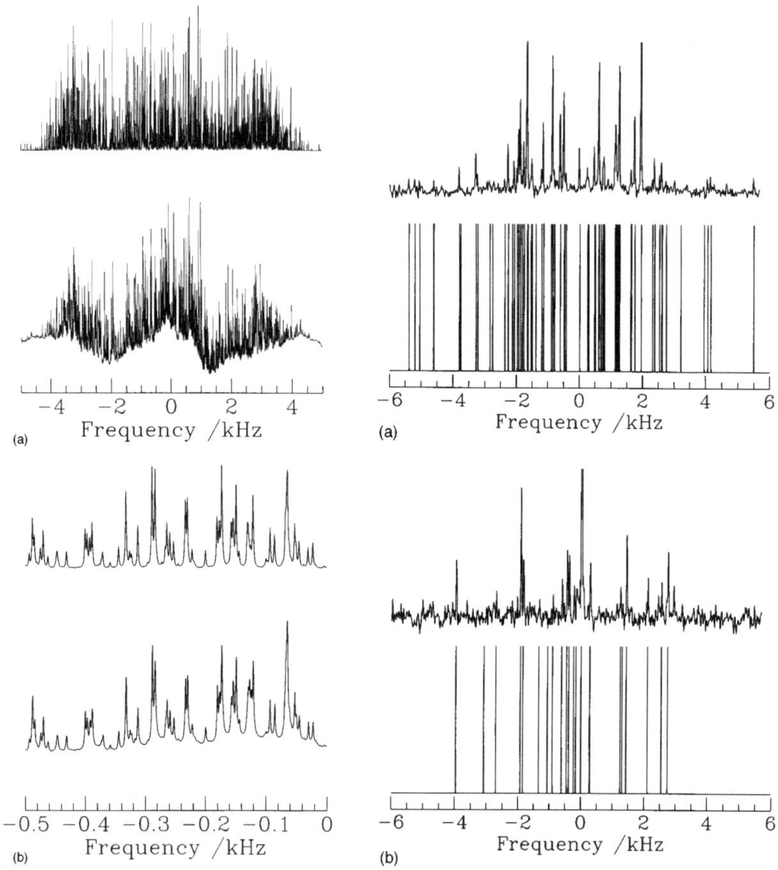

Fig. 1.9. Left: (a) Experimental (bottom) and simulated (top) spectra of partially oriented butane in 55 wt% ZLI 1132 / EBBA. (b) Expanded region of the spectra in (a). Right: (a) Experimental (top) and simulated (bottom) seven-quantum spectra of partially oriented butane. (b) Experimental (top) and simulated (bottom) eight-quantum spectra of partially oriented butane. For the simulated spectra of (a) and (b), the line intensities have been arbitrarily set equal since the intensity of each MQ transition is a complicated function of the preparation time τ and the parameters in the spin Hamiltonian, and does not provide any information for the present study. (Reproduced from Ref. [16] with permission from the American Institute of Physics.)

the experiments are time consuming, and the analysis of complicated spin systems requires considerable effort and NMR-expertise. While it is very powerful, it is certainly not an automated spectral analysis tool.

1.4. Automated analysis using evolutionary strategies

Attempts to develop automated computer routines to efficiently search parameter space in order to find unique values for the spectral parameters that reproduce an observed high-resolution NMR spectrum were made at an early stage, initially for isotropic liquids. Such an alternative approach to the more conventional methods of spectral analysis was first suggested by Diehl *et al.*[17,18] In a similar spirit Stephenson and Binsch developed a basic algorithm which relies on a matrix method derived from a general formulation of the least-squares problem.[19] This approach was implemented in the computer programme DAVINS (Direct Analysis of Very Intricate NMR Spectra),[20] and was reasonably successful. Subsequently the more challenging task of applying the method to the NMR spectra of solutes in anisotropic liquids was undertaken. The programme DANSOM (Direct Analysis of NMR Spectra of Oriented Molecules) was first used in the analysis of the spectra of a number of allyl halides[21] and cyclopentene.[22] However, the limitations of the method in these applications became apparent, because background corrections proved troublesome and for the more complicated cases the intervention of skilled operators during the fitting procedure was usually required. Later the Cosenza group in Italy achieved some notable successes, despite the need for significant operator intervention to avoid the problem of trapping in local minima.[23,24]

In a separate development the use of a Genetic Algorithm (GA) for the analysis of NMR spectra in integrated form of solutes in liquid crystals was reported in a short paper that gave little detail.[25] It appears that trapping in local minima during the convergence process was problematic. In a follow-up paper[26] it was made clear that GA fitting methods were employed to obtain a first approximation to the experimental NMR spectrum. The best GA fit obtained was then used as a starting point for the assignment of individual lines and subsequent least-squares adjustment and refinement. Apart from these two papers we have found no further applications of this method in the literature.

In this section we discuss the implementation of a sophisticated and robust Genetic Algorithm to fit highly complex anisotropic NMR spectra that avoids problems commonly encountered in the automated analysis of NMR spectra of solutes in liquid-crystalline solvents. In particular, the GA-method is inherently capable of circumventing situations where the convergence procedure gets trapped in local minima, without ever reaching the desired global minimum. Our routine is extremely efficient, requires minimal operator interference, and constitutes essentially an intelligent parameter search method. In the absence of reasonable initial guesses for order parameters or dipole-dipole couplings and chemical shielding parameters convergence is still reached and excellent fits are obtained. Sometimes additional information about order or spectral parameters is available which can then be used to narrow the search range and accelerate the convergence process. Owing to the enormous advances in computer technology and parallel processing the analysis of complex anisotropic NMR spectra now becomes eminently feasible. We shall deal with several convincing examples, *viz.* the solutes azulene, biphenylene, *p*-bromo-biphenyl and butane.

Despite the successes that we obtained by employing the GA-method, a weakness of the procedure is that it represents a global optimizer without any strategy. This becomes a limitation when the parameter search space gets very extensive. Currently the development of Evolutionary Algorithms (EA's) is an active and fruitful area of research. From the literature[27] it is known that novel Evolutionary Strategies (ES's) can be very successful in improving the search for an optimal solution. Hence, we decided to explore and implement different EA's as well. We shall demonstrate how successful this approach is when studying *n*-pentane as a solute in different nematic phases.

1.4.1. *Limiting parameter search space*

With a large number of spectral parameters it is important to limit the parameter search space in the fitting procedure as much as one can. Any knowledge available from independent sources about possible parameter values and their uncertainties is highly desirable. Since in practice the experimental spectra are dominated by the direct dipolar couplings, we shall focus on restricting their possible values by any means available.

1.4.2. *The GA-method*

A fit using genetic algorithms (GA) mimics the concepts of natural reproduction and selection processes. For a detailed description of the GA the reader is referred to the original literature on evolutionary or genetic algorithms.[27–29] A detailed description of the GA used in this investigation can be found in Refs. 30–32.

The molecular parameters are encoded in binary or real type, each parameter to be optimized representing a gene. The vector of all genes, which contains all molecular parameters, is called a chromosome. In an initial step the values of all parameters are set to random values between lower and upper limits which are chosen by the user. The quality of the solutions then are evaluated by a fitness function.A proper choice of this fitness function is of vital importance for the success of the GA convergence. In Ref. 30 and 31 the fitness function F_{fg} has been defined as:

$$F_{fg} = \frac{(\boldsymbol{f}, \boldsymbol{g})}{\|\boldsymbol{f}\| \, \|\boldsymbol{g}\|}. \tag{1.5}$$

Here $\boldsymbol{f}$ and $\boldsymbol{g}$ are the vector representations of the experimental and calculated spectrum, respectively. The inner product $(\boldsymbol{f}, \boldsymbol{g})$ is defined with the metric $\boldsymbol{W}$ which has the matrix elements $W_{ij} = w(|j - i|) = w(r)$ as:

$$(\boldsymbol{f}, \boldsymbol{g}) = \boldsymbol{f}^T \boldsymbol{W} \boldsymbol{g}, \tag{1.6}$$

and the norm of $\boldsymbol{f}$ as $\|\boldsymbol{f}\| = \sqrt{(\boldsymbol{f}, \boldsymbol{f})}$; similar for $\boldsymbol{g}$. For $w(r)$ we used a triangle function[30] with a width of the base of Δw:

$$w(r) = \begin{cases} 1 - |r| \,/\, \left(\frac{1}{2}\Delta w\right) & \text{for } |r| < \frac{1}{2}\Delta w \\ 0 & \text{otherwise.} \end{cases} \tag{1.7}$$

The above defined fitness function is able to smooth in a controlled way the error landscape and therefore allows the GA to locate the global minimum. The width of the function $w(r)$ critically determines the ability of the GA to converge to the global minimum and also the speed of convergence. The smoothing of the error landscape allows to sense regions far from the minimum. The GA convergence is obtained in a well-defined procedure. At first the function $w(r)$ should be chosen relatively broad; $\Delta w \approx 15 - 20$ times the line width of an individual NMR transitions in the spectrum. In this way, a first set of parameters is obtained, which still has to be refined. This is done by decreasing Δw and narrowing the limits of the parameter space to be searched in the fit. Decreasing Δw improves the accuracy of the molecular parameters obtained from the fit, while narrowing the parameter space leads to an improved sampling in the region of the minimum. This of course is critical in the procedure, but can in most cases be done automatically. In a final calculation Δw is set to zero. Usually full GA convergence to the best set of parameters is achieved by narrowing Δw to zero in one or two steps.

One optimization cycle, including evaluation of the fitness of all solutions, is called a generation. Pairs of chromosomes are selected for reproduction and their information is combined via a crossover process. Since crossover combines information from the parent generations, it basically explores the error landscape. The values of a small number of bits in the binary encoding of the molecular parameters is changed randomly by a mutation operator. Mutation can be viewed as exploration of the fitness surface. The best solutions within a generation are excluded from mutation. This elitism prevents already good solutions from being degraded. Mutation prevents the calculation from being trapped in local minima, as is often the case with more conventional fitting routines.

The GA described above together with the library PGAPack version 1.0[33] are very suited for massive parallel computation. The larger the starting population for a given problem, the faster is the convergence in terms of generations. On the other hand a large starting population tends to slow the algorithm down due to computational demands. A large number of fast processors will circumvent this problem and will lead to fast and straightforward assignments of the spectra. The computational speed is inversely proportional to the number of processors. Therefore this kind of algorithm is perfectly applicable for parallel processing. Modern computer cluster systems make these calculations very feasible, even for complex spectra.

The automatic GA fitting of anisotropic NMR spectra represents one step up in complexity compared to the fitting of high-resolution Laser Induced Fluorescence (LIF) spectra[30–32] that was performed with the same GA procedure before. The reason is that the LIF spectra often show some kind of regularity. This is in general completely absent in anisotropic NMR spectra because of the complex nature of the spin Hamiltonian (Eq. (1.1)).

1.5. Results obtained with the GA-method

In order to test the validity of the GA-method for very complicated ^{1}H NMR spectra of solutes dissolved in nematic phases, we decided to consider molecules of increasing size and flexibility requiring strategies for spectral analysis of increasing sophistication. We started with the relatively simple case of the planar 'rigid' molecules azulene ($C_{10}H_8$) and biphenylene ($C_{12}H_8$), see Fig. 1.1, each possessing C_{2v} symmetry. Both solutes were dissolved in a 'magic mixture' composed of 55 wt% Merck ZLI 1132 / EBBA. The orientational order of each solute can be described by two independent order parameters. Both solutes had been studied before and their eight-spin ^{1}H NMR spectra analyzed. Azulene had been studied in the nematic liquid crystal EBBA and its ^{1}H spectrum analyzed by conventional means.[34] Biphenylene had been studied in the 55 wt% Merck ZLI 1132 / EBBA 'magic mixture' by employing MQ methods.[35] The results of these previous studies can be used as guidance.

In our GA-fits on azulene and biphenylene a two-step procedure was implemented. To start with, a relatively crude geometrical structure of the solute was assumed. In the first step a GA-fitting procedure was performed in which two independent order parameters and the chemical shielding parameters were varied until the best possible correspondence with the experimental spectrum was obtained. Owing to the approximations about the structure made in this first step, complete agreement could of course not be achieved. However, in practice the choice of geometry does not have to be perfect in order to obtain sufficient common features between calculated and experimental spectra to proceed from there. After the best but far from perfect fit of step one was obtained, approximate dipolar couplings were calculated. In a second GA-procedure, these approximate dipolar couplings were then used as starting values for a fit in which all dipole-dipole couplings and chemical shielding parameters were varied until convergence was reached. For both solutes exceptionally good fits to the experimental spectra were obtained after step two, because no constraints were placed on the dipolar coupling and chemical shielding parameters. When we compare the dipolar couplings to those obtained in earlier work which was performed under slightly different experimental conditions, the correspondence is pleasing, giving us confidence that the GA-method can be of great value for the analysis of very complicated high-resolution NMR spectra.

Next, we turned to a solute that possesses more than one configuration, with rapid interconversion between them on the NMR time scale. As an illustrative example we focused on the nine-spin system *p*-bromo-biphenyl (see Fig. 1.2) dissolved in three different nematic phases, *viz.* Merck ZLI 1132, EBBA, and the 55 wt% Merck ZLI 1132 / EBBA 'magic mixture'. This solute interconverts rapidly between two symmetry-related conformers and had not been studied before. The orientational order of one conformer can be described by the three S-parameters S_{xx}, S_{zz} and S_{yz} where the molecular axes are defined with the x-axis along the inter-ring CC-bond and the y and z-axes in planes bisecting the dihedral angles between the rings. Apart from a change of sign in S_{yz}, the orientational order of the second conformer can be described by the same parameters. Since a description of the orientational order of the related biphenyl only requires two order parameters by symmetry, S_{yz} is taken to be close to zero. Inter-ring couplings are averaged by rotation about the central CC bond and thus the molecule has 16 independent dipolar couplings and 5 chemical shielding parameters.

Again a two-step procedure was performed. In the first step an estimated geometrical structure of the solute was taken and two independent order parameters, chemical shielding parameters and the dihedral angle between the two rings were varied in a GA-fitting procedure until the best correspondence with the experimental spectrum was obtained. With the approximations made, complete agreement could not be expected. However, the correspondence was still good enough for approximate dipolar couplings to be calculated. These then served as starting values in a second GA-calculation in which all dipole-dipole couplings and chemical shielding parameters were varied until convergence was reached. In all three liquid-crystal solvents employed exceptionally good agreement between experimental and fitted ^{1}H spectra was obtained.[36]

For this problem a modern cluster of PC's was used. The CPU time required amounted to $\sim$ 64 minutes per GA-run, which with 16 processors came to 4 minutes wall clock time per run. A typical result is presented in Fig. 1.10.

The next level in complexity consists of a solute that undergoes conformational change between symmetry-unrelated conformers. In order to see whether GA-fitting would be successful under these challenging conditions, we addressed the example of butane (C_4H_{10}, see Fig. 1.3) dissolved in the 55 wt% Merck ZLI 1132 / EBBA 'magic mixture'. As in the case of biphenylene, butane had been studied before employing MQ methods.[16] In fact, for our GA-analysis we used the good-quality single-quantum spectrum of butane in 'magic mixture' (Fig. 1.9) measured in this earlier work.

The conformers of butane can be approximated with the rotational isomeric state (RIS) model.[37] In this model each C–C torsion bond is assumed to exist in three local arrangements, *trans* (t), *gauche* plus (g_+) and *gauche* minus (g_-), with dihedral angles of 0°, and roughly +112° and −112°, respectively, corresponding to the angles at the minima of the rotational potential profile. These angles are close to 120°. In principle, butane can exist at other dihedral angles, with the extremes being the eclipsed and staggered forms. However, since steric hindrance is larger in the eclipsed configuration, the staggered forms are strongly favoured. In butane the possible staggered conformations are anti or t (C_{2h} symmetry), g_+ (C_2 symmetry) and g_- (C_2 symmetry), the latter two being related by symmetry. Each independent conformer requires three independent order parameters, leading to a total of six. There is an energy difference E_{tg} between *trans* and *gauche* conformers. The rapid interconversion between these conformers on the NMR time scale leads to a total of seven observable direct dipolar cou-

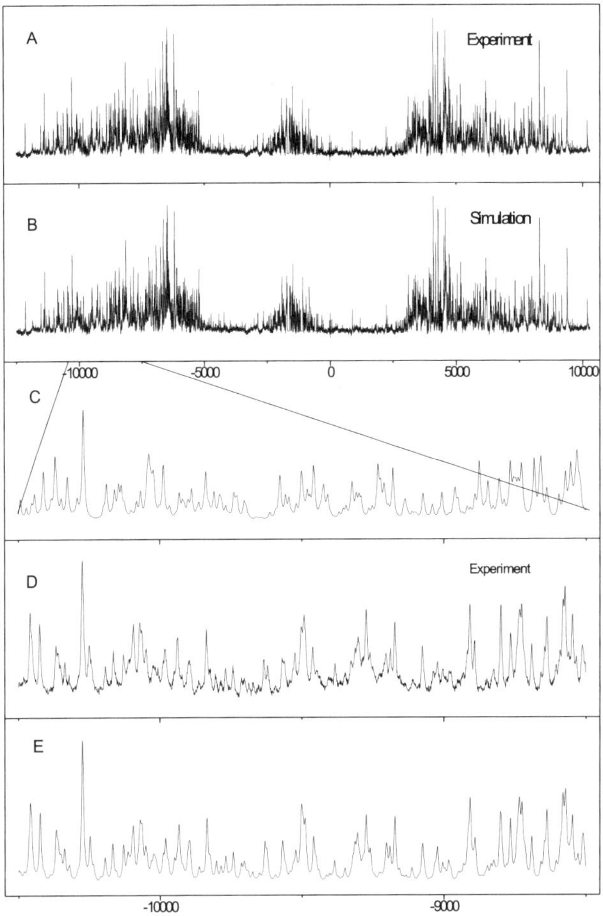

Fig. 1.10. ^{1}H NMR spectrum of *p*-bromo-biphenyl dissolved in 'magic mixture'. The experimental spectrum (A) is compared with the final fit obtained by varying 16 dipolar couplings and 5 chemical shieldings (B); in the next frames the spectrum is enlarged to show the fit obtained by varying 2 S-parameters, 5 chemical shieldings and the dihedral angle (C), the experimental spectrum (D), and the fit obtained by varying 16 dipolar couplings and 5 chemical shieldings (E). All horizontal scales are in Hz. (Reproduced from Ref. [36] with permission from Elsevier Science Publishers B.V.)

plings that are averaged over the internal motions, as well as two chemical shielding parameters.

Since conformational probabilities and order parameters only occur in Eq. (1.3) in product form, we need independent information to achieve a separation between these quantities. For solutes dissolved in a 'magic mixture' the degree of orientational order depends primarily on the size and shape of the solute molecule. The choice of size and shape model is not essential, and we employ the so-called Circumference-Integral (CI) approach that has been used successfully in a study of 46 unrelated solutes in the magic mixture. We fix the ratio of the two interaction parameters of the model to the value 23.529 found in an earlier study of 46 solutes in the magic mixture.[38] The value chosen for the one remaining parameter (a force constant in the CI-model) k is the one that leads to a calculated butane spectrum that possesses the same width as the observed one. In order to proceed, we make a reasonable estimate of the geometries of the different butane conformers and we use the CI-model to obtain relatively crude starting values for the order parameters associated with the different butane configurations.

With a two-step approach similar to that discussed above, with order tensors obtained with the CI-model for every conformer, and with an E_{tg} energy difference estimated from other sources, a set of trial dipolar couplings was established for the first step of the fitting procedure. First, a GA-fit was performed on chemical shieldings and order parameters until some resemblance between experimental and calculated spectra was obtained. From this tentative fit for the order tensors, dipolar couplings were estimated that served as starting parameters for the next step. In step two in a new GA-fitting procedure the chemical shieldings and all the dipolar couplings were varied until excellent agreement between experimental and calculated spectra was obtained (see Fig. 1.11). The dipolar couplings and chemical shieldings obtained from our GA-fitting procedure proved to be virtually identical to those derived previously from the same experimental spectrum following an extensive MQ-approach.[16] This pleasing result lends additional credence to the power of GA-fitting.

Considering the ease with which spectra of butane dissolved in a liquid-crystal solvent can now be reliably analyzed by employing the GA-method, we have recently performed an extensive study of butane in an anisotropic solvent as a function of temperature. In a detailed computer simulation effort in which the solute-solvent interaction is carefully taken into account the degree of orientational order of the various solute conformers can be predicted and the results compared to our experiments. It can be expected that such studies will lead to a better understanding of conformational change in the condensed phase than is currently available.[39]

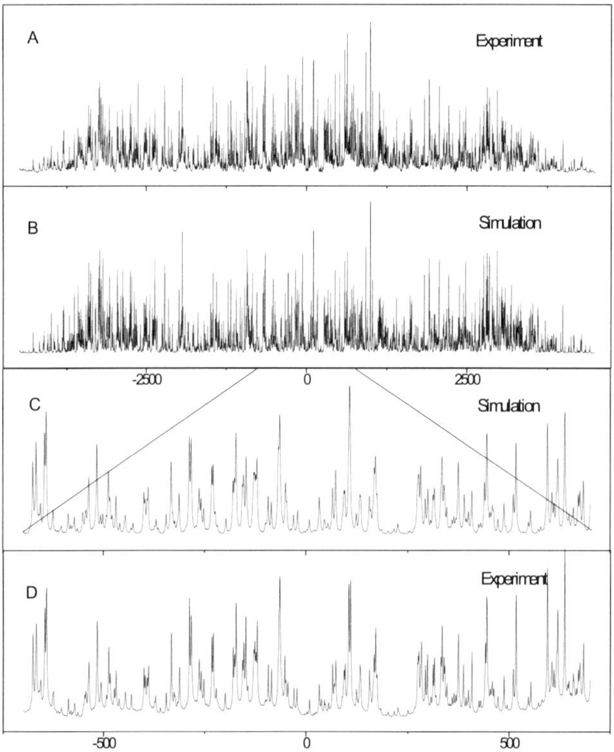

Fig. 1.11. ^{1}H NMR spectrum of butane dissolved in 'magic mixture'. The experimental spectrum (A) is compared with the final fit obtained by varying 7 dipolar couplings and 2 chemical shieldings (B); in the next frames the spectrum is enlarged to show the experimental spectrum (D), and the fit (C). All horizontal scales are in Hz.

In summary, the robustness of the GA-fitting method has been demonstrated for solutes that belong to various categories of increasing complexity:[40] rigid solutes with a single conformation (*e.g.* the eight-spin systems azulene and biphenylene), solutes interconverting between a number of symmetry-related conformers (*e.g.* *p*-bromo-biphenyl), and solutes undergoing conformational change between non-symmetry-related conformers (*e.g.* the ten-spin system butane). In the following section the even more challenging example of *n*-pentane dissolved in different nematic phases will be treated. This will require the use of Evolutionary Algorithms with a higher degree of sophistication than the relatively simple GA-method.

1.6. Spectral analysis using Evolutionary Algorithms

The ^{1}H NMR spectrum of pentane (see Fig. 1.3) dissolved in Merck ZLI 1132 (shown in Fig. 1.12) has no easily recognizable patterns that could aid the spectral analysis. A visual inspection of the spectrum gives no clue whatsoever of the magnitudes of the dipolar couplings. When applying an EA to such a problem it should be realized that the computational time increases very steeply with the search ranges used for every parameter. It is therefore prudent to choose initial values and ranges for each parameter that are as narrow as possible, but still wide enough to include the correct value. In the following we shall describe a strategy to do just that.

High-resolution NMR spectral values were used for indirect spin-spin couplings which were initially assumed to remain unchanged in the spectrum of the orientationally ordered solute. High-resolution NMR values were also used as initial estimates of chemical shifts and the search ranges were chosen to be consistent with centering the calculated with the experimental spectrum.

Pentane exists in several conformers that interchange rapidly on the NMR time scale. As with butane, initial estimates of the eleven independent dipolar couplings were quite problematic because they depend on conformer structures, order tensors and conformer probabilities. Again, we assumed that the conformer problem can be described adequately by the rotational isomeric state (RIS) approximation.[37] For n-pentane there are then nine RIS conformers, some being related by symmetry. The four independent conformers are tt (C_{2v}, singly degenerate), tg (C_1, four-fold degenerate), g_+g_+ (C_2, two-fold degenerate) and g_+g_- (C_s, two-fold degenerate). It is necessary to assign values for conformer probabilities. The extra energy involved in a *gauche* conformer E_{tg} was set to the value 650 cal, with the weight of the g_+g_- conformer set to zero because of the additional energy involved with the steric interaction of the two methyl groups in this configuration (the 'pentane' effect).[37] For the structure of each conformer we used the same bond angles and bond lengths that were used for a previous analysis of butane.[16]

Finally, we required estimates of the pentane conformer order tensors in the two liquid-crystal solvents 'magic mixture' and 1132. Solute orientational order in 'magic mixture' is dominated by a single orientational mechanism that depends on solute size and shape alone.[4,5] Hence, phenomenological size and shape models are appropriate for the prediction of conformer order parameters in this case. For 1132 the situation is some-

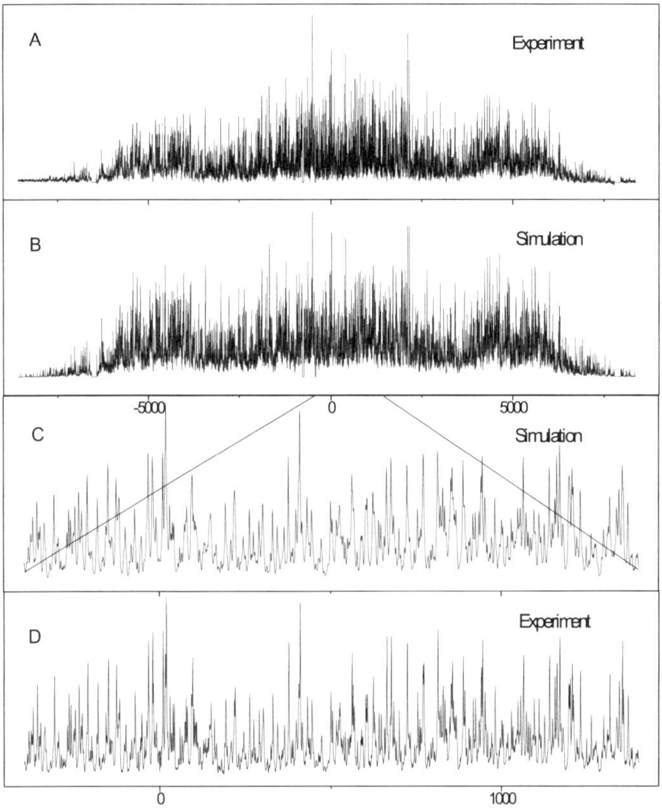

Fig. 1.12. ¹H NMR spectrum of pentane dissolved in 1132. The experimental spectrum (A) is compared with the final fit obtained by varying 11 dipolar couplings, 9 indirect spin-spin couplings and 3 chemical shieldings (B); in the next frames the spectrum is enlarged to show the experimental spectrum (D), and the fit obtained by varying 11 dipolar couplings, 9 indirect spin-spin couplings and 3 chemical shieldings (C). All horizontal scales are in Hz. The 11 indirect spin-spin couplings were varied only in the final fitting procedure; in earlier iterations they were given their isotropic values. (Reproduced from Ref. [41] with permission from the American Institute of Physics.)

what different, because at least one additional mechanism that depends on the anisotropic electrostatic properties of the solute contributes to the orientational potential. However, in the case of alkanes these additional mechanisms are considered to be of minor importance and can be neglected.[42] Thus the moniker 'magic solute' is often used for alkanes. Hence, size and shape models (with the interaction constant adjusted to account for changes in liquid-crystal solvent and temperature) are probably also appropriate to describe the orientational order of pentane in 1132. These considerations are supported by literature data on propane[43] for which the ratios of the two propane order parameters are approximately equal in the liquid crystals 'magic mixture' and 1132. As in the case of butane, we employ the so-called CI-model[38] to obtain order parameters for the various pentane conformers to probably better than 10%.[4,5] Again, we fix the ratio of the two interaction parameters of the model to the value 23.529.[38] The value chosen for the one remaining parameter k is the one that leads to a calculated pentane spectrum that possesses the same width as the observed one.

Spectra calculated using the order parameters estimated above bore resemblance to the experimental ones in the sense that differences were appreciable, but the overall underlying features were reasonable. An initial, unsuccessful attempt with the GA fitting procedure was performed using (and varying within reasonable limits) the above chemical shifts, order parameters for each conformer, and conformer probabilities. Such parameters cannot be expected to give an exact solution because they are based on models and guesses, and hence may have significant uncertainties. The search was abandoned in favour of one that varied the eleven independent dipolar couplings and three chemical shifts that could, in principle, yield an exact match to the experimental spectrum. Indeed there are fewer independent dipolar couplings than conformer order parameters, giving further justification for basing the second attempt on the dipolar couplings.

In order to estimate reasonable initial values and ranges for the dipolar couplings, spectra were calculated for various values of order parameters ($\pm 10\%$ of the values above), E_{tg} (between 400 and 1000 cal/mol) and the dihedral angle for a *gauche* rotation (between 112° and 120°). The highest and lowest values of each dipolar coupling resulting from these calculations provided the search ranges for the dipolar couplings. These ranges seemed consistent with spectra that reproduce reasonably the overall features and width of the experimental one.

We then started a fit using the GA-method which was successful in unraveling the spectrum of the nine-spin system *p*-bromo-biphenyl[36] and the ten-spin system butane.[40] Initially this was unsuccessful for the spectrum of pentane in 1132 for a number of reasons. The uncertainties in some of the larger dipolar constants were still quite large. As a consequence the GA procedure was not able to converge. Probably the main reason for this failure comes from the fact that the GA procedure is a global optimizer without any strategy. If the search ranges become too large there is insufficient coverage of the parameter space to locate the global minimum. We solved this problem initially by breaking up the uncertainty regions for a number of parameters into regions so small that we were certain the GA would converge. The size of the smaller search regions was determined with test spectra. This was done for the dipolar couplings with the largest uncertainties. Now a grid search was set up in which the GA was applied in the different subregions, until a subregion was found for which the GA succeeded in reproducing the experimental spectrum. In this way we obtained a calculated spectrum that is essentially a perfect fit to the experimental one.[41]

It is clear that this grid search process rapidly will become very expensive. If for example n uncertainty regions are divided into m subregions the size of the grid will be m^n, a number that increases steeply with n and m. We therefore decided to implement different Evolutionary Algorithms. From the literature[27] it is known that Evolutionary Strategies (ES's) can be very successful in improving the search for an optimal solution.

The ES algorithm starts with one or more parents. A parent is a trial solution that corresponds to a vector of parameters like in the genetic algorithm, with the difference that each vector has an extra set of parameters corresponding to the type of strategy used. This parent generates offspring. The quality or performance of these children is checked and, depending on the strategy, the next parent is generated. There are several different strategies for the generation of the offspring as well as the generation of the next parent. For a more detailed description see the literature.[27]

The offspring is generated from the parents in a mutative step-size fashion. The magnitude and direction of the steps are dependent on previous history and on the quality of offspring generated. A drawback of the standard ES is the fact that the mutations of the decision and the strategy parameters are subject to independent random processes. If for example an individual with a large step size undergoes only a very small change in the decision parameters and this small change turns out to yield a high fitness, the large step size will be inherited by the next generation. As a result, the fitness in the next mutations may worsen. This problem is resolved in derandomized (DR) algorithms which make the random mutations in decision and strategy parameters dependent on each other. This idea was implemented first as DR1 and soon improved by the concept of accumulated information,[44] which is called DR2. The history of the optimization is recorded and the evolution of the mutation ellipsoid is partially governed by past successful mutations.

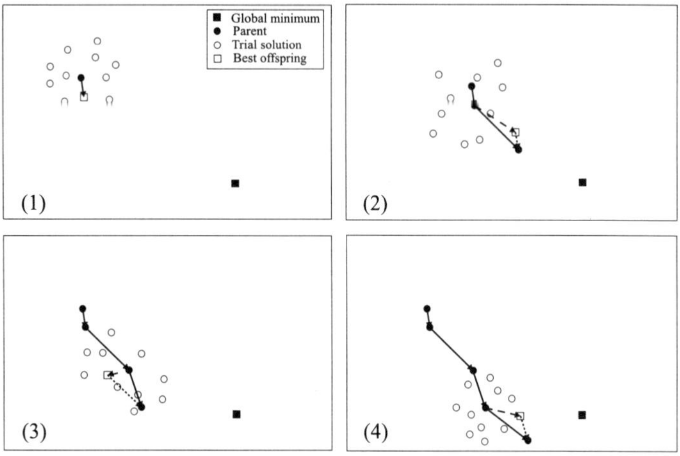

Fig. 1.13. The first four generations of an Evolutionary Strategy: (1) An initial population is generated, and the best offspring is used as the next parent. (2) The offspring is spread over a larger area in the second generation due to the relatively large step made in the previous generation. The vector from the parent to the best offspring (dashed line) is combined with the (shortened) mutation vector of the last generation (dotted line) to generate the new parent (solid line). (3) Due to the correlation between the past two mutations the search range has been extended again in the general direction of both mutations while it has been limited in the perpendicular direction. The best offspring is now a local minimum. The memory effect of the evolutionary algorithm, which incorporates past mutation vectors into the calculation of the next parent, helps to overcome the local minimum and the next parent is still closer to the global minimum. (4) The barrier between the local and global minima has been overcome, and the optimization is progressing towards the global minimum. (Reproduced from Ref. [41] with permission from the American Institute of Physics.)

A further improvement was achieved by Hansen and Ostenmeier[45] with the Covariance Matrix Adaptation Evolution Strategy (CMA-ES). It turns out to be a particularly reliable and highly competitive evolutionary algorithm for local optimization and, surprisingly at first sight, also for global optimization.[46] The CMA-ES does not leave the choice of strategy parameters open to the user — only the population size can be set. Finding good strategy parameters is considered to be part of the algorithm design.

Figure 1.13 depicts the first four generations of an evolutionary strategy and demonstrates the effect of the chosen strategy. In general the Evolution Strategies converge faster and are more robust than the genetic algorithm.

Next, we implemented this algorithm into our programme, selected the estimated uncertainty regions discussed above for the dipolar couplings, and performed a fit of the spectrum of pentane in 'magic mixture'. This spectrum was even harder to deal with than that of pentane in 1132, because the density of lines is larger (smaller dipolar couplings) and the signal-to-noise ratio was smaller, while at the same time the background removal was more difficult. This background is due to an abundance of unresolved dipolar splittings from the liquid-crystal molecules themselves that typically contain of order 20 ^{1}H spins. It turned out that the CMA-ES procedure was immediately successful in obtaining an essentially perfect fit of the observed spectrum and determining the parameters unambiguously. At that point there was only one small glitch left. Due to the large density of lines it was relatively difficult to remove completely the background from the observed spectrum. In spite of this the CMA-ES succeeded in fitting all parameters and transitions. Improvement of the accuracy in the final values of the parameters can be realized by a better way of dealing with the background.[41]

With our initial fit we were in the position to carry out an almost perfect removal of the background from the experimental spectrum. This was performed in the following way. The calculated spectrum resulting from the CMA-ES was subtracted from the experimental one. To remove the high-frequency noise in this difference a degree of smoothing was applied. Finally this smoothed difference was subtracted from the original experimental spectrum. In this way an experimental spectrum was obtained which is almost background-free. The result for part of the spectrum is shown in Fig. 1.14. Using this approach we were able to improve the parameters in a final CMA-ES fit. In addition, it was possible to determine accurately the small indirect spin-spin couplings $J_{\mu\nu}$ between nuclei μ and ν.[41]

The background removal procedure as described in the previous paragraph was repeated for the spectrum of pentane in 1132 (Fig. 1.12) and again we were able to fit the indirect spin-spin coupling constants. Within estimated error the results for the $J_{\mu\nu}$ for pentane in 1132 and in magic mixture were identical, as expected.

What was left to calculate were the statistical errors in the parameters. This is in general a difficult task for EA's. However, in our case it was relatively simple. The converged result of the EA not only produces the best values for the fit constants, but at the same time the quantum numbers of the individual transitions are assigned. This allows a classical least-squares fit using the assigned frequencies. In Meerts and Schmitt[32] this is called

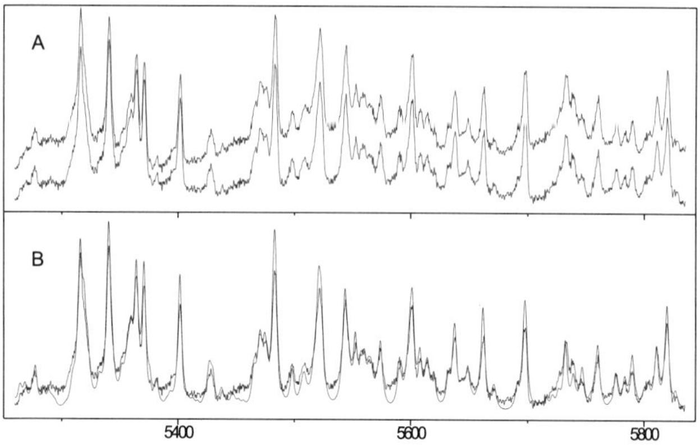

Fig. 1.14. A small part of the ^{1}H NMR spectrum of pentane dissolved in a magic mixture. This shows the effect of the background removal method discussed in the text. (A) The upper and lower traces show the experimental spectra without and with background removal, respectively. (B) The noise free trace is the calculated spectrum and the other the experimental spectrum after removal of background. All horizontal scales are in Hz. (Reproduced from Ref. [41] with permission from the American Institute of Physics.)

an 'assigned fit' and the definitions of the statistical errors and correlation coefficients are discussed in Appendix B of that paper.

1.7. Conclusions

NMR-studies of solutes dissolved in anisotropic solvents provide a rich source of information about molecular structure in the condensed phase. In particular, the direct dipolar couplings in liquid-crystal environments are not averaged to zero, and usually dominate the NMR-spectra. Since these direct dipolar couplings depend on distances between nuclear spins, obtaining these quantities from experiment is an essential step. Application of the liquid-crystal technique to solutes with low symmetry and with more than approximately 8 spins is seriously hampered by the fact that the single-quantum NMR-spectrum shows a multitude of often overlapping transitions without any apparent structure or pattern. Analyzing such spectra and extracting the dipolar couplings from them is a challenging task that for many years appeared to present insurmountable problems. The difficulties encountered were exacerbated in the case of solutes that interconvert among several symmetry-unrelated conformers. For many years the extension of the liquid-crystal method to larger solutes was therefore considered 'impossible'.

In this chapter we focus both on recent experimental and novel theoretical methods to address the central problem of spectral analysis. In recent years the development of multiple-quantum NMR has significantly increased the possibilities of liquid-crystal NMR. High-order multiple-quantum spectra are much simpler than single-quantum ones, and their analysis provides an approximate determination of the direct dipolar couplings that contain most of the relevant information. These estimates can then be employed as a starting point for the detailed analysis of the high-resolution single-quantum spectrum. Various sophisticated versions of multiple-quantum NMR have been developed and successfully applied to the study of relatively large and complicated solutes dissolved in liquid-crystals. All these advanced techniques have the drawback that they require considerable effort and NMR-expertise.

Recently a novel approach has been developed in which mathematical algorithms from evolutionary biology (Evolutionary Algorithms, EA's) were applied to NMR of solutes in anisotropic environments. These methods comprise applications of a relatively simple Genetic Algorithm (GA), as well as the use of more sophisticated Evolutionary Strategies (ES's). These techniques have been described in some detail and have been applied to a number of representative solutes, ranging from 'rigid' solutes with eight spins to the flexible butane and even the pentane molecules. The availability of enormous computer power, and the fact that dealing with the Hamiltonian problem in liquid-crystal NMR lends itself to extensive parallel programming, are the key elements why these spectral fitting techniques are eminently successful. The convergence progress can be accelerated considerably if the search ranges for the spectral parameters can be limited in a sensible fashion. *A priori* estimates of the degree of orientational order of solutes in liquid crystals to a level of better than 10%, based on simple phenomenological size and shape models, are very effective in this context.

In summary, through the application of modern experimental multiple-quantum methods and novel spectral analysis techniques based on evolutionary strategies, the liquid-crystal NMR-method can now be extended to much larger and more complicated solutes than previously possible. Introduction of these methods has therefore given liquid-crystal NMR a new lease of life.

Acknowledgments

E.E.B. acknowledges financial support from the Natural Sciences and Engineering Research Council of Canada. W.L.M. thanks the National Computer Facilities of the Netherlands Organization of Scientific Research for a Grant on the Dutch supercomputing facility SARA.

References

1. P. Diehl and C. L. Khetrapal, *NMR Basic Principles and Progress*, vol. *1*, (Springer-Verlag, Berlin, 1969, p. 1).
2. J. W. Emsley and J. C. Lindon, *NMR Spectroscopy using Liquid Crystal Solvents*. (Pergamon Press, Oxford, 1975).
3. E. E. Burnell and C. A. de Lange, *Chem. Phys. Letters*. **76**, 268, (1980).
4. E. E. Burnell and C. A. de Lange, Prediction from molecular shape of solute orientational order in liquid crystals, *Chemical Reviews*. **98**, 2359, (1998).
5. E. E. Burnell and C. A. de Lange (Eds.), *NMR of Ordered Liquids*. (Kluwer Academic, Dordrecht, The Netherlands, 2003).
6. D. P. Weitekamp, *Adv. Magn. Reson.* **11**, 111, (1983).
7. J. C. T. Rendell, Ph. D. thesis, University of British Columbia, 1987.
8. R. T. Syvitski, N. Burlinson, E. E. Burnell, and J. Jeener, *J. Magn. Reson.* **155**, 251, (2002).
9. D. N. Shykind, J. Baum, S. B. Liu, and A. Pines, *J. Magn. Reson.* **76**, 149, (1988).
10. L. W. Field and M. L. Terry, *J. Magn. Reson.* **69**, 176, (1986).
11. XWIN-NMR Software Manual, Bruker Analytik GmbH, 2000.
12. J. C. T. Rendell and E. E. Burnell, *J. Magn. Reson.* **A 112**, 1, (1995).
13. J. C. T. Rendell and E. E. Burnell, *Mol. Phys.* **90**, 541, (1997).
14. G. A. Morris and R. Freeman, *J. Magn. Reson.* **29**, 433, (1978).
15. T. Chandrakumar, J. M. Polson, and E. E. Burnell, *J. Magn. Reson.* **A 118**, 264, (1996).
16. J. M. Polson and E. E. Burnell, *J. Chem. Phys.* **103**, 6891, (1995).
17. P. Diehl, S. Sýkora, and J. Vogt, *J. Magn. Reson.* **19**, 67, (1975).
18. P. Diehl and J. Vogt, *Org. Magn. Reson.* **8**, 638, (1976).
19. D. S. Stephenson and G. Binsch, *J. Magn. Reson.* **37**, 395, (1980).
20. D. S. Stephenson and G. Binsch, *J. Magn. Reson.* **37**, 409, (1980).
21. D. S. Stephenson and G. Binsch, *Org. Magn. Reson.* **14**, 226, (1980).
22. D. S. Stephenson and G. Binsch, *Mol. Phys.* **43**, 697, (1980).
23. F. Castiglione, M. Carravetta, G. Celebre, and M. Longeri, *J. Magn. Reson.* **132**, 1, (1998).
24. F. Castiglione, G. Celebre, G. De Luca, and M. Longeri, *J. Magn. Reson.* **142**, 216, (2000).
25. H. Takeuchi, K. Inoue, Y. Ando, and S. Konaka, *Chem. Lett.* **11**, 1300, (2000).
26. K. Inoue, H. Takeuchi, and S. Konaka, *J. Phys. Chem. A.* **105**, 6711, (2001).

27. I. Rechenberg, *Evolutionsstrategie - Optimierung technischer Systeme nach Prinzipien der biologischen Evolution.* (Frommann-Holzboog, Stuttgart, 1973).
28. J. H. Holland, *Adaption in Natural and Artificial Systems.* (MI: The University of Michigan Press, Ann-Arbor, 1975).
29. D. E. Goldberg, *Genetic Algorithms in search, optimisation and machine learning.* (Addison-Wesley, Reading Massachusetts, 1989).
30. J. A. Hageman, R. Wehrens, R. de Gelder, W. L. Meerts, and L. M. C. Buydens, *J. Chem. Phys.* **113**, 7955, (2000).
31. W. L. Meerts, M. Schmitt, and G. Groenenboom, *Can. J. Chem.* **82**, 804, (2004).
32. W. L. Meerts and M. Schmitt, *Int. Rev. Phys. Chem.* **25**, 253, (2006).
33. D. Levine. PGAPack V1. 0, PgaPack can be obtained via anonymous ftp from: ftp://ftp.mcs.anl.gov/pub/pgapack/pgapack.tar.z, (1996).
34. A. D. Hunter, E. E. Burnell, and T. C. Wong, *J. Molec. Structure.* **99**, 159, (1983).
35. J. M. Polson and E. E. Burnell, *J. Magn. Reson.* **A 106**, 223, (1994).
36. W. L. Meerts, C. A. de Lange, A. C. J. Weber, and E. E. Burnell, *Chem. Phys. Lett.* **441**, 342, (2007).
37. P. J. Flory, *Statistical Mechanics of Chain Molecules*, Wiley-Interscience, New York, 1969.
38. D. S. Zimmerman and E. E. Burnell, *Mol. Phys.* **78**, 687, (1993).
39. R. Berardi, E. E. Burnell, C. A. de Lange, W. L. Meerts, A. C. J. Weber, and C. Zannoni. Work in progress.
40. A. C. J. Weber, E. E. Burnell, W. L. Meerts, and C. A. de Lange. To be published.
41. W. L. Meerts, C. A. de Lange, A. C. J. Weber, and E. E. Burnell, *J. Chem. Phys.* **130**, 044504, (2009).
42. A. F. Terzis, C. D. Poon, E. T. Samulski, Z. Luz, R. Poupko, H. Zimmermann, K. Müller, H. Toriumi, and D. J. Photinos, *J. Am. Chem. Soc.* **118**, 2226, (1996).
43. E. E. Burnell, L. C. ter Beek, and Z. Sun, *J. Chem. Phys.* **128**, 164901, (2008).
44. A. Ostermeier, A. Gawelczyk, and N. Hansen, *Step-Size Adaptation Based on Non-Local Use of Selection Information. In Parallel Problem Solving from Nature (PPSN3).* (Springer, 1994).
45. N. Hansen and A. Ostermeier, Completely derandomized self-adaptation in evolution strategies, *Evolutionary Computation.* **9**(2), 159–195, (2001).
46. N. Hansen and S. Kern. Evaluating the CMA evolution strategy on multimodal test functions. In ed. X. Yao, *et al.*, *Parallel Problem Solving from Nature PPSN VIII*, vol. 3242, *LNCS*, pp. 282–291. Springer, (2004).

Chapter 2

Analytical Potentials of Natural Abundance Deuterium NMR Spectroscopy in Achiral Thermotropics and Polypeptide Chiral Oriented Solvents

Philippe Lesot[1] and Christie Aroulanda[2]

[1]*RMN en Milieu Orienté, Université de Paris Sud 11, ICMMO, UMR CNRS 8182, Bat. 410, 91405 Orsay, France*
email: philesot@icmo.u-psud.fr
[2]*Méthodologie RMN, UMR CNRS-UHP 7036, Institut Jean Barriol, Nancy-Université, BP 70239, 54506 Vandoeuvre-lès-Nancy, France*

Deuterium NMR spectroscopy in oriented solvents is a powerful methodology that can be efficiently used for various analytical applications. The majority of studies involved isotopically enriched analytes so far. However as deuterium nuclei are naturally present at the level of ca. 0.011 % on average in all (bio)organic compounds, the natural abundance deuterium NMR (NAD-NMR) is possible and has been explored very early in the NMR history. Actually this quadrupolar nucleus (I = 1) with a small nuclear electric quadrupole moment is a remarkable tracer that can be easily detected with modern high-field FT NMR spectrometers. This chapter presents fundamentals of NAD-NMR spectroscopy in achiral and chiral liquid-crystalline environments and its various possible contributions to the chemical analysis. This includes some theoretical aspects and a survey of the most useful applications. The potentialities of NAD-NMR using liquid crystals with low and high degree of ordering are reviewed. A special emphasis will be given on the development of NAD-NMR in polypeptide chiral liquid crystals that provides an original method for enantiomeric analysis and stereochemical investigations. Finally the most recent analytical progresses of anisotropic NAD-NMR using spectrometers equipped with deuterium cryogenically cooled probes will be presented.

2.1. Introduction

The analytical use of magnetically active nuclei with low isotopic natural abundance is an exciting challenge for the Nuclear Magnetic Resonance (NMR) spectroscopy. Among the most remarkable nuclei of the Periodic Table, deuterons (^{2}H) at natural abundance level, noted thereafter NAD (Natural Abundance Deuterium) has a particular position for three main reasons: (i) as isotope of hydrogen atom, it is naturally present in all organic molecules; (ii) it provides the same NMR content as proton (^{1}H) in terms of chemical shifts with no ^{2}H-^{2}H coupling that can sometimes be cumbersome for analyzing large molecules or for some specific analytical applications; (iii) this isotope of hydrogen atom possesses a quadrupolar moment, Q_D.[1] This particular property of spins I > 1/2 leads to an interesting anisotropic NMR observable through the quadrupolar interaction, that can be very insightful for diverse analytical studies as will be shown in this chapter. From an NMR viewpoint, deuterons are among the less receptive nuclei at natural abundance level such as exotic nuclei ^{40}K, ^{57}Fe or ^{187}Os. Actually its reference natural abundance is equal to 1.55×10^{-2} % (according to the Vienna Standard Mean Ocean Water value or V-SMOW), while relative receptivity compared to ^{1}H and ^{13}C nuclei at natural abundance are about 1.11×10^{-6} and 6.52×10^{-3}, respectively.[1-3]

In practice, the intrinsic low sensitivity of deuterons has never been a real obstacle to the exploration of NAD-NMR spectroscopy. Indeed in 1964, namely in the early years of NMR history, Diehl and Leipert demonstrated first the feasibility of recording a ^{2}H signal at natural abundance level when observing the NAD signal of acetone using a frequency-sweep NMR (CW-NMR) spectrometer operating at 7.6 MHz for deuterons.[4] However the first major analytical application of NAD-NMR has really began nine years later when Randall *et al.* reported, in 1973, the first well-resolved proton-decoupled NAD spectra (^{2}H-{^{1}H}) of various small organic molecules using a Fourier Transform NMR (FT-NMR) spectrometer.[5] This key-study proved the invaluable interest of accumulating the signal of a low abundant nucleus combined with FT techniques. It demonstrated also the possibility to separate the single resonance associated with each non-equivalent ^{2}H site in molecules, and hence to measure accurately their ^{2}H chemical shifts ($\delta_{^2H}$) and subsequently evaluate the proton chemical shifts ($\delta_{^1H}$). This approach was introduced to simplify the analysis of overcrowded ^{1}H 1D spectra. This idea was later re-used to give reliable assignment of $\delta_{^1H}$ for a series of polysubstituted cyclohexane.[6] Since these pioneering studies, significant technological progresses in NMR have been achieved. In particular, the advent of modern FT-NMR instrumentations operating at very high magnetic field together with advanced quality electronic circuitries has considerably extended the application domains of the NAD-NMR in liquids as well as in liquid crystal or solid phases. More recently the development of cryogenically cooled probes has really revolutionized the world of NMR for the analysis of low naturally abundant nuclei.[7]

Disregarding the sensitivity problem, ^{2}H nuclei (spin I = 1) possess various interesting spectral properties that make possible recording NAD-NMR spectra in all states of the condensed matter with reasonable experimental conditions. The main one is a rather small nuclear electric quadrupole moment, ($Q_D = 2.86 \times 10^{-31}$ m^2),[1,2] that can interact with the electric field gradient (*EFG*) caused by the surrounding electrons such as those along the C-D bond. The existence of the ^{2}H quadrupolar interaction has direct consequences both on the analytical content of anisotropic ^{2}H spectra (liquid crystals and solids) as well as NMR relaxation mechanisms.[8] Actually the quadrupolar splittings measured in anisotropic media is a relevant source of detailed data for studying molecular properties of ordered solutes or mesogenic molecules.[9] As far as spectral analysis is concerned, NAD-NMR spectra appear rather simple to analyze because ^{2}H-^{2}H couplings are not observed due to the very low probability to have two deuterons in the same isotopomer (2.4×10^{-6} %). Hence, when the protons are decoupled from deuterons, the NAD-NMR spectra contain a superposition of more or less resolved independent signals corresponding to all non-equivalent mono-deuterated isotopomers in the sample.

This chapter presents an overview of the NAD-NMR spectroscopy in achiral thermotropic and chiral lyotropic liquid-crystalline solvents. The analytical possibilities of these tools will be emphasized by highlighting both the major applications and the most recent advances. The Chapter is organized as follows. After a general introduction, Sec. 2 gives a brief theoretical description of the anisotropic NAD-NMR. Sections 3 and 4 focus on the potentialities of NAD-NMR in thermotropic liquid crystals (LC) and in weakly ordering chiral liquid crystals (CLC). Finally, Sec. 5 presents the latest advances of NAD-NMR in CLCs using selective ^{2}H cryogenically cooled probes.

2.2. Theoretical Framework of NAD-NMR in Anisotropic Solvents

To simplify the theoretical description of spin Hamiltonian useful in NAD-NMR, we do not consider any homo- and heteronuclear dipolar and scalar interactions for two reasons: (i) the very low probability at natural abundance level to have two deuterons in the same isotopomer prevents the ^{2}H-^{2}H couplings, (ii) all heteronuclear ^{2}H-^{1}H coupling interactions are experimentally removed by ^{1}H decoupling.

2.2.1. *Brief description of spin Hamiltonian involved in NAD-NMR*

In liquid state, all anisotropic spin interactions of spin Hamiltonian are averaged to zero due to molecular isotropic Brownian motions.[9,10] As a consequence, the ^{2}H NMR signal of an isolated deuteron originates from two degenerated transitions, thus producing the same resonance frequency in isotropic media (see Fig. 2.1a). In practice, isotropic NAD-NMR spectra consist of independent, single resonance for each chemically non-equivalent ^{2}H site in the molecule, presuming that the magnetic field strength is large enough to resolve all resonances. Indeed the relatively low magnitude of ^{2}H magnetogyric ratio ($\gamma_{^2H} = \gamma_{^1H}/6.515$) reduces strongly the frequency dispersion of ^{2}H peaks in the spectra. This situation can be unfavorable for large molecules when numerous signals overlap, but this drawback is less important when operating with high-field NMR spectrometers. Interestingly, deuterons have electronic clouds comparable to protons, and hence ^{2}H chemical shifts (in ppm) are essentially identical to the ^{1}H chemical shifts of the same sites.

In a static magnetic field of 'sufficient' strength, molecules of liquid crystals as well as solutes dissolved inside have a homogeneous

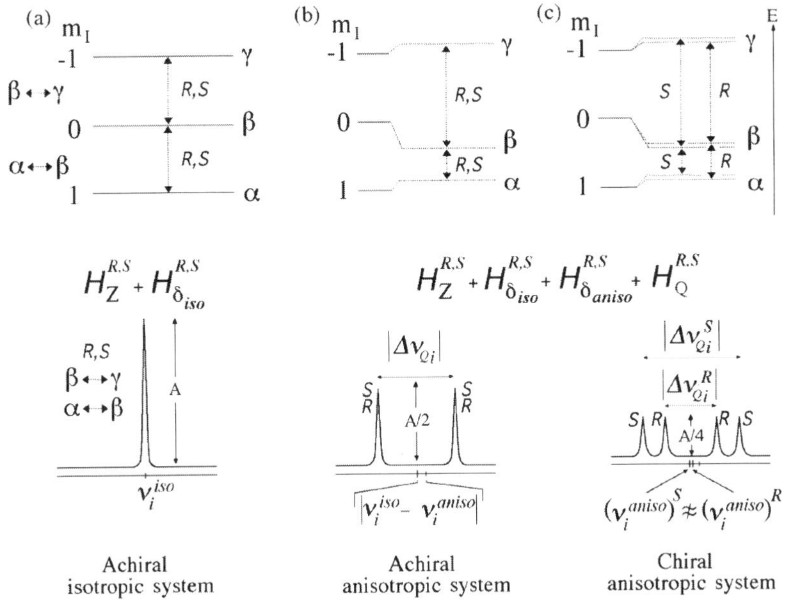

Fig. 2.1. Schematic ^{2}H-{^{1}H} signals expected to be observed for monodeuterated enantiomers dissolved in (a) liquid solvents, (b) in achiral and (c) chiral nematic mesophases. On top are depicted the diagrams of energy levels under H_Z, H_δ and H_Q interactions. All spectra are plotted at the same scale.

macroscopic orientation.[9] The existence of orientational order, denoted *S*, for molecules (mesophase or solute) has important consequences on the NMR Hamiltonian, because all order-sensitive intramolecular interactions are no longer motionally averaged to zero. Consequently, disregarding any ^{2}H-^{2}H couplings, the proton-decoupled NAD-NMR spectra in partially ordered media are governed by the spin Hamiltonian, H_{tot}^{aniso}, that is the sum of the Zeeman interaction (H_Z), the isotropic (H_δ^{iso}) and anisotropic (H_δ^{aniso}) part of the electronic shielding and the quadrupolar interaction H_Q. Therefore, one can write:[9]

$$H_{tot}^{aniso} = H_Z + H_\delta^{iso} + H_\delta^{aniso} + H_Q. \tag{2.1}$$

In practice the Hamiltonian sum, $H_Z + H_\delta^{iso} + H_\delta^{aniso}$, can be written in frequency units (Hz) as:[9,10]

$$H_Z + H_\delta^{iso} + H_\delta^{aniso} = -\frac{\gamma_D B_o}{2\pi} \sum_i \left[1 - \sigma_i^{iso} - \Delta\sigma_i^{aniso}\right] \mathbf{I}_{zi} \tag{2.2}$$

with

$$\sigma_i^{iso} = \frac{1}{3}\left(\sigma_{iaa} + \sigma_{ibb} + \sigma_{icc}\right) \text{ and } \Delta\sigma_i^{aniso} = \frac{2}{3} \sum_{\alpha,\beta=a,b,c} \sigma_{i\alpha\beta} S_{\alpha\beta}\,. \tag{2.3}$$

In these equations, γ_D is the ^{2}H magnetogyric ratio (in rad.s^{-1}.T^{-1}), and σ_i^{iso} and $\Delta\sigma_i^{aniso}$ are respectively the isotropic and anisotropic part of the electronic shielding second-rank tensor whose elements are denoted $\sigma_{i\alpha\beta}$. Here the indices α, β (α, β = a, b, c) are associated to the axes of a molecule-fixed reference frame. $\Delta\sigma_i^{aniso}$ is also defined as chemical shift anisotropy or CSA, while $S_{\alpha\beta}$ are the second-rank order parameters:[9,11]

$$S_{\alpha\beta} = \frac{1}{2}\left\langle 3\cos\theta_{z\alpha}\cos\theta_{z\beta} - \delta_{\alpha\beta}\right\rangle. \tag{2.4}$$

In Eq. (2.4), $\theta_{z\alpha}$ and $\theta_{z\beta}$ are the angles between the z axis (z axis along the director, i.e. magnetic field axis, $\mathbf{B_o}$) and the α and β axes. $\delta_{\alpha\beta}$ is the Kronecker symbol (equals to unity if $\alpha = \beta$ and zero otherwise). The angular brackets denote a statistical ensemble average over all possible molecular orientations. Actually, Eq. (2.4) defines the elements of the alignment tensor, ***S***, (often called the Saupe's order matrix), that is a (3 × 3) symmetric and traceless tensor. ***S*** contains at most five non-zero independent order parameters. From the components, $S_{\alpha\beta}$, we can derive the internuclear order parameter, S_{ij}, associated with any ij direction in the molecule,[9]

$$S_{ij} = \sum_{\alpha,\beta=a,b,c} \cos\theta_{ij\alpha}\cos\theta_{ij\beta} S_{\alpha\beta}, \tag{2.5}$$

where $\theta_{ij\alpha}$ and $\theta_{ij\beta}$ are the angles between the ij direction and the molecular axis α and β respectively (α,β = a, b, c).

Disregarding any solvent effects, the ^{2}H CSA can lead to a shielding or deshielding effect compared to the isotropic situation. This effect is however relatively small for ^{2}H nuclei, and hence the chemical shifts of signals in isotropic and anisotropic phases are mostly identical. As in isotropic NMR, anisotropic NAD-NMR suffers also from the small range of ^{2}H chemical shifts. However, the presence of the quadrupolar interaction, H_Q, in H_{tot}^{aniso} yields a wide range of quadrupolar splittings that may eliminate peak overlaps observed in isotopic ^{2}H spectra.

Assuming that the quadrupolar contribution is much smaller than the Zeeman term, the truncated ^{2}H quadrupolar Hamiltonian, H_Q, can be written in Hertz as:[9,12]

$$H_Q = \frac{1}{h}\sum_i \frac{eQ_D eq_{izz}}{4}\left[3\mathbf{I}_{zi}^2 - \mathbf{I}_i^2\right]. \tag{2.6}$$

In Eq. (2.6), eQ_D is the electric quadrupole moment of ^{2}H and eq_{izz} is the principal component of the *EFG* tensor along $\mathbf{B_o}$ at the deuterium i (denoted also V_{izz}). The product, eq_{izz}, is related to the elements of the *EFG* tensor in the molecular frame, $eq_{i\alpha\beta}$, using:[9,12]

$$V_{izz} = eq_{izz} = \frac{2}{3}\sum_{\alpha,\beta=a,b,c} eq_{i\alpha\beta} S_{\alpha\beta}. \tag{2.7}$$

As displayed in Fig. 2.1b, the first-order quadrupolar interaction influences the Zeeman energy levels. Under this effect, the two possible transitions no longer possess the same resonance frequency. The NMR signal of a single ^{2}H nucleus in ordered systems is therefore a doublet centered on the ^{2}H anisotropic chemical shift. The separation between the two components of the doublet, $\Delta\nu_{Q_i}$, is referred to as the residual quadrupolar coupling or the quadrupolar splitting. Experimentally, in homogeneous uniaxial oriented systems, NAD-NMR spectra consist of a sum of independent quadrupolar doublets each corresponding to an inequivalent ^{2}H site in the molecule studied.

In the case of deuterons, the *EFG* is usually assumed to be axially symmetric along the C-D bond. Consequently, the asymmetry parameter of the *EFG* tensor is supposed to be equal to zero (i.e $\eta_{EFG} = (V_{xx} - V_{yy}) / V_{zz} = 0$). Neglecting η_{EFG}, the quadrupolar splitting associated with a given C-D bond is given by:[9,11]

$$\Delta\nu_{Qi} = \frac{3}{2}C_Q^D \times S_{C-Di} = \frac{3}{2}\left(\frac{eQ_DV_D}{h}\right)\times\left(\frac{1}{2}\left\langle 3\cos^2\left(\theta_{C-D_i}\right)-1\right\rangle\right). \tag{2.8}$$

In Eq. (2.8), θ_{C-D_i} is the angle between the $\mathbf{B_0}$ magnetic field axis and the C-D_i bond axis, C_Q^D is the ^{2}H quadrupolar coupling constant and S_{C-Di} is the order parameter of the C-D_i bond axis (see Eq. (2.5)). The C_Q^D value can vary from one deuterium site to another one depending on the hybridization state of the bonded carbon. C_Q^D is approximately equal to 170 ± 5 kHz, 185 ± 5 kHz and 210 ± 5 kHz, for sp^3, sp^2 and sp carbons, respectively.[9,13] The uncertainty of 5 kHz on C_Q^D is mainly due to small *EFG* variations depending on the electronic vicinity. Although the C_Q^D term is large, the $\Delta\nu_{Qi}$ value can be equal to zero if $S_{C-Di} = 0$, namely when the average orientation of a C-D inter-nuclear axis points along the so-called 'magic angle' direction ($\theta_{C-D_i} = 54.7° = \theta_m$) (see Fig. 2.2b or 2.2e).

2.2.2. *NAD-NMR in chiral oriented solvents*

NMR spectroscopy is totally blind to chirality unless the external environment already discriminates between enantiomers. This is because mirror-image compounds have identical magnetic properties along $\mathbf{B_0}$ axis. In the last decade, an alternative to classical strategies used for analyzing chiral molecules (chiral derivatizing agents, chiral complexing agents or chiral solvating agents) has been intensively developped.[14-16] It consists of using chiral ordering agents, such as CLCs as enantiomeric selectors. This concept is quite old since it was proposed first by Sackmann and Snyder in 1968.[17-18] However, it was revisited and really exploited from the mid 90s by Courtieu's group.[19-21]

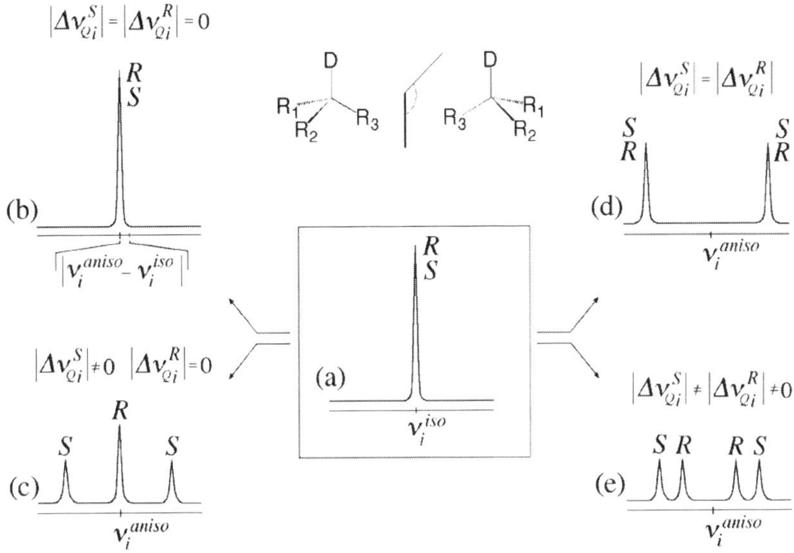

Fig. 2.2. Example of ^{2}H-{^{1}H} spectral patterns associated with two monodeuterated enantiomers (racemic mixture) dissolved in an achiral liquid (a) and a CLC (b-e): (a) isotropic spectrum, (b) $|\Delta\nu_Q^S| = |\Delta\nu_Q^R| = 0$, (c) $|\Delta\nu_Q^S| \neq 0$ and $|\Delta\nu_Q^R| = 0$, (d) $|\Delta\nu_Q^S| = |\Delta\nu_Q^R| \neq 0$, (e) $|\Delta\nu_Q^S| \neq 0$ and $|\Delta\nu_Q^R| \neq 0$. The difference of the ^{2}H CSA between *R* and *S* is assumed to be zero. The various spectra are plotted to scale. The *R*/*S* assignment is arbitrary.

The key point of this approach is the ability of chiral liquid crystals to orient differentially on average two enantiomers inside the mesophase in order to be detected by NMR. In other words, spectral enantiodiscrimination in CLCs is possible when $S^S \neq S^R$ (see Eq. (2.4)).[22,23] So far, the best results were obtained with CLCs made of chiral homopolypeptides such as the poly-γ-benzyl-L-glutamate (PBLG), poly-ε-carbobenzyloxy-L-lysine (PCBLL) or poly-γ-ethyl-L-glutamate (PELG) dissolved in various organic solvents ($CHCl_3$, DMF, THF, ...).[24] In contrast to thermotropics, these ordering systems possess a surprising ability to dissolve a wide variety of molecules, ranging from charged organometallic complexes to apolar compounds.[25,26]

Differences in intermolecular interactions between *R*/*S* isomers and the polypeptide α-helices lead to differences in the ordering effect. This orientational ordering difference can be sufficiently large to be revealed by NMR through the measurement of anisotropic interactions, such as dipolar or quadrupolar interactions.[25,26] Consequently, the choice of nuclei to observe and the related anisotropic observables are crucial since these latter do not exhibit the same sensitivity towards the ordering difference.[25] Compared to other anisotropic interactions (CSA or dipolar), the quadrupolar interaction is the most sensitive to small variations in the ordering. In the case of *S* and *R*-enantiomers, Eq. (2.8) can be rewritten as:[25,26]

$$\Delta\nu_{Q_i}^{S\,or\,R} = \frac{3}{2} C_Q^D \times S_{C-Di}^{S\,or\,R} = \frac{3}{2} C_Q^D \left(\frac{1}{2} \left\langle 3\cos^2\left(\theta_{C-D_i}^{S\,or\,R}\right) - 1 \right\rangle \right) \quad (2.9)$$

where C_Q^D at site *i* is assumed to be the same for the *S* and *R* isomers.

The spectral enantiodiscrimination at site i is detected when $|\Delta\nu_{Q_i}^S| \neq |\Delta\nu_{Q_i}^R|$, *i.e.* $|S_{C-D_i}^S - S_{C-D_i}^R| \neq 0$. Generally, two doublets are observed (see Fig. 2.2e) but one of the splittings can be averaged to zero (see Fig. 2.2c) when the *C-D* axis lies along the 'magic angle' direction (see above). More rarely is encountered the case where $|\Delta\nu_Q^R| = |\Delta\nu_Q^S| = 0$ (see Fig. 2.2b). The relatively large magnitude of C_Q^D is advantageous because it implies that the splittings may be sufficiently different to provide enantiomeric separation even though the difference in local orientation ($S_{C-D_i}^{S\,or\,R}$) between the *S* and *R* isomers is small. Besides, Eq. (2.9) indicates that order parameters, $S_{C-D_i}^{S\,or\,R}$ for both isomers are composite quantities as they depend both on the angle $\theta_{C-D_i}^{S\,or\,R}$ and on the motional averaging (<···>). Consequently, whenever two order parameters are different, one generally ignores if this happens from a purely geometrical reason ($\theta_{C-D_i}^S \neq \theta_{C-D_i}^R$) or from a difference in motional averaging, or even both.[22,25] Assuming that the motional averaging for enantiomers should be more or less the same, we deduce that the magnitude of the enantiodifferentiation should depend mainly on the relative orientations of the *EFG* direction for a given C-D bond relative to $\mathbf{B_o}$.

2.2.3. *^{2}H relaxation and consequences on NAD-NMR spectra*

Both longitudinal (spin-lattice) (T_1) and transversal (spin-spin) relaxation time (T_2) contain information on molecular motion, including internal motion in flexible molecules.[27] Extensive theoretical descriptions of the ^{2}H relaxation mechanisms can be found in the literature.[27-29] Here only qualitative presentation of the main consequences of ^{2}H relaxation on NAD-NMR spectra is given.

Due to the relatively small quadrupole moment of deuterons,[1,2] the ^{2}H linewidths are narrow enough to give high-resolution spectra both in isotropic and LC solvents of low viscosity samples. In a first approximation, the quadrupolar relaxation process is sensitive to molecular ordering, *S*, because the associated spectral density possesses a dependence on *S*.[2,30,31] When the ordering is large (mesogenic molecules), complex anisotropic motional models must be developed to reach desired molecular information.[28] In contrast when the ordering is small ($|S_{\alpha\beta}| < 10^{-3}$), the S^2 dependency of the ^{2}H quadrupolar relaxation can be neglected, and so the understanding of the ^{2}H relaxation mechanisms can be achieved with classical isotropic models.[10] Generally, $T_1(^2H)$ and $T_2(^2H)$ relaxation times of solutes measured in weakly ordering mesophases are smaller than values measured in liquids (at the same temperature), mainly due to the liquid crystal viscosity.[32] Experimentally for weakly oriented solute molecules in organic solutions of polypeptide, ^{2}H nuclei have T_1 and T_2 values ranging from approximately 0.1 to 2 s, leading to linewidths of less than 5 Hz for orientationnally homogeneous samples. The T_1 values are short enough to acquire many FIDs with almost fully relaxed magnetization without using long time-consuming relaxation delays. This property is highly valuable for NAD-NMR.

The dominance of the quadrupolar relaxation mechanism means that ^{1}H irradiation produces a negligible intramolecular nuclear Overhauser effect (nOe). Thus ^{1}H decoupling yields an improvement of signal-to-noise (S/N) ratio through the elimination of ^{2}H-^{1}H scalar and dipolar couplings. Generally these couplings are too small to be resolved but do contribute to the ^{2}H linewidths.[8] Their elimination significantly narrows the linewidths, thus improving the S/N ratio. The absence of nOe is, however, an advantage since it means that NAD-NMR can be used for quantitative measurements when spectra are recorded under 'quantitative' conditions (recyling time: $T_R > 5T_1$). Thus, disregarding any effects due to relaxation or the specific natural isotopic fractionation (see below), the integrals of NAD resonances are proportional to the number of equivalent deuterons contributing to the signal.

In anisotropic environments, order parameters depend on temperature (T), and so any gradient of T can lead to a line broadening, whose size is proportional to the ^{2}H doublet splitting. A dependence of linewidths on the magnitude of quadrupolar splittings may also occur if there is a non-homogeneous alignment of the mesophase directors (**n**) by the magnetic field of NMR spectrometer. In practice we observe that the larger the splitting is, the larger the linewidth is. This effect is a consequence of the 'molecular disorder' inside the mesophase, and reflects that mesogenic molecules or solutes are not uniformly aligned with respect to the director **n**.[9]

2.3. NAD-NMR Spectroscopy of Strongly Aligned Molecules

The liquid crystal state possesses anisotropic physical properties similar to those of crystals, and fluidity close to those of liquids.[33,34] Contrary to liquids where molecules are freely tumbling (Brownian motion), the molecules in LCs (mesogen or solute) acquire some orientational order. As a consequence, the order-dependent NMR interactions are no longer averaged to zero,[9] and hence anisotropic NMR spectra contain numerous information about molecular geometry and orientation as well as conformational dynamics.[35,36] The presence of anisotropic interactions such as dipolar couplings makes the analysis of oriented systems more complex than in isotropic liquids. Thus ^{1}H NMR spectra of bulk liquid crystals show generally only poorly resolved resonances with linewidths up to several tens of kHz, while those of highly ordered solutes are composed of tens to thousands of well-resolved peaks.[9,34] This spectral complexity strongly limits the range of applications of anisotropic ^{1}H NMR as a routine analytical method in organic chemistry. As we will see below, NAD-NMR provides an interesting solution to this problem mainly due to the absence of ^2H-^2H dipolar and scalar couplings.

2.3.1. *Analysis of mesogenic molecules through NAD-NMR*

^2H-{^1H} NMR is a spectroscopic method of choice for examining microscopic properties (C-D order parameters, dynamical properties, molecular geometry and conformation...) of isotopically enriched mesogenic molecules in the nematic or even smectic phase.[37,38] The efficiency of this tool originates from the fact that ^2H spectra and relaxation rates are mainly dominated by the nuclear quadrupolar interaction. Hence the extraction of desired information in terms of orientational order or dynamic properties is facilitated. Thus numerous perdeuterated 4,4'-cyanobiphenyl derivatives such as 4-pentyl-4'-cyanobiphenyl (5CB) or 4-hexyloxy-4'-cyanobiphenyl (6-OCB) in their nematic phase have been extensively investigated.[36-39]

Although numerous chemical syntheses exist, the preparation of selectively deuterated, polydeuterated or perdeuterated mesogenic compounds is often an arduous task.[40,41] This isotopic enrichment step has limited the application of ^2H NMR to investigating LC properties.

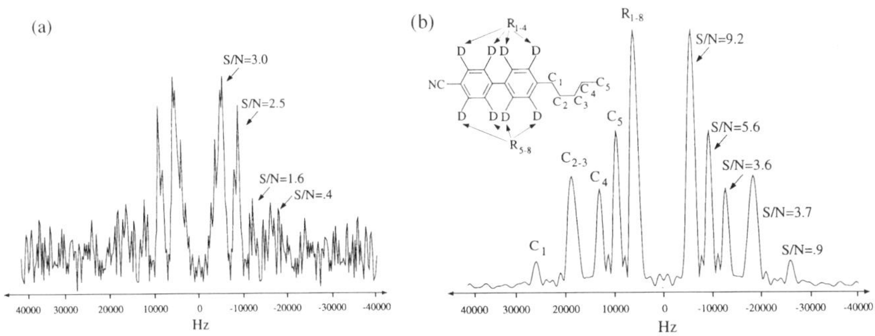

Fig. 2.3. 46.1 MHz NAD-NMR spectra of 5CB in the nematic phase (301 K) recorded under the same condition using (a) a conventional quadrupolar echo sequence (18,000 scans) and (b) a CPMG pulse train consisting of ten refocusing pulses followed by a FT of the sum of the autocorrelation functions associated with ten echoes. (Courtesy of Prosser *et al.* [42]. Reprinted with permission from Academic Press).

Obviously NAD-NMR applied for investigating such materials appears to be a direct tool to obtain the desired information. This solution was pioneered by Prosser, Heaton and Kothe in 1996.[42] As the number of dideuterated isotopomers is statistically too low to be detected, the NAD-NMR spectra of uniaxial liquid crystals are made of independent quadrupolar doublets with a range of splittings that varies generally from tens to hundreds kHz. Generally, rather symmetrical NAD 1D spectra are recorded where the lines can be several hundred Hertz wide. From a S/N point of view, the significant broadening of lines is not a favorable situation. Thus in the example shown, only the doublet associated to the aromatic deuterons of 5CB are observable (see Fig. 2.3a). Although the investigation of nematogens in their nematic phase requires a neat sample, the sensitivity of these NAD experiments remains basically rather poor. To improve the sensitivity of NAD-NMR 1D experiments, in particular for the measurement of spin-lattice relaxation times, these authors have proposed the application of the Carr-Purcell-Meiboom-Gill (CPMG) pulse trains inserted inside ^{2}H pulse sequences such as the quadrupolar echo or Jeener-Broekaert sequences.[42] Taking advantage of the pulse spacing dependence of CPMG relaxation time (T_2^{CP}), it becomes possible to collect several echoes and then add their corresponding autocorrelation functions prior to FT. The gain in sensitivity is significant since the S/N ratios are enhanced by a factor varying between 2 and 3 (Fig. 2.3b). Actually the loss in resolution generated by the echo-train acquisition is not dramatic as signals of all non-equivalent deuterons are well separated. In this example, the analysis of NAD spectrum allows the determination the $\Delta\nu_Q$'s of almost all sites of 5CB, and also the calculation of order parameters, $S_{C\text{-}D}$, for each C-D bond.

Whereas Prosser *et al.* have obtained the NAD spectra of 5CB using a 7.1 T NMR spectrometer designed for solids with a 10-mm solenoid probe, Tabayashi and Akasaka in 1997 have reported the possibility of recording the NAD spectrum of 6-OCB in its nematic phase using a conventional high-resolution spectrometer designed for liquids together with a standard 5-mm saddle coil probe.[43] Actually recording NAD-NMR spectra of a mesogenic compound in the nematic phase with a

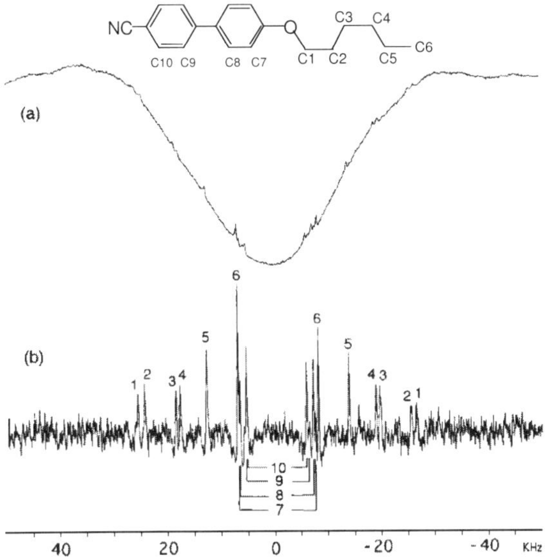

Fig. 2.4. 61.2 MHz NAD-NMR spectrum (180,000 scans) of 6-OCB in its nematic phase (333 K) with no filtering (a) and with trapezoidal filtering (b). The spectrum was recorded using a single rf-pulse sequence under ^{1}H decoupling. (Courtesy of Tabayashi *et al.* [43]. Reprinted with permission from American Chemical Society).

5-mm probe instead of a 10-mm is no longer a real problem nowadays. Besides, small diameter samples are valuable for spectral resolution because a better homogeneity of the mesophase and the magnetic field are generally reached in that case.

Taking advantage of the higher sensitivity of high-field NMR equipments, it is possible to greatly reduce the total amount of the solute to be studied. The NAD 1D spectrum shown in Fig. 2.4 was recorded by adding 180,000 scans during 20 hours on a 9.4 T spectrometer, but only 20,000 scans and two hours have been sufficient to record the same spectrum by us with a satisfactory S/N ratio using a 18.8 T spectrometer (ν_D = 115.1 MHz). This spectrum was obtained with a broadband decoupling pulse scheme[44] at a radiofrequency (RF) field strength of 100 kHz. This multiple-pulse sequence for ^{1}H decoupling is recommended in order to reduce a possible noticeable increase of the sample temperature. As a result of an efficient ^{1}H decoupling, ^{1}H-^{2}H spin-spin couplings are eliminated, thus significantly narrowing the linewidth of the doublets, and clearly increasing the S/N ratio of each resonance peak (see above).

The strongly distorted baseline seen in Fig. 2.4a originates from acoustic ringing effect that is usually eliminated by using a quadrupolar echo sequence.[45] In this case, however, the use of an echo was not feasible due to long ^{2}H pulse durations (over 20 μs). Other pulse sequences taking advantage of differences in T_1 and T_2 relaxation times between the ^{2}H signals could be proposed to cancel this phenomenon, but so far this method has only been applied to the cases where the quadrupolar splittings are rather small. Finally a simpler solution consists in filtering the rapidly decaying ringing component using a trapezoidal function applied on the FID prior to FT.[43] As seen in Fig. 2.4b, such post-acquisition treatment leads to a reasonably flat baseline and signals with a sufficient S/N ratio. In this work, the authors could evaluate the order parameters for each C-D direction with respect to $\mathbf{B_o}$.[43]

Using the same approach, they have also studied the 4-propyl-4'-(3,4-diflurophenyl)bicyclohexane in its nematic phase and examined whether the ^{2}H signals measured in specifically deuterated sites allowed to monitor safely the microscopic ordering of the LC, as in case of the non-enriched materials.[46] Two major difficulties exist when recording the NAD spectrum of mesogens in a nematic phase: (i) the low sensitivity of ^{2}H nuclei; (ii) the magnitude of ^{2}H-^{1}H dipolar couplings that can be particularly large in nematogens, thus requiring high power to decouple ^{2}H-^{1}H dipolar interactions. The first problem can be avoided using high-field NMR spectrometers ($\mathbf{B_o} \geq 14.1$ T) equipped with a selective deuterium cryogenic probe. Such a combination has been tested by recording the 92.1 MHz NAD spectrum of 5CB in the nematic phase at 305 K.[47] Although all the quadrupolar doublets are observed in the NAD 1D spectrum, the NMR results were quite disappointing. Indeed in this experiment, the protons have been decoupled with the classical low-power WALTZ-16 composite pulse decoupling (CPD) sequence and the RF field strength did not exceed 10 kHz. This RF value is clearly not large enough to properly decouple the ^{2}H-^{1}H dipolar interactions, and hence the linewidths are unusually large (between 250 and 400 Hz). Although partly compensated by the sensitivity of the ^{2}H cryoprobe,

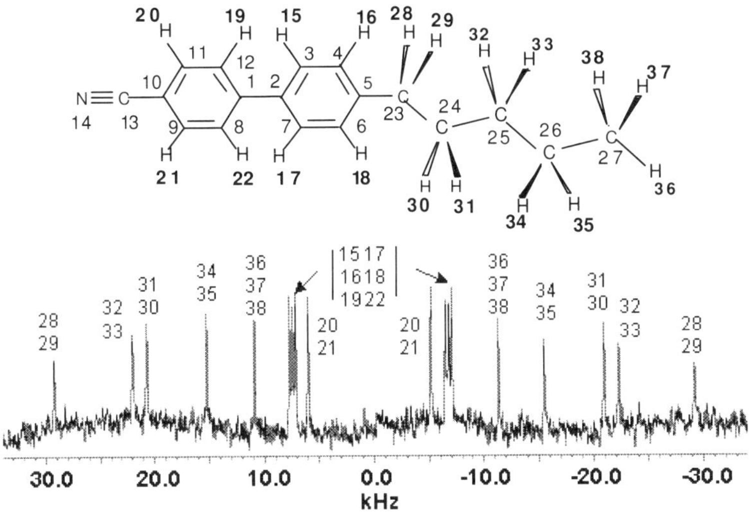

Fig. 2.5. 107.5 MHz NAD-NMR spectrum of 5CB (16,000 scans) using a 4-mm CPMAS probe and no sample spinning (293 K). The recycle delay was set at 2s.

this low resolution strongly reduces the S/N ratio of each doublet. Unfortunately for technological reasons, high-power ^{1}H decoupling is currently not allowed with cryogenic probes in order to protect the probehead cooled at around 15-20 K.

An alternative to cryogenic probes is the use of high-field solid-state NMR spectrometers (17.5 T) coupled with a CPMAS probe as proposed by Emsley *et al.*[47] This combination appears to be a promising solution as seen in Fig. 2.5. To prevent the averaging out of the quadrupolar interaction, the NAD spectrum of 5CB is recorded without sample spinning. Besides the ^{1}H SPINAL-64 decoupling scheme is applied during the acquisition of signal at a RF field strength of *ca.* 50 kHz. In these conditions the ^{1}H decoupling is efficient enough to produce reasonable and uniform linewidths ($\approx$ 100 Hz) throughout the spectrum, thus allowing the NAD spectrum to be recorded with an experimental time smaller than with the cryoprobe. In addition, the measurement of ^{2}H splittings can be performed with a better accuracy.

2.3.2. *Studies of solutes oriented in thermotropic liquid crystals*

Solutes embedded in thermotropics are generally strongly aligned, and so their ^{1}H spectra are very complex to analyze for two reasons: (i) the magnitude of long and short range ^{1}H-^{1}H dipolar couplings leads to second order spectral patterns; (ii) the number of interacting nuclei yields a substantial number of single quantum transitions. In practice, the ^{1}H spectrum of a solute bearing more than ten protons is usually undecipherable due to the considerable number of resonances in the ^{1}H spectra (up to few thousands).[9] To overcome these problems, miscellaneous approaches such as selective or partial isotopic substitution,[9,48] multiple-quantum NMR,[49,50,51] spinning near magic angle experiments[52] or coherent reduction of the spin Hamiltonian by multiple-pulse sequences[53,54,55] were explored as very helpful aids in providing starting data for analyzing very complex anisotropic ^{1}H spectra.

An alternative of the above solutions, based on NAD-NMR, was described by Khetrapal *et al.* in 1998. Indeed the authors proposed to record both ^{1}H and ^{2}H spectra at natural abundance level at the same temperature.[56] The idea consists of using quadrupolar information extracted from NAD spectra in order to determining the molecular order parameters of the solute, and subsequently for estimating the ^{1}H-^{1}H dipolar couplings, with the geometry of the guest molecule being known. Figure 2.6 reports the 61.4 MHz NAD-NMR 1D spectrum obtained for benzene dissolved in ZLI-1114 and recorded in 18 hours of experimental time. In this spectrum, both the NAD-NMR signals of solute and solvent are superimposed, but their respective assignment is possible because the solute signals are much sharper than the ^{2}H lines from the LC itself.

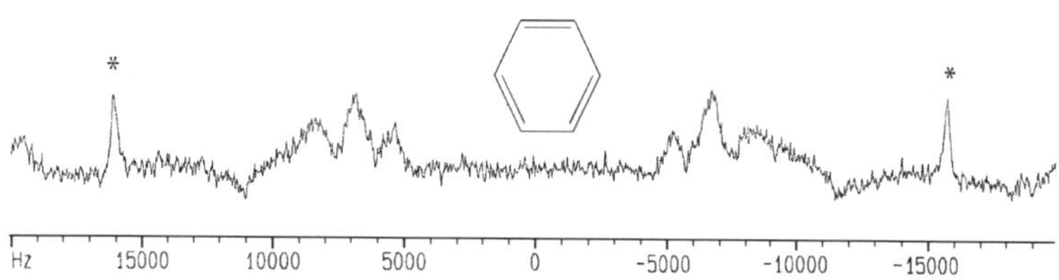

Fig. 2.6. 61.4 MHz NAD-NMR spectrum of benzene in ZLI-1114 and recorded using a 10-mm probe and single pulse sequence (64,000 scans) (303 K). The doublet of benzene ($\Delta\nu_Q$ = +31,952 Hz) is marked by a star while the broad signals originate from the LC. (Courtesy of Khetrapal *et al.* [56]. Reprinted with permission from Elsevier).

From the experimental value of $\Delta\nu_Q$, the authors have calculated the unique order parameter of benzene (S = +0.108), and also evaluated the dipolar coupling between the *ortho* protons (D_{HH} = -852 Hz). From this value, the ^{1}H spectrum of benzene was simulated with only small deviations between experimental and calculated line positions.

2.4. NAD-NMR in Weakly Orienting Polypeptide Solvents

Various efforts have been devoted to design new LCs that orient solutes much less strongly than thermotropic systems.[57-60] Though many systems for aligning hydro-soluble molecules were described, the range of weakly orienting LCs available for organo-soluble compounds is rather small and inadequate for NAD-NMR due to their weak ability to dissolve a large amount of solute.[24,61,62] So far, LCs made of organic solutions of homopolypeptide gave the best results in particular for analytical applications using NAD-NMR as first proved in 1998,[63] and then intensively developed by Lesot and co-workers (see below). In these rather fluid systems, the degree of solute orientation is low ($|S_{\alpha\beta}|$ = 10^{-3} to 10^{-5}).[22,23] This range of S values is slightly larger than in amphiphilic ordered systems but much smaller compared to values measured in thermotropics ($|S_{\alpha\beta}|$ = 0.1 to 0.2). Interestingly, these polypeptide oriented systems may be chiral, thus providing an unique opportunity to spectrally discriminate between enantiomers of chiral solutes,[20,21] including compounds that are chiral by virtue of substitution as reported by Meddour *et al.*[64] or enantiotopic elements in prochiral molecules as first established by Czarniecka and Samulsky,[65] and further demonstrated in various studies.[66,67] Another advantage of polypeptide organic solutions is the great capacity to dissolve a large amount of solute (up to 30 % w/w) without losing the liquid crystalline properties. This specificity of polypeptidic systems is highly valuable from the sensitivity viewpoint because it partly compensates the larger number of resonances when enantiomers or enantiotopic directions are discriminated. In practice, the spectral enantiodiscrimination in CLCs reduces the S/N ratio of ^{2}H NMR signals by a factor of 4 (racemic mixtures, same linewidths) compared to isotropic achiral solvents (see

Figs. 2.1 and 2.2). Finally due to the small magnitude of the ^1H-^2H couplings, the power required to decouple ^1H is similar to that used in isotropic NMR, and so proton-decoupled NAD-NMR spectra can be performed quite easily on routine 'liquid' NMR spectrometers

2.4.1. *Spectral discrimination of enantiomers using NAD-NMR*

As an illustrative example, Fig. 2.7 presents the first NAD spectrum of a solute ((±)-phenethyl alcohol) dissolved in the PBLG/$CHCl_3$ mesophase. This spectrum was recorded on an 'old' 250 MHz spectrometer (AM 250) in 1998.[63] Contrary to NAD spectra recorded in nematic thermotropics, the NAD spectra obtained in PBLG mesophases have not a symmetrical aspect. This is primarily due to the smaller range of quadrupolar splittings (generally less than several hundred of Hertz) compared to the range of $\delta_{^2H}$. As another consequence of low ordering, the average linewidths of the peaks is about 2-5 Hz. Such a situation is highly valuable because a significant gain in S/N ratio for each resonance is obtained. Note also that the NAD signals of the polypeptide helices are rarely observed,[21] and generally do not interfere with those of the solute. From the chiral analysis viewpoint, it appears that the methyl group exhibits two well-resolved doublets, indicating that the

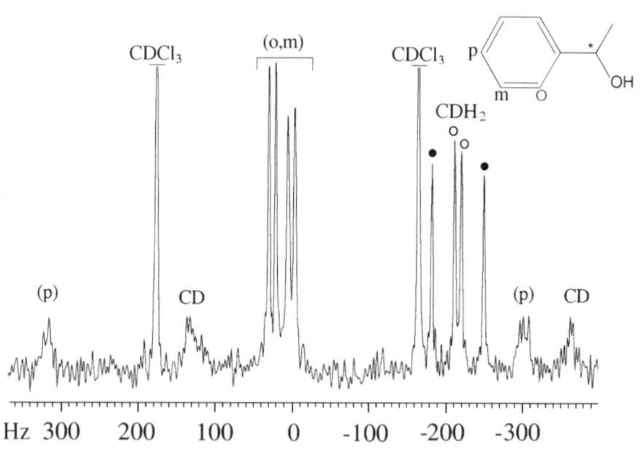

Fig. 2.7. 38.4 MHz NAD-NMR spectrum of (±)-phenethyl alcohol oriented in PBLG/$CHCl_3$ (298 K). This spectrum (150,000 scans) was recorded using a 5-mm BBI probe with a single-pulse NMR sequence and the proton WALTZ-16 pulse scheme.

enantiomers are spectrally discriminated, while other sites do not show any differentiation. This situation illustrates the interest of NAD-NMR for which all possible ^{2}H sites in the molecule are simultaneously probed.

2.4.2. *Design of 2D pulse sequences dedicated to NAD-NMR*

The ability to observe the enantiodifferentiation on each ^{2}H site of a chiral solute is really advantageous, but also makes the analysis of NAD spectra more complex. Disregarding the doublets originating either from the co-solvent or from exchangeable site (-OD), 2*n* doublets are expected for a racemic mixture of enantiomers possessing *n* non-equivalent deuterons, assuming that all monodeuterated isotopomers are discriminated and no line overlaps or null quadrupolar splittings occur. As the NAD-NMR spectra recorded in PBLG have not symmetrical aspect, the pairing the two components of each quadrupolar doublet is not always trivial. To solve this problem, several ^{2}H-{^{1}H} 2D NMR experiments referred to as QUOSY (Quadrupole Ordered SpectroscopY) were designed (see Fig. 2.8).[68] Their aim is to autocorrelate between the two components of each doublet in a 2D map. Among them, the 2D

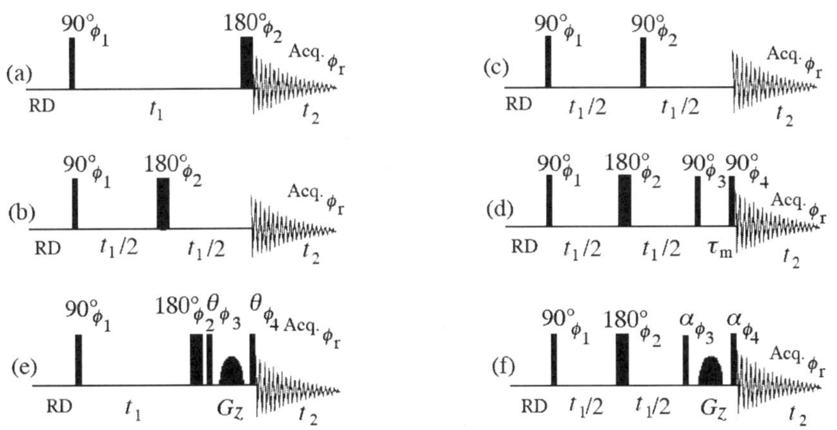

Fig. 2.8. (a-d) Schematic diagrams of basic QUOSY experiments: (a) *Q*-COSY, (b) *Q*-resolved, (c) δ-resolved and (d) *Q*-Double Quantum (*Q*-DQ) pulse sequence. (e and f) pulse diagrams of *Q*-COSY Fz (θ_m = 54.7°) and *Q*-resolved Fz (α = 60.0° (odd exp.) and α = 45.0° (even exp.)) where spin-1 z-filter gradient is added. The full phase cycling of these sequences can be found in refs. 68 and 70.

autocorrelation experiments named *Q*-COSY and *Q*-resolved have been found to be the most suitable 2D sequences in terms of the signal sensitivity and possibility of tilting the 2D spectrum.[68,69] Disregarding all relaxation terms and phase factors, the signal expression, $S(t_1, t_2)$, of the *Q*-COSY sequence after one single scan, is:[68,69]

$$S(t_1,t_2) = A\Big\{ \exp\big[-i(2\pi\nu_D + \pi\Delta\nu_Q)t_1\big]\times\exp\big[i(2\pi\nu_D - \pi\Delta\nu_Q)t_2\big] + \exp\big[-i(2\pi\nu_D - \pi\Delta\nu_Q)t_1\big]\times\exp\big[i(2\pi\nu_D + \pi\Delta\nu_Q)t_2\big]\Big\} \quad (2.10)$$

where $2\pi\nu_D$ and $\pi\Delta\nu_Q$ (expressed in rad.s^{-1}) denote the deuterium offset frequency and the quadrupolar splitting (see Eq. (2.8)), respectively.

For the *Q*-resolved experiment, the expression after one single scan, is

$$S(t_1,t_2) = A\Big\{ \exp\big[i(\pi\Delta\nu_Q)t_1\big]\times\exp\big[i(2\pi\nu_D + \pi\Delta\nu_Q)t_2\big] + \exp\big[-i(\pi\Delta\nu_Q)t_1\big]\times\exp\big[i(2\pi\nu_D - \pi\Delta\nu_Q)t_2\big]\Big\}. \quad (2.11)$$

Although the first four 2D sequences (Figs. 2.8a-d) facilitate the analysis of complex NAD spectra, they yield phase-twist signals, and so the 2D spectra have to be displayed in the magnitude mode.[68] Due to the line broadening associated with this spectral mode, these sequences are not convenient for analyzing very congested NAD spectra. This drawback is eliminated using *Q*-COSY-Fz and *Q*-resolved-Fz sequences that have been developed in 2004 to obtain phased 2D spectra.[70] In these sequences, a modified z-gradient filter (adapted for spin I = 1) was inserted before the acquisition period (Figs. 2.8e and 2.8f). This filter allows the separation of sine and cosine parts of the ^{2}H signal evolving during t_1. After a double FT, the resulting 2D spectrum can be phased in pure absorption in both spectral dimensions. The presence of the z-gradient filter leads to the *Q*-COSY-Fz and *Q*-resolved-Fz experiments being less sensitive (-33% and -25%) compared to the initial sequences. However these modified sequences remain a substantial achievement to analyze the NAD spectra of large molecules in particular when spectra

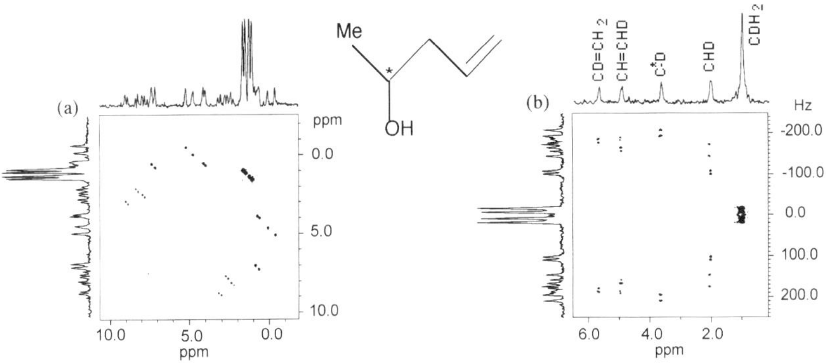

Fig. 2.9. (a) 61.4 MHz NAD Q-COSY and (b) tilted Q-resolved spectrum of (±)-4-penten-2-ol oriented in PBLG/$CHCl_3$ and recorded with a 5-mm BBI probe (298 K).

are recorded with ^{2}H cryoprobes (see below).[70] Indeed the loss in sensitivity is greatly compensated by the gain due to the ^{2}H cryoprobe and by the remarkable increase in resolution due to the refocusing of magnetic and sample inhomogeneities. As in the case of the Q-COSY 2D experiments, the tilting procedure is still applicable on the Q-COSY Fz maps due to the absence of diagonal peaks.[70]

To exemplify the usefulness of NAD QUOSY experiments, Fig. 2.9 presents the Q-COSY and tilted Q-resolved 2D spectra of (±)-4-penten-2-ol dissolved in PBLG/$CHCl_3$ (16 hours) recorded with a routine NMR spectrometer (9.4 T).[69] In both 2D spectra, seven pairs of chiral isotopomers (disregarding the -OH group) are discriminated, thus leading to 14 visible doublets for the solute. The diastereotopicity of methylenic deuterium sites produces two pairs of quadrupolar doublets centered on two distinct chemical shifts, *i.e.* 2.21 and 2.23 ppm.

2.4.3. *Analysis of chiral rigid and flexible hydrocarbons*

Enantiomeric discrimination observed by NAD-NMR is due to a selective orientational ordering of two mirror-image isomers inside the CLC. *A priori* this differential ordering effect depends on both electronic density profile and geometrical factors.[25,26] Hence the difference in ordering does not require a specific chemical function in the chiral solutes to be efficient.

To assess the limits of the enantioselectivity of polypeptide mesophases, a series of rigid and flexible enantiomers devoid of any polar functions has been experimentally investigated.[26,71] Thus it was demonstrated by Sarfati *et al.* that signals of chiral flexible alkanes such as (±)-3-methylhexane could be efficiently differentiated using NAD-NMR in PBLG mesophases.[71] Figure 2.10b displays the NAD *Q*-COSY map of (±)-*trans*-1,3-dimethylcyclohexane dissolved in PBLG/$CHCl_3$. The inversion of the chair conformation (*a,e* ↔ *e,a*) in this flexible cycle leads at room temperature to a single isomer with six inequivalent sites, that are not superposable with its mirror image (see Fig. 2.10a). As seen in Fig. 2.10b, six pairs of doublets centered on six distinct δ_{2H} are observed, thus indicating that all enantiomeric isotopomers are discriminated in the 2D spectrum. To confirm the NMR results obtained in PBLG, it can be very helpful to record the NAD spectrum of a solute in a racemic mixture of PBLG and PBDG (its enantiomer). In this achiral compensated mixture denoted 'PBG',[72] the spectral enantiodiscrimination disappears while spectral diastereodiscrimination is still active. Note that the spectral differentiation of enantiotopic elements (pro-*R*/pro-*S*) in prochiral molecules also vanishes in the PBG mesophase. In this example, the use of PBG mesophase (see Fig. 2.10c) confirms the observed enantiodifferentiation, and that no other physical effect arises in this example.[26]

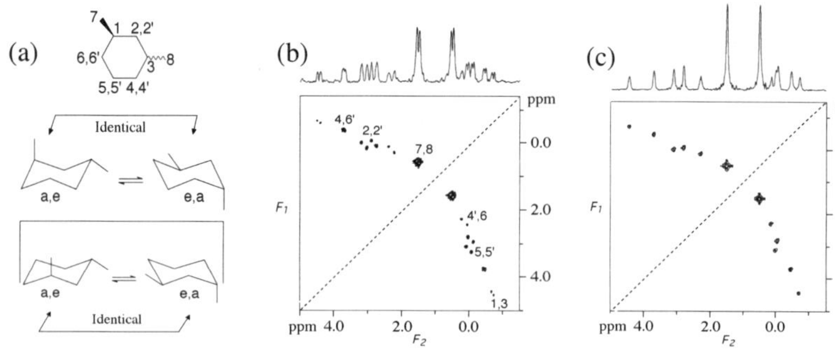

Fig. 2.10. (a) Structure, numbering and stereochemical relationship of (±)-*trans*-1,3-dimethylcyclohexane. 61.4 MHz NAD *Q*-COSY 2D spectrum obtained in (b) PBLG/$CHCl_3$ and (c) PBG/$CHCl_3$ mesophases (298 K).

The excellent NMR results obtained for apolar chiral hydrocarbons suggest that electrostatic solute-solvent interactions are not dominant in the chiral recognition of apolar or weakly polar solutes by these polypeptide solvents. In contrast, enantioselective contributions originating from short-range repulsive intermolecular forces, correlated with the molecular shape anisotropy, seem to play an important role in the enantiomeric discrimination mechanisms of such molecules.[26]

2.4.4. *Applications of NAD-NMR in asymmetric synthesis*

Small organic chiral building blocks play an essential role in synthesis of bioactive products and the control of their enantiomeric purity is of crucial importance. NAD-NMR in polypeptide CLC using routine NMR equipment is very well adapted for analysing such chiral molecules for two reasons: (i) the molecular weight of chiral building blocks are generally rather low (MW ~ 150); (ii) during the initial stages of a total synthesis, a large amount of chiral compound is available, thus allowing the preparation of PBLG samples with a large amount of solute.[73]

In the framework of the total synthesis of dolastatine-14 (Fig. 2.11a), the hept-6-yn-2-ol (**2**) was identified as an interesting chiral building block for preparing the dolatrienoic acid (the south fragment). The specific question addressed here was the control of the enantioselectivity of the alkyne zipper reaction leading to compound **2** from hept-3-yn-2-ol (**1**). Figures 2.11b and 2.11c present the NAD signals of the methyl group of **1** and those associated with one of the diastereotopic deuterons of the methylene group 3 for compound **2**, both in racemic and enantioenriched series. The presence of two quadrupolar splittings centered on the same chemical shift indicates that the chiral discrimination occurs for both sites shown in solutes **1** and **2**. Using an enantioenriched mixture prepared by an enantioselective reaction, only one quadrupolar doublet is observed in NAD-NMR spectra (Figs. 2.11d and 2.11e). Considering the NMR results obtained with **2**, the absence of NAD signals for the *S*-(+)-**2** isomer enables an estimate of the enantiomeric excess, *ee*, ($ee = |m^R - m^S| / (m^R + m^S)$) at least equal to the *ee* of *R*-(+)-**1**. This experimental NMR results assess unambiguously that the alkyne zipper reaction investigated in this example is a racemisation-free process.

The choice of sites for measuring the enantiomeric excess depends on both the S/N ratio and the difference between the *R*/*S* quadrupolar splittings. The best deuterium sites for quantification are the sites that possess the largest number of magnetically equivalent spins that contribute to a given doublet and have the strongest difference in quadrupolar splittings. Methyl groups are interesting sites but the spectral enantiodiscriminations observed can be sometime small or null. In practice, working with routine NMR spectrometers (9.4 T) equipped with standard 5-mm probes, it is possible to evaluate the enantiomeric purity in PBLG mesophases within an accuracy of 5-10% depending on the S/N ratios of ^{2}H site on which the *ee* measurement is performed, using either a deconvolution or integration method of signals.

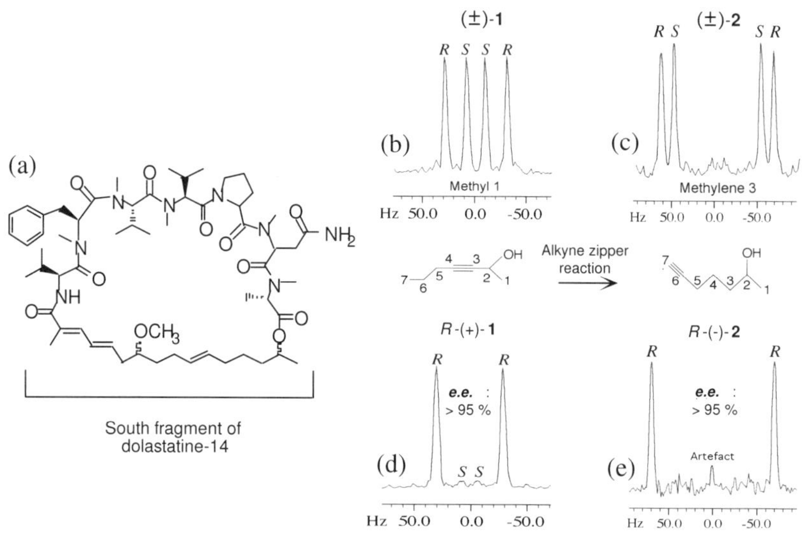

Fig. 2.11. (a) Structure of dolastatine-14. Right panel: 61.4 MHz NAD signals of (b and c) methyl group 1 of **1** and (d and e) one of two deuterons in methylene group 3 of **2** in racemic (top) and enriched series (bottom) in PBLG/$CHCl_3$ (300 K). These doublets (sum of columns) are extracted from the tilted *Q*-COSY 2D experiment.

2.4.5. *Spectral discrimination of enantiotopic elements*

Prochiral molecules possess enantiotopic elements (nuclei, groups or internuclear directions) that are only exchanged by an improper operation of symmetry, e.g. plane of symmetry.[74] Under the effect of enantioselective solute/solvent interactions, enantiotopic elements can be distinguished through NMR in CLCs using various nuclei.[75,76] From a theoretical point of view, the spectral discrimination of enantiotopic elements is brought about by a reduction of the symmetry in the CLC mesophase. In other words, the chirality of the orientating system causes the elimination of all improper symmetry elements. Hence a solute can have an effective symmetry, as far as ordering is concerned, which may be different from that of its molecular symmetry point group as demonstrated by Merlet *et al.* using arguments based on group theory.[77] Thus, for rigid prochiral compounds belonging to C_s, C_{2v}, D_{2d} or S_4 symmetry groups, the symmetry properties of the chiral ordering solvent will make the apparent symmetries respectively C_1, C_2, D_2 and C_2, thus leading to the spectral differentiation between enantiotopic elements.[65,77,78] This reduction in the apparent solute symmetry can change the location of the principal axes system of the Saupe's order matrix of the solute, which means that C-H or C-D enantiotopic directions in prochiral molecules (R-CD_2-R' or R-CH_2-R') are no longer equivalent. A direct consequence is that the NAD-NMR signals of the two possible enantio-isotopomers (R-C*HD-R') of prochiral molecules of the structure (R-CH_2-R') will also be different.[79]

In conclusion the spectral discrimination of molecule that are chiral by virtue of the isotopic substitution (isotopic enantiomers) does not originate from a difference in the molecular ordering, but is a consequence of the difference in the order parameters of enantiotopic directions in a CLC. This result is very noticeable because only few examples of spectral discrimination between enantiotopic nuclei or isotopic enantiomers using 'classical' methods in the liquid state have been described so far.[80,81,82]

Figure 2.12 presents the NAD *Q*-COSY spectrum of norbornene, a rigid C_S-symmetry prochiral molecule.[83] As seen, all C-H directions are differentiated since two ^{2}H doublets (with same δ) are observed for each

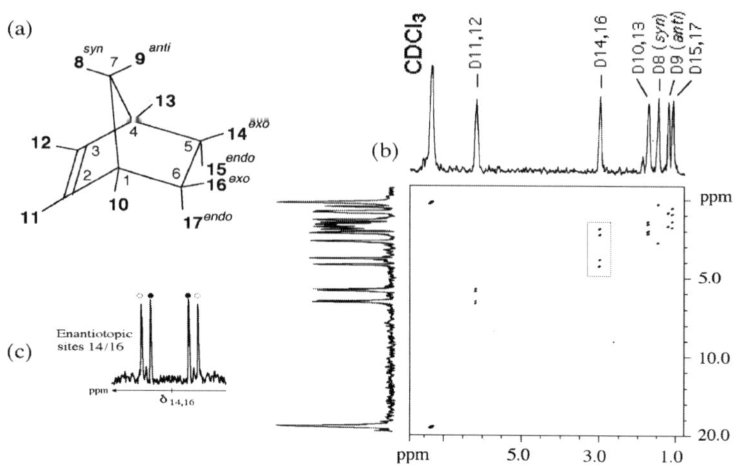

Fig. 2.12. (a) Structure and numbering of norbornene. (b) 61.4 MHz tilted *Q*-COSY NAD spectrum of norbornene in PBLG/$CHCl_3$ recorded in 15 hours (selective ^2H probe). The doublets of exo-deuterons (sites 14 and 16) are shown in c.

enantiotopic pair, whereas the diastereotopic *syn* and *anti* deuterons exhibit also two splittings but centered on two chemical shifts. In this example, the assignment of NAD signals for these sites were obtained by exploiting the dependence of local order parameter, S_{ij}, the molecular geometry (angles) and the elements of the Saupe's order matrix, $S_{\alpha\beta}$ (see Eq. (2.5)). As the S_{ij} parameters are derived from spectral information such as quadrupolar splittings, there exists an unique relationship between the anisotropic interactions, the orientational order parameters of a solute and its geometry as proposed by Aroulanda *et al.*[83] Using the magnitude and the sign of experimental quadrupolar splittings, $\Delta\nu_Q^{exp}$, as relevant anisotropic information, we can write by combining Eqs. (2.5) and (2.8):[83]

$$\frac{2}{3}\frac{\Delta\nu_{Qi}^{\exp}}{C_{Q\,i}^{D}} = \sum_{\alpha,\beta=a,b,c} \cos\theta_{C-D_i}^{\alpha}\cos\theta_{C-D_i}^{\beta} S_{\alpha\beta}\,. \qquad (2.12)$$

A very good agreement between experimental data, $\Delta\nu_Q^{exp}$, and back-calculated data, $\Delta\nu_Q^{calc}$, derived through $S_{\alpha\beta}$ calculations, indicates that the molecular structure and the anisotropic data are compatible; otherwise the proposed assignment or the structure is wrong, and hence a new combination 'assignment/geometry' must be tested.[83]

2.5. NAD-NMR in CLCs Using the Technology of Cryogenic Probes

As the spin manipulation ($^2H/^1H$ polarization transfers) fails in increasing the sensitivity of ^2H experiments, one simplest solution is to record NAD spectra with spectrometer operating at very high magnetic fields. The reason is that the NMR sensitivity is approximately proportional to $B_o^{3/2}$, and the experimental time (t_{exp}) needed to reach a given S/N ratio scales as $1/B_o^3$. Another way of improving the sensitivity consists of replacing the standard NMR probes by cryogenically cooled probes.

In these cryoprobes, both the receiver coil and the preamplifier electronics are cooled to very low temperature with helium gas, thereby reducing greatly the thermal noise and enhancing at least four fold the sensitivity compared with selective conventional probes.[7] Ideal NMR combination involves spectrometers operating at high fields with ^2H cryoprobes. The considerable gain in sensitivity and the current possibilities of NAD-NMR in anisotropic solvents have led to new analytical frontiers. Here the recent advances using a 14.1 T NMR spectrometer equipped with a selective ^2H cryoprobe are reviewed.[84] The advantages of the ^2H cryogenic probe are: (i) the amount of solute may be strongly diminished; (ii) the experimental time is reduced by a factor of over two to reach similar S/N ratios; (iii) the ratio 'gain in sensitivity/price' of cryoprobes is very advantageous compared to the price of an NMR spectrometer operating at a higher magnetic field but equipped with standard NMR probes.

2.5.1. *NAD-NMR and ^2H cryoprobe: new analytical frontiers*

As discussed above, NAD-NMR in CLCs is a powerful method for investigating the small chiral synthons that are key precursors in the total synthesis of bioactive chiral compounds. When large quantities of solute are available, NAD-NMR spectra can be recorded in few hours with satisfactory S/N ratios.[84] Another objective of NMR in CLC is to determine the enantiomeric purity in mixtures. Figure 2.13 shows a typical example of a small chiral building block (tridec-3-yn-2-ol) involved in the synthesis of derivatives in the dolastatine family[85] and synthetised both in racemic and enantioenriched series using two different synthetic pathways. The NAD *Q*-resolved Fz spectrum was obtained in only 5 hours (m_{solute} = 36 mg) with an excellent S/N ratio on the methyl site 1; spectral separation between the methyl doublets (> 80 Hz) improved the quality of the integration or the deconvolution of peak area. Here the enantiomeric enrichment measured by NAD-NMR is around 97%.

Another advantage of the ^2H cryoprobe is obviously the possibility of working with very small quantities of solute, in particular when the solubility of the solute into the mesophase is low. This situation was found with the (±)-nonamethoxycyclotriveratrylene ((±)-nonamethoxy-CTV), a chiral crown molecule without stereogenic carbon atom (see Fig. 2.14a).[78] For this compound, only four milligrams of each enantiomer could be dissolved in the PBLG/DMF mesophase. The

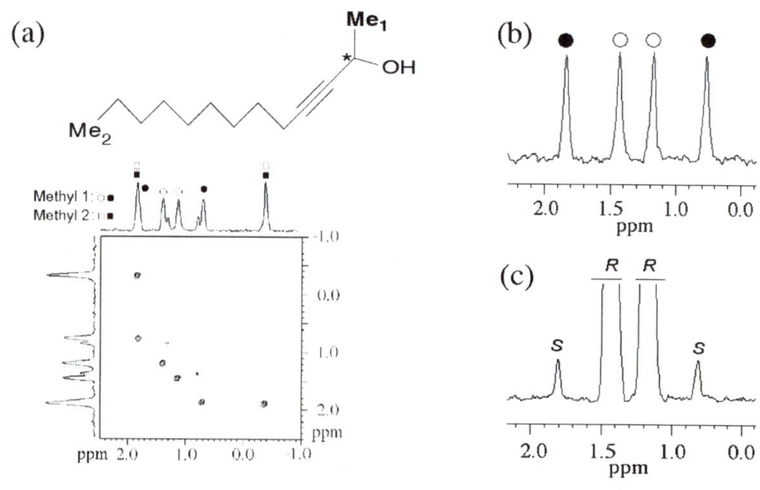

Fig. 2.13. (a) 92.1 MHz NAD *Q*-COSY Fz map of tridec-3-yn-2-ol recorded in PBLG/$CHCl_3$ (300 K). On right panel, doublets of methyl group 1 extracted from 2D maps are shown for the (b) racemic and (c) enantioenriched mixture (*ee* ~ 97%).

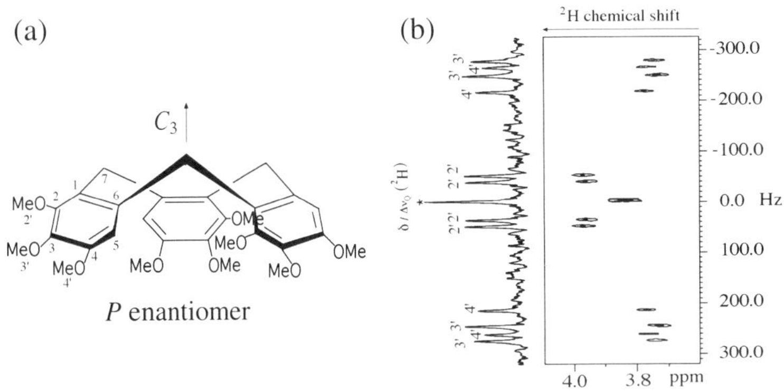

Fig. 2.14. (a) Structure of nonamethoxy-CTV (*P*-enantiomer). (b) Methyl region of the tilted 92.1 MHz NAD *Q*-resolved Fz of (±)-nonamethoxy-CTV in PBLG/DMF (320 K). The peak labeled with an asterisk is an artefact.

concentration in monodeuterated *M* or *P* enantio-isotopomers for the three equivalent methine sites (C_3-symmetry) that contribute to a unique doublet, $[CD]^e$, is around 4.4 μmol/L.[78]

The NAD 2D *Q*-resolved Fz map in Fig. 2.14b shows that only doublets associated with the methoxy groups ($[CDH_2]^e = 3 \times [CD]^e \sim 14$ μmol/L) emerge distinctly from the noise. Thus using a 14.1 T spectrometer with a 5-mm ^2H cryoprobe, the minimal concentration in monodeuterated isotopomer contributing to a given analyzable signal on NAD spectra is estimated between 4 to 10 μmol/L.

The sensitivity of a ^2H cryoprobe is sufficiently high to detect the heteronuclear couplings, ^2H-X, when the solute possesses other magnetically active heteroatoms such as fluorine-19 nuclei (Abund. = 100%, I = 1/2). Consequently it becomes possible to analyze the coupling fine structures between ^{19}F and ^2H atoms at natural abundance level as well as the ^2H quadrupolar doublets on the same NAD QUOSY map. This situation is illustrated with the (±)-2-(fluoromethyl)oxirane dissolved in PCBLL/DMF (see Fig. 2.15).[86] The *Q*-COSY Fz 2D spectrum recorded in around 5 hours (m_{solute} = 60 mg) shows that deuterons coupled with the ^{19}F nucleus (methylene group 1,1') are characterized by a quadrupolar doublet whose components are doubled by the ^2H-^{19}F coupling. As seen, this last splitting appears parallel to the main diagonal of the map (dotted line). The separation of spectral

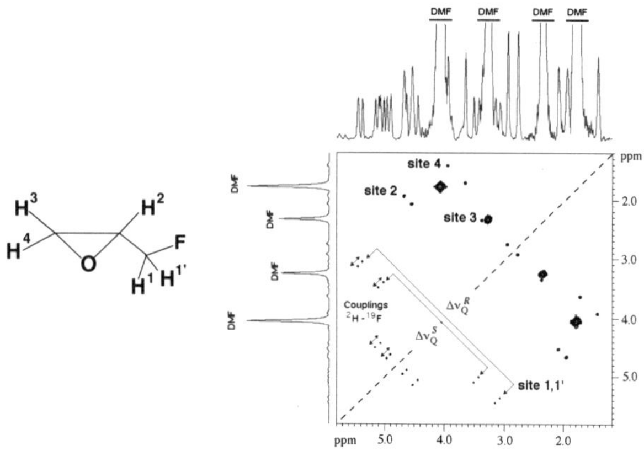

Fig. 2.15. 92.1 MHz NAD *Q*-COSY Fz 2D spectrum of (±)-2-(fluoromethyl)oxirane dissolved in the PBLG/DMF phase (310 K). The *R*/*S* assignment is arbitrary.

information allows a direct measurement of the total $^{2}H^{19}$-F spin-spin couplings ($^{n}T_{DF} = {}^{n}J_{DF} + 2^{n}D_{DF}$). Here, it appears that the two bond ^{2}H-^{19}F dipolar couplings are slightly different for the enantiomers whereas those between the ^{19}F nucleus and the deuteron in stereogenic carbon (site 2) is null by compensation with scalar couplings ($^{3}J_{DF} = -2^{3}D_{DF}$).

2.5.2. *From 2D to 3D NAD-NMR experiments*

The interest of 3D NMR approaches is to split NMR information between three spectral dimensions in order to increase the resolving power of the 2D experiments and/or to correlate information that are not accessible using a sum of 2D experiments.[93] This strategy was also applied to assign quadrupolar doublets in overcrowded ^{2}H spectra recorded in CLCs. In 2005, three new NMR 3D sequences designed for analyzing overcrowded ^{2}H spectra of weakly oriented perdeuterated solutes were described by Lafon *et al.*[94] These 3D sequences, derived from 2D QUOSY experiments, are entitled 3D *Q*-DQ, 3D *δ*-resolved/*Q*-COSY and 3D *δ*, *Q*-resolved experiments. All of them lead to pure absorption mode peaks in three dimensions, and the small number of RF-pulses reduces the amount of magnetization lost due to pulse imperfections.[95] They differ from each other by both the coherence pathway and the NMR information contained in the three dimensions. In practice, the two last sequences lead to 2D projections equivalent to tilted *Q*-COSY or *Q*-resolved experiments without post-acquisition data treatment (tilting procedure).

Figure 2.16 reports the first NAD 3D *Q*-DQ experiment obtained in 2008 at 14.1 T using a 5-mm ^{2}H cryoprobe in the case of (±)-hept-3-yn-2-ol dissolved in PBLG/$CHCl_3$.[95] Here the F_2 spectral dimension corresponds to the evolution of the double quantum coherence of ^{2}H nuclei during t_2. The 3D cube and the associated 2D projections allow for a facilitated analysis of the quadrupolar doublet network. This simple example illustrates that NAD 3D NMR is possible in a CLC, and permits the assignment of quadrupolar doublets of weakly oriented compounds.

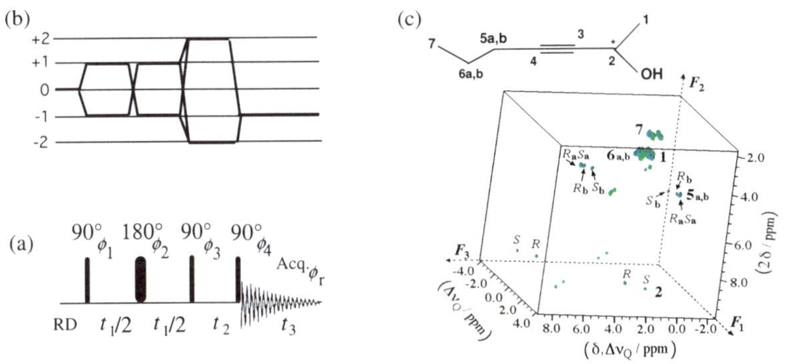

Fig. 2.16. (a) Pulse scheme of the *Q*-DQ 3D experiment and (b) associated coherence pathway diagram. (c) 92.1 MHz NAD *Q*-DQ 3D spectrum of (±)-hept-3-yn-2-ol in PBLG/$CHCl_3$ (14 hours of acquisition). The 3D spectrum was recorded with 8 scans *per* FID and 90 × 110 × 832 data points in t_1, t_2, t_3 dimensions, respectively.

2.5.3. *Anisotropic extension of the SNIF-NMR® method*

The investigation of the natural ^{2}H distribution by quantitative ^{2}H-{^{1}H} NMR in isotropic solvents is the method of choice for elucidating the biosynthesis pathways of natural products or for studying kinetic isotopic effects (KIE).[87] This well-known NMR method developed by Martin's group in 1981 is denoted SNIF-NMR®.[88] Although its application domains are large, its efficiency is, however, limited by two factors: (i) the small ^{2}H chemical shift dispersion; (ii) the inability to measure the (^{2}H/^{1}H) ratios for enantiotopic hydrogen sites in isotropic solvents.[81] Actually these two drawbacks can be overcome by combining quantitative NAD-NMR 2D experiments and the use of polypeptide CLCs as enantioselective oriented solvents.[81] Such approach was pioneered in 2004 by Lesot, Aroulanda and Billault who have investigated the isotopic profile of 1,1'-bis-(phenylthio)hexane (BPTH), a fragment of the linoleic acid (an unsaturated fatty acid) in a CLC. Thus, contrary to isotropic NMR,[89] it was demonstrated that in a PBLG/$CHCl_3$ system, all ^2H sites was spectrally discriminated on the basis of their δ_{2H} and quadrupole splittings, including the enantiotopic sites of methylenic groups using a 9.4 T spectrometer with a 5-mm BBI probe. Thus it was possible to estimate the ($^2H/^1H$) isotope ratios (in ppm) for some

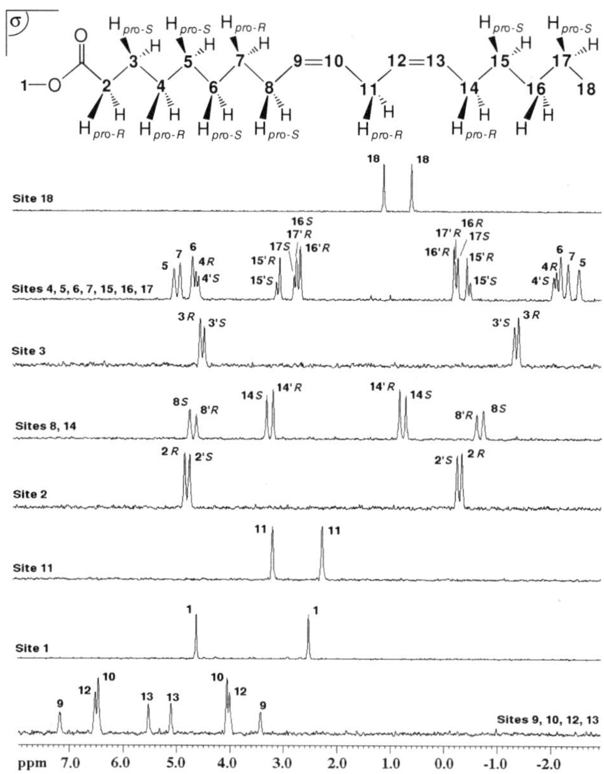

Fig. 2.17. Series of NAD sub-spectra extracted from the 92.1 MHz tilted *Q*-COSY Fz 2D spectrum of methyl linoleate dissolved in PBLG/$CHCl_3$ and recorded in 15 h (300 K).

methylene pairs and it was shown for the first time, a marked difference in the $(^2H/^1H)_i$ ratio for enantio-isotopomers of a prochiral molecule.

More interestingly, the assignment of the stereodescriptors, *R/S*, relative to each quadrupolar doublet observed on the NAD spectra of BPTH was achieved and it was shown that the *R* enantio-isotopomers in both odd and even methylene sites are naturally enriched in deuterons compared to the *S* enantio-isotopomers.[90] These two findings provide a new type of SNIF-NMR information in terms of the origin of the non-statistical distribution of deuterons in methylene groups along the chain of fatty acids and KIE associated with enzymatic reactions.

In 2008, taking again advantages of ^{2}H cryoprobe sensitivity, the NAD-NMR in a CLC system for the determination of ^{2}H isotopic profiles of bio-products was again applied to study four C-18 unsaturated fatty acid methyl esters, the methyl -oleate, -linoleate, -vernoleate and -linolenate.[91] In these examples several ^{2}H resonances are not resolved in the isotropic NAD spectra, thus prohibiting a measurement of the $(^2H/^1H)$ ratios for all hydrogen sites. Figure 2.17 reports a series of 1D NAD sub-spectra for methyl linoleate dissolved in PBLG/$CHCl_3$. Each spectrum displayed are a sum of columns extracted from the tilted *Q*-COSY Fz 2D map obtained using 100 mg of solute (instead of 200 mg as in case of BPTH) and the 5-mm ^{2}H cryoprobe (ν_D = 92.1MHz). The analysis of the various NAD sub-spectra shows that almost all non-equivalent ^{2}H sites are spectrally discriminated, including the four ethylenic sites and 70 % of enantiotopic methylene sites. Thus it was possible to plot the overall evolution of average $(^2H/^1H)$ ratios *versus* ^{2}H sites, establishing the well-known ^{2}H depletion/enrichment effect on the odd/even methylene sites along the aliphatic chain in fatty acids.[92] Finally, as the methyl linoleate and the BPTH investigated initially had the same botanical origin, the most intense doublets for each methylene group (except for the site 8) of the former was assigned to the *R*-enantio-isotopomer (correspond to the pro-*R* direction in the fully protonated molecule).[92]

The contribution of NAD-NMR in CLCs for the study of the isotope fractionation is a noteworthy achievement because it provides pieces of information that are inaccessible in isotropic environment but are very essential in order to improve the understanding of the enantioselective enzymatic mechanisms at the origin of numerous bioproducts.[92]

2.5.4. *Empirical assignment of the absolute configuration of enantiomers using NAD-NMR*

The determination of the absolute configuration of enantiomers of chiral compounds or enantiotopic elements in prochiral molecules using data extracted from NMR is a continuous challenge.[18,96] In CLCs, the spectral enantiodiscrimination originates in a difference of orientational ordering, and the assignment of absolute configurations through NMR data is only imaginable if the prediction of order parameters is possible by *ab initio* and/or molecular dynamic simulations.[98]

Waiting for such a prediction, empirical approaches for assigning absolute configurations were explored as reported by Ziani *et al.*[86] The method is based on the fact that in a closely related homologous series of iso-structural chiral compounds the relative magnitude of ordering for each enantiomer and their difference in ordering when dissolved in the same CLCs should be similar. Using NAD-NMR 2D experiments, this

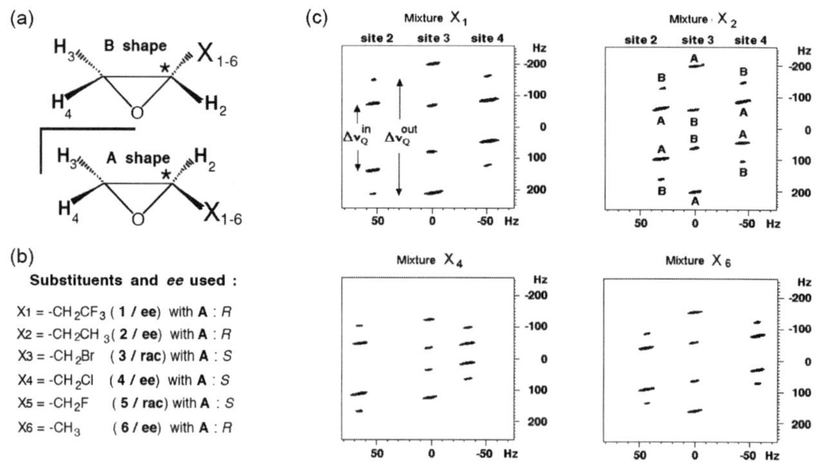

Fig. 2.18. (a) Structure of enantiomeric shapes (A and B) and nature of substituents (X_1 to X_6). (b) Correspondence between the *R*/*S* stereodescriptors and the A shape. (c) 92.1 MHz NAD spectral 2D fingerprints (*Q*-COSY Fz experiment) associated with deuterium sites of the epoxide ring in compounds 1, 2, 4 and 6 (*ee* = 33%).

concept was applied to determine the unknown absolute configuration of the major enantiomer in a prepared enantio-enriched mixture of 1,1,1-trifluoromethyl-2,3-epoxypropane (compound X_1) which had been synthetised both in racemic and enantiopure series (see Fig. 2.18).[86] The step-by-step strategy of the method is as follows: (i) the optimisation of spectral enantiodiscrimination for X_1; (ii) the analysis of at least the NAD QUOSY 2D spectra of a series of enantio-enriched mixtures where the absolute configuration of the major enantiomer (compounds X_2, X_4 and X_6) is known; (iii) the assignment of the absolute stereochemistry of NAD signals of X_1 from the NMR results of reference molecules.

For a set of homologous compounds, a correlation between the stereochemical descriptors, *R*/*S*, for each enantiomer and the quadrupolar splittings (for a given site) is not directly observed. Indeed, in agreement with the priority rules based on the atomic number of atoms (the CIP rules) used to define the stereochemistry of enantiomers[76,99] the *R* (or *S*) stereodescriptors do not always correspond to the same molecular topology in the reference molecules (the substituent X can be behind or in front of the epoxide ring, see Fig. 2.18a). To avoid such inversions, two iso-structural enantiomeric shapes denoted A and B were defined in which the substituent X is always either in front of (form A) or behind (form B) the epoxide cycle when the oxygen atom is pointing down.[100] The *ee* in the reference mixtures (2, 4 and 6) were prepared with the same major topological enantiomer (form A). From the analysis of spectral 2D fingerprints, it appears that the doublet with the smallest magnitude or inner doublets ($\Delta\nu_Q^{in}$) for ^{2}H sites 2 and 4, and the doublet with the largest magnitude or outer doublet ($\Delta\nu_Q^{out}$) for the site 3 (see Fig. 2.18b) correspond to enantiomers A (major), namely the *R* isomer for X_1. Thus it was concluded that the mixture X_1 is 'most probably' enriched in the *R* isomer according to the CIP rules.

The key point in this empirical methodology is the choice of the substituents, X_i, in the reference molecules. Interestingly this first experimental study has raised two crucial theoretical questions: (i) how does one define the degree of homology between two molecular objects when talking about molecular ordering? (ii) what pertinent criteria (electronic distributions, Van der Waals volumes, hydrogen bonds, ...) should be adopted to achieve this aim?

2.6. Conclusion

In this chapter, we have attempted to present the various facets of NAD-NMR spectroscopy in oriented systems. The fundamental interest of NAD-NMR compared to other NMR involving nuclei with low natural abundance is that deuterium nuclei are present in all organic molecules. Therefore, it provides a remarkable spy for a wide range of analytical applications. Since the first experiments, significant technological progresses have been made in the area of NAD-NMR. In particular, spectrometers operating at high magnetic fields together with cryogenically cooled NMR probes have considerably extended the application domain of NAD-NMR. The combination of NAD-NMR with the use of CLC solvents has opened up wide prospects in various distinct analytical domains such as: (i) the stereochemical analysis (configuration studies); (ii) the investigation of isotopic fractionation in natural organic compounds dedicated to the elucidation or the understanding of biochemical mechanistic pathways.

Acknowledgments

The authors thank all the co-authors cited in their own references as well as all copyright holders for Figs. 2.3, 2.4 and 2.6 used. They are also grateful to Prof. A. Loewenstein for his original idea to use organic solutions of polypeptides, and Prof. J. Courtieu for his numerous helpful discussions.

References

1. C. Brevard and P. Granger, *Handbook of High Resolution Multinuclear NMR*, J. Wiley & Sons, New York, (1981).
2. J. R. De Laeter, J. K. Böhlke, P. De Bievre, H. Hidaka, H. S. Peiser, K. J. R. Rosman and P. D. P. Taylor, *Pure Appl. Chem.* **75** 683 (2003).
3. R. K. Harris and E. D. Becker, *J. Magn. Reson.* **156** 323 (2002).
4. P. Diehl and T. Leipert, *Helv. Chim. Acta* **47** 545 (1964).
5. J. M. Briggs, L. F. Farnell and E. W. Randall, *Chem. Commun.* (1973) 70.
6. J. Curtis, D. K. Dalling and D. M. Grant, *J. Org. Chem.* **51** (1986) 136.
7. H. Kovacs, D. Moskau and M. Spraul, *Prog. Nucl. Magn. Reson. Spectrosc.* **46** (2005) 131.

8. D. Neuhaus and M. P. Williamson, *The Nuclear Overhauser Effect in Structural and Conformational Analysis*, (2nd ed.), Wiley-VCH, 2000.
9. J. W. Emsley and J. C. Lindon (Eds.) in Chapter 2, *NMR Spectroscopy Using Liquid Crystal Solvents*, Pergamon Press, Oxford, (1975).
10. R. R. Ernst, G. Bodenhausen and A. Wokaun, *Principles of Nuclear Magnetic Resonance in One and Two Dimensions*, Clarendon Press, Oxford, (1987).
11. C. Zannoni in Chapter 1, *Nuclear Magnetic Resonance of Liquid Crystals*, Ed. J.W. Emsley, NATO ASI Series, (1985).
12. C. P. Schlichter, in *Principles of Magnetic Resonance* (2nd ed.), Harper & Row, New York, 1978.
13. H. Jensen and K. Schaumburg, *Acta Chem. Scand.* **25** 663 (1971).
14. D. Parker, *Chem. Rev.* **91** (1991) 1441.
15. R. Rothchild, *Enantiomer* **5** (2000) 457.
16. T. J. Wenzel, *Discrimination of Chiral Compounds using NMR Spectroscopy*, J. Wiley & Sons, (2007).
17. E. Sackmann, S. Meiboom and L. C. Snyder, *J. Am. Chem. Soc.* **90** (1968) 2183.
18. E. Sackmann and L. C. Snyder, *Science* **162** (1968) 3860.
19. J. Courtieu, E. Lafontaine, J.-M. Péchiné and C. L. Mayne, *Liquid Crystals* **7** (1990) 293.
20. J.-P. Bayle, J. Courtieu, E. Gabetty, A. Loewenstein and J.-M. Péchiné, *New J. Chem.* **16** (1992) 837.
21. I. Canet, J. Courtieu, A. Loewenstein, A. Meddour and J.-M. Péchiné, *J. Am. Chem. Soc.* **117** (1995) 6520.
22. P. Lesot, D. Merlet, Y. Gounelle and J. Courtieu, *J. Phys. Chem. A.* **99** (1995) 14871.
23. P. Lesot, D. Merlet, J. Courtieu, J. W. Emsley, R. R. Rantala and J. Jokisaari, *J. Phys. Chem.* **101** (1997) 5719.
24. C. Aroulanda, M. Sarfati J. Courtieu and P. Lesot, *Enantiomer* **6** (2001) 281.
25. M. Sarfati, P. Lesot, D. Merlet and J. Courtieu, *Chem. Commun.* (2000) 2069, and references therein.
26. P. Lesot, M. Sarfati and J. Courtieu, *Chem. Eur. J.* **9** (2003) 1724, and references therein.
27. R. Y. Dong in Chapter 16, *NMR of Ordered Liquid*, Kluwer Academic Publishers, Drodrecht, (2003).
28. R. R. Vold in Chapter 11, *Nuclear Magnetic Resonance of Liquid Crystals*, Ed. J. W. Emsley, NATO ASI Series, Dordrecht, (1985).
29. R. Y. Dong, *Ann. Rep. on NMR Spect.* **53** (2004) 67.
30. T. C. Lubensky, *Phys. Rev.* **A** (1970) 2497.
31. P. Pincus, *Solid State Commun.* **7** (1969) 415.
32. P. Lesot, O. Lafon, C. Aroulanda and R. Y. Dong, *Chem. Eur. J.* **14** (2008) 4082.
33. The Molecular Dynamics of Liquid Crystals, Ed. G. R. Luckhurst, C. A. Veracini, NATO ASI series, Kluwer Academic Publishers, Dordrecht, (1993).
34. P. Diehl in Chapter 7, *Nuclear Magnetic Resonance of Liquid Crystals*, Ed. J.W. Emsley, NATO ASI Series, Drodrecht, 1985.
35. B. M. Fung, Prog. Nucl. Magn. Reson. Spectrosc. **41** (2002) 171.

36. R. Y. Dong, *Nuclear Magnetic Resonance of Liquid Crystals*, Spring-Verlag, New York, (1994).
37. R. Y. Dong, in pp 2752-2759, Encyclopedia of NMR, Eds. R. K. Harris, D. M. Grant, J. Wiley & Sons, Chichester, (1996).
38. R. Y. Dong, *Ann. Rep. on NMR Spect.* **53** (2004) 68.
39. E. T. Samulski and R. Y. Dong, *J. Chem. Phys.* **77** (1982) 5090.
40. J. Atzrodt, V. Derdau, T. Fey and J. Zimmermann, *Angew. Chem. Int. Ed.* **45** (2007) 7744.
41. H. Zimmermann, *Liquid Crystals* **4** (1989) 591.
42. R. S. Prosser, N. J. Heaton and G. Kothe, *J. Magn. Reson. B* **112** (1996) 51.
43. K. Tabayashi and K. Akasaka, *J. Phys. Chem. B* **101** (1997) 5108.
44. J. S. Waugh, *J. Magn. Reson.* **49** (1982) 517.
45. I. P. Gerothanassis, *Prog. Nucl. Magn. Reson. Spectrosc.* **19** (1987) 267.
46. K. Tabayashi and K. Akasaka, *Liquid Crystals* **26** (1999) 127.
47. J. W. Emsley, P. Lesot, G. de Luca, A. Lesage, D. Merlet and G. Pileo, *Liquid Crystals* **35** (2008) 443.
48. J. W. Emsley, J. C. Lindon, J. M. Tabony and T. H. Wilmshurst, *Chem. Commun.* (1971) 1277.
49. M. Longeri, G. Celebre in pp 2774-2781, *Encyclopedia of NMR*, Eds. R. K. Harris, D. M. Grant, J. Wiley & Sons, Chichester, (1996).
50. E. E. Burnell, C. A. de Lange in Chapter 10, *NMR of Ordered Liquids*, Eds. E. E. Burnell, C. A. de Lange, Kluwer Academic Publishers, Dordrecht (2003).
51. L. D. Field in Chapter 4, *NMR of Ordered liquids*, Eds. E. E. Burnell, C. A. de Lange, Kluwer Academic Publisher, Dordrecht, (2003).
52. J. Courtieu, J.-P. Bayle and B. M. Fung, *Prog. Nucl. Magn. Reson. Spectrosc.* **26** (1994) 141.
53. J.-M. Ouvrard, B. N. Ouvrard, J. Courtieu, C. L. Maine and D. M. Grant, *J. Magn. Reson.* **93** (1991) 225.
54. P. Lesot, J.-M. Ouvrard, B. N. Ouvrard and J. Courtieu, *J. Magn. Reson. A* **107** (1994) 141
55. P. Lesot, J. W. Emsley, J.-M. Ouvrard and E. Curzon, *J. Magn. Reson. A* **133** (1998) 166.
56. C. L. Khetrapal, K. V. Ramanathan, N. Suryaprakash and S. Vivekanandan, *J. Magn. Reson.* **135** (1998) 265.
57. E. de Alba and N. Tjandra, *Prog. Nucl. Magn. Reson. Spectrosc.* **40** (2002) 175.
58. J. H. Prestegard, C. M. Bougault and A. I. Kishore, *Chem. Rev.* **104** (2004) 3519, and references therein.
59. A. S. Tracey and K. Radley, *J. Phys. Chem.* **88** (1984) 6044.
60. K. Baczko, C. Larpent and P. Lesot, *Tetrahedron: Asymmetry* **15** (2004) 971.
61. K. Kobzar, H. Kessler and B. Luy, *Angew. Chem., Int. Ed.* **44** (2005) 3145.
62. R. R. Gil, C. Gayathri, N. V. Tsarevsky and K. Matyjaszewski, *J. Org. Chem.* **73** (2008) 840.
63. P. Lesot, D. Merlet, A. Loewenstein and J. Courtieu, *Tetrahedron: Asymmetry* **9** (1998) 1871.

64. A. Meddour, I. Canet, A. Loewenstein, J.-M. Péchiné and J. Courtieu, *J. Am. Chem. Soc.* **116** (1994) 9652.
65. K. Czarniecka and E. T. Samulski, *Mol. Cryst. Liq. Cryst.* **63** (1981) 205.
66. C. Aroulanda, D. Merlet, J. Courtieu and P. Lesot, *J. Am. Chem. Soc.* **123** (2001) 12059.
67. P. Lesot, D. Merlet, M. Sarfati, J. Courtieu, H. Zimmermann and Z Luz, *J. Am. Chem. Soc.* **124** (2002) 10071.
68. D. Merlet, B. Ancian, J. Courtieu and P. Lesot, *J. Am. Chem. Soc.* **121** (1999) 5249.
69. D. Merlet, M. Sarfati, B. Ancian, J. Courtieu and P. Lesot, *Phys. Chem. Chem. Phys.* **2** (2000) 2283.
70. O. Lafon, P. Lesot, D. Merlet and J. Courtieu, *J. Magn. Reson.* **171** (2004) 135.
71. M. Sarfati, J. Courtieu and P. Lesot, *Chem. Commun.* (2000) 1113.
72. C. Canlet, D. Merlet, P. Lesot, A. Meddour, A. Loewenstein and J. Courtieu, *Tetrahedron: Asymmetry* **11** (2000) 1911.
73. A. Parenty, J.-M. Campagne, C. Aroulanda and P. Lesot, *Org. Lett.* (2002) 1663.
74. E. E. Eliel, S. H. Wilen, *Stereochemistry of Organic Compounds,* J. Wiley & Sons, INC, New York, (1994).
75. P. Lesot, O. Lafon, J. Courtieu and P. Berdagué, *Chem. Eur. J.* **10** (2004) 3741.
76. P. Lesot, O. Lafon, H. Zimmermann, Z. Luz, *J. Am. Chem. Soc.* **130** (2008) 9754.
77. D. Merlet, J. W. Emsley, P. Lesot and J. Courtieu, *J. Chem. Phys.* **111** (1999) 6890.
78. O. Lafon, P. Lesot, H. Zimmermann, R. Poupko and Z. Luz, *J. Phys. Chem. B* **111** (2007) 9453.
79. P. Lesot, C. Aroulanda and I. Billault, *Anal. Chem.* **76** (2004) 2827.
80. R. R. Fraser, M. A. Petit and M. Miskow, *J. Am. Chem. Soc.* **94** (1972) 3253.
81. A. Bilz, T. Stork and G. Helmchen, *Tetrahedron: Asymmetry* **8** (1997) 3999.
82. J. Huskens, R. Goddard and M. T. Reetz, *J. Am. Chem. Soc.* **120** (1998) 6617.
83. C. Aroulanda, P. Lesot, D. Merlet and J. Courtieu, *J. Phys. Chem. A* **107** (2003) 10911.
84. P. Lesot, *Spectra Analyse* **260** (2008) 16.
85. M. Georgy, P. Lesot and J.-M. Campagne, *J. Org. Chem.* **75** (2007) 3543 and references therein.
86. L. Ziani, P. Lesot, A. Meddour and J. Courtieu, *Chem. Commun.* (2007) 4737.
87. D. A. Singleton and A. A. Thomas, *J. Am. Chem. Soc.* **117** (1995) 9357 and references cited therein.
88. J.-R. Duan, I. Billault, F. Mabon and R. J. Robins, *ChemBioChem* **3** (2002) 52.
89. M. L. Martin and G. J. Martin, *NMR Basic Principles and Progress*, Ed. P. Diehl, 23 (1990), pp 1-61.
90. I. Billault, R. J. Robins and S. Akoka, *Anal. Chem.* **74** (2002) 5902.
91. V. Baillif, I. Billault, R. J. Robins and P. Lesot, *J. Am. Chem. Soc.* **128** (2006) 11180.
92. P. Lesot, V. Baillif and I. Billault, *Anal. Chem.* **80** (2008) 2963.
93. G. W. Vuister, R. Boelens, A. Padilla, G. J. Kleywegt and R. Kaptein, *Biochemistry* **29** (1990) 1829, and references therein.
94. O. Lafon and P. Lesot, *Chem. Phys. Lett.* **404** (2005) 90.
95. P. Lesot and O. Lafon, *Chem. Phys. Lett.* **458** (2008) 219.

96. J. M. Seco, E. Quiñoá and R. Riguera, *Chem. Rev.* **104** (2004) 17.
97. E. Alvira, J. Breton and J. Plata, *Chem. Phys.* **155** (1991) 7.
98. J. Helfrich, R. Hentschke and U. M. Apel, *Macromolecules* **2** (1994) 472.
99. R. S. Cahn, C. K. Ingold and V. Prelog, *Angew. Chem., Int. Ed.* **5** (1966) 385.
100. A. Meddour, A. Loewenstein, J.-M. Péchiné and J. Courtieu, *Tetrahedron: Asymmetry* **8** (1997) 485.

Chapter 3

Noble Gas Probes in NMR Studies of Liquid Crystals

Jukka Jokisaari

Department of Physical Sciences, University of Oulu,
P.O. Box 3000, FI-90014 University of Oulu, Finland
email: Jukka.Jokisaari@oulu.fi

The use of noble gas probes in NMR studies of the properties of liquid crystals is a very practical alternative to detecting spectra of nuclei in liquid crystal molecules. In particular, joint analysis of NMR spectra of spin-1/2 and quadrupolar noble gas nuclei recorded over temperature ranges, covering all the mesophases, makes feasible the derivation of phase transitions, all orientational order parameters and tilt angle. Complementary information can be derived from noble gas self-diffusion and relaxation time measurements. ^{129}Xe NMR is also a very effective and informative method in studies of the behavior of liquid crystals in restricted spaces.

3.1. Introduction

The first successful NMR experiments were carried out in late 1945 at Harvard and Stanford universities and only about 5 years later NMR was applied to measure nuclear magnetic moments of the xenon isotopes ^{129}Xe (spin ½) and ^{131}Xe (spin 3/2).[1] Consequently, noble gas NMR is almost as old as NMR in general.

A significant milestone in noble gas NMR spectroscopy was the observation that ^{129}Xe NMR may give valuable information on pore dimensions in microporous materials.[2] Since then, ^{129}Xe NMR spectroscopy has been applied to numerous different materials, such as microporous solids (carbon nanotubes, zeolites, molecular sieves, nanotubes formed by dipeptides, *etc*.), mesoporous solids (porous glasses, aerogels, wood, concrete, *etc*.), polymers, proteins, and so on, and, of course, liquid crystals.[3] To the author's best knowledge, the first Xe NMR experiment of xenon in a liquid-crystalline environment is the ^{131}Xe NMR spectrum recorded from xenon in a mixture of poly-γ-benzyl-L-glutamate (PBLG) and deuterated chloroform[4]. Such a spectrum consists of three resonance lines which arise from the fact that ^{131}Xe is quadrupolar (spin 3/2) and its electric quadrupole moment interacts with the electric field gradient at the nuclear site created by the environment, although also another contribution is effective as will be discussed in Sec. 3.4. Consequently, ^{131}Xe NMR as well as the NMR of the other quadrupolar noble gases, ^{21}Ne and ^{83}Kr, give information of the electric field gradient arising from liquid crystal molecules.

The first ^{129}Xe NMR experiments in a thermotropic liquid crystal were performed by Bayle *et al.* in 1988.[5] They studied xenon in a so-called critical mixture of two nematic liquid crystals with opposite diamagnetic anisotropy. Since then, noble gas NMR has been applied to various thermotropic and lyotropic liquid crystals, as will be discussed in the next chapters. Earlier reviews on noble gas NMR in isotropic liquids and liquid crystals can be found in Refs. [6] and [7].

Noble gases have many properties that favor their use in materials studies. One of the most important properties is their inertness, though of very great importance is the fact that their polarization can be enhanced by several orders of magnitude compared to thermal polarization using optical pumping (OP).[8] In the literature, different acronyms are used for optically pumped noble gases, such as OP, HP (hyperpolarized) and LP (laser-polarized). In this chapter, the acronym HP will be used. The principle of the HP technique is discussed in Sec. 3.3.

3.2. NMR Properties of Noble Gases

Only five isotopes of all the stable noble gases possess a non-zero spin quantum number, although, for example, xenon alone has nine stable isotopes. Three of these five isotopes are quadrupolar, as mentioned above. Natural abundance and thus NMR sensitivity of ^{3}He and ^{21}Ne are very low and those of ^{83}Kr fairly modest, meaning that NMR experiments require use of gases enriched with these isotopes. Table 1 lists the NMR properties of the noble gas isotopes.

The number of resonance lines in a spectrum recorded in a liquid-crystalline environment is determined by the spin quantum number; the number of lines is $2I$. Consequently, ^{3}He and ^{129}Xe give only a single peak, ^{21}Ne and ^{131}Xe three peaks with relative intensity ratios 3:4:3, and ^{83}Kr nine peaks with relative intensity ratios 9:16:21:24:25:24:21:16:9. Particularly, in the case of ^{83}Kr the distribution of intensity within nine peaks lengthens acquisition times. Other problems in this case are the low Larmor frequency and acoustic ringing which may cause severe background disturbances. In all, if possible, recording NMR spectra of both spin-1/2 nuclei and quadrupolar nuclei gives complimentary information about liquid crystals. Table 3.1 shows that ^{129}Xe has the highest NMR sensitivity. It is also the most sensitive to environmental effects; its chemical shift range in isotropic and anisotropic liquids is about 300 ppm. Due to these facts, the following discussion will concentrate on the use of ^{129}Xe NMR. Another point that favors the xenon-129 probe is that its resonance signal can be detected even from natural abundance on a single scan, and the line width of a few Hz is easily obtainable. A disadvantage, on the contrary, is its long spin-lattice relaxation time, tens of seconds in liquid-crystalline phases, often making particularly diffusion and relaxation measurements very time consuming.

3.3. Enhancement of Polarization by Optical Pumping

Likely the weakest aspect in NMR spectroscopy is its inherently low sensitivity. This arises from the fact that the population differences, which determine the polarization, between various spin states are very small. For example, for ^{129}Xe at room temperature and magnetic field of 9.39 T, the ratio of the population difference of the two possible spin states to the total number of spins, $\Delta N/N$, is only ~9 ppm. By applying optical pumping (OP) this can be increased to a value close to 1.[8] Optical pumping leads to a non-equilibrium state, which in the literature is often called hyperpolarized (HP) state. When time passes hyperpolarization approaches thermal polarization because of relaxation processes.

Table 3.1. NMR properties of noble gases.

Property	^{3}He	^{21}Ne	^{83}Kr	^{129}Xe	^{131}Xe
Spin	½	3/2	9/2	½	3/2
Natural abundance (%)	$1.34 \cdot 10^{-4}$	0.27	11.5	26.4	21.2
Gyromagnetic ratio (10^{7} radT^{-1}s^{-1})	-20.379	-2.113	-1.033	-7.452	2.209
Electric quadrupole moment (fm^{2}) [a)]		10.155	25.9		-11.4
NMR sensitivity [b)]	$3.48 \cdot 10^{-3}$	$3.91 \cdot 10^{-2}$	1.28	33.5	3.51
NMR frequency (MHz) [c)]	304.817	31.587	15.395	111.277	32.986

[a)]Ref. [9].
[b)]With respect to ^{13}C.
[c)]At 9.3798 T. (Proton frequency 400.130 MHz)

The OP technique is based on interaction of optically pumped rubidium and xenon atoms. In an OP apparatus, mixture of rubidium vapor and xenon gas, placed in a weak magnetic field, is illuminated by circularly polarized laser light (wave length 795 nm) which pumps rubidium atoms to an excited electronic state. When rubidium and xenon atoms collide, spin exchange between electrons and xenon nuclei takes place, and xenon becomes hyperpolarized. In order to maintain high polarization in experiments requiring long accumulation times the optical pumping apparatus has to work in a continuous flow mode. Corresponding techniques can be used to polarize ^{3}He, ^{83}Kr and ^{131}Xe.[10] In applications of HP ^{129}Xe the long spin-lattice relaxation time of the isotope turns out to be a positive matter since accumulation may not be required.

3.4. Noble Gases in Bulk Liquid Crystals

3.4.1. *Chemical shift*

In this chapter ^{129}Xe chemical shift and shielding in thermotropic and lyotropic liquid crystals is emphasized. This is derived from the fact that ^{129}Xe chemical shift is most sensitive to physical and chemical changes in its environment, as mentioned above. Figure 3.1 displays, as an example, the ^{129}Xe NMR spectrum at variable temperatures of xenon dissolved in thermotropic FELIX-R&D (consisting mostly of phenylpyrimidine derivatives) ferroelectric liquid crystal which possesses nematic N*, smectic A and smectic C* phases besides isotropic and crystal phases.[11]

Figure 3.1 immediately reveals that the ^{129}Xe NMR of xenon in a liquid-crystalline solution is able to distinguish the various mesophases. A particularly clear shielding (chemical shift) jump can be detected at the isotropic-nematic phase transition, and this is a general feature. The direction of this jump indicates the sign of the anisotropy of diamagnetic susceptibility tensor, $\Delta\chi_d$; if the jump is toward smaller shielding, $\Delta\chi_d > 0$, whereas if the jump is toward higher shielding, $\Delta\chi_d < 0$. Other experiments have shown that there is no abrupt jump in shielding at the N - SmA transition but the shielding is curved to a higher value (for example, see Fig. 3.4 below). This behavior has been interpreted to arise from redistribution of xenon atoms during the formation of layer structure in the smectic A phase; part of xenon atoms is expelled from the denser core region to the less dense chain region. Figure 3.1 also shows that in the isotropic phase the shielding constant (and chemical shift) is a linear function of temperature. Curve shape in the nematic phase indicates that the shielding is affected by the degree of orientational order of the liquid crystal. It has appeared that indeed this is the case, and even the translational and mixed translational-orientational order parameters of a smectic A phase contribute to the shielding. The shielding can be modeled with a function[12,13]

$$\sigma(T) - \sigma_o = \rho(T)\Big\{\sigma_d^{'}[1-\epsilon(T-T_o)][1+2c\gamma_1^2(T)] + \frac{2}{3}P_2(cos\theta)\Delta\sigma_d^{'}[1-\Delta\epsilon(T- T_o)][S(T)+ 2c\sigma_1(T)\gamma_1(T)]\Big\} \quad (3.1)$$

where σ_0 is the shielding of an external gas reference and $\rho(T)$ is the density of liquid crystals. The quantities σ_d' and $\Delta\sigma_d'$ are the isotropic shielding constant and shielding anisotropy, respectively, both in units ppm/g^{-1}cm^3. $S(T)$ is the normal second rank orientational order parameter, $\gamma_1(T)$ is the translational order parameter, and $\sigma_1(T)$ is the mixed translational-orientational order parameter. $P_2(\cos\theta)$ is the second order Legendre polynomial with θ being the angle between the liquid crystal director and the external magnetic field. The parameters ϵ and $\Delta\epsilon$ describe temperature dependence of the shielding constant and shielding

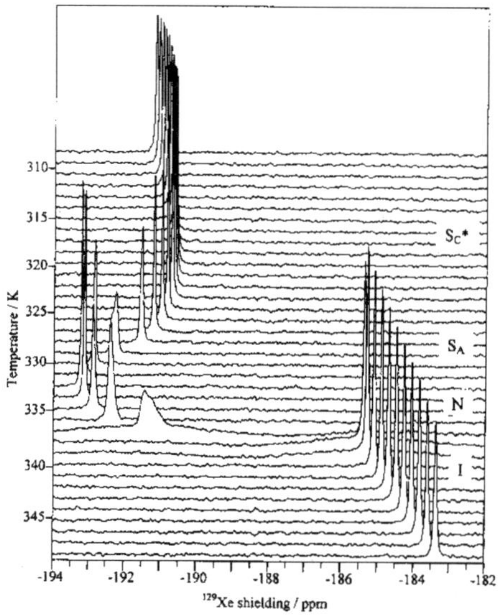

Fig. 3.1. ^{129}Xe NMR spectrum of xenon in FELIX-R&D liquid crystal at variable temperatures. Note that the horizontal axis is a shielding axis and the values are referenced to an external low pressure xenon gas sample. The symbols on the right refer to isotropic (I), nematic (N), smectic A (S_A) and smectic C* (S_{C*}) phases. The nematic phase is actually a chiral phase, N*. The helix orients in low magnetic fields so that its axis is perpendicular to the external magnetic field. The field strength in this particular case is so strong that it causes untwisting of the helical structure and a transition from chiral nematic phase to the uniformly oriented nematic phase takes place. The director of the N phase is along the external magnetic field, i.e. $\Delta\chi_d > 0$. (Reproduced from Ref. [11] with permission. Copyright (2001), Taylor and Francis.)

anisotropy, respectively. T_o is a chosen reference temperature, often the nematic-isotropic transition temperature T_{NI}. The parameter c measures the redistribution of probe atoms or molecules during the formation of smectic layers.

When applying Eq. (3.1) experimental results have to be corrected for bulk susceptibility, if an external reference is used. For a long cylindrical liquid-crystalline sample (sample tube axis parallel with the external magnetic field) the correction is[14]

$$\sigma_b(T) = -\frac{1}{3}[\chi_d + \frac{2}{3}\Delta\chi_d S(T)]\frac{\rho(T)}{M} \tag{3.2}$$

where χ_d is the isotropic average of the diamagnetic susceptibility tensor and $\Delta\chi_d$ is the anisotropy of the tensor and M is molar mass of liquid crystal. Temperature-dependence of the orientational order parameters in Eqs. (3.1) and (3.2) can be modeled with functions[12]

$$S(T) = (1 - y_S\frac{T}{T_o})^{z_S} \tag{3.3}$$

$$\sigma_1(T) = (1 - y_\sigma\frac{T}{T_o})^{z_\sigma} \tag{3.4}$$

$$\gamma_1(T) = \left(1 + x\frac{T}{T_o}\right)(1 - y_\gamma\frac{T}{T_o})^{z_\gamma}. \tag{3.5}$$

The application of Eq. (3.1) proceeds phase by phase. In the isotropic phase, all the orientational/translational order parameters are set equal to zero. In nematic (N) phases, $S(T)$ is included whereas all the other order parameters vanish, and finally in smectic A (SmA) phases all the orientational/translational order parameters are included in the shielding function. In nematic and smectic A phases, $P_2(\cos\theta)$ is either 1 (in case the liquid crystal director is parallel with the external magnetic field) or -1/2 (in case the director is perpendicular to the external field). For smectic C (SmC) and C* (SmC*) phases, $P_2(\cos\theta)$ is assigned varying values due to the temperature dependence of the tilt angle of the director (see below). The shielding jump at the isotropic - nematic phase transition arises from the anisotropic term in Eq. (3.1) and a possible density change.

All experiments in liquid-crystalline solutions until now have shown that the ^{129}Xe shielding anisotropy is negative. Consequently, when the director points along the external magnetic field $P_2(\cos\theta) = 1$ ($\Delta\chi_d > 0$) and the anisotropic term is negative, leading to a *decrease* in total shielding. On the contrary, if the director is perpendicular to the external field ($\Delta\chi_d < 0$), $P_2(\cos\theta) = -1/2$ and the anisotropic term is positive, this leads to an *increase* in total shielding. (Note that the term including order parameters in Eq. (3.1) is always positive.)

Figure 3.2 displays ^{129}Xe shielding as a function of temperature in 1-methylheptyl-4'-(4-n-decyloxybenzoyloxy)biphenyl-4-carboxylate (10B1M7) which possesses isotropic, smectic A, and ferroelectric, ferrielectric and antiferroelectric smectic C* phases.[15] The curve clearly reveals the I - SmA and SmA - SmC*(ferroelectric) phase transitions. On the contrary, no clear transition points are visible for the transitions from SmC*(ferroelectric) to SmC*(ferrielectric) nor from SmC*(ferrielectric) to SmC* (antiferroelectric). In this particular case, Eq. (3.1) was used to derive all the order parameters as a function of temperature. However, certain assumptions had to be made in order to keep the problem overdetermined. In this case the tilt angle of the liquid crystal director was derived from the ^{129}Xe NMR shielding data for the first time. This also required some assumptions, such as that density and orientational/translational order parameters change smoothly at phase transitions. These assumptions appeared to be quite reasonable approximations and are verified by comparing the results to those derived from ^{2}H NMR experiments of partially deuterated liquid crystal and from optical measurements. This comparison is shown in Fig. 3.3. Temperature-dependence of the tilt angle was modeled with the function[16]

$$\theta(T) = \theta_0 \left(1 - \frac{T}{T_{AC*}}\right)^{\beta} \tag{3.6}$$

where θ_0 and β are adjustable parameters in least-squares fit, resulting in the values 33.38° and 0.142 in 10B1M7.

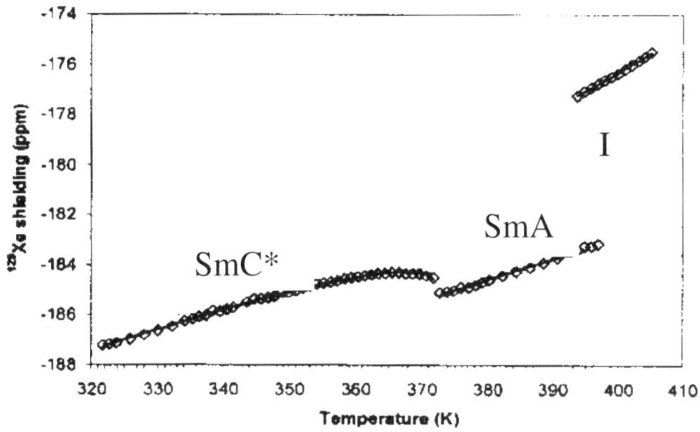

Fig. 3.2. ^{129}Xe shielding (relative to low-pressure xenon gas) at variable temperature for xenon in 10B1M7 liquid crystal. The solid lines are results of the least-squares fits to Eq. (3.1). The change in curvature at the SmA – SmC* phase transition was interpreted to arise from the tilt angle. (Reproduced from Ref. [15] with permission. Copyright (2004), American Chemical Society.)

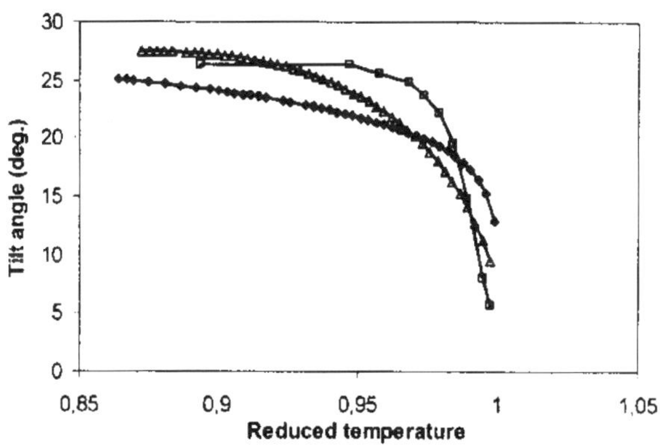

Fig. 3.3. Tilt angle of the director in the smectic C* phases of 10B1M7 as a function of reduced temperature, T/T_{AC^*}, from the ^{129}Xe shielding values (♦), from the ^{2}H NMR experiments (Δ) in Ref. [17], and from optical measurements (□). (Points were estimated from Fig. 3.17 of Ref. [18].) (Reproduced from Ref. [15] with permission. Copyright (2004), American Chemical Society.)

The theoretical model described above has been applied to the ^{129}Xe shielding in HAB (4,4'-diheptylazoxybenzene).[19] Experimental results are shown in Fig. 3.4. Interestingly, a xenon signal can be also detected in solid state, and the curve behaves quite smoothly at the SmA to solid phase transition. In many cases, xenon is expelled from the solid during freezing. Obviously, in HAB xenon remains partly in the solid matrix. (^{129}Xe resonance signals have also been detected in the solid state of Merck Phase 4 liquid crystals as is shown below in Sec. 3.5.) The parameter *c*, indicating the redistribution of xenon atoms during the formation of smectic layers was analyzed to be -0.0779 in HAB. For comparison, the corresponding value in NCB84 (1-butyl-*c*-4-(4'-octylbiphenyl-4-yl)-*r*-1-cyclo-hexan-carbonitrile) was found to be -0.05215,[12] and in 10B1M7 -0.0325.[15] These results mean that the deviation from uniform distribution is of the order of a few percent and that the magnitude of the deformational contribution to σ reduces. In a liquid-crystalline environment, the electron cloud is deformed by the anisotropic forces, leading to non-zero shielding anisotropy and change in the isotropy of the shielding tensor. The same forces are responsible for the dipolar couplings measured for symmetric methane and tetramethylsilane molecules in liquid crystals.[20] Earlier on, this phenomenon was called "apparent orientation", the basic reason, however, being the interference of molecular vibrational and reorientational motions.[21] The shielding can be divided into isotropic and anisotropic parts:

$$\sigma_{iso}(T) = \sigma_o[1-\alpha(T-T_o)][1-\varepsilon(T-T_o)](1+2c\gamma_1^2) \quad (3.7)$$

$$\Delta\sigma_{aniso}(T) = \Delta\sigma_o[1-\alpha(T-T_o)][1-\Delta\varepsilon(T-T_o)][S(T)+2c\sigma_1\gamma_1]. \quad (3.8)$$

In Eqs. (3.7) and (3.8), $\sigma_o = \rho_o\sigma_d'$ and $\Delta\sigma_o = \rho_o\Delta\sigma_d'$, *i.e.* they are isotropic shielding constant and shielding tensor anisotropy at the reference temperature T_0. It is assumed, which in fact is valid in most cases, that the density of liquid crystals is a linear function of temperature, $\rho(T) = \rho_o[1-\alpha(T-T_o)]$, with α being isobaric thermal expansion coefficient. When moving from one phase to another one has to consider a possible density jump term, $\Delta\rho/\rho_o$, in $\rho(T)$ as well. The σ_{iso} and $\Delta\sigma_{aniso}$ for xenon in HAB are shown in Fig. 3.5 whereas Fig. 3.6 displays the order parameters as a function of temperature.

Figure 3.5 reveals an interesting behavior of the isotropic shielding constant at the N-SmA phase transition; σ_{iso} starts to increase (chemical shift decreases). This arises from the negative $2c\gamma_1^2$ term in Eq. (3.7). Consequently, during the formation of the SmA phase the translational order parameter is switched on and xenon atoms redistribute between the dense core region and less dense chain region.

The discussions above concentrate exclusively on xenon shielding (chemical shift). Similar information, as from ^{129}Xe NMR spectra, can, of course, be obtained from the spectra of other noble gases as well. Figure 3.7 shows as an example the comparison of ^{129}Xe and ^{21}Ne shielding in d_2-EBBA (4-ethoxybenzyldine-2,6-dideutero-4'-n-butylaniline).[22] It appeared that the ratios $\sigma_{iso}(Xe)/\sigma_{iso}(Ne)$ and $\Delta\sigma(Xe)/\Delta\sigma(Ne)$ are practically independent of temperature and equal to about 15 and 10, respectively. Consequently, one could say that the ^{129}Xe shielding (chemical shift) parameter is about 15 times more sensitive than that of ^{21}Ne. A similar factor, *ca.* 16, has been measured for the two isotopes in isotropic liquids.[23]

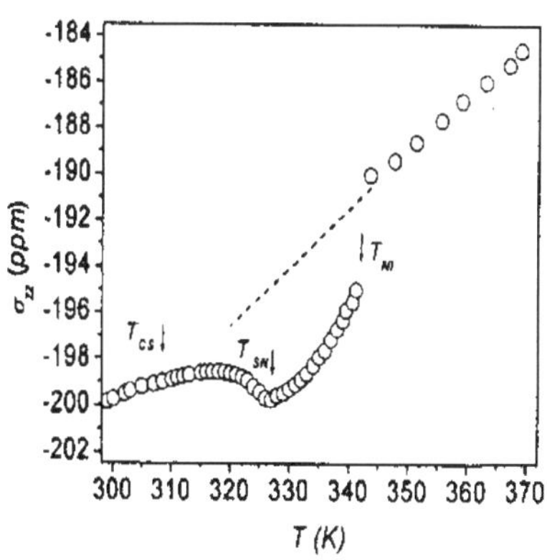

Fig. 3.4. ^{129}Xe shielding (with respect to an external low-pressure gas sample) as a function of temperature for xenon in HAB. The dashed line indicates the variation of shielding if nly temperature-dependence of density is considered. (Reproduced from Ref. [19] with permission. Copyright (2004), American Chemical Society.)

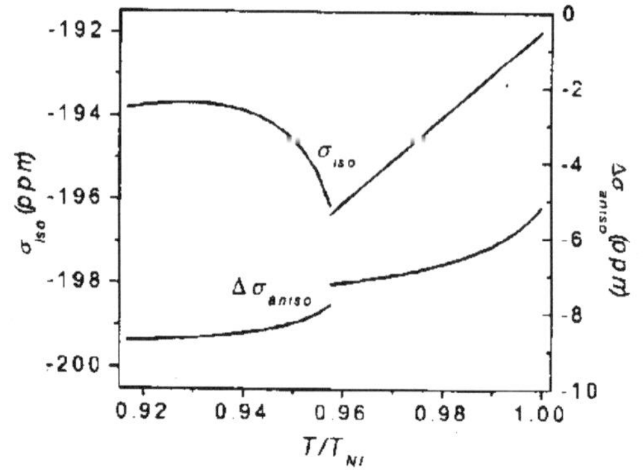

Fig. 3.5. σ^{iso} and $\Delta\sigma^{aniso}$ of xenon in HAB liquid crystal as a function of reduced temperature calculated from Eqs. (3.7) and (3.8). The discontinuities appear at the nematic -smectic A phase transition. (Reproduced from Ref. [19] with permission. Copyright (2004), American Chemical Society.)

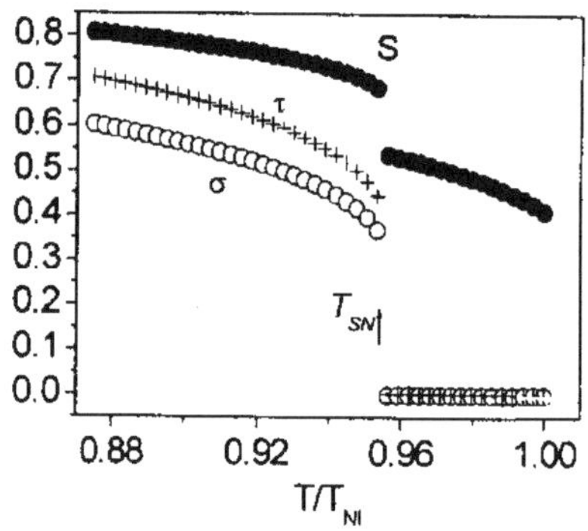

Fig. 3.6. Variation of the theoretically estimated orientational/translational order parameters as a function of reduced temperature. In nematic phase, γ_1 (τ in the figure) and σ_1 (σ in the figure) are zero. The results were calculated using parameters corresponding to those of HAB. (Reproduced from Ref. [19] with permission. Copyright (2004), American Chemical Society.)

The NMR of noble gas probes in liquid-crystalline systems has mostly been applied to thermotropic liquid crystals, although the first ^{131}Xe NMR study, in fact the first one conducted in a liquid crystal as mentioned in Sec. 3.1, was carried out in a lyotropic solution of PBLG and d-chloroform.[4] One reason may be, the low solubility of noble gases in water and thus in water-based lyotropics. A more systematic study than that in Ref. [4] dealt with a mesophase formed by a mixture of cetyltrimethylammonium bromide (CTAB) and formamide (FA).[24] This system is known to form micelles at low concentrations of CTAB and hexagonal, cubic and lamellar mesophases at higher concentrations and elevated temperatures. ^{129}Xe chemical shift measured over a wide temperature range indicated one clear phase transition. The transition appeared at 341, 316 and 297 K when the CTAB concentration was 78, 60 and 25 wt%, respectively. The shielding jumped several ppm as it does in thermotropic liquid crystals. Joint analysis of ^{129}Xe chemical shift data and ^{131}Xe quadrupole coupling data allowed conclusions about the existence of various phases.

During the last decade, lyotropic liquid crystals have been used to orient biological macromolecules.[25] Obviously this has motivated scientists to study, with the aid of $^{129/131}$Xe NMR spectroscopy, properties of a system comprising DMPC (dimyristoyl-phosphatidylcholine) and DHPC (dihexanoyl-phosphatidycholine), which in aqueous solution form liquid-crystalline phases just above room temperature.[26] Figure 3.8 shows, apart from ^{2}H NMR spectra of D_2O, ^{129}Xe NMR spectra in such a solution.

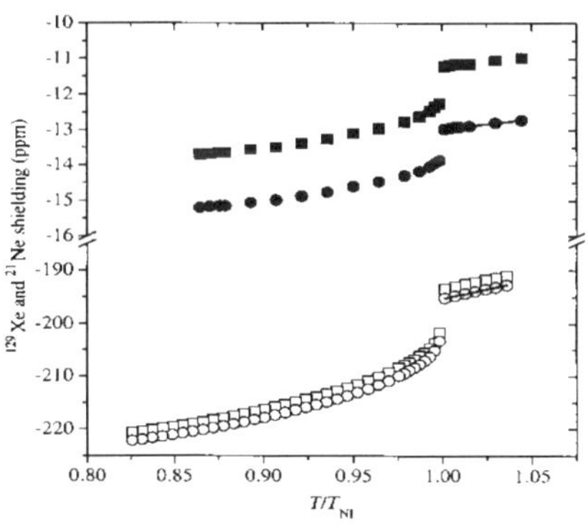

Fig. 3.7. ^{129}Xe (open symbols) and ^{21}Ne (solid symbols) shielding in EBBA with (circles) and without (squares) the bulk susceptibility correction. The solid lines show the least-squares fits of the isotropic part in Eq. (3.1). (Reproduced from Ref. [22] with permission. Copyright (1999), Elsevier.)

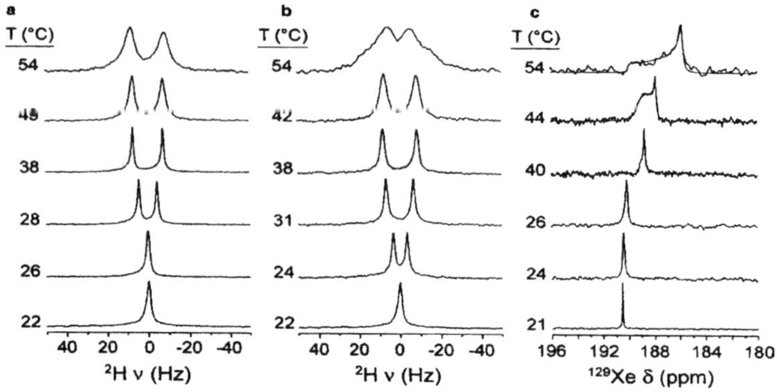

Fig. 3.8. ^{2}H and ^{129}Xe NMR spectra from DMPC/DHPC bicelle solutions (q = 3.5%, 7.5% w/v; q is the molar ratio [DMPC]/[DHPC]) at variable temperatures. (a) No xenon in the sample, (b,c) sample with about 4.7 atm of xenon. Comparison of spectra in (a) and (b) reveals the effect of xenon on the phase behavior. ^{129}Xe chemical shift is referenced to in situ Xe gas capillary. (Reproduced from Ref. [26] with permission. Copyright (2005), Elsevier.)

Comparison of ^{2}H NMR spectra in solutions with and without xenon (Fig. 3.8) reveals interesting behavior: xenon causes a reduction in the onset temperature of alignment and a reduction in the onset of collapse in the liquid crystal phase. The ^{2}H splitting is small and can be explained with the aid of a two-site model: water molecules exchange between the anisotropic and isotropic sites but at 28 °C and higher temperatures averaging is incomplete resulting in a doublet structure in the spectrum. The ^{129}Xe spectrum consists of a relatively sharp single peak in the isotropic phase. In the nematic phase, the peak shifts to higher shielding (smaller chemical shift) due to the anisotropic contribution (see, Eqs. (3.1) and (3.8)). This indicates that the director orients perpendicular to the external magnetic field, and consequently the anisotropy of diamagnetic susceptibility is negative. At the transition from the liquid crystal phase to the extended lamellae[27] the ^{129}Xe line shape starts to resemble a CSA powder spectrum of an axially symmetric shielding tensor. This is interpreted to arise from a situation where xenon resides predominantly within bilayer pores of axial symmetry.[26] At the highest temperature in Fig. 3.8, the shielding tensor anisotropy was found to be -1.55 ppm. The sign is negative as in thermotropic liquid crystals but the magnitude is about one order smaller. This may be due to the idea that the deforming forces in lyotropic liquid crystals are weaker than in thermotropic liquid crystals and that xenon exchanges between two sites, one of them being isotropic. This system was also studied with applying ^{131}Xe NMR, as will be discussed below.

As indicated above, ^{129}Xe shielding constant (chemical shift) is a very informative parameter in studies of properties of liquid crystals. When measured as a function of temperature, it reveals phase transitions, gives information on density (including thermal expansion coefficient), and allows the determination of orientational/translational order parameters (S, σ_1 and γ_1) as well as tilt angle in SmC/C* phases. Complementary information may be gained from ^{129}Xe self-diffusion and relaxationn time measurements.

3.4.2. *Diffusion*

^{129}Xe NMR self-diffusion experiments are usually based on the PGSE (pulsed gradient spin echo) method[28]. The self-diffusion coefficient is determined from the echo amplitude attenuation as a function of the strength of the magnetic field gradient (along the external field) pulse. The amplitude of the echo decays according to Eq. (3.9)[29]

$$A = A_o \exp\left\{-(\gamma G)^2 \left[\delta^2\left(\Delta - \frac{3}{\delta}\right) + \frac{d^2}{30} - \frac{\delta d^2}{6}\right] D_{||}\right\} \tag{3.9}$$

where $D_{||}$ is the self-diffusion constant in the direction of the external magnetic field, Δ is the diffusion time, d is the increment (decrement) time of the gradient and $\delta = d+h$ with h being the duration of the constant gradient G. Figure 3.9 displays the ^{129}Xe self-diffusion coefficient as a function of reciprocal temperature for xenon in FELIX-R&D ferroelectric liquid crystal.[11] The figure clearly shows the different mesophases in FELIX-R&D and the slopes of the straight lines within the phases allow determining diffusion activation energies with the help of the logarithmic form of the Arrhenius equation

$$ln(D) = -\frac{E_a}{RT} + lnD_o \qquad (3.10)$$

where E_a is the activation energy, R is the universal gas constant and D_o is a constant which is assigned different values in different phases and does not possess any physical meaning.

Figure 3.9 also shows a few points of the ^{129}Xe self-diffusion constant in the perpendicular direction to the external magnetic field, $D_\perp$, and consequently in the perpendicular direction to the director. These values were determined by applying constant magnetic field gradients created by shim coils. The results in the isotropic phase are, of course, independent of direction because of the random orientation of liquid crystal molecules. On the contrary, in oriented phases $D_{||}$ and $D_\perp$ differ

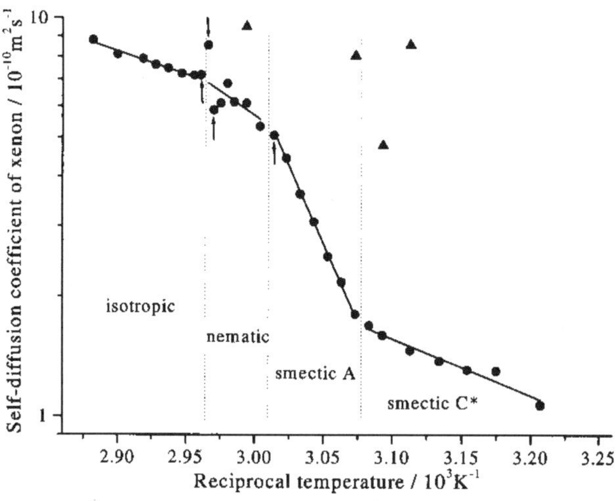

Fig. 3.9. Parallel, $D_{||}$ (•) and perpendicular, $D_\perp$ (▲) self-diffusion coefficient of xenon with respect to the external magnetic field in FELIX-R&D. The vertical dashed lines indicate the phase transitions whereas the solid lines indicate fits to the Arrhenius Eq. (3.10). Note the logarithmic scale in the vertical direction. (Reproduced from Ref. [11] with permission. Copyright (2001), Taylor and Francis.)

from each other, meaning that the ^{129}Xe self-diffusion tensor is anisotropic. The ratio $D_\perp / D_{||}$, which is commonly used as a measure of diffusion tensor anisotropy, increases from the nematic phase toward SmA and SmC* phases. This indicates that xenon atoms diffuse more freely in the perpendicular direction to the director, in other words along the layer surface in the interlayer space, than through the denser core region. Moreover, the diffusion constant along the field and director direction decreases monotonically in the order I > N > SmA > SmC*. The slowest diffusion and the decrease with decreasing temperature along the field direction is most likely a consequence of increasing tilt angle.

Similar results to those for FELIX-R&D have been measured for xenon in 10B1M7.[15] These are shown in Fig. 3.10. The ^{129}Xe shielding values (see Fig. 3.2) do not reveal transitions between the ferroelectric, ferrielectric and antiferroelectric SmC* phases. Similarly, not very clear changes at these phase transitions can be detected in the D vs. $1/T$ graph. It, however, appeared that the slopes, *i.e.* activation energies are slightly different for each phase.

^{129}Xe self-diffusion measurements provide complementary information to that obtainable from chemical shift and shielding data on the structure and behavior of liquid crystals. A generally valid result seems to be that the ^{129}Xe self-diffusion tensor anisotropy increases when moving from the nematic phase to smectic A and smectic C phases. One should, however, point out that the diffusion experiments are usually very time-consuming because at each temperature one has to measure at least 10 echo amplitudes with varying gradient strengths. When measuring diffusion constants (or relaxation times) one should make sure that convection does not start when cooling/warming a sample.

3.4.3. *Relaxation*

The ^{129}Xe spin-lattice relaxation time, T_1, in pure gas is very long. A recent study has shown that the *intrinsic* T_1 is due to the formation of Xe_2 van der Waals dimers and then two mechanisms are effective: spin-rotation and chemical shift anisotropy (CSA).[30] The latter depends upon the strength of the external magnetic field and thus is the dominating

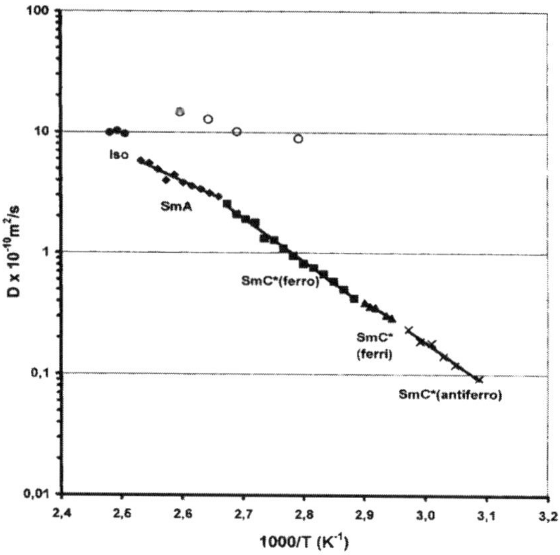

Fig. 3.10. ^{129}Xe self-diffusion coefficients of xenon in 10B1M7 as a function of reciprocal temperature. The points (o) indicate $D_{\perp}$ whereas the other points indicate $D_{||}$. (Reproduced from Ref. [15] with permission. Copyright (2004), American Chemical Society.)

mechanism at high magnetic fields, whereas spin-rotation interaction dominates at low fields. The intrinsic T_1 at room temperature is several hours. The dominating extrinsic relaxation mechanism is due to collisions of xenon atoms with the walls of a glass container. In solutions, ^{129}Xe T_1 is much shorter, a few tens of seconds, the dominating relaxation interaction being the dipole-dipole coupling between xenon and protons of a solvent molecule.

It has appeared that ^{129}Xe T_1 and T_2 (spin-spin relaxation time) are sensitive parameters in studies of various phases of liquid crystals. An example of the dependence of T_1 on phases is shown in Fig. 3.11. The figure illustrates $ln(T_1)$ as a function of reciprocal temperature for xenon in HAB.[19] (The shielding behavior of xenon in HAB is shown in Fig. 3.4.) Contrary to the shielding *vs.* temperature curve, the $ln(T_1)$ vs. $1/T$ indicates clear jumps at both N – I and SmA – N transitions. In the isotropic phase, T_1 is approximately 53 s and the activation energy is 1.64 ± 0.03 kJmol^{-1}. At the isotropic-nematic transition T_1 jumps to 43 s and decreases to 28 s with decreasing temperature, and jumps to 48 s in the transition to the smectic phase. Furthermore, the slope of the straight line in the smectic phase gives the activation energy of 4.0 ± 0.1 kJmol^{-1} which is smaller than 14 kJmol^{-1} in the nematic phase. In accordance with the ^{129}Xe shielding data, the change in T_1 and activation energy can be interpreted to arise from xenon redistribution during the development of the smectic phase.

Similar information as recorded from ^{129}Xe T_1 can be derived from ^{129}Xe T_2 measurements. Figure 3.12 shows ^{129}Xe ln(T_2) and ln(T_1) as a function of reciprocal temperature for xenon in two liquid crystal mixtures possessing nematic, smectic and reentrant nematic phases.[31] ^{129}Xe chemical shift jumps abruptly at the isotropic-nematic transition but surprisingly the shift changes smoothly at the nematic-smectic transition. As pointed out above, usually a clear change in the ^{129}Xe chemical shift behavior can be detected in the latter transition because of xenon redistribution and onset of the order parameters of a smectic phase. A slight change in the curvature can be seen at the smectic-reentrant nematic transition. On the contrary, the plots ln(T_2) and ln(T_1) vs. 1000/T reveal clear changes in the slopes at each phase transition.

The spin-lattice relaxation time ^{129}Xe T_1 was measured in the DMPC/DHPC bicelle solution (D_2O as buffer solvent) using hyperpolarized ^{129}Xe, polarization being 5-8%. Figure 3.13 displays the normalized ^{129}Xe signal intensity as a function of time in the isotropic and liquid crystal phase of the solution. The analysis of the results shown in Fig. 3.13 gives ^{129}Xe T_1 in the isotropic phase (15 °C) of 112 ± 4 s and in the liquid crystal phase (35 °C) of 116 ± 7 s. These are significantly longer than those measured in thermotropic liquid crystals. It is obvious that the ^{129}Xe relaxation times T_1 and T_2 are sensitive parameters in studies of liquid crystal phases and give complimentary information with shielding data.

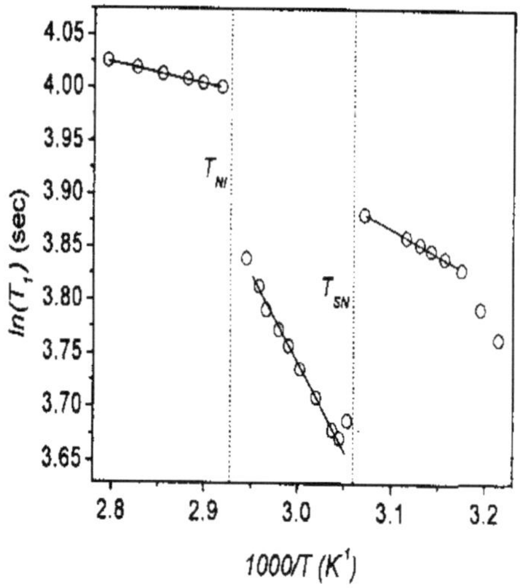

Fig. 3.11. ^{129}Xe ln(T_1) as a function of reciprocal temperature for xenon in HAB. The solid lines are results of the fits to the Arrhenius equation whereas the vertical dashed lines indicate phase transition temperatures. (Reproduced from Ref. [19] permission. Copyright (2004), American Chemical Society.)

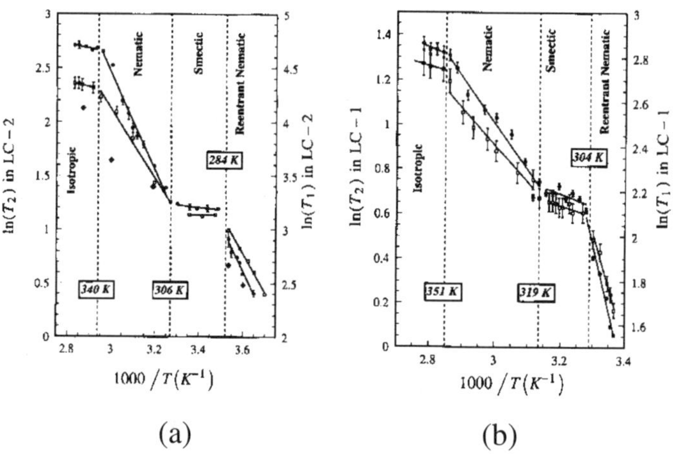

Fig. 3.12. $ln(T_2)$ and $ln(T_1)$ as a function of reciprocal temperature in (a) LC-1 (27.16 wt% 6OCB in 8OCB) and (b) LC-2 (73% 7CB, 8% 8OCB and 9% 5CT). (Reproduced from Ref. [31] with permission. Copyright (1999), American Chemical Society.)

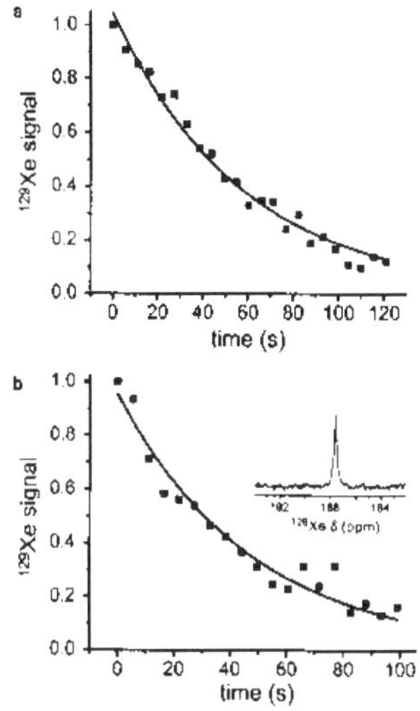

Fig. 3.13. Decay of the hyperpolarized ^{129}Xe signal in (a) isotropic phase, and (b) liquid crystal phase of the DMPC/DHPC bicelle solution. (Reproduced from Ref. [26] with permission. Copyright (2005), Elsevier.)

3.4.4. *Quadrupole coupling*

Quadrupolar noble gas nuclei probe electric field gradients (EFGs) which in liquid-crystalline systems originate from the charge distribution in the liquid crystal molecules. The theoretical description of the temperature dependence of the quadrupole coupling, B, can be derived in a similar way as the temperature dependence of shielding. The main difference, however, is the fact that the electric field gradient tensor is traceless, and consequently only the anisotropic part needs to be considered. The function for B is [12]

$$B(T) = \frac{2}{3}P_2(\cos\theta)[1-\alpha(T-T_o)](\Delta B_d + \Delta B_{ext}^{o} + \Delta B_{ext}^{1}\frac{T}{T_{NI}})[S(T) + 2c\sigma_1(T)\gamma_1(T)] \quad (3.11)$$

where ΔB_d arises from the distortion of the electron cloud from spherical symmetry and $\Delta B_{ext} = \Delta B_{ext}^{o} + \Delta B_{ext}^{1}\, T/T_{NI}$ is due to the permanent electric moments of neighboring liquid crystal molecules and is assumed to be a linear function of temperature. The other symbols have the same meaning as in Eq. (3.1). A joint analysis of shielding and quadrupole coupling data allows the determination of the isotropic and anisotropic shielding terms, the quadrupole coupling terms, all three order parameters as a function of temperature if a smectic A phase is present, and finally temperature dependence of the tilt angle in smectic C* phases.

Figure 3.14 shows both the quadrupole coupling of ^{21}Ne and chemical shift of ^{129}Xe in FELIX-R&D. The former averages out in the isotropic phase while the latter averages to a non-zero value which increases linearly with temperature. The two curves possess similar features in the nematic and smectic A phases.

As stated above, there are two contributions to the quadrupole coupling of noble gases in liquid crystals: one from the deformation of an electron cloud and the other from the electric charge distribution in liquid crystal molecules. Experiments have shown that these may have the same or opposite sign. This is illustrated in Fig. 3.15.

The curves in Fig. 3.15a display a maximum. This is an indication that the two contributions to the field gradient must have opposite signs in this particular liquid crystal. The slight shift of the maxima stems from the fact that the two EFGs contribute in different proportions. The EFG due to distortion of the electron cloud is different in each case. An example of a case in which the two gradients possess the same sign is given in Fig. 3.15b which shows $B(^{21}\mathrm{Ne})$ as a function of reduced temperature for neon in EBBA.

An example of the utilization of ^{131}Xe NMR in a study of the lyotropic DMPC/DHPC system is shown in Fig. 3.16. The figure shows both ^{2}H and ^{131}Xe NMR spectra of water and xenon, respectively, in the DMPC/DHPC bicelle solution. The ^{131}Xe spectrum displays a triplet due to quadrupole coupling in the liquid-crystalline phase, $B(^{131}\mathrm{Xe})$ being a few kHz. The B value is about one order of magnitude smaller than that observed in thermotropic ZLI 1167 and decreases with increasing temperature. At the same time the ^{2}H splitting of D_2O increases. Outside the liquid crystal region, satellite transitions ($-3/2 \leftrightarrow -1/2$ and $1/2 \leftrightarrow 3/2$

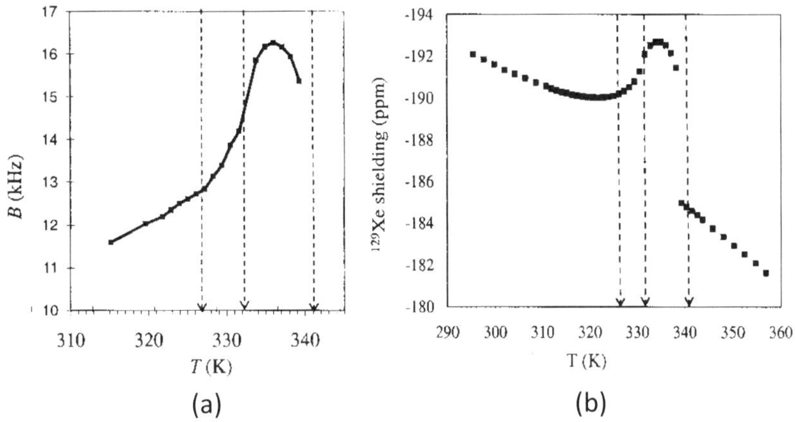

Fig. 3.14. (a) Absolute value of the quadrupole coupling B of ^{21}Ne (spectrum does not yield the sign) and (b) chemical shift of ^{129}Xe in ferroelectric FELIX-R&D liquid crystal as a function of temperature. The vertical dashed lines indicate phase transition temperatures of pure liquid crystal: $T_{S_{C^*}S_A} = 54$ °C, $T_{S_AN} = 59$ °C, and $T_{NI} = 68$ °C. The chemical shift data in (b) correspond to the spectra shown in Fig. 3.3. (Unpublished results from the research group of the author.)

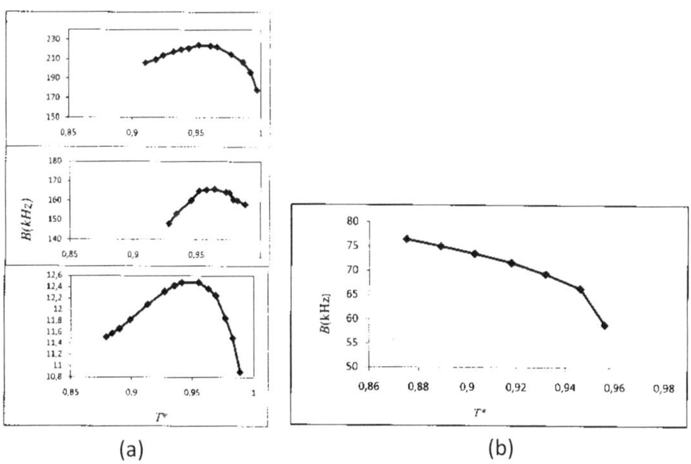

Fig. 3.15. (a) Absolute value of the quadrupole coupling constant B (in kHz) as a function of reduced temperature $T^*=T/T_{NI}$ for krypton (top), xenon (middle) and neon (bottom) dissolved in ZLI 1167 liquid crystal. ZLI 1167 possesses a negative anisotropy of diamagnetic susceptibility and thus its director orients perpendicular to the external magnetic field. (b) Absolute value of $B(^{21}Ne)$ for neon in EBBA as a function of reduced temperature. (Unpublished results from the research group of the author.)

for an $I = 3/2$ nucleus) are not observable, most likely because of a wide distribution of orientational order parameter values which causes the satellites to become so broad that they disappear into the background and noise. Although the two contribution ΔB_d and ΔB_{ext} (see, Eq. (3.11)) cannot be separated (this would require a lot more experimental points than given in Fig. 3.16) it is quite obvious that EFGs are about one order of magnitude smaller in lyotropic than in thermotropic liquid crystals.

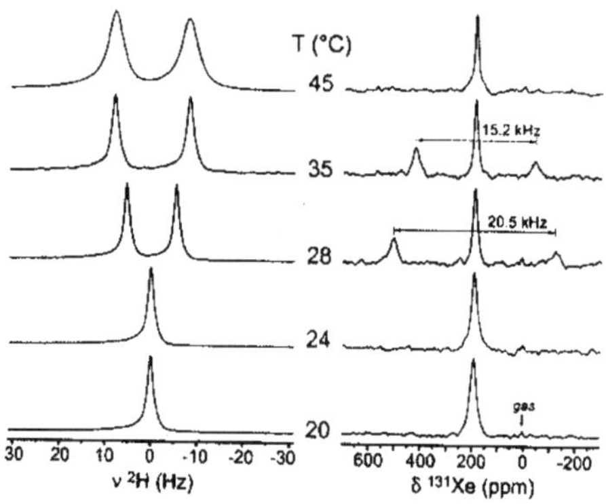

Fig. 3.16. ^{2}H (left) and ^{131}Xe (right) NMR spectra of D_2O and xenon in DMPC/DHPC bicelle solution. The small peak marked by "gas" was interpreted to come from outside the coil region. (Reproduced from Ref. [26] with permission. Copyright (2005), Elsevier.)

3.4.5. *Simulations*

In deriving Eqs. (3.1) and (3.2), pairwise additive approximation was applied. As a consequence, the shielding, for example, reduces to a sum of two terms, the isotropic and anisotropic terms. If the pair additive approximation is not used, the solvent effect on the shielding is a linear combination of products of the liquid crystal orientation parameters of varying rank.[13]

A constant-pressure Monte Carlo simulation of ^{129}Xe shielding and ^{131}Xe quadrupole coupling in a model thermotropic liquid crystal was carried out in Ref. [33]. The liquid crystal solvent was described by Gay-Berne (GB) molecules. The model exhibits isotropic, nematic, smectic A and smectic B/molecular crystal phases. The Xe NMR parameters were calculated using the pairwise additive approximation. Figure 3.17a shows temperature dependence of simulated order parameters for the GB model with xenon solutes. The curves reveal discontinuities at certain T^* values ($T^* = k_B T/\epsilon_o$, where k_B is the Boltzmann constant and ϵ_o is a parameter in Gay-Berne potential and defines the energy scale). They correspond to the nematic - isotropic, smectic A - nematic and crystal - smectic A phase transitions. The simulation results in a small shift in transition temperatures as compared to "pure" liquid crystal when xenon was added to the system. Experiments have yielded shifts of a few degrees as a consequence of adding 1-5 atm of xenon in a sample.[34] The number density (see Fig. 3.17b) shows discontinuity at the crystal - SmA and N - I transitions but not at the SmA - N transition.

The simulation produces qualitatively similar results (see Fig. 3.18) as obtained experimentally but, for example, changes in the NMR parameters and density in phase transitions are clearly exaggerated. However, temperature dependence of these parameters can be interpreted in terms of density and order parameters of the GB model liquid crystal. This is in complete agreement with the theoretical model presented above and experimental results.

3.5. Xenon in Confined Liquid Crystals

Properties of liquid crystals confined in porous materials have been studied with numerous techniques, such as neutron scattering,[35] Raman spectroscopy,[36] differential scanning calorimetry,[37] and NMR.[38-46] NMR is especially useful because of the sensitivity of a spectrum to molecular order, director configurations, and molecular dynamics. A conventional approach in NMR experiments is to utilize partially deuterated liquid crystals and detect their ^{2}H NMR resonances. Utilization of a xenon probe has, however, appeared as an applicable alternative.

^{129}Xe and ^{131}Xe NMR have been applied to probe the behavior of ZLI 1132 (mixture of *trans*-4-*n*-alkyl-(4-cyanophenyl)-cyclohexanes) confined to the cavities of anopore membrane.[38] The average diameter of the cavities is 0.2 μm (200 nm), the diameter distribution is narrow, and the cavities are parallel, straight cylinders which are not connected.

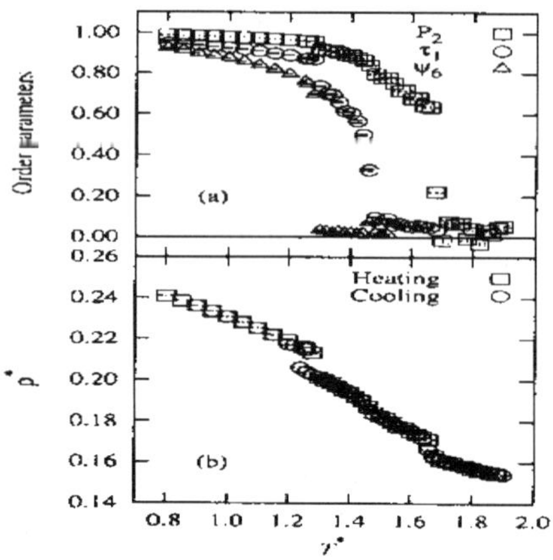

Fig. 3.17. (a) Orientational order parameter P_2 (= S in eq. (3.1)), the translational order parameter τ_1 (= γ_1 in eq. (3.1)), and the bulk orientational order parameter ψ_6 derived from the simulation with Gay-Berne model liquid crystal. ψ_6 differentiates between the smectic A and smectic B phases. (b) Temperature dependence of the average number density ρ. (Reproduced from Ref. [33] with permission. Copyright (2007), American Physical Society.)

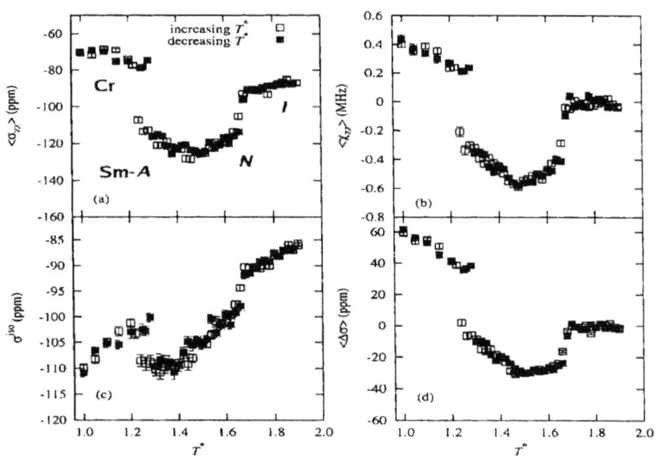

Fig. 3.18. Simulated Xe total shielding, $<\sigma_{zz}>$ (a), isotropic shielding constant, σ^{iso} (b), ^{131}Xe quadrupole coupling, $<\chi_{zz}>$ (c), and shielding tensor anisotropy, $<\Delta\sigma>$ (d) for xenon in Gay-Berne model liquid crystal as a function of temperature. (Reproduced from Ref. [33] with permission. Copyright (2007), American Physical Society.)

Figure 3.19 shows experimental and simulated ^{129}Xe NMR spectra on xenon dissolved in ZLI 1132 liquid crystal and confined to the cavities of an anopore membrane. The channel axis is perpendicular to the plane of the pile of membrane sheets, each being 60 μm thick.

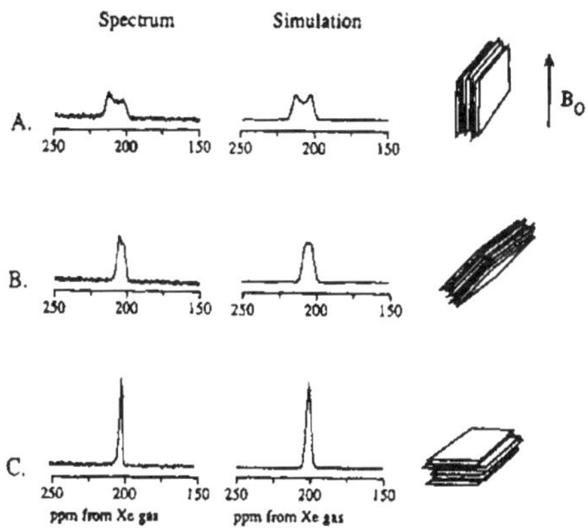

Fig. 3.19. ^{129}Xe NMR spectra of xenon in ZLI 1132 confined to anopore membranes at 19 °C and at various orientations with respect to the external magnetic field. Xenon pressure is 5 atm. The channel axis forms an angle of A) 90°, B) 45°, and C) 0° with the external field. (Reproduced from Ref. [38] with permission. Copyright (1995), American Chemical Society.)

The orientation of liquid crystals inside a confined space is a consequence of the competition between the elastic forces within the liquid crystal, the molecular interactions with the cavity surface, and the effect of the external magnetic field. The anisotropy of diamagnetic susceptibility of ZLI 1132 is positive, and consequently the magnetic field tends to orient the director to its own direction. ^{2}H NMR studies have shown that ZLI 1132 orients in lecithin coated cavities so that the molecular long axes are perpendicular to the surface.[46] (In untreated anopore, the director is aligned along the pore axis.) Such a molecular arrangement is also supported by the ^{129}Xe NMR results shown in Fig. 3.19. When the channel axis is parallel with the external magnetic field, as in Fig. 3.19C, a single, relatively narrow line at 201 ppm is observed. This is possible in two cases: (a) the liquid crystal director is parallel with the field direction, and (b) directors are spread in the plane perpendicular to the field direction. When the channel axis forms an angle of 45° and 90° with the field direction (Figs. 3.19B and A), a powder spectrum is observed. In particular, the powder spectrum in case A is characteristic of a random distribution of directors in the plane perpendicular to the external field direction. The simulated powder spectrum is in fair agreement with the experimental one which can be interpreted to be proof that the external magnetic field (11.7 T) affects the orientation distribution only very little if at all. Interestingly, these experiments predict that the isotropic nematic phase transition temperature *increases* from 74 ± 1 °C measured in bulk ZLI 1132 to 96 ± 3 for ZLI 1132 confined with xenon gas (6 atm) to anopore channels. It was also concluded on the basis of ^{131}Xe NMR measurements that the electric field gradient experienced by xenon in confined ZLI 1132 is slightly higher than in the bulk sample.

Other examples of the application of ^{129}Xe NMR to studies of liquid crystals in confined geometries are cases in which the solid matrix is formed by mesoporous controlled pore glasses (CPGs)[44]. The CPG materials differ from anopore membranes in that the porous network consists of randomly oriented and connected channels. Experiments were carried out using CPGs with different average pore diameters, extending from 81 to 2917 Å and liquid crystal Merck Phase 4 (eutectic mixture of *p*-methoxy-*p'*-*n*-butylazoxybenzenes). The melting point and clearing point of pure Phase 4 are 20 °C and 74 °C, respectively. In the following, porous materials are referred to with the abbreviation CPG followed by pore diameter in ångströms, for example CPG 81. For comparison, Fig. 3.20 displays ^{129}Xe spectra of xenon in bulk Phase 4 (there is no CPG material in the sample). Spectra were taken from high to low temperature. The labels attached to resonance signals in Fig. 3.21 have the following meaning: I_B, I refers to isotropic and B to bulk, N to nematic, S to solid, L to liquefied, and G to gas. Chemical shift is given with respect to an external xenon gas sample with a pressure of 6.16 atm. Chemical shift behaves as a function of temperature as described in Sec. 3.4.1. There is a linear behavior in the isotropic phase, a jump to a larger shift (smaller shielding) at the I-N phase transitions, and a curved behavior in the nematic phase. Interestingly, when the liquid crystal freezes a weak signal can be detected from the solid matrix (most of the xenon is expelled from the solid but a small amount is occluded in the solid) and at the same time a signal at 0 ppm emerges. The signal arises from fairly large gas bubbles developed during the freezing process. On the other hand, during freezing the density of Phase 4 increases, and as a consequence some free space in between the solid liquid crystal and sample tube wall is created. Part of the signal at 0 ppm may arise from this volume. At the lowest temperatures xenon starts to liquefy and the resonance appears at the highest chemical shift.

Some differences in the ^{129}Xe shielding behavior are observed if the experiments are performed in the opposite direction to those in Fig. 3.20. Fig. 3.21 shows the spectra taken from the same sample as those in Fig. 3.20 but recorded from low to high temperature. The sample was frozen in liquid nitrogen outside the magnet and set inside the magnet to stabilize to the lowest temperature shown in Fig. 3.21. The main difference between the two series of spectra can be seen at the lowest temperatures. Without knowing the details of the solid phase(s) of Phase 4, the resonances labeled S1, S2 and S3 were interpreted to be indications of three solid phases, each having different symmetry.

In a sample including CPG particles the space in between the particles (size 125-177 μm) is so large that liquid crystals occupying the space behave as if in bulk. Therefore, the information derived from the previous experiments is of great importance when interpreting the spectra from samples with xenon, liquid crystal and CPG materials. Figure 3.22 shows the ^{129}Xe spectra recorded from a sample containing CPG 81, Phase 4 liquid crystal and xenon gas. The mean pore diameter of CPG 81 is comparable with the length of liquid crystal molecules. The solid matrix of CPGs is nearly independent of temperature within the temperature range studied. Therefore, all the temperature-dependent effects in the following discussion can be attributed to the change in physical properties of liquid crystal. Although the experiments were performed from low to high temperature, the resonances are assigned from high to low temperature. In the isotropic phase, two signals with slightly different chemical shift can be resolved: I_B and I_C. They arise from the bulk phase in between the CPG particles and from confined liquid crystal/xenon solution, respectively (Subscript C refers to confined). The chemical shift of I_C is about 1.5 ppm larger than that of I_B.

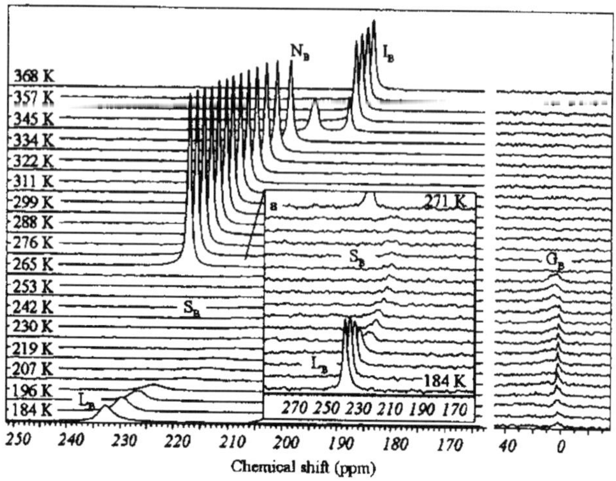

Fig. 3.20. ^{129}Xe NMR spectra of xenon in bulk Phase 4 liquid crystal at variable temperatures. Experiments were carried out from the isotropic phase to solid phase. For the assignment of resonance signals, see text. (Reproduced from Ref. [44] with permission. Copyright (2006), American Chemical Society.)

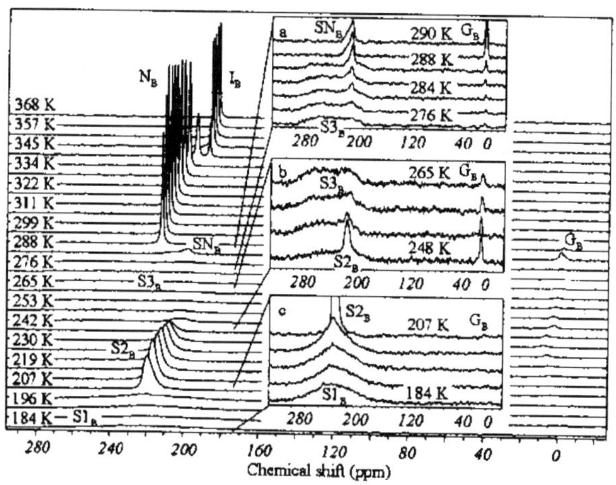

Fig. 3.21. ^{129}Xe NMR spectra of xenon in bulk Phase 4 liquid crystal at variable temperatures. Experiments were carried out from the solid phase to isotropic phase, i.e. from low to high temperature. (Reproduced from Ref. [44] with permission. Copyright (2006), American Chemical Society.)

At the transition from the isotropic to nematic phase (the phases are classified according to the bulk liquid crystal) I_B disappears and N_B emerges; N_B arises from the space in between the CPG particles where the liquid crystal behaves as in bulk. Interestingly, I_C remains and indicates that xenon experiences on average an isotropic environment in the pores. At the nematic-solid transition N_B disappears and a gas peak G emerges similarly to as in bulk Phase 4. Here one has to note that liquids in general behave differently in bulk and confined spaces. For example, the melting point of normal isotropic liquids depresses in confinements. The melting point depression is inversely proportional to the pore radius R_p according to the Gibbs-Thompson equation[47]

$$\Delta T = T_o - T = \frac{k_p}{R_p} \tag{3.12}$$

where T_o and T are the melting points of the bulk and confined substances, respectively, and k_p is a characteristic constant for each substance. Thus I_C remains to be observed still below the melting point of bulk liquid crystal. The results of ^{129}Xe experiments can be interpreted so that liquid crystal in the pores solidifies at about 38 degrees below the bulk melting point, and therefore the signal is labeled with S_C. This resonance signal is not observable in the spectra recorded from the bulk Phase 4 sample (see Figs. 3.20 and 3.21). Therefore, it is obvious that it arises from inside the pores. When the mean pore size of the CPG material increases to 2917 Å the ^{129}Xe NMR spectrum from inside the pores changes drastically when compared to the previous ones. The pore diameter is of the same magnitude as the magnetic coherence length. The spectra for the sample with CPG 2917 are shown in Fig. 3.23.

In the isotropic phase, two separate signals, I_B and I_C, are detected as in the case of smaller pores. The signal, N_B, from the space in between the CPG particles behaves naturally similarly as in the previous case. On the contrary, the N_C signal from inside the pores resembles a powder pattern, indicating orientational order distribution inside the pore. Near to the pore surface, liquid crystal molecules are oriented along the surface (along the pore axis) though particularly at higher temperatures near the

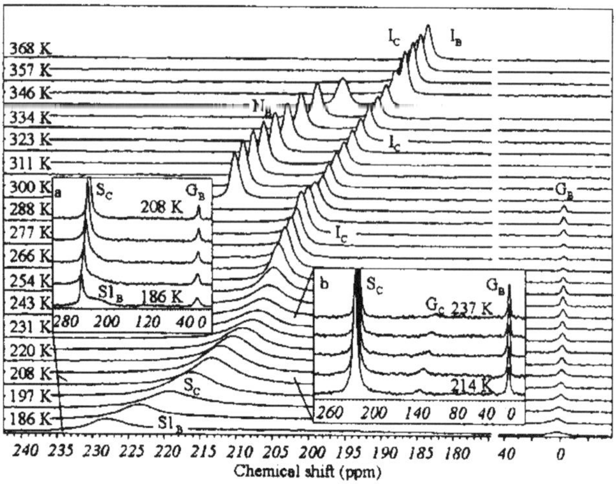

Fig. 3.22. ^{129}Xe NMR spectra from the sample containing CPG 81, Phase 4 liquid crystal and xenon gas. (Reproduced from Ref. [44] with permission. Copyright (2006), American Chemical Society.)

clearing point of bulk liquid crystal, molecules are oriented almost along the external field direction. At these higher temperatures also the ^{129}Xe chemical shifts from the bulk and confined liquid crystals approach each other.

Figure 3.24 shows the ^{129}Xe chemical shift data collected for samples containing Phase 4 liquid crystal, CPG with different mean pore diameters and xenon gas. The results can be condensed into the following items: (a) The ^{129}Xe chemical shift changes linearly in the smallest CPG materials over the whole temperature range, 285-370 K. This means that xenon experiences on average isotropic environment in the pores. (b) The linear temperature dependence signifies that there is no isotropic - nematic phase transition inside the small pores, such as in CPG 81. (c) In the medium-sized pores, such as CPG 156, the nematic phase builds up gradually. (d) The discontinuity in the ^{129}Xe chemical shift in the neighborhood of the isotropic-nematic phase transition of bulk liquid crystal means that a bulk-like first-order isotropic - nematic phase transition takes place inside the large pores, from CPG 538 up, and the transition temperature is close to that of bulk liquid crystal. (e) In the

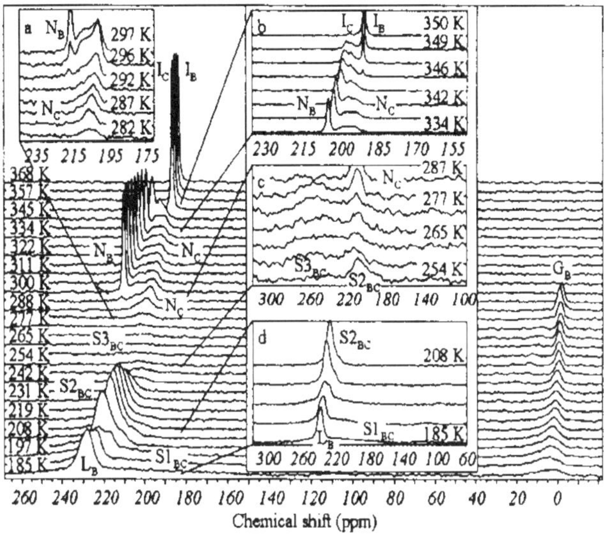

Fig. 3.23. ^{129}Xe NMR spectra taken from the sample containing CPG 2917, Phase 4 liquid crystal and xenon gas. (Reproduced from Ref. [44] with permission. Copyright (2006), American Chemical Society.)

largest pores, in CPG 2917, the pore diameter is close to magnetic coherence length, and therefore the liquid crystal director deviates from the pore axis direction to the magnetic field direction. The magnetic coherence length decreases with increasing temperature and therefore the ^{129}Xe chemical shift approaches that of bulk liquid crystal which in the present case orients so that its director is parallel with the external magnetic field.

As described above, xenon resides in two different sites in the xenon/liquid crystal/porous material systems. The sites are: (a) xenon dissolved in the bulk liquid crystal between the CPG particles, and (b) xenon in the liquid crystal confined inside the pores. The interpretation of xenon shielding and quadrupole coupling results in confined spaces requires the knowledge of a possible exchange between the sites. In these particular systems, a practical means to investigate exchange is 2-dimensional ^{129}Xe exchange spectroscopy (2D ^{129}Xe EXSY).[48] Figure 3.25 shows the spectra recorded in the systems consisting of xenon, Phase 4 liquid crystal and CPG 81 as a function of mixing time[49].

The spectra show that off-diagonal signals, which indicate exchange of xenon between the two sites, begin to gradually build up and increase at a mixing time of about 80 ms. After a mixing time of 2 s, the intensity of cross-peaks remains practically constant. This means that the majority of xenon atoms have exchanged their state during that time. The evolution of the cross-peaks with mixing time makes feasible the determination of the exchange rate constants between the two sites. They appeared to be 0.46 s^{-1} from bulk to confined space and 1.09 s^{-1} in the reverse direction. In the case where the bulk liquid crystal is in isotropic phase, two resonance signals can be observed, I_C and I_B, the former arising from inside the pores and the latter from the interparticle space.

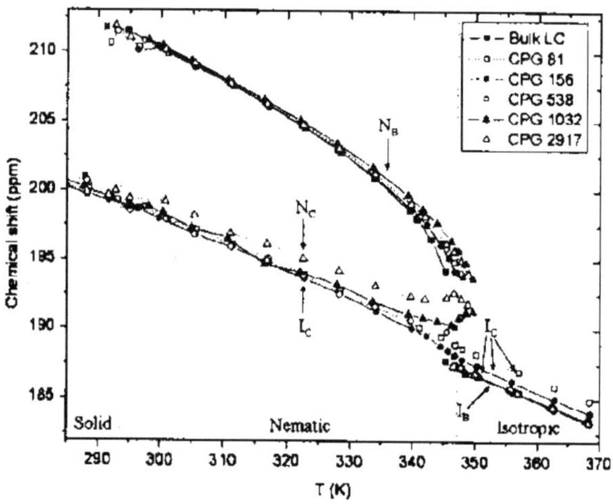

Fig. 3.24. [129]Xe chemical shift of xenon dissolved in Phase 4 liquid crystal confined to five different CPG materials. For comparison, also the data points for xenon in bulk Phase 4 are included in the figure. The lines between the points are only to guide the eye. The vertical dashed line shows the nematic-isotropic phase transition temperature of bulk Phase 4. Subscripts B and C refer to bulk and confined states. All the measurements were carried out from low to high temperature. (Reproduced from Ref. [44] with permission. Copyright (2006), American Chemical Society.)

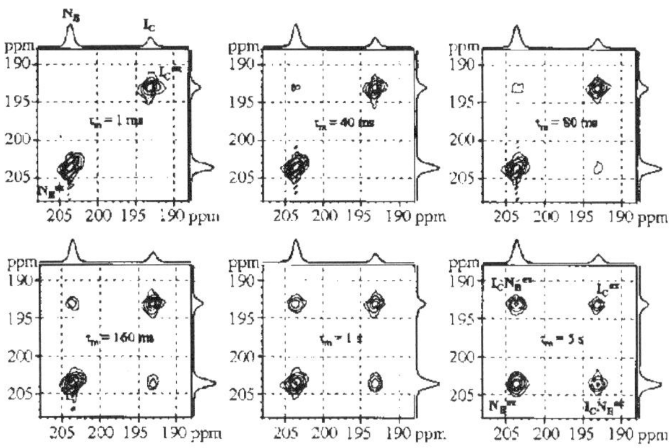

Fig. 3.25. 2D ^{129}Xe EXSY spectra from the sample containing CPG 81, Phase 4 liquid crystal and xenon gas. T = 322 K. Mixing time is shown in each spectrum. N_B is the resonance from xenon in bulk liquid crystal between the CPG particles and I_C is the resonance from xenon in confined liquid crystal. N_B^{ex} (see the lower left corner in the first spectrum) illustrates xenon exchange between different states in the interparticle space whereas I_C^{ex} represents the exchange of xenon in CPG pores. (Reproduced from Ref. [49] with permission. Copyright (2006), PCCP Owner Societies.)

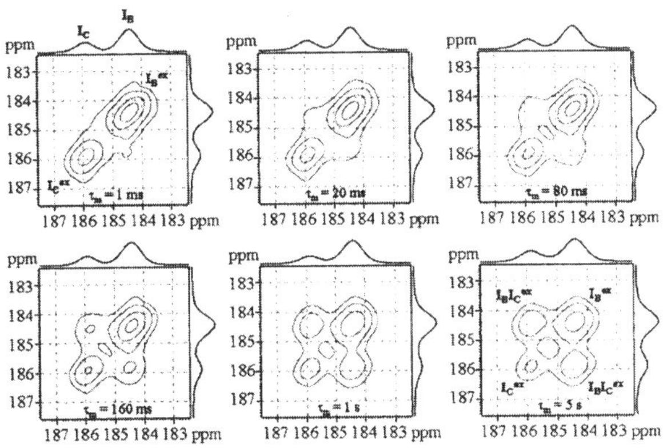

Fig. 3.26. 2D ^{129}Xe EXSY spectra from the same sample as in Fig. 3.27 at 357 K where liquid crystal Phase 4 is in isotropic state. (Reproduced from Ref. [49] with permission. Copyright (2006), PCCP Owner Societies.)

Figure 3.26 displays 2D ^{129}Xe EXSY spectra from such a situation. It appears that now the exchange is much faster than in the previous case where liquid crystal is in nematic phase. Cross-peaks start to emerge already at a mixing time of 20 ms, and they reach a maximum at the mixing time of 500 ms, about four times faster than in the nematic phase. This can be explained on the basis of faster diffusion of xenon in the isotropic phase than in nematic phase.

3.5. Conclusions

The use of noble gas probes in NMR investigations of liquid crystals is a useful means to derive very versatile information. A particularly sensitive probe is ^{129}Xe. ^{129}Xe NMR spectra recorded at variable temperatures make feasible assignment of various phases, derivation of temperature dependence of all the orientational order parameters and tilt angle. Moreover, because ^{129}Xe chemical shift depends upon liquid crystal density, it is possible to determine thermal expansion coefficients in various phases. Additional information about liquid crystal properties can be obtained by NMR relaxation time and diffusion constant measurements. A drawback in such experiments, however, is the fact that they are quite time consuming even though enriched xenon-129 gas is used. In some case, hyperpolarized xenon-129 may offer a solution to this problem. The use of quadrupolar noble gas isotopes gives information on electric field gradients created by liquid crystal molecules. Furthermore, because quadrupole splitting is similarly dependent upon orientational order parameters as in ^{129}Xe chemical shift, joint analysis of shift data and quadrupole coupling data allows derivation of a very complete picture of the properties of liquid crystals. In the case of confined liquid crystals, ^{129}Xe NMR is a workable alternative and a source of additional information to ^{2}H NMR of partially deuterated liquid crystals. In the case of complex pore networks, analysis of powder spectral shapes may reveal director distribution in the pores.

Acknowledgments

This work was supported by the Academy of Finland. The author thanks J. Lounila, P. Tallavaara and V.-V. Telkki for co-operation.

References

1. W.G. Proctor and F.C. Yu, *Phys. Rev.*, 81, 20 (1951).
2. J. Fraissard and T. Ito, *Zeolites*, 8, 350 (1988); J. Fraissard, *Z. Phys. Chem.*, 269, 657 (1988).
3. K. Bartik, P. Choquet, A. Constantinesco, G. Duhamel, J. Fraissard, J.-N. Hyacinthe, J. Jokisaari, E. Locci, T.J. Lowery, M. Luhmer, T. Meersman, I.L. Moudrakovski, G.E. Pavlovskaya, K.L. Pierce, A. Pines, J.A. Ripmeester, V.-V. Telkki and W.S. Veeman. *L'actualité chimique*, 287, 16 (2005).
4. A. Loewenstein and M. Brenman *Chem. Phys. Lett.*, 58, 435 (1978).
5. J.P. Bayle, J. Courtieu and J. Jullien, *J. Chim. Phys.*, 85,147 (1988).
6. J. Jokisaari, *Progr. Nucl. Magn. Res. Spectr.*, 26, 1 (1994).
7. J. Jokisaari, in *NMR of Ordered Liquids*, Ed. E.E. Burnell and C.A. de Lange, (Kluwer, Dordrecht, 2003), p. 109.
8. W. Happer, E. Miron, S. Schaefer, D. Schreiber, W.A. van Wijngaarden and X. Zeng, *Phys. Rev. A*, 29, 3092 (1984); D. Raftery, H. Long, T. Meersman, P.J. Grandinetti, L. Reven and A. Pines, *Phys. Rev. Lett.*, 66, 584 (1984).
9. P. Pyykkö, *Molec. Phys.*, 106, 1965 (2008).
10. T.G. Walker and W. Happer, *Rev. Mod. Phys.*, 69, 629 (1997); K.F. Tupic, Z.I. Cleveland, G.E. Pavlovskaya and T. Meersman, *Solid State Nucl. Magn. Reson.*, 29, 79 (2006) and reference therein ; J. Luo, X. Mao, J. Chen, M. Zhao, L. Fu and X. Zeng, *Appl. Magn, Reson.*, 17, 587 (1999).
11. J. Ruohonen, M. Ylihautala and J. Jokisaari, *Mol. Phys.*, 2001, 99(9), 711-719.
12. J. Lounila, O. Muenster and J. Jokisaari, *J. Chem. Phys.*, 97, 8977 (1992).
13. M. Ylihautala, J. Lounila and J. Jokisaari, *J. Chem. Phys.*, 110, 6381 (1999.)
14. A.D. Buckingham and E.E. Burnell, *J. Am. Chem. Soc.*, 89, 3341 (1967).
15. M. Cifelli, J. Saunavaara, J. Jokisaari and C.A. Veracini, *J. Phys. Chem. A*, 108, 3973 (2004).
16. S. Chandrasekhar, *Liquid crystals*, 2nd ed., (Cambridge University Press, New York, 1992).
17. D. Catalano, M. Cavazza, L. Chiezzi, M. Geppi and C.A. Veracini, *Liq. Cryst.*, 27, 621 (2000).
18. J.W. Goodby, S. Patel and E.J. Chin, *J. Mater. Chem.*, 2, 197 (1992).
19. B. Jagadeesh, A. Prabhakar, M.H.V. Ramana Rao, C.V.S. Murty, V.G.K.M. Pisipati, A.C. Kunwar and C.R. Bowers, *J. Phys. Chem. B*, 108, 11272 (2004).
20. A. Loewenstein, *Chem. Phys. Lett.*, 38, 543 (1976); L.C. Snyder and S. Meiboom, *J. Chem. Phys.*, 44, 4057 (1966).
21. J. Lounila, *Molec. Phys.*, 58, 897 (1986).
22. M. Ylihautala, J. Lounila and J. Jokisaari, *Chem. Phys. Lett.*, 301, 153-158 (1999).
23. R. Seydoux, P. Diehl, R.K. Mazitov and J. Jokisaari, *J. Magn. Reson.* A, 101, 78 (1993).

24. M. Ylihautala, P. Ingman, J. Jokisaari and P. Diehl, *Appl. Spectroscopy*, 50, 1435 (1996).
25. See for example, N. Tjandra and A. Bax, *Science* 278, 1111 (1997).
26. X. Li, C. Newberry, I. Saha, P. Nikolaou, N. Whiting and B.M. Goodson, *Chem. Phys. Lett*. 419, 233-239 (2006).
27. T.A. Harroun, M. Koslwsky, M.-P. Nieh, C.-F. de Lannoy, V.A. Raghunathan and J. Katsaras, *Langmuir*, 21, 5356 (2005).
28. E.O. Steijskal and J.E. Tanner, *J. Chem. Phys*., 42, 288 (1965).
29. W.S. Price and P.W. Kuchel, *J. Magn. Reson*., 94, 133 (1991).
30. B.C. Anger, G. Schrank, A. Schoek, K.A. Butler, M.S. Solum, R.J. Pugmire and B. Saam, *Phys. Rev. A*, 78, 043406 (2008).
31. J. Bharatam and C.R. Bowers, *J. Phys. Chem. B*, 103, 2510 (1999).
32. M.P. Augustine and K.W. Zilm, *Chem. Phys. Lett*., 280, 24 (1997).
33. J. Lintuvuori, M. Straka and J. Vaara, *Phys. Rev. E*, 75, 031707 (2007).
34. J. Jokisaari and P. Diehl, *Liq. Cryst*., 7, 739 (1990).
35. R. Guégan, D. Morineau, C. Loverdo, and W. Guendouz, *Phys. Rev. E*, 73, 011707 (2006).
36. Y. Shao and T.W. Zerda, *J. Phys. Chem. B*, 102, 3387 (1998).
37. M.D. Dadmun and M. Muthukumar, *J. Chem. Phys*., 98, 4850 (1993).
38. H.W. Long, M. Luzar, H.C. Gaede, R.G. Larsen, J. Kritzenberger and A. Pines, *J. Phys. Chem*., 99, 11989 (1995).
39. G.S. Iannacchione, G.P. Grawford, S. Žumer, J.W. Doane and D. Finotello, *Phys. Rev. Lett*., 71, 2595 (1993).
40. A. Golemme, S. Žumer, D.W. Allender, J.W. Doane, *Phys. Rev. Lett*., 61, 2937 (1988).
41. G.P. Crawford, D.K. Yang, S. Žumer, D. Finotello and J. Doane, *Phys. Rev. Lett*., 66, 723 (1991).
42. T. Jin, B. Zalar, A. Lebar, M. Vilfan, S. Žumer and D. Finotello, *Eur. Phys. J*., 16, 159, (2005).
43. G.S. Iannacchione, S. Qian, D. Finotello and F.M. Aliev, *Phys. Rev. E*, 56, 554 (1997).
44. P. Tallavaara, V.-V. Telkki and J. Jokisaari, *J. Phys. Chem. B*, 110, 21603 (2006).
45. G.P. Crawford and S. Žumer (eds.), *Liquid Crystals in Complex Geometries* (Taylor&Francis, London), 1996.
46. D.W. Allender, G.P. Crawford and J.W. Doane, *Phys. Rev. Lett*., 67, 1442 (1991).
47. C.L. Jackson and G.B. Mc Kenna, *J. Chem. Phys*., 93, 9002 (1990).
48. J. Jeener, B.H. Meier, P. Bachmann and R.R. Ernst, *J. Chem. Phys*., 71, 4546 (1979).
49. P. Tallavaara and J. Jokisaari, *Phys. Chem., Chem. Phys*., 8, 4902 (2006).

Chapter 4

Bicelles – A Much Needed Magic Wand to Study Membrane Proteins by NMR Spectroscopy

Ronald Soong, Jiadi Xu, and Ayyalusamy Ramamoorthy*

Biophysics and Department of Chemistry, University of Michigan,
Ann Arbor Michigan, 48109-1055, USA
**email: ramamoor@umich.edu*

High-resolution structures, images, and dynamics are essential in understanding the function of membrane associated molecules such as proteins, peptides, polysaccharides, lipids, cholesterol, etc. Recent studies have shown that the preparation of a suitable model membrane, that closely mimics the native cell membrane and behaves well for NMR experiments, is the major hurdle for studies using NMR spectroscopy. In this context, bicelles, a combination of a lipid and a detergent, render unique advantages for a variety of studies using solid-state and solution-state NMR techniques. For example, an easy preparation of aligned bicelles with different sizes without the need for a mechanical support has been exploited in several NMR studies. A summary of recent NMR studies on bicelles, with a specific emphasis on the structural studies of membrane proteins using solid-state NMR spectroscopy under static and magic angle spinning condition, is presented in this chapter.

4.1. Introduction

The structural biology of membrane proteins is rapidly becoming an exciting new frontier of science. Remarkably, these proteins are intimately involved in various essential regulatory processes, ranging from the complicated network of cellular communications to the metabolism and breakdown of unwanted substances in our body. However, this class of vital proteins continues to remain unknown in regards to their intrinsic structure-function relationships due to lack of their 3D high-resolution structures. One of the obstacles of studying membrane proteins is the need of a native bilayer environment, which represents a significant challenge to most biophysical techniques due to the non-crystalline and insoluble nature of membrane systems. A proper bilayer environment is crucial in the study of this class of proteins since their secondary, tertiary, and quaternary structures, dynamics, folding, and refolding are closely connected to the bilayer environment, which significantly influence their functions.[1,2] Previous studies have demonstrated the influence of bilayer compositions on the function of membrane proteins and peptides. Biophysical studies have shown that an improper bilayer environment can render a protein inactive, especially for various membrane-associated enzymes, amyloid peptides, antimicrobial/anticancer peptides, toxins, and fusion proteins/peptides.[3,4] Therefore, finding a suitable bilayer environment continues to remain a considerable challenge in this field of research. Fortunately, the inception of bicelles as a model membrane has led to a giant leap forward in resolving this issue. This class of bilayer mimics, first described by Sanders et al,[5] bridges the gap between liposomes and micelles; in fact, they are robust, versatile and stable under a variety of sample conditions, making them the ideal system for functional reconstitutions of membrane proteins for *in vitro* studies.

One of the intrinsic properties of bicelles is their propensity to spontaneously align in the presences of an external magnetic field.[6-8] This alignment can be maintained over a wide range of physiological conditions, making bicelles an appealing biomembrane mimics for solid-state NMR (SS-NMR) studies of membrane proteins. SS-NMR is a powerful technique capable of providing detailed structural information of even the most complex and disordered system.[2] Unlike liquid state NMR, whose resolution is heavily dependent on the molecular size of the system under investigation, SS-NMR, under static conditions, relies only on the degree of sample alignment and the ability to implement various multiple-pulse line-narrowing decoupling techniques to achieve the desired high-resolution spectrum. Recently, SS-NMR has flourished in the field of structural biology of membrane proteins due to the advent of higher magnetic fields, various advances in instruments and NMR methodologies, and the availability of well-studied model membrane systems.[9] Numerous high-resolution multidimensional RF pulse sequences developed specifically for the measurement of NMR parameters such as isotropic and anisotropic chemical shifts (of nuclei like ^{15}N, ^{13}C, and even protons), heteronuclear (such as ^{1}H-^{15}N, ^{1}H-^{13}C, ^{31}P-^{13}C, ^{31}P-^{15}N, ^{31}P-^{1}H, ^{19}F-^{13}C, ^{19}F-^{15}N, ^{19}F-^{31}P, and ^{13}C-^{15}N) and homonuclear (^{13}C-^{13}C) dipolar couplings, and quadrupole couplings of selective nuclei like ^{2}H from model membranes containing a membrane protein are major stepping stones in the structural studies of membrane proteins.[10] While these amazing developments enabled the applications of static and magic angle spinning (MAS) NMR techniques, probing the dynamical structures of molecules embedded in fully-hydrated lipid bilayers is still a major challenge for NMR spectroscopy.[11,12] In this context, studies on aligned lipid bilayers are valuable as they have provided structural and dynamical information in physiologically-relevant experimental conditions and also been successful in addressing biological questions related to the function of membrane-associated peptides.[2]

Polarization Inversion Spin Exchange At the Magic Angle (PISEMA)[13-15] and Heteronuclear Isotropic Mixing Spin Exchange via Local Field (HIMSELF)[16] are two major techniques that have been commonly used in the measurement of structural and orientational constraints from aligned samples. These rotating-frame separated-local-field (SLF) experiments provide ultra-high resolution spectral lines by suppressing line broadening homonuclear proton dipolar couplings and also the long-range heteronuclear dipolar couplings. However, one of the challenges of these experiments is proper sample alignment during lengthy multidimensional experiments and bilayer hydrations. While studies on mechanically-aligned samples have provided insights into the function of membrane-associated peptides, they are not suitable to study the structure of membrane proteins containing a large soluble domain which requires bulk water for proper native-like folding in a membrane environment.[12] Consequently, the application of bicelles has virtually eliminated most of these problems associated with sample preparations and facilitates the acquisition of high-resolution SS-NMR spectra of a functional membrane protein in a suitable bilayer environment. Recent NMR studies on cytochrome-b5 embedded in bicelles are presented below in this Chapter.

4.2. Bicelles: Simple, Elegant, Yet Controversial

Bicelles are a fascinating category of lyotropic liquid crystals whose popularity in the field of structural biology has considerably grown over the years (Fig. 4.1). Bicelles mainly consist of a simple binary mixture of long chain and short chain lipids, such as 1,2-Dimyristoyl-sn-Glycero-3-Phosphocholine (DMPC) and 1,2-Dihexanoyl-sn-Glycero-3-Phosphocholine (DHPC), and exhibit a variety of controversial, yet interesting, morphologies, ranging from perforated lamellae to giant worm like micelles, as dictated by sample conditions, such as temperatures, *q* ratios (q = [DMPC]/[DHPC]), and lipid concentrations. One of the often-exploited properties of bicelles is their tendency to align when placed in a magnetic field, such as that of an NMR spectrometer.[17] The bicelles spontaneously orient with their bilayer normal perpendicular to the applied magnetic field (B_0), which is a consequence of the interaction between the negative magnetic susceptibility anisotropy ($\Delta\chi$) of the acyl chains of the lipids and B_0. Interestingly, bicelles can be reoriented, with their bilayer normal parallel to B_0, via the addition of specific lanthanide ions, such as Yb^{3+}, Eu^{3+}, Er^{3+} and Tm^{3+}, whose positive magnetic susceptibility anisotropy is large enough to overcome that of the lipids.[18]

The morphology of bicelles has been a subject of great controversy for many years and only recently has this issue been resolved via small angle neutrons scattering (SANS) and NMR diffusion studies (Fig. 4.1).[19,20] In the past, it was believed that bicelles exist as disc-like aggregates with the DMPC occupying the planar regions and the DHPC are sequestered to the rim of the disc. This morphology was consistent with ^{31}P chemical shift (Fig. 4.2) and ^{2}H quadrupole coupling NMR spectra of magnetically aligned bicelles; yet, the idea that individual discs align with the magnetic field seems unrealistic. Interestingly, a series of diffusion and SANS studies have provided new insights into

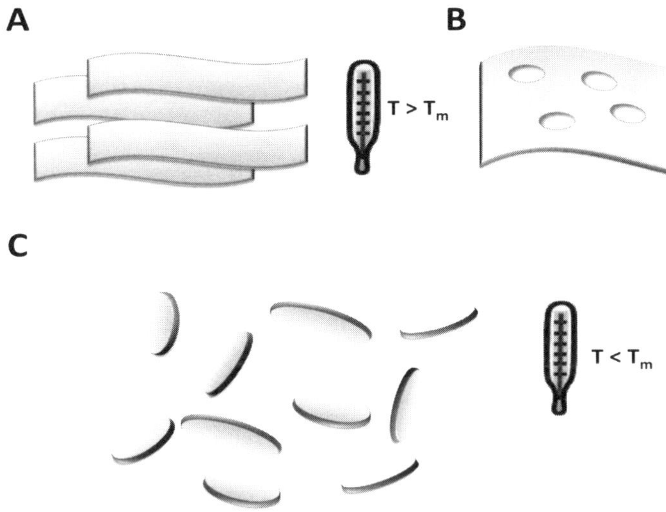

Fig. 4.1. A range of morphologies exhibited by Bicelles as a function of temperatures. At temperatures above the phase transition of DMPC, bicelles display morphology similar to that of ribbons (A) or perforated lamellae (B) dependent on the composition of bicelles. While the planar layer is populated by DMPC, toroidal pores are due to the detergent DHPC. At temperatures below the phase transition of DMPC, bicelles exist as discs as shown in (C).

the fascinating morphology of bicelles.[21,22] From diffusion studies, an amphiphilic probe molecule, embedded in bicelles, is able to undergo free diffusion covering distances in the range of microns, which is greater than the expected radius of an ideal bicelle disc, while the SANS data shows a series of Bragg peaks indicative of large lamellae structures. Hence, these results are inconsistent with the proposed disc like morphology of bicelles. The current accepted concensus is that bicelles exist as perforated lamellae in which the DMPC reside in the planar regions while the DHPC line the high curvature regions of the toroidal-pore like defects. NMR spectra of unaligned multilamellar vesicles, mechanically aligned bilayers, and magnetically aligned bicelles are given in Fig. 4.2 to illustrate the effect of alignment. As shown in Fig. 4.2 (top), ^{31}P chemical shift spectrum of unaligned MLVs is a motionally averaged powder pattern. On the other hand, ^{31}P chemical shift spectra of mechanically aligned lipid bilayers or

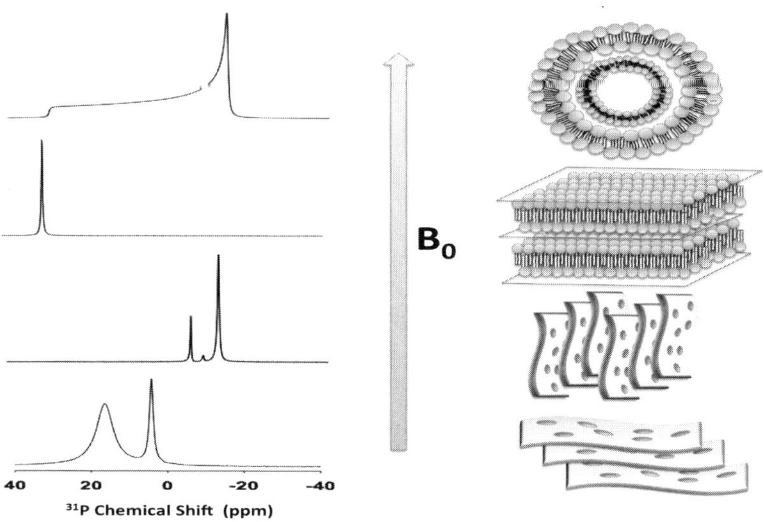

Fig. 4.2. Phosphorus-31 chemical shift spectra of lipid bilayers and bicelles (from top to bottom): multilamellar vesicles, mechanically aligned bilayers, negatively aligned bicelles with the bilayer normal perpendicular to the external magnetic field (B_0) direction, and positively aligned bicelles (or called as flipped) with the bilayer normal parallel to B_0. As mentioned in the text, bicelles can be flipped by the addition of paramagnetic ions.

magnetically aligned bicelles consists of narrow spectral lines (Fig. 4.2). We have recently systematically analyzed the effects of the *q* ratio, extent of hydration, and temperature on bicelle properties. In this study, we utilized ^{31}P chemical shifts to probe the changes in the head group region and ^{1}H-^{13}C dipolar couplings in the acyl region of bicelles.

A sample 2D NMR spectrum used to measure ^{1}H-^{13}C dipolar couplings at the natural abundance of ^{13}C nuclei is given in Fig. 4.3.[23-25] Resonances in the 2D spectrum are well-resolved because of the magnetic alignment of bicelles. The C-H dipolar couplings measured for each carbon site of the lipid or detergent can be used to measure the order (or a ligand-induced disorder) in different regions of bicelles. Such experiments have been successfully applied to determine the effects of antimicrobial peptides[24] and curcumin.[26]

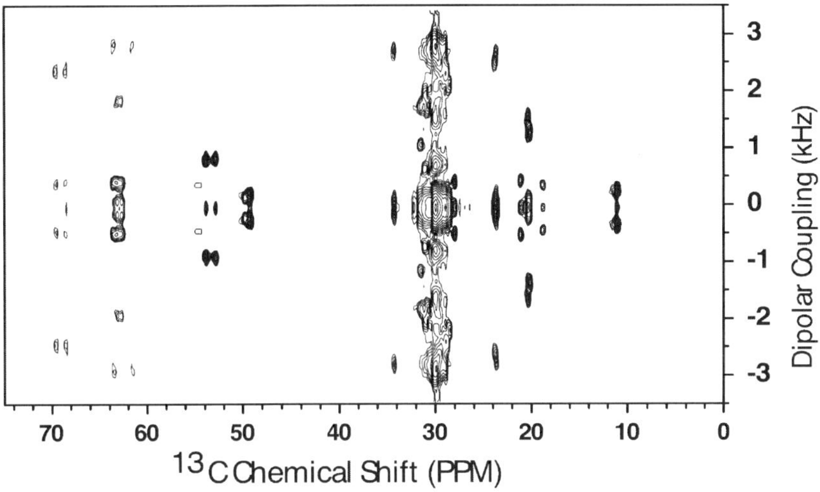

Fig. 4.3. A 2D SLF spectrum of magnetically aligned bicelles at natural abundance of ^{13}C. The data was obtained using a 2D proton-evolved local field (PELF, also known as PDLF) pulse sequence that correlates ^{13}C chemical shifts with ^{13}C-^{1}H dipolar couplings. Narrow spectral lines observed in the 2D spectrum are due to magnetic alignment of bicelles and the decoupling of ^{1}H-^{1}H and ^{1}H-^{13}C dipolar couplings. As demonstrated in the literature, the 2D PELF experiment is efficient for the measurement of long-range or weak ^{1}H-^{13}C dipolar couplings especially in semi-solids like liquid crystalline materials and bicelles.

4.3. Applications of Bicelles in Structural Biology

Bicelles have become a universal platform for the functional reconstitution of a variety of membrane proteins. The spontaneous alignment of bicelles, in the presence of a magnetic field, facilitates the acquisition of high-resolution solid-state NMR (SS-NMR) spectra of numerous membrane proteins in a true bilayer environment. In an aligned medium, various anisotropic interactions, such as dipolar couplings and chemical shift anisotropy (CSA), are retained and manifest themselves in a NMR spectrum. These anisotropic interactions are rich in orientational information, providing insights into the structure of a membrane protein at the atomic-level. The determination of a membrane protein structure, via SS-NMR, in an aligned medium, relies on the ability to measure different orientational constraints for each residue via SLF spectroscopy such as PISEMA or HIMSELF (see Fig. 4.4).

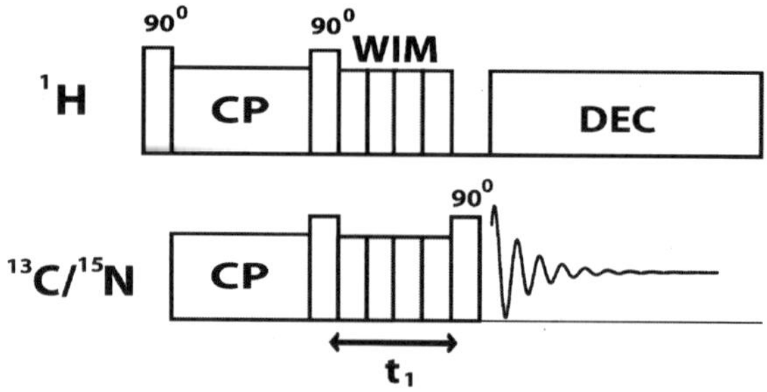

Fig. 4.4. A schematic of a 2D HIMSELF pulse sequence composed using the WIM (windowless isotropic mixing) proton homonuclear decoupling multiple pulse sequence.

These techniques have shown tremendous success in providing detailed structural information on a great variety of membrane proteins, especially for those with helical secondary structures.[27-31] In a typical 2D SLF spectrum, a helical membrane protein often exhibits a characteristic PISA (Polarized Index Slant Angle) wheel pattern in which the size and position of the wheel are dependent on the tilt angle of the helix with respect to the bilayer normal.[32,33] For a typical PISA wheel, no assignment is needed to determine the tilt of a helix. The assignment of the PISA wheel can be done using a combination of computer simulations and selective labeling of specific residues. Therefore, this simple 2D experiment provides various critical parameters necessary for determining membrane protein structures.

Recently, a full-length cytochrome b_5 (Cyt-b_5) protein was reconstituted in a solution of bicelles.[27] This is an important membrane protein found in the endoplasmic reticulum of the liver cells whose intimate interactions with cytrochrome P450 facilitate a variety of essential oxidative processes.[2] This protein consists of three critical domains: a highly mobile water-soluble region, 14-residues linker region, and a transmembrane segment. In the past, a truncated (with its transmembrane region removed), inactive version of this protein was extensively studied using solution state NMR and X-ray crystallography. However, the structure of the full-length cyt-b_5 continues to remain largely unknown; in fact, only recently has this issue been addressed using solid-state NMR experiments. A series of 1D ^{15}N NMR spectra, shown in Fig. 4.5, of cyt-b5 were recorded as a function of cross polarization (CP) contact time. At a short contact time, mainly the immobile region of the protein, the transmembrane helix, is observed and as the contact time increases, other mobile regions of the proteins begin to become visible, such as the linker region and different regions of the the water-soluble domain. These simple experiments indicate that this protein consists of regions with vastly different dynamics, which is a reflection of its overall structure. A 2D SLF spectrum of cyt-b_5 was also recorded using the HIMSELF pulse sequence and a characteristic PISA wheel pattern was observed, which is indicative of an alpha helical structure of the transmembrane region of the protein as shown in Fig. 4.5.[27] Remarkably, this is the first glimpse into the secondary structure of a membrane bound cyt-b_5 under physiological conditions and an important step forward in resolving the interaction between cyt-b_5 and P450.

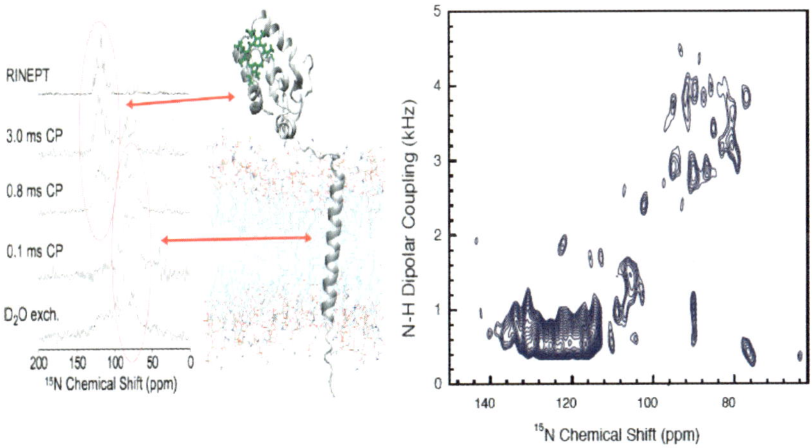

Fig. 4.5. NMR spectra of magnetically-aligned bicelles containing ^{15}N-labeled cytochrome b5. Chemical shift spectra of Cyt-b_5 in aligned bicelles (left) were acquired at various contact times and different regions of the protein are displayed based on the experimental conditions. A 2D SLF spectrum of the transmembrane region of Cyt-b_5 (right) in aligned bicelles shows a distinctive wheel pattern indicative of alpha helical secondary structure.

Our recent investigation of bicelles under MAS conditions has provided a new avenue of attaining high-resolution SS-NMR spectra of membrane proteins in a bilayer environment. Under the MAS condition, various anisotropic interactions are eliminated and only isotropic chemical shifts remain. One of the drawbacks of MAS is the effect of mechanical strain on the system since it is often the case that a spinning rate of greater than 10 kHz is required to eliminate various unwanted anisotropic interactions in a liposome system. However, molecules embedded in bicelles often experience a high degree of motions. Therefore, a reduced spinning rate can be used to achieve the desired resolution. A series of MAS spectra of Cyt-b_5 reconstituted in bicelles was recorded and shown in Fig. 4.6. The MAS spectra show significant remarkable resolution and partial assignments can be made on some of the amino acids as indicated in the sample spectrum (Fig. 4.6).[12] While these experiments are only preliminary, however, MAS of bicelles holds great promise in achieving high-resolution NMR spectra of integral membrane proteins compared to other bilayer systems. Other MAS recoupling sequences were also applied and our results suggested that the performance of these sequences depended on the dynamical influence of the protein on the pulse sequences. Performing MAS experiments at physiologically-relevant temperature provides biologically meaningful insights into the system. Therefore, bicelles will be the key in enabling structural studies using MAS experiments on membrane proteins without freezing them.

Within their relatively short history, bicelles have managed to play an important role in the structural determination of a great variety of membrane proteins. Bicelles containing bulk water offer a proper bilayer environment for functional reconstitution and an aligned medium for solid-state NMR studies. These unique properties have made bicelles an attractive medium for solid-state NMR studies of integral membrane proteins. Interestingly, with the advent of new pulse sequences and hardware, we are now beginning to realize the full potential of these fascinating lipid assemblies as they are increasingly used in various MAS studies. Therefore, bicelles will no doubt be a standard model membrane system for structural biology of membrane proteins as we strive to answer questions regarding their structure-function relationship.

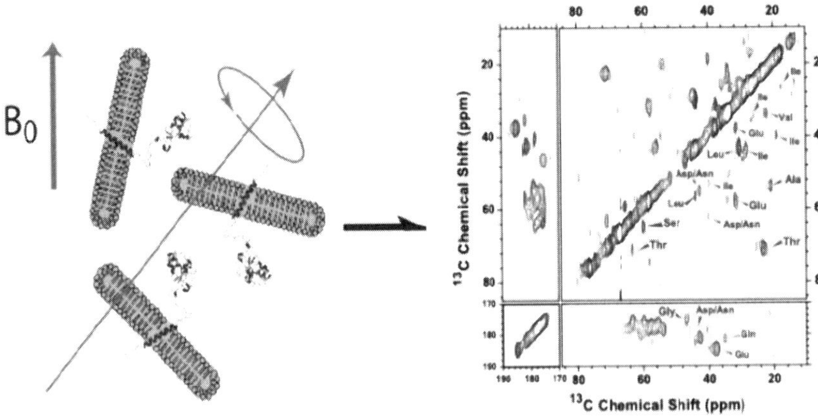

Fig. 4.6. Under the MAS condition, the bicelles are randomly oriented such that they behave as if they tumble isotropically as shown in the schematic (left). A well-resolved 2D ^{13}C-^{13}C (DARR or RAD-Mixing) correlation spectrum (right) of cyto-b_5 in bicelles was recorded under MAS conditions showing significant improvement in resolution and sensitivity.

Acknowledgement

We thank Kazutoshi Yamamoto, Dr. Sang-Choul Im, Dr. Ulrich Dürr, and Dr. Lucy Waskell for their contributions to this project. This research was supported by funds from NIH (GM084018, AI054515 and RR023597 to A.R.).

References

1. T.A. Cross and S.J. Opella, *Curr. Opin. Struct. Biol.* **4**, 574 (1994).
2. U.H.N. Dürr, L. Waskell and A. Ramamoorthy, *Biochim. Biophys. Acta. Biomembr.* **1768**, 3235 (2007).
3. U.H.N. Dürr, U.S. Sudheendra and A. Ramamoorthy, *Biochim. Biophys. Acta. Biomembr.* **1758**, 1408 (2006).
4. V. Dhople, A. Krukemeyer and A. Ramamoorthy, *Biochim. Biophys. Acta. Biomembr.* **1758**, 1499 (2006)
5. C.R. Sanders, B.J. Hare, K.P. Howard and J. H. Prestegard, *Prog. Nucl. Magn. Reson. Spectrosc.* **26**, 421 (1994).
6. C.R. Sanders and R.S. Prosser, *Structure.* **6**, 1227 (1998).
7. C.R. Sanders and J.P. Schwonek, *Biochemistry.* **31**, 8898 (1992).
8. R.S. Prosser, F. Evanics, J.L. Kitevski and M.S. Al-abdul-Wahid, *Biochemistry.* **45**, 8453 (2006).

9. A. Ramamoorthy, editor. *NMR spectroscopy of biological solids.* (Taylor & Francis Group, New York, 2006)
10. M. Baldus, *Curr. Opin. Struct. Biol.* **16**, 618 (2006).
11. O.C. Andronesi, S. Becker, K. Seidel, H. Heise, H.S. Young and M. Baldus, *J. Am. Chem. Soc.* **127**, 12965 (2005).
12. J.D. Xu, U.H.N. Durr, S.C. Im, Z.H. Gan, L. Waskell and A. Ramamoorthy, *Angew. Chem. Int. Ed.* **47**, 7864 (2008).
13. C.H. Wu, A. Ramamoorthy and S.J. Opella, *J. Mag. Reson.* **109**, 270 (1994).
14. A. Ramamoorthy, Y.F. Wei and D.K. Lee, *Annual report on NMR spectroscopy.* **52**, 1 (2004).
15. K. Yamamoto, D.K. Lee and A. Ramamoorthy, *Chem. Phys. Lett.* **407**, 289 (2005).
16. K. Yamamoto, S.V. Dvinskikh and A. Ramamoorthy, *Chem. Phys. Lett.* **419**, 533 (2006).
17. S.J. Opella and F.M. Marassi, *Chem. Rev.* **104**, 3587 (2004).
18. R.S. Prosser, JS.. Hwang and R.R. Vold, *Biophys. J.* **74**, 2405 (1998).
19. T.A. Harroun, M. Koslowsky, M.P. Nieh, C.F. de Lannoy, V.A. Raghunathan and J. Katsaras, *Langmuir.* **21**, 5356 (2005).
20. M.P. Nieh, C.J. Glinka, S. Krueger, R.S. Prosser and J. Katsaras, *Langmuir.* **17**, 2629 (2001).
21. R. Soong, and P.M. Macdonald, *Biophys. J.* **88**, 255 (2005).
22. R. Soong and P.M. Macdonald, *Biophys. J.* **89**, 1850 (2005).
23. S.V. Dvinskikh, U.H.N. Dürr, K. Yamamoto and A. Ramamoorthy, *J. Am. Chem. Soc.* **129**, 794 (2007).
24. S.V Dvinskikh, K. Yamamoto, U.H.N. Dürr and A. Ramamoorthy, *J. Am. Chem. Soc.* **128**, 6326 (2006).
25. S.V. Dvinskikh, K. Yamamoto, U.H.N. Dürr and A. Ramamoorthy, *J. Mag. Reson.* **184**, 228 (2007).
26. J. Barry. F. Michelle, J.R. Brender, P.E.S. Smith, D.K. Lee, and A. Ramamoorthy, *J. Am. Chem. Soc.* (in press).
27. U.H.N. Durr, K. Yamamoto, S.C. Im, L. Waskell and A. Ramamoorthy, *J. Am. Chem. Soc.* **129**, 6670 (2007).
28. F.A. Kovacs and T.A. Cross, *Biophys. J.* **73**, 2511 (1997).
29. J.F. Wang, S. Kim, F. Kovacs and T.A. Cross, *Protein. Sci.* **10**, 2241(2001).
30. K.A.H. Wildman, D.K. Lee and A. Ramamoorthy, *Biochemistry* **42**, 6545 (2003).
31. N.J. Traaseth, R. Verardi, K.D. Torgersen, C.B. Karim, D.D. Thomas, and G. Veglia, *Proc. Nat. Acad. Sci. U.S.A.* **104**, 14676 (2007).
32. F.M. Marassi and S.J. Opella,. *J. Mag. Reson.* **144**, 150 (2000)
33. J. Wang, J. Denny, C. Tian, S. Kim, Y. Mo, F. Kovacs, Z. Song, K. Nishimura, Z. Gan, R. Fu, J.R. Quine and T.A. Cross,. *J. Mag. Reson.* **144**, 162 (2000).

Chapter 5

Advances in Proton NMR Relaxometry in Thermotropic Liquid Crystals

Pedro J. Sebastião*, Carlos Cruz, and António C. Ribeiro

Departamento de Física, Instituto Superior Técnico,
TU Lisbon Av. Rovisco Pais 1049-001, Lisboa, Portugal
Centro de Física da Matéria Condensada, Av. Prof. Gama Pinto 2,
1649-003 Lisboa, Portugal

Advances on the investigation of molecular dynamics in liquid crystals using both standard and fast field-cycling proton NMR relaxometry techniques are presented. Results obtained on systems of increasing complexity are discussed in perspective with respect to the first works in the field. Investigations addressed in this review include various liquid crystalline systems, from the more standard low molecular weight calamitics and discotics to polycatenar molecules, and dendrimers of different generations. The influence of molecular structure, local molecular organization, mesophase structure, and confinement conditions on the molecular dynamics will be addressed.

5.1. Introduction

The investigation of molecular self-organization in soft matter is of crucial importance both for the understanding of specific systems and for the achievement of technological skills of probably unsuspected potential. In fact, when the first liquid crystals were discovered at the end of the nineteenth century, it would have been hard to grasp the future application of those materials in the display industry or their implications in the biological world. The combination of anisotropic physical properties of liquid crystals (e.g. dielectric constants, magnetic susceptibilities, refractive indices, viscosities, elastic constants) with their flow capacity, characteristic of common liquids, is the key to their technological usefulness and their omnipresence in biological systems. The possibility of controlling the align-

*E-mail: pedro.jose.sebastiao@ist.utl.pt

ment of such molecular materials by the interaction with specially prepared surfaces is of major importance. The competition between surface forces and the delicate anisotropic responses of these materials to electromagnetic fields are fundamental to many practical applications including the prolific display technology.

The structural characterization and the study of physical properties of liquid crystal (LC) phases involve many experimental techniques including differential scanning calorimetry, polarizing optical microscopy, x-ray diffraction, neutron scattering, dielectric relaxation, nuclear magnetic resonance, dynamic light scattering, electro-optical and mechano-optical measurements, rheology, and others. In particular, proton NMR relaxometry is a powerful tool for the study of molecular dynamics in LC phases. As will be discussed hereafter, the molecular dynamics depend on the symmetry of the phase and on the molecular organization that gives rise to specific phase structures. Moreover, the molecular architecture, especially in complex systems, also influences in a crucial way the molecular movements within the mesophases. NMR relaxometry of LCs has been the subject of previous reviews where different types of mesophases have been considered.[1–6] The perspective of the present review is to address the advances on the field, with particular emphasis on investigations regarding the effects of molecular properties and specific molecular organization of peculiar LC phases on the NMR relaxation results.

On the one hand, the works revisited here allow for a comparison between phases of similar symmetry but having molecules of completely different sizes, shapes and internal structures (e.g. lamellar phases of calamitic and dendritic molecules). On the other hand, investigations comparing phases of very different structures but with the same molecules will be discussed (e.g. lamellar and columnar phases of biforked molecules). Proton NMR relaxometry studies, in a wide frequency range, have been performed on a number of systems of increasing complexity, from conventional calamitic (rod-like) and discotic (disk-like) molecules to dendrimers (tree-like molecules) of several generations and branching types.[7–9] These investigations give a comprehensive impression on how the molecular architecture and phase structure affect the molecular dynamics in LC phases. Moreover, the peculiar structure of certain mesophases, like frustrated smectics and TGB phases (that will be mentioned in the next sections), induces particular molecular motions with characteristic NMR relaxation contributions, which may be regarded as structural signatures. Therefore, in a way, proton NMR relaxation studies are also important for the investigation of the structure of these LC phases. In the following sections, a brief description of a wide variety of LC phases and the types of molecular movements generally observed by proton NMR relaxation in those phases will be presented. The latest developments of the fast field-cycling NMR relaxometry technique will be reviewed. Finally, the dependence of the molecular dynamics of thermotropic LC on the molecular architecture, the local molecular organization and the corresponding phase structure will be discussed on the basis of available experimental evidence.

5.2. Mesophases of Thermotropic Liquid Crystals

Liquid crystals appear in many forms. Their building blocks vary from small elongated or plate-like molecules of about two or three nanometers in length to macromolecules of polymeric LC materials. In between, the progression in complexity and shape diversity of the LC molecules gives rise to a corresponding variety of phase structures and molecular arrangements.[10] As in other thermodynamic systems, the phase transitions of LC may be controlled by different physical parameters. LC systems formed by pure compounds are called thermotropic, whose phase transitions are controlled by varying the temperature. LC systems resulting from solutions of amphiphilic molecules (partially hydrophilic, partially hydrophobic) in appropriate solutes (mainly water), where the phase transitions are determined (at given temperature and pressure) by concentration, are called lyotropic. The latter are the most similar to the complex structures present in living matter and are useful for the preparation of model systems with biological implications. Thermotropic LCs are used for technological purposes with particular emphasis on the production of electro-optical devices for displays and photonics applications. This chapter concerns NMR relaxometry of thermotropic LCs.

Liquid crystals may be described as partially ordered systems, whose phases (mesophases) are structurally more ordered than common isotropic liquids (with short range positional order and no orientational order of their molecules) and less ordered than crystals (with long range orientational order and tridimensional positional order). Long range orientational order is common to all LC phases and may be defined by the Saupe order tensor:[11]

$$S^{ij}_{\alpha\beta} = \left\langle \frac{1}{2}\left(3\cos\theta_{i\alpha}\cos\theta_{j\beta} - \delta_{\alpha\beta}\right)\right\rangle \tag{5.1}$$

where the brackets stand for ensemble average and $\theta_{i\alpha}$, $\theta_{j\beta}$ represent the

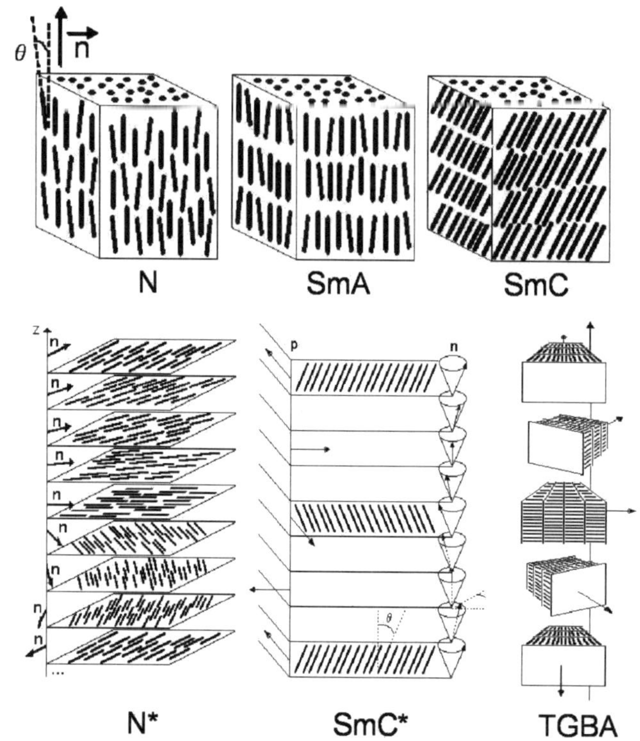

Fig. 5.1. Schematic representation of examples of thermotropic LC phases of calamitic molecules: nematic (N), smectic A (SmA), Smectic C (SmC), chiral nematic (N*), chiral smectic C (ferroelectric) (SmC*), and Twist Grain Boundary A (TGBA) phase.

angles between the laboratory frame axis $i, j = (X, Y, Z)$ and a fixed molecular frame axis $\alpha, \beta = (x, y, z)$. In the most general case $S^{ij}_{\alpha\beta}$ has 25 independent parameters (by definition it is a symmetric and traceless tensor). In the case of the uniaxial nematic phase, the simplest (less ordered) LC phase, Eq. (5.1) is reduced to a single independent element, $S = S^{ZZ}_{zz}$, the nematic order parameter:

$$S = \left\langle \frac{1}{2}\left(3\cos^2\theta - 1\right)\right\rangle . \tag{5.2}$$

In Eq. (5.2), θ represents the angle between the longitudinal axis of a given molecule of cylindrical symmetry and a single direction of common preferential molecular alignment, characteristic of the phase, defined by a unit vector $\boldsymbol{n}$ called *director*[a] (see Fig. 5.1). The nematic phase exhibits long-range orientational order but only short range positional order, i.e. the centers of mass of the molecules are randomly distributed as in ordinary (isotropic) liquids. This phase may be composed by LC molecules of many different kinds like calamitic (elongated), discotic (disc-like), sanidic (lath-like), polycatenar (rod-like core with several peripheral flexible chains), dendritic (monodisperse tree-like), polymeric, and others, which acquire on average cylindrical symmetry for the physical parameters characteristic of the phase due to the fast motions of the molecules (or mesogenic units in the case of complex molecules like polymers or dendrimers). For instance, in the case of calamitic molecules, a long molecular axis (sometimes, but not necessarily, the main axis of the molecular inertia tensor) tends to align along $\boldsymbol{n}$. In the case of discotic molecules, the disk-like molecular cores tend to remain approximately parallel to a common plane and the director is defined by the direction perpendicular to this plane (see Fig. 5.2).

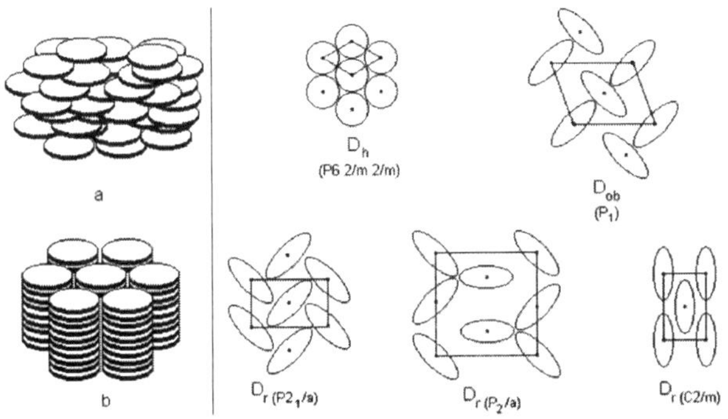

Fig. 5.2. Schematic representation of columnar phases of discotic molecules. Nematic discotic (ND left a), columnar (left b); two dimensional molecular organizations in the plane perpendicular to the columnar axis (right).

Liquid crystals in general and nematics in particular exhibit *curvature elasticity*. This type of elasticity derives from the tendency of the director field $\boldsymbol{n}(x, y, z)$ to remain uniform in the absence of external forces. Any distortion from this uniform situation represents an energy cost, which may be expressed, in the case of nematics, by the Frank's free energy density

[a]In fact, $\boldsymbol{n}$ is not a unit vector (the term was used for simplicity), since both directions $\boldsymbol{n}$ and $-\boldsymbol{n}$ are equivalent in the nematic phase. The director may be associated to the direction of the principal axis of a second-rank tensor (the tensor order parameter) which may be defined in terms of any second-rank tensorial physical property of the phase (e.g. the magnetic susceptibility).

function given by:

$$F = \frac{1}{2}K_1 \left[\nabla \cdot \boldsymbol{n}\right]^2 + \frac{1}{2}K_2 \left[\boldsymbol{n} \cdot (\nabla \times \boldsymbol{n})\right]^2 + \frac{1}{2}K_3 \left[\boldsymbol{n} \times (\nabla \times \boldsymbol{n})\right]^2 \qquad (5.3)$$

where K_1, K_2 and K_3 are elastic constants for the three elementary types of nematic distortions: splay, twist and bend, respectively (see Fig. 5.3). In the presence of electromagnetic or surface force fields, additional coupling terms are added to the free energy function. The solution for the resulting director field is given by the minimization of the free energy, which corresponds to the balance between the different forces present (for instance the electrostatic force and the elastic force associated with the alignment imposed by a given surface confinement).

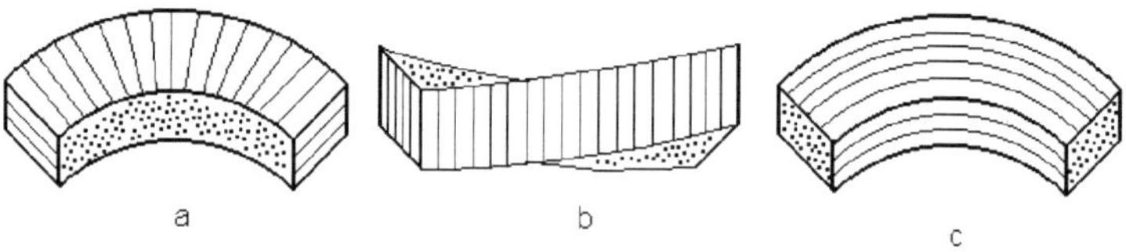

Fig. 5.3. Elementary modes of nematic elastic distortions: splay (a), twist (b) and bend (c).

Some calamitic molecules with an element of chirality exhibit the so-called chiral nematic or cholesteric phase. This phase is locally similar to the nematic and exhibits a helical structure with the helix axis perpendicular to the local director and a characteristic pitch p (see Fig. 5.1). The Frank free energy for the cholesteric phase is given by an expression similar to Eq. (5.3) where the term corresponding to the contribution of the twist distortion is now given by $1/2\, K_2[\boldsymbol{n}.(\nabla \times \boldsymbol{n}) + q]^2$. The parameter $q = 2\pi/p$ expresses the natural pitch of the phase ensuring that F equals zero for the spontaneously twisted configuration of the cholesteric.

Another important variant of the nematic phase is the *Biaxial Nematic* (N_B). This phase, theoretically predicted by Freiser in 1970,[12] exhibits additional orientational order characterized by a secondary director in the plane perpendicular to the principal director $\boldsymbol{n}$ defined before. In recent years this phase has been observed in different systems of thermotropic liquid crystals namely, bent core mesogens,[13–16] bent-rod dimers,[17] side-chain polymers,[18,19] and organosiloxane tetrapodes.[20–23] Reviews on NMR studies in N_B phases may be found in Refs. [24] and [25].

Besides orientational order, some LC phases present positional order.

Smectic phases exhibit long range positional order in one dimension which means that the molecules are disposed in layers of average constant thickness[b]. In smectics of disordered layers, the molecular positional order in the plane of the layers is of short range. The centers of mass of the molecules (or mesogenic elements in the case of more complex systems like polymers) are randomly distributed in the plane of the layers, forming two-dimensional liquid-like systems. In uniaxial smectics of disorderd layers (generally formed by rod-like mesogens), such as smectic A (SmA), the positional order may be described by the function

$$\rho = \rho_0 + Re\left\{\psi \cdot e^{i\frac{2\pi}{d}z}\right\} \tag{5.4}$$

corresponding to the first order Fourier series approximation of the periodic distribution of density in the direction perpendicular to the layers, z (see Fig. 5.1). ρ_0 is the average density, d is the layers' spacing and the smectic order parameter, $\psi = |\psi|e^{i\varphi}$, is a complex quantity, where $|\psi|$ represents the amplitude of the density variation in the direction perpendicular to the layers and φ determines the layers position.[8] Furthermore, the long molecular axis (considering molecules or mesogenic units of cylindrical simmetry) may be perpendicular to the plane of the layers (SmA) or tilted with respect to the layers' normal. In the latter case, the phase is called smectic C (SmC) and an additional order parameter (the SmC order parameter) is introduced to describe the orientation of the tilting direction in the layers' plane:[8]

$$\chi = \alpha e^{i\phi} \tag{5.5}$$

where α is the tilt angle, defined with respect to the layers' normal, and ϕ is the azimuthal angle, which defines the orientation of the tilting in the layer's plane. Here, a director $\vec{c}$ may be defined in the layer's plane, which becomes analogous to a two-dimensional nematic system.

Positional order in LCs requires the introduction of different descriptions of the free energy associated with elastic deformations in various LC phases. In the case of the SmA phase, twist and bend deformations are incompatible with a lamellar structure, and therefore only the splay term (K_1) remains. Considering that $u(r)$ describes the dislocation of a point with respect to

[b]In fact positional order in smectic phases must be more precisely defined as quasi-long range as the effect of Landau-Peierls instability (logarithmic increase of thermal fluctuations with distance in one dimensional systems) forbids the existence of long range positional order in one dimension. In practical terms, however, the number of layers in a smectic domain is high enough to ensure macroscopic size of the domains.

the unperturbed position, we have in a first order approximation:

$$n_x = -\frac{\partial u_z}{\partial x}; n_y = -\frac{\partial u_z}{\partial y}; n_z \simeq 1$$

$$F = \frac{1}{2}B\left(\frac{\partial u_z}{\partial x}\right)^2 + \frac{1}{2}K_1\left(\frac{\partial^2 u_z}{\partial x^2} + \frac{\partial^2 u_z}{\partial y^2}\right)^2 \tag{5.6}$$

where B is the elastic constant related to the compression of the layers.[26]

Certain smectic phases exhibit some degree of order in the plane of the layers. The so called *hexatic phases* (SmB_H, SmF, SmI, and SmL) exhibit no long range positional order in the layer plane but long-range *bond orientational order.* This means that the orientation of the axes defining the local distribution of neighboring molecules (typically a hexagonal lattice) exhibits long range orientational order (but the positional order of the molecules in the plane of the layer is still short range). Lamellar phases exhibiting long range positional order in the plane of the layers were historically classified as smectics (SmB, SmE, SmG, SmH, SmJ, SmM and SmK). These phases, should be more properly called soft crystals since they show tridimensional long range positional order with some degree of local orientational disorder, and should just be termed B, E, etc..[27]

Calamitic mesogens with strong terminal electric dipoles, that form smectic phases of disordered layers, exhibit special variations of molecular arrangements with different layer spacings (SmA_1, SmA_2, SmA_d, $Sm\tilde{A}$, SmC_1, SmC_2, SmC_d, $Sm\tilde{C}$). The peculiar molecular packing in these phases may be explained by the competition between the tendency of formation of single layered structures (characterized by the smectic order parameter) with the propensity of molecules to form dimers by dipolar interaction between terminal dipoles (described by a anti-ferroelectric order parameter). These systems are generally called *frustrated smectics* due to the frustration mechanism that produces the peculiar variation of layer spacing. This mechanism is also responsible for the appearance, in some cases, of the so-called *reentrant* nematic (N_{re}) and *reentrant* smectic phases.[28]

Chiral calamitic molecules with transverse dipoles may also show very interesting phase structures and polymorphisms, including the ferroelectric chiral smectic C phase (SmC*) and the *Twist Grain Boundary* (TGB) phases. Besides the interest on the applications of ferroelectric phases in electro-optical cells (e.g. *surface stabilized liquid crystal cells* - SSFLC), the fundamental scientific interest of this type of systems has culminated with the discovery of the *Twist Grain Boundary* A (TBGA) phase, as predicted by de Gennes based on the theoretical analogy between liquid crystals and superconductors.[29,30] The molecular dynamics of frustrated smectics and peculiar smectic phases exhibited by chiral molecules with lateral dipoles (e.g. SmC*, TGBA) present a variety of specific features related to their phases' structures. These effects could be detected through NMR relaxation studies that will be addressed in this review.

The synthesis of bent-core ('banana shaped') mesogenic molecules, brought in a very interesting research topic to the LC field. The introduction of additional elements of molecular orientation (namely the bent direction) with respect to the calamitic molecules sets off new possibilities of molecular arrangements. This particularity, especially when associated to the presence of dipoles with a component transverse to the plane of the bent molecular core, lead to a rich variety of polymorphisms including the biaxial nematic phase[13,14] and the new banana phases, B_1-B_7. These new phases are generally lamellar and show different combinations of synclinic and anticlinic with ferro and anti-ferroelectric molecular arrangements.[31] NMR investigations are also of major importance in the study of these new phases and have been the subject of a recent comprehensive review on this topic.[32]

Liquid crystalline phases, which exhibit long range positional order in two dimensions, are called Columnar phases. The molecules in these phases are stacked into parallel columns that are organized, in the plane perpendicular to the columns axis, in one of five possible two dimensional crystalline lattices[c]. These phases (hexagonal Col_h, rectangular Col_r $P2_1/a$, $P2/a$, $C2/m$ or oblique Col_{ob}) may have ordered or disordered columns.[33] A schematic representation of the columnar phases of discotic molecules is presented in Fig. 5.2. In oblique or rectangular columnar phases of discotic molecules, the planes of the molecular rigid cores are tilted with respect to the plane perpendicular to the columns' axes. In more complex LC systems, columnar phases may be obtained by different molecular arrangements. In some cases, the transverse section of the columns may be composed of several molecules (e.g. in polycatenar mesogens), while in other systems, mesogenic segments belonging to different macromolecules may pile up to form each of the columns (like in columnar phases of LC polymers). Columnar phases of dendrimers also present peculiar molecular arrangements.[9] The molecular dynamics had been investigated by NMR relaxation on columnar mesophases of discotic, polycatenar and dendritic molecules. The influence of the molecular organization in the columnar

[c]Additional rotation, reflection and glide reflection symmetry operations lead to 17 possible two dimensional arrangements corresponding to the 17 two-dimensional space groups.

phases on their respective molecular dynamics will be addressed hereafter. The order parameter for the different columnar mesophases may be obtained from the expression:[34]

$$\rho = \rho_{\perp} f(x,y) + \chi \tag{5.7}$$

where $\rho_{\perp} f(x,y)$ describes the density variation in the plane perpendicular to the column's axis (it tends toward a constant at the transition from a columnar to a discotic nematic phase) and χ describes the density variation along the column's axis (χ is zero for disordered phases and varies periodically for ordered columnar phases). The free energy density that describes the elastic deformations in columnar phases is given by:[34]

$$F = \frac{B}{2}\left(\frac{\partial u_x}{\partial x} + \frac{\partial u_y}{\partial y}\right)^2 + \frac{D}{2}\left[\left(\frac{\partial u_x}{\partial x} - \frac{\partial u_y}{\partial y}\right)^2 + \left(\frac{\partial u_x}{\partial y} + \frac{\partial u_y}{\partial x}\right)^2\right] + \frac{K_3}{2}\left[\left(\frac{\partial^2 u_x}{\partial z^2}\right)^2 + \left(\frac{\partial^2 u_y}{\partial z^2}\right)^2\right] \tag{5.8}$$

where u_x and u_y represent dislocations in the plane perpendicular to the columns' axes, B and D are constants that characterize the deformation of the two-dimensional lattice in the plane perpendicular to the columns, and K_3 is the Frank constant associated to bending of the columns.

In this very brief and general description of LC phases we look for some key aspects assumed in the description of molecular movements that are detected by NMR relaxometry. The main issues, related to the structure and elastic properties, are especially important in defining molecular movements in LC phases. More details on the molecular organization and phyisical properties of these phases will be mentioned in the text whenever needed for the discussion of particular aspects of relaxation mechanisms.

5.3. Proton NMR Relaxation in Liquid Crystals

Proton NMR relaxation describes the energy exchange between the proton spins of the hydrogen nuclei and the energy exchange between this spin system and the surrounding lattice. The lattice represents all degrees of freedom associated with molecular motions and acts as a large heat reservoir. These energy exchanges are characterized by two relaxation rates: the *spin-spin* relaxation rate, T_2^{-1}, and the *spin-lattice relaxation rate*, T_1^{-1}.[35–37] The most important interactions regarding proton NMR are the Zeeman coupling, i.e. direct interaction between the spins and the external magnetic field, and the dipolar coupling (direct dipolar interaction) between a pair of spins. The movements of the hydrogen nuclei in the system modulate the spatial part of the dipolar coupling Hamiltonian and are reflected in the values of the relaxation rates T_1^{-1} and T_2^{-1}. Therefore, through the study of these relaxation rates it is possible to characterize the motions that can produce effective relaxation mechanisms.[3] In the following we will focus on the spin-lattice relaxation.

In the classical Abragam's notation[35] the proton spin-lattice relaxation rate is given by

$$T_1^{-1} = \frac{9}{8}\left(\frac{\mu_0}{4\pi}\right)^2 \gamma^4 \hbar^2 \left[J_1(\omega) + J_2(2\omega)\right] \tag{5.9}$$

where γ is proton gyromagnetic ratio and $J_k(\omega)$ are the spectral densities which depend on the Larmor frequency ω, a quantity proportional to the magnetic field B, $\omega = \gamma B$. In uniaxial LCs, the spectral densities $J_k(\omega)$ are calculated usually for the orientation of the director parallel to the magnetic field. If the director is turned by an angle Δ away from the magnetic field, the relaxation rate is determined by the spectral densities $J_i(\omega, \Delta)$, which are linear combinations of the original $J_k(\omega)$:[38]

$$J_i(\omega, \Delta) = \sum_{k=0}^{2} f_{ik} J_k(\omega) \tag{5.10}$$

with

$$f_{ik}(\Delta) = \begin{bmatrix} 1 - 3\sin^2\Delta + \frac{9}{4}\sin^4\Delta & 18(\sin^2\Delta - \sin^4\Delta) & \frac{9}{8}\sin^4\Delta \\ \frac{1}{4}(\sin^2\Delta - \sin^4\Delta) & 1 - \frac{5}{2}\sin^2\Delta + 2\sin^4\Delta & \frac{1}{4}\sin^2\Delta - \frac{1}{8}\sin^4\Delta \\ \frac{1}{4}\sin^4\Delta & 2(2\sin^2\Delta - \sin^4\Delta) & 1 - \sin^2\Delta + \frac{1}{8}\sin^4\Delta \end{bmatrix}. \tag{5.11}$$

For a sample consisting of domains with a given angular distribution of director's orientation, it is necessary to calculate the averages $\overline{f_{ij}(\Delta)}$ over the 4π solid angle. Generally, proton NMR relaxation rate T_1^{-1} in LCs is expressed as a sum of the contributions of all effective relaxation mechanisms in the system, that is

$$T_1^{-1} = \sum_{n=1}^{N} \left(T_1^{-1}\right)_n . \tag{5.12}$$

This assumption is supported by the fact that the characteristic time-scales associated with the different relaxation mechanisms present in the studied systems are considerably different.

5.3.1. *Relaxation Mechanisms: Theoretical Models*

5.3.1.1. *Local molecular rotations/reorientations (R)*

The most basic relaxation mechanism introduced by Bloembergen, Purcel and Pound (BPP)[39] considers the isotropic rotational diffusion of molecules with a correlation time τ_R and the associated intra-molecular interaction between two spins at a distance r. The spectral densities $J_0(\omega)$, $J_1(\omega)$, and $J_2(\omega)$, are proportional to $J_1(\omega)$,

$$J_k(\omega, r, \tau_R) = c_k \frac{4}{15r^2} \frac{\tau_R}{1+\omega^2\tau_R^2} \tag{5.13}$$

$$c_k = \begin{cases} 6, \; k=0 \\ 1, \; k=1 \\ 4, \; k=2. \end{cases} \tag{5.14}$$

The relaxation rate in Eq. (5.9) is then given by

$$T_1^{-1} = \frac{3}{10}\left(\frac{\mu_0}{4\pi}\right)^2 \frac{\gamma^4\hbar^2}{r^6}\left(\frac{\tau_R}{1+\omega^2\tau_R^2} + \frac{4\tau_R}{1+4\omega^2\tau_R^2}\right). \tag{5.15}$$

The anisometric character of the LC molecules requires more elaborated models for the rotational/reorientational motions. One of these models is the extended Woessner model for the relaxation induced by reorientations of rigid elongated molecules.[3,40–46] The spectral densities are given by:

$$J_k(\omega) = \frac{4}{3}c_k \sum_{m=-2}^{2} \overline{\frac{|d_{k0}^2(\alpha_{ij})|^2}{a_{ij}^6}} \left\langle |D_{km}^2(\theta)|^2 \right\rangle \frac{\tau_{|m|}}{1+\omega^2\tau_{|m|}^2} \tag{5.16}$$

where a_{ij} denotes the distance between the proton pair and α_{ij} the angle between the interproton vector and the long molecular axis. D_{km}^2 is the second rank Wigner rotation matrix and the averages $\left\langle |D_{km}^2(\theta)|^2 \right\rangle$ can be expressed in terms of the second and fourth rank Legendre polynomials of the angle θ between the long molecular axis and the nematic director, i.e. the nematic order parameter S and $\langle P_4 \rangle$,[3,45] respectively. $d_{km}^2(\alpha_{ij})$ is the second rank reduced Wigner rotation matrix and the expressions for the factors $\overline{|d_{k0}^2(\alpha_{ij})|^2/a_{ij}^6}$ can be found in Refs. [42,45]. The correlation times $\tau_{|m|}$ can be expressed in terms of two correlation times corresponding to molecular reorientations with respect to directions parallel and perpendicular to the long molecular axis, τ_L and τ_S, respectively.[45,46]

More detailed models can be found in the literature for the spin-lattice relaxation due to rotations/reorientations.[3] In most cases they are expressed by different sets of correlation times and require more detailed descriptions of the molecules and of their rotational diffusion motions.

Conformational changes are usually not considered when analyzing proton spin-lattice relaxometry data. On the one hand, the proton T_1^{-1} is not site specific (due to the very large number of proton pairs involved in dipolar interaction in the molecules) and the contribution of internal motions to relaxation is masked by the rotations/reorientations contribution of the whole molecule. On the other hand, the correlation times corresponding to these motions are very short, typically of the order of $\sim 10^{-12}$s, a time scale that is accessible only when T_1^{-1} is measured at frequencies above hundreds of MHz. Some LC systems require, however, an approach where internal motions are partially considered. For instance, in the case of polycatenar molecules, presented in Sec. 5.2, molecular reorientations might be treated in such a way that different sets of correlation times are assigned to motions of the chains (internal motion relative to a frame fixed to the core) and of the core (corresponding to the global molecular rotations).[47]

5.3.1.2. *Molecular translational diffusion (SD)*

Molecular self-diffusion is present in isotropic and LC phases. The translational self-diffusion can be detected by proton NMR relaxation since they modulate intermolecular spin interactions. These interactions are also affected by the details of molecular structure and local molecular organization. Therefore, different models have to be considered for this relaxation mechanism depending on the phase structure and/or confinement conditions of the molecular system. In bulk samples, the spectral densities can be written as

$$J_k(\omega)_{SD} = \frac{n\tau_D}{d^3} j_k^{SD}(\omega\tau_D) \tag{5.17}$$

where n is the density of spins, τ_D is the mean jump time, d is the distance of closest approach between molecules, and $j_k^{SD}(\omega\tau_D)$ are dimensionless functions that depend on the phase structure, e.g. isotropic or LC phase.[48]

For the isotropic phase the spectral densities had been calculated analytically by Torrey[48,49] for spherical molecules, $j_k^{SD}(\omega\tau_D) = j_k^{SD-Torrey}(\sigma, \omega\tau_D)$, with $\sigma = \langle r^2 \rangle /(12d^2)$, and $\langle r^2 \rangle = 6\tau_D D$, the mean square jump distance, where D is the self-diffusion coefficient. This model can also be used in low molecular mass LCs as long as the molecules can be treated approximately as spherical objects as far as concerning molecular translational displacements.

For the N, SmA, SmB and columnar phases the spectral densities required to calculate T_1^{-1} in Eq. (5.9) must be obtained numerically using the models proposed by Žumer and Vilfan.[50–54] The spectral densities are functions of several parameters besides mean jump time τ_D. For the N,[51] SmA,[52] and SmB[54] phases

$$j_k^{SD} = j_k^{SD_{N,SmA,SmB}}\left(\omega\tau_D, \frac{\langle r_\perp^2\rangle}{d^2}, \frac{D_\parallel^0}{D_\perp^0}\right) \tag{5.18}$$

where $D_\parallel^0/D_\perp^0$ is the ratio for the self-diffusion coefficients of the perfectly ordered phase ($S = 1$), and $\langle r_\perp^2\rangle = 4\tau_D D_\perp$ the lateral mean square jump distance. In the nematic phase, the relaxation rate given by Eq. (5.18) corresponds approximately to the analytical Torrey's expression divided by 1.4 for frequencies below 10 MHz.

In the case of columnar phases, the self-diffusion process takes into account inter-columnar molecular jumps.[53] The spectral densities are still given by an equation similar to Eq. (5.18) but, in this case, d is the average distance between the centers of neighboring columns in the plane perpendicular to the columnar axes. These dimensionless functions

$$j_k^{SD} = j_k^{SD_{Col}}\left(\omega\tau_D, \frac{\ell}{d}, \frac{D_\parallel^0}{D_\perp^0}, \epsilon\right) \tag{5.19}$$

further depend on the average distance ℓ between centers of neighboring molecules in the columns, and on the distribution of spins in the molecule parameterized by ϵ.

When liquid crystals are confined into nano-scale cavities, molecular translational (self-diffusion) displacements must be thoughtfully analyzed since the translational jumps can induce rotations/reorientations of the molecules. In these systems it is necessary to consider a specific relaxation mechanism that considers the effect of reorientations mediated by translational diffusion (RMTD).[55–58] The spectral densities for the RMTD relaxation mechanism may have different expressions depending on the orientational structure factor of the director field inside the confining cavities. A detailed discussion can be found in Refs. [46,59]. For an isotropic distribution of pores' orientations, the spectral densities are written as

$$J_k^{RMTD}(\omega) = c_k \int_{q_{min}}^{q_{max}} Q(q)\frac{2\tau_q}{1+\omega^2\tau_q^2}\,dq \tag{5.20}$$

with c_k given by Eq. (5.14). $Q(q)$ is the orientational structure factor which depends on the wave number q of a diffusive mode that exponentially decays with a characteristic time τ_q. Depending on the structure factor $Q(q)$, different frequency dependencies can be observed for the spin-lattice relaxation rate.[46] In the case of a constant $Q(q)$, $\left(T_1^{-1}\right)_{RMTD}$ is proportional to $\omega^{-1/2}$ in a frequency range between a low and a high cut-off frequencies.[46,59]

5.3.1.3. *Collective motions*

Due to the specific structural properties of the LC medium, elastic deformations involving the collective motion of a large number of molecules may be observed. These collective motions can be associated to *order director fluctuations* (ODF) in nematic phases, *layer undulations* in liquid-like layered phases, and *column undulations* in columnar phases, to mention the most common types.

In the case of the N phase, the spectral densities for the ODF contribution to the relaxation rate may be written as:.[60–64]

$$\begin{aligned} J_k^{ODF}(\omega, \Delta) &= f_{k1}(\Delta)\, \frac{\overline{|d^2_{00}(\alpha_{ij})|^2}}{a_{ij}^6}\, \frac{k_B T S^2 \eta^{1/2}}{\pi\sqrt{2}\, K^{3/2}}\, \omega^{-1/2} \\ &\quad \times \left[f\left(\frac{\omega_{cM}}{\omega}\right) - f\left(\frac{\omega_{cm}}{\omega}\right)\right] \end{aligned} \tag{5.21}$$

where

$$f(x) = \frac{1}{\pi}\left[\tan^{-1}(\sqrt{2x}+1) + \tan^{-1}(\sqrt{2x}-1) - \tanh^{-1}\frac{\sqrt{2x}}{1+x}\right]. \tag{5.22}$$

In Eq. (5.21), besides the previously introduced physical parameters, T is temperature, η is the effective viscosity, K the elastic constant (in a one constant approximation), k_B is the Boltzmann constant, and $f(x)$ is the cut-off function, while $\omega_{cM} = Kq_{max}^2/\eta$ and $\omega_{cm} = Kq_{min}^2/\eta$ are the high and low cut-off frequencies, respectively. q_{max} and q_{min} are the highest and the lowest wave number of the fluctuation modes, respectively. The $T_1^{-1} \propto \omega^{-1/2}$ law that is obtained from Eqs. (5.9) and (5.21) for the ODF relaxation mechanism in N phases was first introduced by Pincus.[65] Translational self-diffusion can contribute to the damping of the fluctuation modes. When SD is not considered as an independent relaxation mechanism, this effect can be included in the spectral densities by modifying the ratio $K/\eta \to K/\eta + D$.[3,65,66]

In the SmA and lamellar phases of lyotropic systems, the collective motions associated with director fluctuations give a contribution from molecular displacements that preserve the layer's structure - *layer undulations*.

In this case the spectral densities are given by[46,62,67–69]

$$J_k^{LU}(\omega,\Delta) = f_{k1}(\Delta)\frac{\overline{|d_{00}^2(\alpha_{ij})|^2}}{a_{ij}^0}\frac{k_BTS^2}{2K_1L}\omega^{-1}\frac{1}{\pi}$$
$$\times\left[\tan^{-1}\left(\frac{\omega_{cM}}{\omega}\right)-\tan^{-1}\left(\frac{\omega_{cm}}{\omega}\right)\right] \quad (5.23)$$

where K_1 is the splay elastic constant, and L is the correlation length of undulations in the direction perpendicular to the layers. For $\omega_{cM} \gg \omega \gg \omega_{cm}$ the relaxation rate is $\left(T_1^{-1}\right)_{LU}$ is proporcional to ω^{-1}.

In the case of the smectic C phase, in-plane fluctuations of the tilt angle direction can occur. These fluctuations have a nematic-like character as they happen for all q wave vector directions. As for the N phase, the contribution of this type of fluctuations to relaxation rate is also proportional to the inverse square root of the Larmor frequency $\left(T_1^{-1}\right)_{DF_N} \propto \omega^{-1/2}$.[69]

In the case of columnar phases, the collective movements give rise to the relaxation mechanism by the columns' undulation described by the Žumer and Vilfan's model:[47,53,70]

$$J_k^{ECD}(\omega,\Delta) = f_{k1}(\Delta)C\,\omega^{-1}$$
$$\times\int_0^1\left[\tan^{-1}\left(\frac{R\omega_c}{u\omega}+u\frac{\omega_c}{\omega}\right)-\tan^{-1}\left(u\frac{\omega_c}{\omega}\right)\right]du \quad (5.24)$$

where C is a constant that depends on the viscoelastic properties of the columnar phase in the LCs, R depends on the elastic constants corresponding to the bending and compression of the columns and ω_c is a cut-off frequency. Eq. (5.24) is obtained using a single constant approximation with respect to the deformation of the two dimensional lattice in the plane perpendicular to the columns.

Although collective motions modulate only the intra-molecular interactions, the specific nature of each mesophase is strongly reflected on the T_1 frequency dispersion as described by its corresponding relaxation model.

5.3.1.4. *Other relaxation models for specific motions*

In the case of SmC, SmC*, SmC$^*_{\mathrm{A}}$, SmC*$_{ferri}$, and TGBA phases, the models presented above for the SmA have been used as an approximation to interpret the experimental T_1 frequency dispersion data. However, due to some particular characteristics of the structure of these mesophases, additional relaxation mechanims have been proposed: e.g. fluctuations of the tilt angle at the transition from the SmA to the SmC phase,[62] fluctuations of the tilt angle in the SmC* phase (*soft-mode*), phase fluctuations (*Goldstone mode*),[71–73] *anti-phase* collective fluctuations in the case of SmC^*_A,[74] rotations induced by the self-diffusion across the TGBA *grain boundaries*,[75] that is rotations induced by diffusion along the helix axis as in the SmC* phase.[76]

5.3.2. *Experimental Techniques*

Different experimental techniques can be used in NMR relaxometry to measure the spin relaxation times.[35,77] Here we will focus on techniques to measure the spin-lattice relaxation rate. In the presence of a magnetic field $\overrightarrow{B}(t) = B_0\overrightarrow{e_z} + \overrightarrow{B_{RF}}(t)$, where B_0 is the static magnetic field and $\overrightarrow{B_{RF}}(t)$ is the time dependent radio frequency (RF) field usually perpendicular to B_0, the time evolution of the magnetization $\overrightarrow{M}$ can be expressed by the phenomenological Bloch equation[35]

$$\frac{d\overrightarrow{M}}{dt} = \gamma\overrightarrow{M} \times \overrightarrow{B} - \frac{M_x\overrightarrow{e_x} + M_y\overrightarrow{e_y}}{T_2} - \frac{M_z - M_0}{T_1}\overrightarrow{e_z} \tag{5.25}$$

where $M_0 = \chi B_0$, and χ is the magnetic susceptibility of the system. Eq. (5.25) shows that T_1 is related with the time evolution of the z-component of the magnetization (longitudinal relaxation time) and T_2 is associated with the evolution of the magnetization on the x, y plane (transverse relaxation time). In classical pulse NMR, T_1 is measured using the *inversion-recovery* sequence (180° - τ - 90°) after which the free induction decay (FID) signal, corresponding to $\overrightarrow{M_\perp} = M_x\overrightarrow{e_x} + M_y\overrightarrow{e_y}$, is detected and $M_z(\tau) = M_\perp(0)$.[35,77] The RF pulse is applied on resonance with the Larmor frequency $\omega_0 = \gamma B_0$, where B_0 is kept constant during the whole measuring cycle. $T_1(\omega_0)$ is then determined by fitting the experimental FID signals according to the expression

$$M_z(\tau) = M_0\left(1 - 2e^{-\frac{t}{T_1(\omega_0)}}\right). \tag{5.26}$$

The signal-to-noise ratio in pulse NMR is S/N$\propto B_0^{3/2}$.[35,78,79] As a consequence, the application of this technique is limited in practice to B_0 fields corresponding to Larmor frequencies typically above 4 MHz. For lower frequencies, the preferred technique is the Fast Field-Cycling (FFC) NMR. Transverse relaxation[5,80,81] and longitudinal relaxation in the rotational frame ($T_{1\rho}$ measurements)[35,42,77] can also be used to study the molecular dynamics at low frequencies. These techniques are, however, less effective (when compared to FFC NMR) due to technical constraints that impose narrower limits in the frequency range.

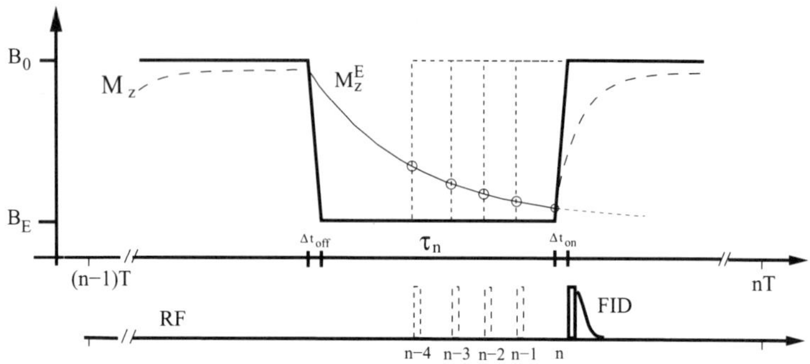

Fig. 5.4. Schematic representation of a fast magnetic field-cycle for a spin-lattice relaxation rate measurement. The evolution of magnetic field and magnetization with time is represented for a repetition period T. The RF pulse is applied after the $B_E \rightarrow B_0$ transition. The detected free induction decay signal (FID) is also schematically represented. The experiment is repeated for different τ_n values. $T_1(\omega_E)$ is estimated ($\omega_E = \gamma B_E$) using a non-linear least-square minimization method, in order to fit Eq. (5.27) to the mesured values of $M_z^E(\tau)$ as explained in the text.

The basics of FCC NMR can be found in several reviews,[78,79,82,83] and its principle is to submit the spin system to a magnetic field, parallel to the z-axis, that cycles between (at least) two different values (see Fig. 5.4): a high field B_0 (the polarization/detection field), and B_E (the evolution field, corresponding to the Larmor frequency at which T_1 is measured). When B changes from one value to the other, the magnetization of the system varies according the Bloch equation [Eq. (5.25)]. This means that, for a cycle of the type $B_0 \overset{\Delta t_{off}}{\longrightarrow} B_E \overset{\tau}{\longrightarrow} B_E \overset{\Delta t_{on}}{\longrightarrow} B_0$, with transition times $\Delta t_{off} \ll T_1(B_E)$ and $\Delta t_{on} \ll T_1(B_0)$, we have, after $B_0 \rightarrow B_E$

$$M_z^E(\tau) = M_\infty^E + \left(M_0 - M_\infty^E\right) e^{-\frac{\tau}{T_1(\omega_E)}} \tag{5.27}$$

with $M_\infty^E = \chi B_E$ (equilibrium magnetization) and $\omega_E = \gamma B_E$. After $B_E \rightarrow B_0$ we get

$$M_z(t) = M_0 + \left(M_z^E(\tau) - M_0\right) e^{-\frac{t}{T_1(\omega_0)}} \tag{5.28}$$

If a 90° RF pulse, with the Larmor frequency on resonance $\omega_0 = \gamma B_0$, is applied just after $B_E \rightarrow B_0$, the initial value of the FID is $M_\perp(0) = M_z(0) = M_z^E(\tau)$. Therefore, the FID is measured at the high B_0 field, which assures that the S/N ratio has a reasonable value. The $M_\perp(0)$ value corresponds to the magnetization of the spin system under the influence of the B_E field, which depends of the value $T_1(\omega_E)$. After a sufficiently long time $[\sim 10T_1(\omega_0)]$ the magnetization reaches its initial value M_0, and by repeating the cycle for increasing values of τ, $T_1(\omega_E)$ can be easily obtained from Eq. (5.27). The same experiment can be made for different values of B_E in order to measure T_1 for a set of low Larmor frequencies. It is important to remark that the detection (high) frequency is always the same. In this way, the S/N ratio (corresponding to the detection B_0 field) is kept at a reasonable value and independent of the measuring frequency ω_E.

In a real FFC cycle the magnet transition times have to fulfill several conditions. On the one hand ΔT_{on} and ΔT_{off} have to be shorter than $T_1(\omega_0)$ and $T_1(\omega_E)$, respectively. On the other hand, they have to assure that the adiabatic switching condition is fulfilled,[78,79] i.e. the local magnetization has to remain always aligned with the magnetic field. Also, the 90° RF pulse is always applied with a delay $\Delta T_{on} + \Delta t_{RF}$, where Δt_{RF} is the time interval between the $B_E \rightarrow B_0$ transition and the RF pulse. Therefore, the solutions for the Bloch equation in the case of a real FFC experiment are slightly different from Eqs. (5.27) and (5.28).[78,79] If ΔT_{on}, ΔT_{off}, and Δt_{RF} are reproducible in every field-cycle during a $T_1(\omega_E)$ measurement, Eqs. (5.27) and (5.28) are only affected by some corrective factors which do not influence the obtained $T_1(\omega_E)$ value.[78,79] The field cycle $B_0 \xrightarrow{\Delta t_{off}} B_E \xrightarrow{\tau} B_E \xrightarrow{\Delta t_{on}} B_0$ corresponds to a simple case. Other cycles where the magnetic field is switched between three or more field levels during a field cycle can be found in the literature.[78,79]

Standard commercial NMR spectrometers use typically solenoid-type superconducting magnets for fields above 2T and iron-core magnets for variable field strengths to cover the range 0.2T-2T. In these systems, a standard $T_1(\omega_0)$ measurement requires the setting of the RF systems of the spectrometer to generate the RF pulses with the selected Larmor frequency. On the contrary, the magnetic field in FFC NMR can change to allow for the $T_1(\omega_E)$ measurement. There is no need to readjust or reset the RF resonance setting, as it is determined by the fixed detection B_0 field.

The development of FFC NMR spectrometers has always been connected with the developments in power electronics and magnet design. In fact, the necessity of fast switching the magnetic field from values above 0.2T to zero (available commercial FFC spectrometers can operate with a maximum field of 1T) has driven the development of optimized low self-inductance solenoid magnets which require high values of electrical current to operate.[78,79,84] These high current (and power) values involve in most cases the use of several power semiconductor devices (MOSFETs, GTO thyristors, or IGBT's) in parallel, to control the current in the magnets.[78,79,85] There are alternative solutions though, that require the use of just a single power semiconductor device.[86]

Solenoid-type magnets have been extensively used in FFC NMR. However, this geometry severely restricts the possibility to perform angular dependent measurements where the aligned sample is rotated about a direction perpendicular to the magnetic field. This type of studies may be important in the case of aligned mesophases. In that case, additional saddle coils may be introduced into the main solenoid in order to obtain a perpendicular magnetic field component and thus produce an effective off z-axis magnetic field. In this case the sample is immobile and it is the magnetic field direction that turns with respect to the sample's alignment direction.[87,88] Generally, this type of FFC magnets requires the use of optimized cooling systems and in some cases the field cycles must include specific cooling periods in which no current is circulating in the magnet. This might be a problem when working with aligned mesophases of LCs.

Recently, a new high permeability FFC magnet has been developed that allows for the use of solenoid RF coils and an easy rotation of the sample perpendicularly to the magnetic field.[89,90] The new FFC magnet design opens the possibility to reduce the size of the whole magnet system and the electrical power necessary to operate at $B_0 = 0.2$T. It involves field transitions with switching times shorter than 2.5 ms. The most recent home-built propotype requires a 200W power supply and operates with a detection field of 0.2T. A simple fan cooling system is enough to allow for a continuous operation of the spectrometer with magnetic field permanently on. In this way the working conditions are ideal for the sample's alignment. The new optimized power supply design ensures a stable FID quadrature detection, in phase sensitive mode, and both real and imaginary parts of the FID signal can be used to obtain the spin-lattice relaxation time.[90]

When phase sensitive detection is not possible, T_1 can be estimated using the FID signal modulus. However, in this case, the average noise level must be thoughtfully considered when analyzing the measured values of $M_z(\tau)$. In fact, although the noise on both real and imaginary parts of the FID signal has a null average, the modulus of the FID is always positive. Therefore, it introduces a bias in the measured magnetization values, which is particularly relevant for week signals. The contribution of this bias can be suppressed using a suitable numerical algorithm.[91] Otherwise, the bias effect may also be attenuated by considering a suitable choice of Δt_{RF}.[78]

It was mentioned above that the switching time Δt_{on} has to be as short as technically possible but must not violate the adiabatic switching condition. The use of arbitrary values for this experimental parameter setting may compromise the analysis of the $T_1(\omega_E)$ dispersion at low frequencies.[92,93] The particular linear frequency dispersion at low frequencies extensively reported in the literature for the SmA phase was at some point questioned and the possibility that this linear dispersion could be associated with technical artifacts was suggested in a work where the smectic A state was disturbed by sound waves.[93] A careful analysis of the results presented for the SmA phase obtained with different home-built and commercial FFC spectrometers[46,67–69] as well as the results obtained with a new LC compound has shown that the low frequency part of the dispersion is not due to technical issues.[94]

The development of superconducting magnets for FFC NMR[95] opens a new area of work with the possibility to raise the detection magnetic field to values above those currently available.

5.4. Relaxometry Results and Molecular Dynamics

The combined used of different NMR spectrometers allows for the assessment of the spin-lattice relaxation dispersion over a broad frequency band from the kHz range to 300 MHz or more. The frequency range above 100 MHz is accessible through NMR spectrometers with superconducting magnets. The range 4-100 MHz is usually covered by standard variable field iron-core magnets, although the more recent commercial FFC spectrometers operate at frequencies up to 20 MHz. The use of FFC spectrometers is imperative for frequencies below 4 MHz. Molecular dynamics can also be studied by measuring the spin-lattice relaxation in the rotating frame.[35,42,77] In the following a brief review on the proton NMR relaxometry results obtained in different phases of LC compounds is presented.

5.4.1. *Isotropic phase*

In the Isotropic phase of a large number of LC systems, the T_1 dispersion is usually characterized by a T_1 increase with frequency for frequencies in the MHz range and a low frequency plateau that extends to the kHz range (see Fig. 5.5). The frequency at which the leveling-off starts depends on

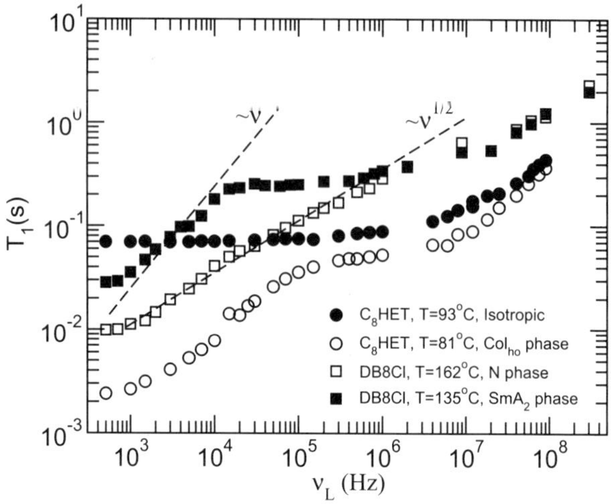

Fig. 5.5. Frequency dependence of the spin-lattice relaxation in different phases of two liquid crystalline compounds DB8Cl and C_8HET.[69,70] The dashed lines are guides-to-the-eye to show the characteristic frequency dependences $\nu^{1/2}$ and ν^1 associated with the order director fluctuations and layer undulations relaxation mechanisms, respectively. Results presented in the plot were obtained for the nematic and bilayer smectic A phase of the DB8Cl (calamitic) and of the isotropic and hexagonal phases of C_8HET (discotic) liquid cristal compounds.

each particular system.[3,46,59,94] The frequency dispersion can be quantitatively analyzed by a sum of two relaxation mechanisms in Eq. (5.12): rotations/reorientations (R), and translational self-diffusion (SD). In a new tetrapode (generation 0 dendrimer[96]) with lateral attached mesogens, the T_1 dispersion in the I phase presents some peculiar differences with respect to more conventional calamitics.[97,98] In fact, in this system it was necessary to consider an additional relaxation mechanism in the analysis of the experimental results. It was found that the particular structure of these molecules (where 4 mesogenic units are attached to a central siloxane centre by the long flexible spacers) induces an enhanced local molecular organization leading to the persistence of cybotatic nematic clusters in the I phase several degrees above the N-I transition temperature. A typical $T_1^{-1} \sim \omega^{1/2}$ contribution was included in the quantitative T_1 dispersion analysis.[97]

The molecular dynamics in the I phase can also be affected by the boundary conditions of the LCs when they are confined in glasses with 10-100 nm pores. The surface interaction induces order in the LC and ad-

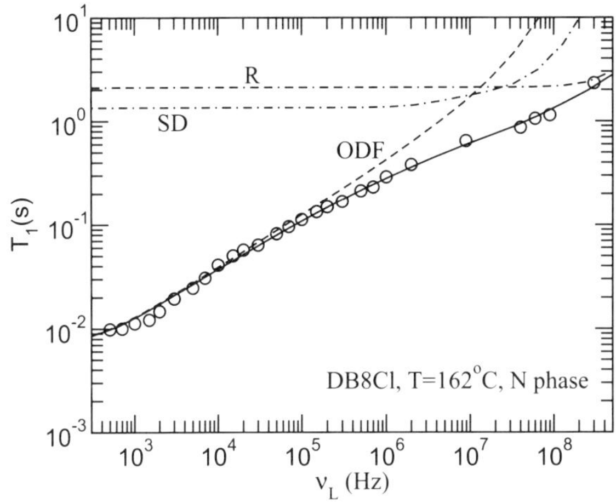

Fig. 5.6. Plot of experimental spin-lattice relaxation time versus frequency in the N phase of DB8Cl and theoretical curves obtained from the fit of Eq. (5.12) to the experimental data (solid line). Three relaxation mechanisms: rotations/reorientations R, translational self-diffusion, SD, and order director fluctuations, ODF.[69] The curves corresponding to each individual contribution to the relaxation are shown in the figure as dashed lines.

ditional contributions to relaxation (e.g. RMTD) become important, thus reflecting a change in the molecular dynamics in the system.[60]

5.4.2. *Nematic phase and Smectic phases with liquid-like layers*

The spin-lattice relaxation in the nematic and smectic phases is completely different from the one observed in the I phase of LCs. The differences between the I and N phases are known since the first FFC relaxometry studies of Noack et al. in the PAA[99] and MBBA[100] liquid crystals. As an example, the distinct characteristics of the T_1 dispersion in the nematic and smectic phase of liquid-like layers are clear when comparing their results for the DB8Cl compound as presented in Fig. 5.5. The results shown in this figure for the bilayer SmA_2 phase of DB8Cl are similar to others obtained from monolayer and partial bilayer SmA_d phases.[67,68,101,102]

Above 20 MHz, T_1 increases with increasing frequency in all mesophases. Below 1 MHz there is a clear deviation between the T_1 dispersion in the N phase and that observed for the SmA phase. In fact, for intermediate frequencies (20kHz-1MHz) T_1 changes very little with increasing frequency in the SmA phase but in the N phase it shows the typical strong increase with increasing frequency. In the SmA_2 phase (see Fig. 5.7), T_1 in the range 1-10 kHz presents also a strong increase with increasing frequency with a slope which is higher than the one observed in the N phases. A leveling-off in T_1 seems to start below 2 kHz for both N and SmA_2 phases. Similar differences have been observed in many other LC systems of calamitic molecules.[46,67,69,98,101,102]

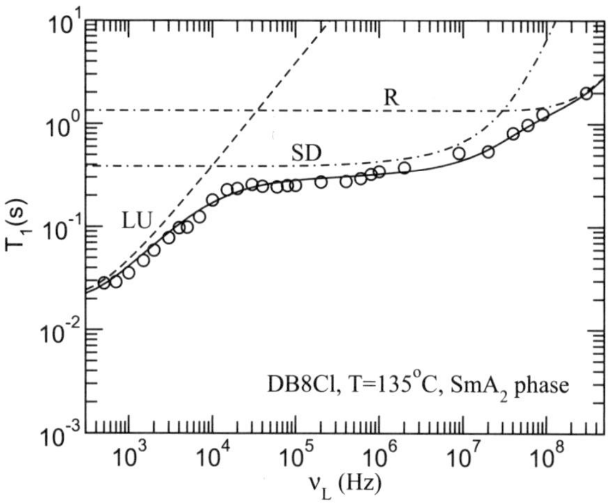

Fig. 5.7. Plot of experimental spin-lattice relaxation time versus frequency in the SmA_2 phase of DB8Cl and theoretical curves obtained from the fit of Eq. (5.12) to the experimental data (solid line). Three relaxation mechanisms were considered: rotations/reorientations R, translational self-diffusion, SD, and layer undulations, LU.[69] The curves corresponding to each individual contribution to the relaxation are shown in the figure as dashed lines.

Three contributions are considered in the model used to fit the T_1 dispersion in the N phases: R, SD, and ODF with a square-root frequency dependence [see Fig. 5.6 and Eq. (5.21)]. The quantitative analysis of the spin-lattice relaxation in the SmA phases is usually performed using a similar combination of relaxation mechanisms except that the slow collective motions are associated with a linear frequency dependence due to layer undulations [see Fig. 5.7 and Eq. (5.23)]. The T_1 dispersion in N phases is dominated by the ODF relaxation mechanism for frequencies up to the 1-10 MHz range in most of the LC systems. Above these frequencies R and SD contributions become more important, although SD is less relevant than rotations/reorientations. In some cases it was possible to estimate the values of the high and low cut-off frequencies for the ODF relaxation mechanism, and to relate these values with the viscoelastic constants of the studied compounds. Conversely, it is possible to use measured or estimated values of the viscoelastic parameters in order to calculate the maximum and minimum wave vectors of the ODF modes. The correlation times for rotations/reorientations obtained from the fits of the theoretical models to the experimental results are typically in the range $10^{-9} - 10^{-11}$s. The mean time between translational self-diffusion jumps are of the order of $10^{-9} - 10^{-10}$s.

In the SmA phases of calamitic liquid crystals (monolayer, partial bilayer and bilayer), the usually detected plateau in the intermediate frequency range for T_1 is associated with molecular translational self-diffusion. At high frequencies the T_1 dispersion is dominated by rotations/reorientations. At low frequencies the dominant mechanism of collective motions is undoubtedly characterized by a linear frequency dependence associated with layer undulations. This contribution has a low frequency cut-off that may be estimated, depending on the particular system studied. The high cut-off frequency is usually not experimentally accessible.

The fit of the T_1 angular dependence is also useful to unravel the contributions of the different relaxation mechanisms and the physical parameters relevant to the molecular dynamics (e.g. the characteristic times associated to each type of motions).[42,67–69,73] Depending on the frequency it is possible to verify that the relaxation rates corresponding to the main relaxation mechanisms, obtained by considering Eqs. (5.10) and (5.11), show different angular dependences. In the case of DB8Cl (Fig. 5.8), in which molecular rotations/reorientations and translational self-diffusion are considered, it is possible to observe that at the 90° angle rotations/reorientations model shows a maximum rate whereas SD exhibits a minimum rate.

In N and SmA phases of calamitic LC composed by molecules with nitrogen atoms, an additional relaxation contribution has been identified associated with the cross relaxation between the protons ($I = 1/2$) and the nitrogen ($I = 1$) spin systems. This relaxation mechanism is characterized by the presence of the so-called 'dips' when T_1 is plotted as a function of the frequency. This contribution is effective at given frequencies below the MHz range but its detection is very much dependent on the fine sampling of the Larmor frequency.[78,82,101,102]

A good quantitative analysis of the T_1 frequency, angular and temperature dependences is possible whenever some additional physical pa-

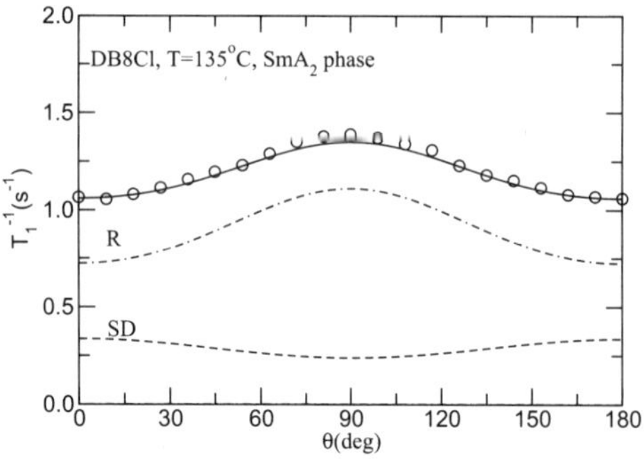

Fig. 5.8. Plot of the experimental angular dependent spin-lattice relaxation time results in the SmA_2 phase of DB8Cl, at a Larmor frequency of 60MHz, and theoretical curves obtained from the fit of Eq. (5.12) to the experimental data (solid line). Three relaxation mechanisms were considered: rotations/reorientations R, translational self-diffusion, SD, and layer undulations, LU.[69] The curves corresponding to the R and SD contributions to the relaxation are shown in the figure as dashed lines. The contribution of LU is negligible at this frequency.

rameters of the system are known or a global analysis can be made when the studied system has several mesophases with known structures.[46,59,67–70,73–75,97,98,103,104] In some cases, this global analysis approach reveals the presence of additional contributions to the spin-lattice relaxation that otherwise would remain hidden by the dominant relaxation mechanisms. This is the case in the molecular dynamics study of the DB8CNSym reported in Ref. [68]. This compound exhibits a partial bilayer SmA_d phase - a smectic phase with a layer between one and two molecular lengths. This layer spacing can be explained as an average from a mixture of single molecules and molecular dimers resulting from the coupling of terminal molecular dipoles.[105] The structure of the SmA_d phase is due to the *frustration* mechanism referred in Sec. 5.2.[28,106] A relaxation mechanism related with the specific nature of the SmA_d phase was identified in this system. This mechanism may be associated to the dynamic process of formation/dissociation of molecular dimers with a mean lifetime of 10^{-8}s.[68]

Recently, the study of molecular dynamics in the SmA phases of tetrapodes with terminally attached mesogens, and dendrimers of higher generations have revealed that layer undulations can be the most effective relaxation mechanism for frequencies up to the MHz range (see Fig. 5.9).

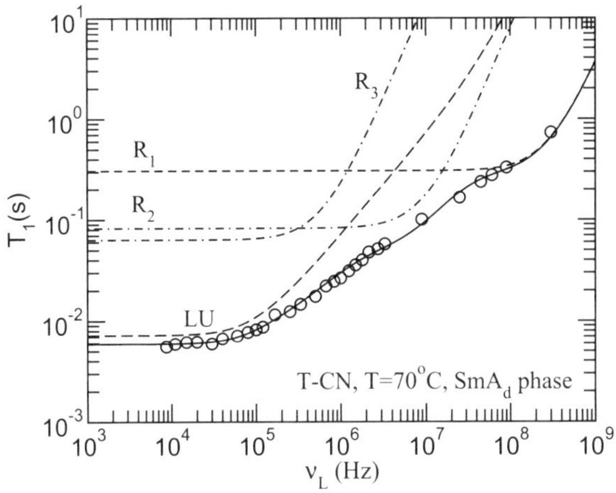

Fig. 5.9. Plot of the experimental spin-lattice relaxation time vs frequency results obtained in the SmA_d phase of T-CN and theoretical curves obtained from the fit of Eq. (5.12) to the experimental data (solid line). Rotations/reorientations, R1,R2,R3 [BPP-type, Eq. (5.15)], and layer undulations, LU[94] were considered as relaxation mechanisms. The curves corresponding to each individual contribution to the relaxation are represented by dashed lines.

This effect is due to the particular local molecular organization, which restricts the translational displacement of the molecules.[94,103,104]

The micro-segregation mechanism is particularly important in the formation of dendrimers' lamellar phases. In the case of the tetrapode T-CN (generation 0 dendrimer), three separate sub-layers (aromatic, aliphatic and siloxane) define the smectic structure. The predominant layer undulations observed in T-CN was related to the mutual decoupling of aromatic sub-layers from the siloxane cores by the long flexible spacers. Also, in the case of the LC dendrimers that exhibit SmA phases (e.g. PAMAM LC dendrimers[9,107]), the covalent links between the mesogenic units and the dendritic cores impose severe restrictions to the motions of the mesogenic units. In this case, the large and dominating contribution of LU detected for frequencies up to 10 MHz might also be related to the reduced size of smectic domains in the polycrystalline medium when compared with what is observed in low molecular mass LCs. In these SmA phases translational self-diffusion of the whole molecules is negligible. The characteristic times associated to the small local translational displacements of the mesogenic peripheral segments are close to the values of the rotational/reorientational correlations times. These similarities inhibit a clear distinction between both types of relaxation mechanisms.[103,104]

The molecular dynamics studies in the SmC phase give qualitatively similar results to those obtained for the SmA phases. Rotations/reorientations, translational self-diffusion, and layer undulations are the most relevant motions in contributing to the modulation of the dipolar spin interactions.[69,108] However, an additional relaxation mechanism associated with in-plane fluctuations of the tilting direction has also been reported for this mesophase.[69]

Studies of the molecular dynamics in the SmC*, anti-ferroelectric $\mathrm{SmC}_{\mathrm{A}}^{*}$, and ferri-electric phases were also reported.[73–76] The T_1 dispersion in these systems was interpretted in terms of the sum of three main contributions: R, SD, and LU. In the case of the SmC* phase, the three relaxation mechanisms could explain well the observed spin-lattice relaxation. Nevertheless, the contribution of other collective motions like the 'Soft-mode' and the 'Goldstone-mode', characteristics of the ferroelectric phases[109]could also be present but hidden by the LU contribution. In the case of the anti-ferroelectric $\mathrm{SmC}_{\mathrm{A}}^{*}$ phase, a contribution associated with anti-phase azimuthal fluctuations of the molecules (important at low frequencies) was also considered in the interpretation of the T_1 relaxation. In that case, the slow motions detected at low frequencies could not be exclusively associated with layer undulations.[74] The T_1 dispersion in the TGBA phase presents some differences with respect to both SmA and SmC* phases as reported for the LC compound 18FBTCO(1)M(7).[75] Indeed, an additional specific relaxation mechanism associated with the self-diffusion of molecules through the grain-boundaries that separate the SmA blocks was necessary to explain the steeper increase observed at low frequencies (5-70 kHz) in the TGBA phase.

NMR relaxometry studies of the molecular dynamics in some mesophases of bent-core molecules have been reported.[110,111] In the N, B_2 and B_6 phases of these LC systems, the T_1 dispersion seems to confirm the existence of reorientational and self-diffusional relaxation mechanisms with correlations times lower than those usually found for normal calamitic LCs. The molecular dynamics in the N phase is also affected by the confinement conditions. In fact, the proton NMR relaxometry study of LC confined in controlled porous glasses (CPG) confirms the importance of surface effects. Besides the bulk-like R and SD contributions typical of N phases, molecular reorientations induced by translational displacements (RMTD) must be included in the relaxation model when analysing the experimental results in confined LCs. Also, the slow collective motions are strongly affected by the reduced size of the cavities.[46]

5.4.3. *Smectic phases with ordered layers*

Proton NMR relaxometry in SmB phases[54,112,113] shows that the T_1 dispersion is explained in terms of translational self-diffusion and rotation/reorientations relaxation mechanisms and that collective motions typical of LCs are not observed in these SmB systems. In the SmB phase of some halogen compounds, the spin-lattice relaxation shows a very strong frequency dependence ($T_1^{-1} \sim \nu^{-2}$) for frequencies below 10 MHz associated with translational self-diffusion.[113] In the high frequency range the T_1 dispersion is explained by the rotations/reorientations of the molecules.

Molecular dynamics studies using standard NMR relaxometry techniques (e.g. T_1, $T_{1\rho}$) in smectic phases of higher in-layer order, such as $\mathrm{SmB_{hex}}$, SmF, and SmG, have been reported for different systems. In the TBDA liquid crystal, a comparative study between the SmC and the SmF phases has shown that the layer undulations mechanism, typical of smectics with liquid like layers, does not give a significant contribution to the relaxation in ordered smectics like the SmF phase.[114]

5.4.4. *Columnar phases*

Proton NMR relaxation measurements have been carried out in columnar phases exhibited by different types of thermotropic LCs. The first liquid crystalline columnar phases were observed in discotic molecules.[115] In this type of columnar phases, the cross section of each column is constituted by a disc-like molecule. The rigid central cores (e.g. triphenylene groups) of the molecules are stacked up, forming the columns that are surrounded by an inter-columnar soft medium of partially molten aliphatic chains (see Fig. 5.2). Relaxometry measurements carried out in the isotropic and columnar hexagonal ordered phases of the compound hexaoctyloxytriphenylene ($\mathrm{C_8HET}$) are explained by the usual contributions of three relaxation mechanisms: R, molecular SD and collective motions at high ($\omega/(2\pi) > 10$MHz), intermediate (10^7 Hz $> \omega/(2\pi) > 10^5$ Hz) and low ($\omega/(2\pi) < 10^5$ Hz) Larmor frequencies, respectively.[70]

The molecular rotations/reorientations mechanism is explained by a generalization of the Woessner model (see Eq. (5.16)) in which overall rotations of the molecules are considered together with faster relative motions corresponding to the reorientation of the molecular chains. Molecular

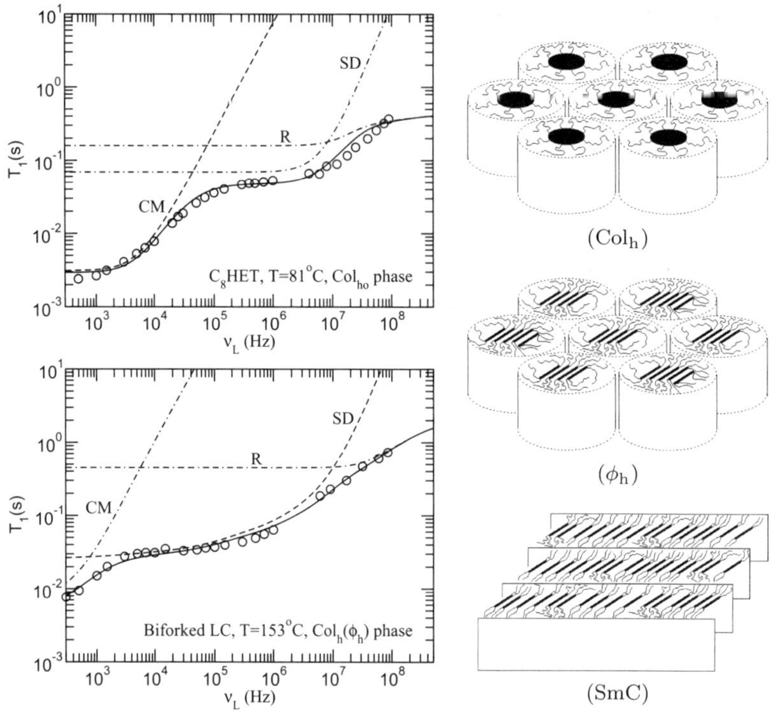

Fig. 5.10. Experimental T_1 data and fits of theoretical models for T_1 frequency dispersion [Eq. (5.9)] in the $\mathrm{Col_h}$ phases of discotic compound $\mathrm{C_8HET}$ (top) (reported in Ref. [70]) and biforked (tetracatenar) compound (reported in Ref. [47]). The structures of the phases, together with the SmC phase of the biforked compound, are schematically represented in the right picture. The columnar phase of the biforked compound is denoted as ϕ_h. In both cases, collective movements (CM on the top Fig.) correspond to column elastic deformations given by Eq. (5.24). In $\mathrm{C_8HET}$ (top graph) molecular translational self-diffusion (SD) is given by Eq. (5.19), associated to a model of random inter-columnar molecular jumps. In the biforked compound, the contribution of molecular translational self-diffusion is given by Eq. (5.18) due to the similarity at local level between the $\mathrm{Col_h}$ (ϕ_h) phase and the SmC phase of biforked molecules. The contribution of molecular rotations/reorientations (ROT) is described in both cases by a generalization of Eq. (5.16) (Woessener model) as described in Ref. [47] and Ref. [70].

translational self-diffusion is described by a *columnar permeation* mechanism, introduced by Žumer and Vilfan (see Eq. (5.19),[53] which results from molecular jumps between adjacent columns. The model considered for the collective motions, also introduced in the same theoretical work, corresponds to elastic oscillations/deformations associated with the bending and compression of the columns. The study presented in Ref. [70] is generally in agreement with the proposed relaxation models, after taking into account the molecular dimensions determined by structural studies.[116,117] The characteristic time-scales for the rotations/reorientations and molecular translational self-diffusion mechanisms ($\sim 10^{-11} - 10^{-12}$s for molecular chains reorientations, $\sim 10^{-9}$s for global molecular rotations, $\sim 10^{-8}$s for self-diffusion) are similar to those found for similar relaxation mechanisms in low molecular weight calamitic LCs. These results are also in agreement with both previous measurements in the same compound[118] and results obtained in the analogue compound C_5HET.[119] Particularly interesting is the experimental result indicating that the molecular reorientaions about the columnar axis are significantly faster (about one order of magnitude) than the transverse (out of the columnar cross-sectional plane) reorientations. The explanation is that the interactions between closely stacked triphenylene rings in molecular cores strongly restrain reorientaions out of the cross-sectional plane, whilst rotations about the columnar axis are practically free due to the soft molten chains in the inter-columnar regions. This was also found in the study on C_5HET.[119]

A proton NMR relaxation study in the columnar hexagonal (Col_h) mesophase of a polycatenar LC with two alkyl end chains in each side of a five ring aromatic core with a central cyclohexane group is presented in Ref. [47]. Some polycatenar mesogens present peculiar columnar phases, whose columns' cross sectional segments are formed by several molecules (instead of one as for discotics). These mesophases were initially observed in *phasmidic* compounds (with an elongated core and six terminal aliphatic chains). They were originally called ϕ_h after their designation. The average number of molecules in the columnar cross-section depends on the number of terminal chains of the polycatenar mesogen. In the case of the biforked compound with four chains, this average number is 4.5, which means that the number of molecules varies (with equal probability) between 4 and 5 along a column and from column to column in the plane perpendicular to the molecular axes (see Fig. 5.10). The rigid molecular aromatic cores that are disposed parallel to each other in a single cross-sectional segment, compose the inner columnar cores. The peripheral, partially molten, aromatic chains constitute a soft intercolumnar medium as in columnar phases of discotic molecules. The referred study shows that both the contributions of molecular reorientations (dominant at high Larmor frequencies) and of collective motions (dominant at low Larmor frequencies) to the relaxation rate may be explained by relaxation mechanisms similar to those considered in the analysis of the Col_h mesophase of the discotic compound C_8HET.

Nevertheless, in the case of molecular R, the relation between correlation times for longitudinal and transverse reorientations is more similar to those observed for calamitics, particularly in the SmC phase of the same compound. On the other hand, molecular self-diffusive motions are explained by the same mechanism used in the SmC phase of this compound (see Eq. 5.18), and not with the one corresponding to a columnar phase of discotic molecules. This may be understood by noting that the structure of the Col_h phase of the biforked molecules is similar at a local level to the SmC phase of the same compound. The fit of the theoretical expressions to the relaxation data of this compound is also presented in Fig. 5.10.

The comparison between the NMR relaxation results on the Col_h phases of C_8HET and of the biforked compound shows that the macroscopic similarity between both phases (both are columnar hexagonal) is reflected by the predominance of the same collective motions mechanism at low frequencies. This mechanism gives asymptotic T_1 behaviors independent of ν and proportional to ν^2 for low and high frequencies, respectively.[47,53] At high frequencies, the reorientation motions are essentially dependent on the molecular structure, and not too much on the phase properties. This type of motions is therefore explained by a mechanism that requires correlation times according to the different molecular structures.[47,70,120] As explained above, the comparison between T_1 results on these compounds also shows that translational self-diffusion is clearly different due to the difference between the two columnar phases at the local level.

In a more recent work, the molecular dynamics on the columnar rectangular (Col_r) phase of a PAMAM LC co-dendrimer of generation 3 is investigated by proton NMR relaxation.[103] This compound also presents a SmA phase, as mentioned in a previous section, and relaxation results in this phase are compared with those in the Col_r phase. Liquid crystalline dendritic molecules, especially those of intermediate generations, may assume different shapes, as the flexible inner dendritic cores can adopt several configurations to give various mesophases, depending on the symmetry of the phase. In the case of PAMAM's Col_r phase, the dendrimer assumes an average rectangular plate like structure as shown by structural studies based on x-ray diffraction data.[9,121] The T_1 fits with the considered theoretical models for PAMAM compound are shown in Fig. 5.11 together with a schematic representation of the Col_r phase. The T_1 relaxation at high Larmor frequency was described by two BPP type contributions [Eq. (5.15)] characterized by two correlation times associated with slow and fast local reorientations, respectively. These contributions are espe-

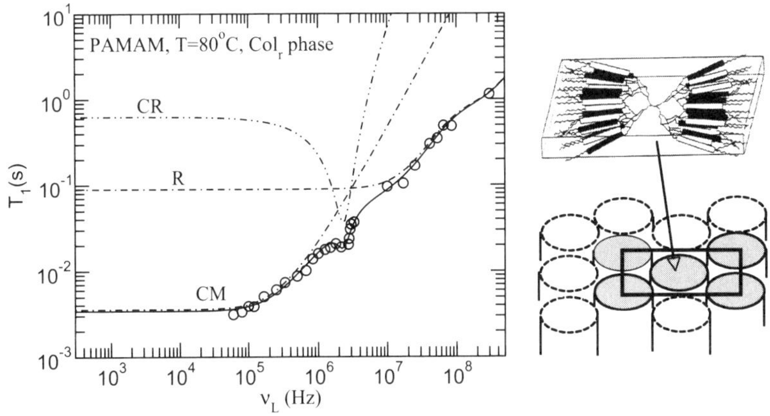

Fig. 5.11. Experimental data and fit of theoretical models for T_1 dispersion [Eq. (5.9)] in the Col_r phase of PAMAM co-dendrimer of generation 3 (left) (reported in Ref. [120]) and schematic reprensentation of the structure of the Col_r mseophase in the plane perpendicular to the columns. Collective movements correspond to column elastic deformations (ECD) given by Eq. (5.24). Local molecular rotations/reorientations is described by two BPP contributions. A cross-relaxation contribution (CR) due to the presence on nitrogen atoms in the PAMAM chemical structure is also included.

cially important for frequencies above 10 MHz. The orders of magnitude of the corresponding correlation times ($\sim 10^{-9}$s and $\sim 10^{-10}$s) are compatible with local molecular reorientations of the mesogenic units being restricted by their covalent links with the dendritic core. At low frequencies, the relaxation model for columnar elastic deformations (ECD) (see Eq. (5.24)) was applied. The resulting parameters are characteristic of rather rigid columns compared to what was found in columnar phases of discotic and biforked molecules. The rigidity of the columns can be understood taking into account the topological constraints imposed by the dendritic core on the peripheral mesogenic units as well as the partial interdigitation of the mesogens belonging to neighboring dendrimers. These severe spacial constraints also explain the negligible contribution of molecular SD in contrast with columnar mesophases of low molecular weight LCs. An additional contribution corresponding to cross-relaxation (CR) due to the presence on nitrogen atoms in the PAMAM structure was also necessary to explain the experimental results.

5.5. Conclusions and Outlook

Since the pioneering work of Bloembergen, Purcell and Pound,[39] widespread developments of the proton NMR relaxation experimental techniques have opened new perspectives to the investigation of molecular dynamics in a variety of materials. Liquid crystals are certainly among the systems where this relaxation technique is most useful. This is particularly clear if one realizes that the diversity of molecular organization in LC phases gives rise to a corresponding variation of relaxation mechanisms accessible and distinguishable through proton NMR relaxometry.

Besides the molecular movements present in common liquids, like molecular reorientations and translational self-diffusion, the specific collective modes, characteristic of the molecular dynamics in different LC phases, are clearly identifiable using this experimental technique. These collective motions give especially important contributions to the relaxation rate at low Larmor frequencies (usually at frequencies below one MHz). Therefore, the development of proton relaxometry at low magnetic fields has been of crucial importance to the molecular dynamics characterization of LCs. The development of the FCC NMR technique represents a breakthrough in that direction. In recent years, the application of novel power electronics techniques to this field has been of essential importance. More recently, the achievements on the design of fast switching power supplies have been associated with technical developments applied to the implementation of small dimension magnets producing highly homogeneous magnetic fields.

As evidenced in this review, the use of FCC NMR in conjunction with standard solid state NMR techniques, allows for the acquisition of proton NMR relaxation results in very wide Larmor frequency ranges. The analysis of the relaxation rate T_1^{-1} frequency dispersion in such a wide Larmor frequency domain provides an effective tool to the identification of molecular movements with distinct characteristic time-scales. Different relaxation mechanisms associated to the structure of specific LC phase have been the object of theoretical models to provide the spectral densities in terms of the physical parameters of the system both at microscopic and macroscopic levels (e.g. molecular dimensions, local molecular structure of the phases, elastic constants, viscosity coefficients, etc.). This type of theoretical studies has been applied in the last decades to many different LC phases of materials of increasing complexity. The fundamental understanding of the molecular dynamics in the mesophases of conventional calamitics (nematics, smectic A, smectic C, cholesteric, etc.) was followed by the investigation of peculiar phases like the frustrated smectics (SmA_2, SmA_d and others), phases of chiral molecules with lateral dipoles (SmC*, SmC_A and TGBA) and mesophases of bent-core molecules. Columnar phases of discotic molecules, polycatenar LCs, and recently more complex systems like dendrimers and nano confined LC phases have been the object of this type of studies. In the case of complex systems, the effect of molecular architecture, phase structure, and confinement conditions on the molecular dynamics is of special importance.

It is worthwhile to stress that the development of experimental relaxometry studies in LC goes together with the formulation of new models, which relate the molecular dynamics with the structure and physical properties (e.g. viscoelastic behavior) of the mesophases, and with complementary experiments, that can measure the physical parameters involved in these models. Future advances on the application of NMR relaxometry to novel LC systems and other soft-matter systems of increasing complexity (e.g. biomaterials) will induce the design of better performing spectrometers, and challenge new theoretical and complementary experimental efforts.

References

1. F. Noack, *NMR Basic Principles and Progress.* (Springer-Verlag Berlin, Heidelberg, New York, 1971).
2. C. G. Wade, *Ann. Rev. Phys. Chem.* **28**, 47, (1977).
3. R. Dong, *Nuclear Magnetic Resonance of Liquid Crystals.* (Springer, New York, 1994).
4. A. Ribeiro, P. J. Sebastião, and C. Cruz, *Mol. Cryst. Liq. Cryst.* **362**, 291–306, (2001).
5. R. Dong, *Prog. Nucl. Magn. Reson. Spectrosc.* **41**, 115–151, (2002).
6. A. C. Ribeiro, P. J. Sebastião, and C. Cruz, *Pramana-J. of Phys.* **61**, 205–218, (2003).
7. S. Chandrasekhar, *Liquid Crystals.* (Cambridge University Press, Cambridge, 1992).
8. P. de Gennes and J. Prost, *The Physics of Liquid Crystals, 2nd Edition.* (Oxford University Press, 1993).
9. J.-M. Rueff, J. Barber, B. Donnio, D. Guillon, M. Marcos, and J. Serrano, *Macromolecules.* **36**, 8368, (2003).
10. C. Tschierske, *J. Mater. Chem.* **8**, 1485–1508, (1998).
11. A. Saupe, *Z. Naturforsch. A.* **19**, 193, (1971).
12. M. J. Freiser, *Phys. Rev. Lett.* **24**, 1041, (1970).
13. L. A. Madsen, T. J. Dingemans, M. Nakata, and E. T. Samulski, *Phys. Rev. Lett.* **92**, 145505, (2004).

14. B. R. Acharya, A. Primak, and S. Kumar, *Phys. Rev. Lett.* **92**, 145506, (2004).
15. M. Lehmann, S.-W. Kang, C. Köhn, S. Haseloh, U. Kolb, D. Schollmeyer, Q. Wang, and S. Kumar, *J. Mater. Chem.* **16**, 4326, (2006).
16. R. Y. Dong, S. Kumar, V. Prasad, and J. Zhang, *Chem. Phys. Lett.* **448**, 54, (2007).
17. C. V. Yelamaggad, S. K. Prasad, G. G. Nair, I. S. Shashikala, D. S. S. Rao, C. V. Lobo , and S. Chandrasekhar, *Angew. Chem. International Edition* **43**, 3429, (2004).
18. K. Severing and K. Saalwächter, *Phys. Rev. Lett.* **92**, 125501, (2004).
19. K. Severing, E. Stibal-Fischer, A. Hasenhindl, H. Finkelmann, and K. Saalwächter, *J. Phys. Chem. B* **110**, 15680, (2006).
20. K. Merkel, A. Kocot, J. K. Vij, R. Korlacki, G. H. Mehl, and T. Meyer, *Phys. Rev. Lett.* **93**, 237801, (2004).
21. J. L. Figueirinhas, C. Cruz, D. Filip, G. Feio, A. C. Ribeiro, Y. Frère, T. Meyer, and G. H. Mehl. *Phys. Rev. Lett.* **94**, 107802, (2005).
22. K. Neupane, S. W. Kang, S. Sharma, D. Carney, T. Meyer, G. H. Mehl, D. W. Allender, S. Kumar, and S. Sprunt, *Phys. Rev. Lett.* **97**, 207802, (2006)
23. C. Cruz, J. L. Figueirinhas, D. Filip, G. Feio, A. C. Ribeiro, Y. Frére, T. Meyer, and G. H. Mehl, *Phys. Rev. E* **78**, 051702, (2008).
24. G. R. Luckhurst, *Thin Solid Films* **393**, 40, (2001).
25. K. Severing and K. Saalwächter, *Phase Biaxiality in Nematic Liquid Crystals, Experimental evidence for phase biaxiality in various thermotropic liquid crystalline systems, in Thermotropic Liquid Crystals, Recent Advances.* (Springer Netherlands, 2007).
26. L. D. Landau, L. P. Pitaevskii, A. M. Kosevich, and E. Lifshitz, *Theory of Elasticity, Third Edition: Volume 7 (Theoretical Physics, Vol 7).* (Butterworth-Heinemann, 1986).
27. D. Demus, J. Goodby, G. Gray, H.-W. Spiess, and V. Vill, *Physical Properties of Liquid Crystals.* (Wiley-VCH, Weinheim, 1999).
28. J. Prost, *Advances in Physics.* **33**, 1–46, (1984).
29. P. de Gennes, *Sol. Stat. Commun.* **10**, 753, (1972).
30. J. W. Goodby, A. M. Waugh, S. M. Stein, E. Chin, R. Pindak, and J. S. Patel, *Nature.* **337**, 449, (1989).
31. G. Pelzl, S. Diele, and W. Weissflog, *Adv. Mater.* **11**, 707, (1999).
32. V. Domenici, M. Geppi, and C. A. Veracini, *Prog. Nucl. Magn. Reson. Spectrosc.* **50**, 1, (2007).
33. D. Guillon, *Columnar Order in Thermotropic Mesophases in Structure and Bonding.* Liquid Crystals II, (Springer Verlag, Berlin Heidelberg, 1999).
34. E. I. Kats, *Sov. Phys. JETP.* **48**, 916, (1978).
35. A. Abragam, *Principles of Nuclear Magnetism.* (Oxford Science Publications, 1961).
36. C. P. Slichter, *Principles of Magnetic Ressonance.* (Springer-Verlag Berlin Heidelberg New York, 1980).

37. D. Wolf, *Spin-Temperature, Nuclear-Spin Relaxation in Matter.* (Academic Press New York London, 1979).
38. P. Ukleja, J. Pirš, and J. W. Doane, *Phys. Rev. A.* **14**, 414, (1976).
39. N. Bloembergen, E. M. Purcell, and R. V. Pound, *Phys. Rev.* **73**, 679, (1948).
40. D. E. Woessner, *J. Chem Phys.* **36**, 1, (1962).
41. V. Rutar, M. Vilfan, R. Blinc, and E. Bock, *Mol. Phys.* **35**, 721, (1978).
42. P. J. Sebastião, M. H. Godinho, A. C. Ribeiro, D. Guillon, and M. Vilfan, *Liq. Cryst.* **11**, 621–635, (1992).
43. P. A. Beckmann, J. W. Emsley, G. R. Luckhurst, and D. L.- Turner, *Molec. Phys.* **50**, 699, (1983).
44. P. A. Beckmann, J. W. Emsley, G. R. Luckhurst, and D. L. Turner, *Molec. Phys.* **59**, 97, (1986).
45. J. Dolinšek, O. Jarh, M. Vilfan, S. Žumer, R. Blinc, J. W. Doane, and G. P. Crawford, *J. Chem. Phys.* **95**, 2154, (1991).
46. M. Vilfan, T. Apih, P. Sebastião, G. Lahajnar, and S. Žumer, *Phys. Rev. E.* **76**, 51708, (2007).
47. C. Cruz, J. Figueirinhas, P. Sebastião, A. Ribeiro, F. Noack, H. T. Nguyen, B. Heinrich, and D. Guillon, *Z. Naturforsch.* **51a**, 155, (1996).
48. H. C. Torrey, *Phys. Rev.* **92**, 962, (1953).
49. J. F. Harmon and B. H. Muller, *Phys. Rev.* **182**, 400, (1969).
50. R. Blinc, M. Burgar, M. Luzar, J. Pirš, I. Zupančič, and S. Žumer, *Phys. Rev. Lett.* **33**, 1192–1195, (1974).
51. S. Žumer and M. Vilfan, *Phys. Rev. A.* **17**, 424, (1978).
52. M. Vilfan and S. Žumer, *Phys. Rev. A.* **21**, 672, (1980).
53. S. Žumer and M. Vilfan, *Mol. Cryst. Liq. Cryst.* **70**, 39, (1981).
54. S. Žumer and M. Vilfan, *Phys. Rev. A.* **28**, 3070–3079, (1983).
55. R. Kimmich and H. W. Weber, *Phys. Rev. B.* **47**, 11788, (1993).
56. S. Stapf, R. Kimmich, and J. Niess, *J. Appl. Phys.* **75**, 529, (1994).
57. T. Zavada and R. Kimmich, *J. Chem. Phys.* **109**, 6929–6939, (1998).
58. T. Zavada and R. Kimmich, *Phys. Rev. E.* **59**, 5848, (1999).
59. P. J. Sebastião, D. Sousa, A. C. Ribeiro, M. Vilfan, G. Lahajnar, J. Seliger, and S. Žumer, *Phys. Rev. E.* **72**, 61702, (2005).
60. R. Vold and R. Vold, *The Molecular Dynamics of Liquid Crystals.* (G.K. Luckhurst, C. A. Veracini, Kluwer Academic Publishers Netherlands, 1994).
61. J. W. Doane, C. E. Tarr, and M. A. Nickerson, *Phys. Rev. Lett.* **33**, 620, (1974).
62. R. Blinc, M. Vilfan, and V. Rutar, *Sol. Stat. Commun.* **17**, 171, (1975).
63. I. Zupančič, V. Žagar, R. Rožmarin, I. Leštik, F. Kogovšik, and R. Blinc, *Sol. State Commun.* **18**, 1591, (1976).
64. J. H. Freed, *J. Chem Phys.* **66**, 4183, (1977).
65. P. Pincus, *Sol. State Commun.* **7**, 415, (1969).
66. T. C. Lubensky, *Phys. Rev. A.* **2**, 2497, (1970).
67. P. J. Sebastião, A. C. Ribeiro, H. T. Nguyen, and F. Noack, *Z. Naturforsch.* **48a**, 851–860, (1993).
68. P. Sebastião, A. Ribeiro, H. T. Nguyen, and F. Noack, *J. Physique II.* **5**, 1707–1724, (1995).

69. A. Carvalho, P. J. Sebastião, A. C. Ribeiro, H. T. Nguyen, and M. Vilfan, *J. Chem. Phys.* **115**, 10484–10492, (2001).
70. C. Cruz, P. Sebastião, J. Figueirinhas, A. Ribeiro, H. T. Nguyen, C. Destrade, and F. Noack, *Z. Naturforsch.* **53a**, 823–827, (1998).
71. R. Dong and J. Sandeman, *J. Chem. Phys.* **78**, 4849, (1983).
72. H. Bender, F. Noack, M. Vilfan, and R. Blinc, *Liq. Cryst.* **5**, 1233, (1989).
73. J. Figueirinhas, A. Ferraz, A. Ribeiro, H. T. Nguyen, and F. Noack, *Ferroelectrics.* **146**, 123, (1993).
74. A. Ferraz, A. Ribeiro, and H. T. Nguyen, *Mol. Cryst. Liq. Cryst.* **331**, 67–74, (1999).
75. J. Figueirinhas, A. Ferraz, A. Ribeiro, F. Noack, and H. T Nguyen, *J. Physique II.* **7**, 79, (1997).
76. D. Pusiol, F. Noack, and C. Aguilera, *Z. Naturforsch. A.* **45a**, 1077, (1990).
77. T. C. Farrar and E. D. Becker, *Pulse and Fourier Transform NMR.* (Academic Press New York London, 1971).
78. F. Noack, *Prog. Nucl. Magn. Reson. Spectrosc.* **18**, 171–276, (1986).
79. R. Kimmich and E. Anoardo, *Prog. Nucl. Magn. Reson. Spectrosc.* **44**, 257–320, (2004).
80. D. Frezzato, G. Kothe, and G. J. Moro, *J. Phys. Chem.B.* **105**, 1281–1292, (2001).
81. M. Vilfan, B. Zalar, and P. J. Sebastião, *to be submitted*, (2009).
82. R. Kimmich, *Bul. Mag. Res.* **1**, 195–218, (1979).
83. S. H. Koenig and R. D. Brown III, *Prog. Nucl. Magn. Reson. Spectrosc.* **22**, 487, (1990).
84. K. H. Schweikert, R. Krieg, and F. Noack, *J. Magn. Reson.* **78**, 77–96, (1988).
85. E. Rommel, K.Mischker, G. Osswald, K. H. Schweikert, and F. Noack, *J. Magn. Reson.* **70**, 219–234, (1986).
86. D. M. Sousa, G. D. Marques, P. J. Sebastião, and A. C. Ribeiro, *Rev. Sci. Instrum.* **74**, 4521–4528, (2003).
87. S. Struppe and F. Noack, *Liq. Cryst.* **20**, 595, (1996).
88. H. Gotzig, S. Grunenberghassanein, and F. Noack, *Z. Naturforsch. A.* **49**, 1179–1187, (1994).
89. D. M. Sousa, G. D. Marques, J. M. Cascais, and P. J. Sebastião. Proceedings of the 5th conference ffc nmr relaxometry. Torino, (2007).
90. D. M. Sousa, G. D. Marques, J. M. Cascais, and P. J. Sebastião, *to be submitted*, (2009)
91. K.-H. Schweikert, *Thesis.* (University of Stuttgart, 1990).
92. P. J. Sebastião, *Thesis.* (Instituto Superior Técnico, Universidade Técnica de Lisboa, 1993).
93. E. Anoardo, F. Bonetto, and R. Kimmich, *Phys. Rev. E.* **68**, 22701, (2003).
94. D. Filip, C. Cruz, P. J. Sebastião, A. C. Ribeiro, M. Vilfan, T. Meyer, P. H. J. Kouwer, and G. H. Mehl, *Phys. Rev. E.* **75**, 11704, (2007).
95. V. Cavaliere, A. Formisano, R. Martone, G. Masullo, A. Matrone, and R. Quarantiello, *IEEE Trans. Applied Superconductivity.* **17**, 1429, (2007).
96. G. Newkome, C. Moorefield, and F. Vögtle, *Dendrimers and Dendrons : Concepts, Syntheses, Applications.* (Wiley-VCH, Weinheim, 2001).

97. D. Filip, C. Cruz, P. J. Sebastião, A. Ribeiro, T. Meyer, and G. Mehl, *Mol. Cryst. Liq. Cryst.* **436**, 17–28, (2005).
98. D. Filip, C. Cruz, P. J. Sebastião, M. Cardoso, A. C. Ribeiro, M. Vilfan, T. Meyer, P. H. J. Kouwer, and G. H. Mehl, *submitted to Phys. Rev. E.* (2009).
99. W. Woelfel, F. Noack, and M. Stohrer, *Z. Naturforsch. A.* **30**, 437, (1975).
100. V. Graf, F. Noack, and M. Stohrer, *Z. Naturforsch. A.* **32**, 61–72, (1977).
101. D. Pusiol and F. Noack, *Liq. Cryst.* **5**, 377, (1989).
102. D. Pusiol, R. Humpfer, and F. Noack, *Z. Naturforsch. A.* **47**, 1105–1114, (1992).
103. A. Van-Quynh, D. Filip, C. Cruz, P. J. Sebastião, A. Ribeiro, J.-M. Rueff, M. Marcos, and J. Serrano, *Eur. Phys. J. E.* **18**, 149–158, (2005).
104. A. Van-Quynh, D. Filip, C. Cruz, P. J. Sebastião, A. C. Ribeiro, J.-M. Rueff, M. Marcos, and J. L. Serrano, *Mol. Cryst. Liq. Cryst.* **450**, 191–201, (2006).
105. D. Guillon and A. Skoulios, *J. Physique.* **45**, 607–621, (1984).
106. J. O. Indekeu and A. N. Berker, *J. Phys. France.* **49**, 353, (1988).
107. J. Barbera, M. Marcos, and J. L. Serrano, *Chem. Eur. J.* **5**, 1834–1840, (1999).
108. T. Mugele, V. Graf, W. Wolfel, and F. Noack, *Z. Naturforsch.* **35a**, 924–929, (1980).
109. I. Musevic, R. Blinc, and B. Zeks, *The Physics of Ferroelectric and Antiferroelectric Liquid Crystals.* (World Scientific Publishing Company, 2000).
110. P. Queiroz, I. Costa, H. T. Nguyen, P. Sebastião, A. Ribeiro, A. Ferraz, and J. Figueirinhas. Ljubljana, Slovenia (July 4-9, 2004). 20th International Liquid Crystal Conference.
111. V. Domenici, T. Apih, and C. Veracini, *Thin Solid Films.* **517**, 1402–1406, (2008).
112. A. Ribeiro, *Mol. Cryst. Liq. Cryst.* **148**, 85–91, (1987).
113. A. Ribeiro, P. J. Sebastião, L. Gonçalves, L. Oswald, and D. Guillon, *Proc Int Disp Workshops.* **12**, 175–176, (2005).
114. J. Figueirinhas, C. Cruz, A. Ribeiro, N. T. Tinh, and M. Vilfan, *Mol. Cryst. Liq. Cryst.* **206**, 45, (1991).
115. S. Chandrasekhar, B. K. Sadashiva, and K. A. Suresh, *Pramana-J. of Phys.* **91**, 471, (1977).
116. J. Billard, J. C. Dubois, H. T. Ngyuen, and A. Zann, *Nouv. J. Chim.* **2**, 535, (1978).
117. C. Destrade, M. C. Mondon, and J. Malthête, *J. Physique.* **40**, C13–C17, (1979).
118. A. Martins and A. Ribeiro, *Port. Phys.* **16**, 117–123, (1985).
119. M. Vilfan, G. Lahajnar, V. Rutar, R. Blinc, B. Topic, A. Zann, and J. Dubois, *J. Chem. Phys.* **75**, 5250, (1981).
120. C. Cruz and A. Ribeiro, *Mol. Cryst. Liq. Cryst.* **331**, 1935–1940, (1999).
121. B. Donnio, J. Barber, R. Gimenez, D. Guillon, M. Marcos, and J. L. Serrano, *Macromolecules.* **35**, 370, (2002).

Chapter 6

Deuterium NMR Study of Magnetic Field Distortions in Ferroelectric Mesogens

Ronald Y. Dong

Department of Physics and Astronomy,
University of British Columbia,
6224 Agricultural Road, Vancouver, BC, Canada, V6T 1Z1
email: rondong@phas.ubc.ca

Ferroelectric liquid crystals (FLCs) have attracted much interest due to their applications in electro-optical devices. Deuterium nuclear magnetic resonance (DNMR) has been used in the past to study soliton-like distortions by observing the spectral line shape as a function of the NMR field direction with respect to the helical axes of aligned deuteriated FLCs. In chiral smectic C* phases, an in-plane electric polarization exists spontaneously in each smectic layer. When the helix is undistorted, the modulation wave along the pitch axis has a linear phase profile. The helicoidal structure may, however, be distorted in the presence of an external electric or magnetic field. Domains are then formed to give a soliton-like lattice where the molecules, mostly aligned by the field, are separated by narrow domain walls in which the molecules twist rapidly. As a result, a non-linear spatial phase modulation profile is seen. If the applied field is large enough, the helices can even be unwound to give a smectic C phase. Spectral simulation based on the Landau theory is used to assess the critical magnetic field strength for unwinding the helices. The structure and degree of distortions can also be monitored from the spectral simulations. In one FLC, the winding/unwinding transitions are studied at different magnetic field strengths by applying the magnetic field along the helical pitch. In other FLCs, the critical field has been determined at different temperatures far away from the Lifshitz point. These results will be discussed and contrasted in this chapter.

6.1. Introduction

Since the discovery by Meyer[1] in the mid-1970's, ferroelectric liquid crystals (FLC) have continued to attract much attention from the viewpoint of fundamental physics, and their applications in electro-optical devices.[2] Following the observation of antiferroelectric smectic C phase (SmC_A^*) in an antiferroelectric liquid crystal (AFLC),[3] many interesting tilted smectic structures (e.g. SmC_γ^*, SmC_α^* and other subphases) have been identified. These arise from combining chirality (caused by the presence of asymmetric carbon(s) in molecules) and polarity (from lateral electric dipole(s) in molecules). With the aid of a theoretical model[4,5] to account for the relevant molecular interactions, numerous experimental studies of FLC, including resonant X-ray,[6–9] optical rotatory power,[10–12] ellipsometry,[13] and dynamic light scattering,[14] have led to a typical phase sequence for these materials: isotropic-N-SmA-SmC_α^*-SmC^*-SmC_{Fi2}^*-SmC_{Fi1}^*-SmC_A^*. In a SmC^* phase molecules in all layers have the same tilt angle θ_0 with respect to the smectic planar normal, while their azimuthal ϕ_i angles remain the same within each layer, but vary continuously and slightly from one layer to the next. SmC_A^* phase has a 2-layer periodicity, and molecules have the same tilt angle, and form an interdigitated double-helix structure with a π difference between the ϕ_i angles within the 2-layer unit. Transition from the synclinic SmC^* to the low temperature anticlinic SmC_A^* phase causes frustrations between two very different helical superstructures, and usually occurs via two intermediate subphases: ferrielectric SmC_{Fi1}^* (or SmC_γ^*) and SmC_{Fi2}^* (or SmC_{AF}^*). The SmC_{Fi1}^* and SmC_{Fi2}^* phases show instead a 3-layer and 4-layer periodicity, respectively.[7,9,15] A variety of experimental probes has been used to obtain the molecular disposition in chiral mesophases. At present, the clock model[16] or asymmetric clock model[17] seems to describe these subphases better than the Ising model.[18]

Deuterium NMR (DNMR) spectroscopy[19] has been proven as a powerful technique to shed light on the structure, ordering and dynamics of LC. It is well known that the quadrupolar interaction of a deuteron (^{2}H) nucleus produces a doublet NMR spectrum with a splitting $\delta\nu_Q$ in aligned LC samples. The quadrupolar splitting can be related to the order parameters[19] of the molecules in the mesophases. Since molecules can carry on fast rotation about their long axes, this motion effectively diagonalizes the electric field gradient (EFG) tensor at the deuteron nucleus, yielding a motionally averaged EFG tensor ($\bar{V}_{xx}, \bar{V}_{yy}, \bar{V}_{zz}$) in a molecular (M) frame, in which $\bar{V}_{zz}$ is along the molecular long (Z_M) axis. However, it is important to note that axial molecular rotations may be hindered in some low temperature LC phases such as smectic C, E, F and G phases. In these phases, the EFG tensors may not be diagonalized in the M frame. To determine the observed DNMR splitting, it is nececessary to transform each EFG tensor in its principal axis system (PAS) to the laboratory (L) frame, or the director ($\vec{n}$) frame. It has been demonstrated that angle dependent DNMR spectral patterns can be used to extract interlayer jump diffusion rate in the SmC^*_A of a chiral FLC.[20] Possible helicoidal distortions by the NMR magnetic field have been satisfactorily treated.[20–22] Recently we have applied the same method to obtain the jump diffusion rate in the SmC^*_A phase of 1-methylheptyl 4'-(4-n-decyloxybenzoyloxy)-biphenyl-4-carboxylate (10B1M7).[23]

If a LC sample is put inside a NMR magnet, the magnetic torque due to the anisotropy of diamagnetic susceptibility $\Delta\chi = \chi_{\parallel} - \chi_{\perp}$,[24] where $\chi_{\parallel}$ and $\chi_{\perp}$ denote principal components of the magnetic susceptibility tensor parallel and perpendicular to $\vec{n}$, can even induce orientational ordering in its isotropic phase.[25] When the constituent molecules contain asymmetric (or chiral) carbons, a twisted nematic (N^*) or cholesteric phase can form. At zero field, the local nematic director $\vec{n}$ in this phase rotates uniformly and continuously (i.e. $d\phi/dz$ is constant) in the direction (z axis) of the pitch. An external magnetic field $\vec{H}$ applied perpendicular to the pitch axis of a N^* phase can cause deformations in the helix by the magnetic torque. Upon increasing the field to a certain critical field H_c, a transition from the N^* to N phase results. This induced transition has been interpretted as a condensation process of solitons.[26] Solitons was first introduced to LC under the influence of a magnetic field by Michelson.[27] A theoretical (H, T) phase diagram is schematized in Fig. 6.1.[28] Variation of the azimuthal ϕ angle in the magnetic field is described by the static Sine-Gordon equation.[29] When chiral molecules form a tilted SmC phase, a spontaneous electric polarization appears in the layer plane and makes a right angle with the director field. In addition, the director field rotates along the layer normal producing a helical superstructure. If the helical structure is unwound by an applied field (either perpendicular or parallel to the pitch), then the spontaneous electric polarization vectors within a helix all point in the same direction, producing a macroscopic polarization, typically about 10 nC/cm^2 for $uSmC^*$ (or unwound SmC^*) phases. For an aligned LC sample with $\Delta\chi > 0$, the helical axes are oriented along $\vec{H}$ with the smectic layer planes perpendicular to it. Typical helical pitch length of a few μm contains several hundreds smectic layers. Significant research has been directed at characterizing the structure of the SmC^* and chiral subphases in newly discovered LC.[11,24,27,28]

The behavior of FLC in an external magnetic[30–32] or electric field[33–35] is important for technical applications[36] such as nonlinear optics, informa-

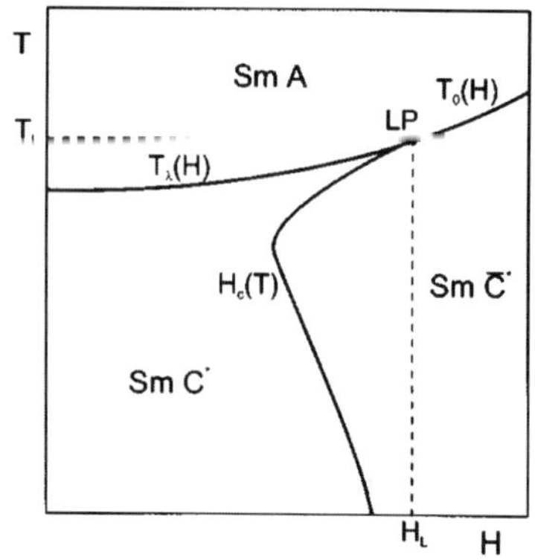

Fig. 6.1. Theoretical (H, T) phase diagram of a ferroelectric liquid crystal.[28] The Sm$\bar{C}^*$ phase is an unwound uSmC* or SmC phase. The field is applied perpendicular to the helix. LP denotes the Lifshitz point (H_L, T_L). (Reproduced with permission of APS)

tion storage, and display devices. The first observation of unwinding SmC* helices by $\vec{H}$ parallel to the helical axis was reported in 1982.[37] This phenomenon has recently been further investigated in several FLC.[38] If $\vec{H}$ is applied normal to the helical structure of a SmC* phase, molecules have a tendency to align along the field, leading to a soliton-like lattice shown in Fig. 6.2, i.e. two uniformly aligned regions are separated by a thin soliton domain. Inside the soliton domain, molecules have their ϕ angles changing rapidly along the pitch. At a certain critical field H_c, the helical structure will be completely unwound, and the sample becomes a spatially homogenous Sm$\bar{C}^*$ (or uSmC*) phase.[27,30] As seen in Fig. 6.1, one property of the SmC* phase in the presence of external $\vec{H}$ is the occurrence of a reentrant phenomenon. That is, there exists a range of H for which the phase sequence SmA $\rightarrow$ SmC* $\rightarrow$ Sm$\bar{C}^*$ $\rightarrow$ SmC* occurs upon decreasing temperature. In the case of an applied electric field, the unwinding of helix is caused by the coupling between the electric field and the electric polarization. The (H, T) phase diagram (Fig. 6.1) is predicted by Landau theory, in which a Landau expansion in powers of variables is used to write down the free energy density f of the LC system.[39] The crossing point of the three phase boundaries SmA $\rightarrow$ SmC*, SmA $\rightarrow$ Sm$\bar{C}^*$ and Sm$\bar{C}^*$ $\rightarrow$ SmC* is called the Lifshitz point (LP). Experimental (H, T) diagrams have been reported for DOBAMBC[37] and for a chiral/racemic CE8 mixture.[28] Among the significant works on the effect of magnetic field, the one by Zalar and co-workers[21] must be emphasized as they have done very careful angular

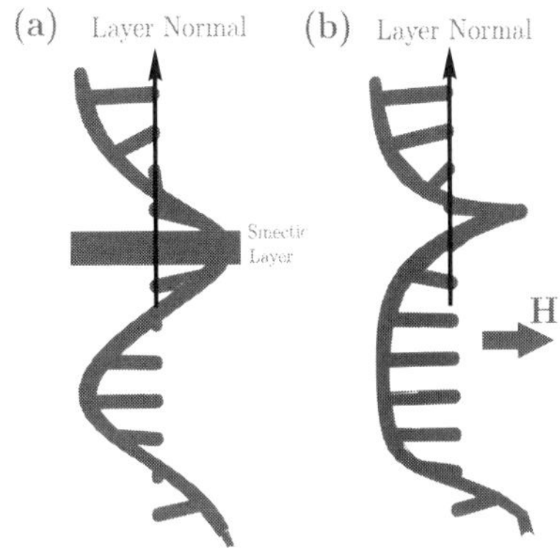

Fig. 6.2. (a) The helical structure of SmC* phase at zero field. (b) Soliton-like distortion of the helical structure when H field is applied perpendicular to the layer normal.

dependent DNMR experiments[40] to obtain spectra of the chiral mesogen DOBAMBC in order to evaluate the anisotropy of H_c. They have found that H_c is not necessarily lower when the field is applied normal to the helix. This is true except for the rotation angle $\theta = 90°$. Recently this type of study has been applied to the chiral mesogen 10B1M7, confirming the real potential of angle dependent ^{2}H NMR as a sensitive method to study the structure and the magnetic properties of its ferrielectric phases.[23] This chapter begins with the theoretical background in Sec. 6.2 to facilitate discussion of NMR data, followed by studies of SmC* phase of two chiral FLC. The anticlinic phases of 10B1M7 are investigated in Sec. 6.5, while in Sec. 6.6 winding and unwinding tranistions of helices by the external NMR field along the pitch axes are described in several FLC. The last section contains a brief conclusion.

6.2. Theoretical background

The theoretical treatment of magnetic distortions and helical unwinding in chiral smectic phases will be surveyed. When a LC sample of $\Delta\chi > 0$ is aligned by slow cooling from the isotropic melt to the chiral smectic phases, a rotation experiment as sketched in Fig. 6.3 can be carried out on its aligned SmC* phase. The treatment of angle dependent NMR spectra is outlined first.

6.2.1. *DNMR*

The observed NMR spectrum represents a sum of frequencies from all the molecules in the helical structure which are uniformly distributed on the surface of the cone shown in Fig. 6.3.[40] The apex of the cone is the tilt angle θ_0 of molecules in the smectic layers. When the sample is rotated by an angle θ w.r.t. $\vec{H}$ field, the geometry of these angles in Fig. 6.3 gives

$$\begin{aligned} \cos\beta &= \sin\theta\sin\theta_0\cos\phi + \cos\theta\cos\theta_0, \\ \sin\beta\cos\alpha &= \cos\theta\sin\theta_0 - \sin\theta\cos\theta_0\cos\phi. \end{aligned} \tag{6.1}$$

The two frequencies of a quadrupolar doublet for molecules in a smectic layer with ϕ angle is expressed in the director frame by[40]

$$\begin{aligned} \nu_\pm(\phi) = \pm\frac{3}{4}\bar{\nu}_Q\{&\frac{3+\bar{\eta}}{2}(\sin\theta\sin\theta_0\cos\phi + \cos\theta\cos\theta_0)^2 \\ &+ \bar{\eta}(\cos\theta\sin\theta_0 - \sin\theta\cos\theta_0\cos\phi)^2 - \frac{1+\bar{\eta}}{2}\} \end{aligned} \tag{6.2}$$

where $\bar{\nu}_Q$ and $\bar{\eta}$ are the average quadrupolar coupling constant and asymmetry parameter in this coordinate frame (whose Z_M axis is along $\vec{n}$). The observed NMR spectrum has contributions from all ϕ, and can be calculated by Fourier transforming (FT) the free induction decay (f.i.d.) $F(t)$

$$F(t) \propto \int_0^{2\pi} \cos(2\pi\nu_\pm(\phi)t)e^{-\sigma^2t^2/2}d\phi \tag{6.3}$$

where σ is a fitting parameter in a Gaussian broadening function $e^{-\sigma^2t^2/2}$.

The spectral intensity $I(\nu)$ at frequency ν is proportional to the number (N) of molecules which generate this particular ν, and[40]

$$I(\nu) = \frac{dN}{d\nu} = (\frac{dN}{d\phi})(\frac{d\phi}{d\nu}) \tag{6.4}$$

where $dN/d\phi$ is the distribution of the molecules along the pitch axis. For an undistorted helical structure, $dN/d\phi$ is a constant. However, when the helical structure is distorted by an external field, $dN/d\phi$ is not constant. The spectral singularities in the frequency domain can be calculated according to Eq. (6.4) by setting $d\nu_\pm(\phi)/d\phi = 0$ This produces three solutions: $\phi = 0$, π and $d\nu/d(\cos\phi) = 0$,[40] corresponding to singularities $s_1^\pm$, $s_2^\pm$

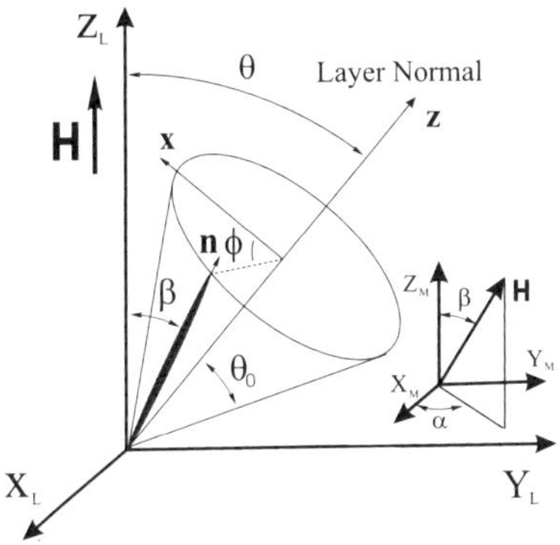

Fig. 6.3. Schematic of coordinate systems in a sample rotation experiment. (X_L,Y_L,Z_L) and (X_M,Y_M,Z_M) are the laboratory and molecular frame, respectively. **n** denotes the director located on the Z_M axis. The (x,y,z) frame is fixed onto the smectic layers. (α, β, 0) are the polar angles of the applied H field in the director frame. θ is the rotation angle, while ϕ and θ_0 are Euler angles of **n** in the (x, y, z) frame.[47] (Reproduced with permission of APS)

and $s_3^{\pm}$:

$$
\begin{aligned}
s_1^{\pm} &= \pm\frac{3}{4}\bar{\nu}_Q[1 - \frac{1}{2}(3-\bar{\eta})\sin^2(\theta-\theta_0)],\\
s_2^{\pm} &= \pm\frac{3}{4}\bar{\nu}_Q[1 - \frac{1}{2}(3-\bar{\eta})\sin^2(\theta+\theta_0)],\\
s_3^{\pm} &= \pm\frac{3}{4}\bar{\nu}_Q[\frac{(3+\bar{\eta})\bar{\eta}\cos^2\theta}{(3-\bar{\eta})\sin^2\theta_0 + 2\bar{\eta}} - \frac{1}{2}(1+\bar{\eta})].
\end{aligned}
\tag{6.5}
$$

The last singularity appears when $\frac{\pi}{2} \geq \theta \geq \tan^{-1}[(3-\bar{\eta})\sin 2\theta_0/((3-\bar{\eta})2\sin^2\theta_0 + 4\bar{\eta})]$. The theoretical spectra and the frequencies of the three pairs of singularities $s_1^{\pm}, s_2^{\pm}, s_3^{\pm}$ are plotted versus θ in Fig. 6.4(a) and Fig. 6.4(b), respectively. $\theta_0, \bar{\nu}_Q$ and $\bar{\eta}$ can be obtained by spectral fitting of angular dependent ^{2}H spectra or fitting their singularities.

To simulate DNMR lineshapes in ferrielectric SmC$^*_{Fi}$ phases, one can use the same method as in the SmC$^*_A$ phase[23] by solving a multi-site jump problem, and the evolution of the magnetization $M(\phi)$ for one base unit (3- or 4-layer) can be calculated using the Bloch-McConnell equation[41]

$$d\mathbf{M}/dt = (\mathbf{\Lambda} + \Gamma)\mathbf{M} \tag{6.6}$$

where $\mathbf{M}$ is a column vector with magnetization components $M(\phi)$ for each ϕ site, $\mathbf{\Lambda} + \mathbf{\Gamma} =$

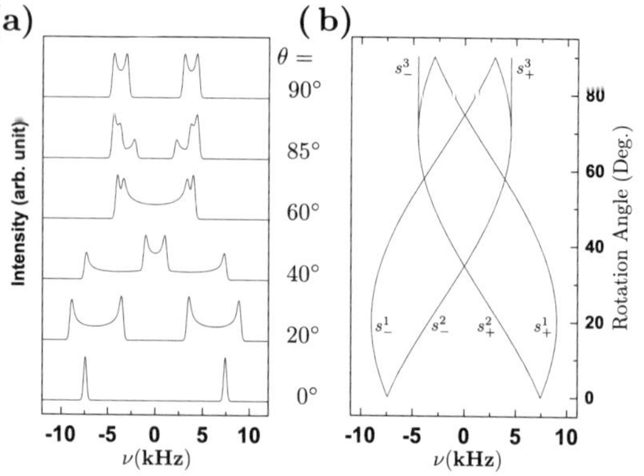

Fig. 6.4. (a) Simulated line shapes for some typical rotation angles θ using $\theta_0 = 20°$, $\bar{\nu}_Q = 12$kHz and $\sigma = 1$kHz. (b) The corresponding angular dependent singularity positions $s_1^{\pm}, s_2^{\pm}, s_3^{\pm}$. are shown.

$$\begin{pmatrix} -i\nu(\phi-\varphi/2)-2K & K & 0 & K \\ K & -i\nu(\phi+\varphi/2)-2K & K & 0 \\ 0 & K & -i\nu(\phi+\pi-\varphi/2)-2K & K \\ K & 0 & K & -i\nu(\phi+\pi+\varphi/2)-2K \end{pmatrix}$$

and

$$\begin{pmatrix} -i\nu(\phi+\pi)-2K & K & K \\ K & -i\nu(\phi-\varphi/2)-2K & K \\ K & K & -i\nu(\phi+\varphi/2)-2K \end{pmatrix} \tag{6.7}$$

for SmC^*_{Fi2} and SmC^*_{Fi1}, respectively, K is the jump rate from one layer to one of the two neighboring layers per unit time, $\nu(\phi)$ is the resonance frequencies for different ϕ sites given by Eq. (6.2) with $\bar{\eta}$ set to zero for simplicity and φ, the incremental azimuthal displacement of the molecule w.r.t the one in the neighbouring layer, as illustrated in Fig. 6.15(b). In the clock model, the disposition of molecules in the 3- and 4-layer structures gives $\varphi = 120°$ and $90°$, respectively. The magnetization formed after a solid echo $(90^\circ_x - \tau - 90^\circ_y - \tau)$ sequence[42] is given by[43]

$$M(t) = Re[\vec{1} \cdot e^{\mathbf{\Lambda}(t+\tau)}(e^{\mathbf{\Lambda}\tau}\mathbf{M}^0)^*]. \tag{6.8}$$

Initially all sites are equivalent, and for convenience one can take the initial condition $\mathbf{M}^0_{Fi2} = [\frac{1}{4}, \frac{1}{4}, \frac{1}{4}, \frac{1}{4}]$ and $\mathbf{M}^0_{Fi1} = [\frac{1}{3}, \frac{1}{3}, \frac{1}{3}]$. The observed NMR spectrum is a linear superposition of magnetization from all the smectic layers within one pitch, and can be obtained by FT the following

$$F(t) = \int_0^{2\pi} (\frac{dz}{d\phi}) M(t) e^{-\sigma^2 t^2/2} d\phi \tag{6.9}$$

where $d\phi/dz$ is given by the distribution of the azimuthal angle along the pitch (z) axis to account for possible distortions in the helical structure. By setting t=0 in Eq. (6.9), $F(0) = I(\theta, 2\tau)$ is the intensity of echo maximum.

6.2.2. *Free energy consideration*

It is convenient to write the free energy density f for chiral SmC* phases using a two-component tilt vector $\vec{\xi} = (\xi_1, \xi_2) = (\sin\theta_0 \cos\phi, \sin\theta_0 \sin\phi)$, which is the projection of $\vec{n}$ onto the xy-plane (the layer plane). Without an external field, f is given by[27,39,44]

$$f = f_A + \frac{1}{2}a_1(\xi_1^2 + \xi_2^2) + \frac{1}{4}b(\xi_1^2 + \xi_2^2)^2 - \Lambda(\xi_1 \frac{\partial \xi_2}{\partial z} - \xi_2 \frac{\partial \xi_1}{\partial z}) + \frac{K_{33}}{2}[(\frac{\partial \xi_2}{\partial z})^2 + (\frac{\partial \xi_1}{\partial z})^2] \tag{6.10}$$

where f_A is the equilibrium free energy density of the SmA phase, Λ is the coefficient of the Lifshitz terms, which are responsible for the helicoidal modulation, and a_1 is used to explain the linear temperature variation in θ_0 angle, while b, Λ and K_{33} are temperature independent. In general, three Frank elastic constants are needed to describe elastic deformations in LC: splay (K_{11}), twist (K_{22}) and bend (K_{33}). Only the bend distortion has a significant effect in the SmC* phase. The bend distortion causes the f to increase by δf given by the last term in Eq. (6.10). When the molecules in the smectic layers change their tilt angles, the thickness of the smectic layers must vary accordingly. This process drastically increases the entropy of the LC system. Therefore, it is hard for molecules to change their tilt angles. Thus, θ_0 does not change significantly, while the azimuthal angle ϕ can vary by the H field to give a nonlinear spatial profile $\phi(z)$. This leads to the so-called "constant amplitude approximation" (CAA). This approximation may not be applicable when θ_0 is small such as near the SmC*-SmA transition. Under the CAA approximation, the nonequilibrium f can be calculated from Eq. (6.10) to give[28,33]

$$f(z) = -\Lambda \sin^2\theta_0 \frac{\partial \phi}{\partial z} + \frac{1}{2}K_{33}\sin^2\theta_0 (\frac{\partial \phi}{\partial z})^2. \tag{6.11}$$

To find the ground state of the SmC* phase, the free energy F per unit volume $F = \int_0^{p_0} f(z)dz/p_0$ (p_0 is the pitch length) is minimized w.r.t. ϕ and $\partial\phi/\partial z$ using the Euler-Lagrange minimization method, i.e.

$$\frac{\partial F}{\partial \phi} - \frac{d}{dz}(\frac{\partial F}{\partial (d\phi/dz)}) = 0, \tag{6.12}$$

This produces $d^2\phi/dz^2 = 0$, which leads to a helix with a pitch length defined by $d\phi/dz = 2\pi/p_0$. Putting this definition back to Eq. (6.11) and minimizing the free energy $F(p_0)$ w.r.t. p_0, the ground state pitch length p_0 is obtained:

$$p_0 = 2\pi K_{33}/\Lambda. \tag{6.13}$$

When $\vec{H}$ is applied to a SmC* phase, one should also include the magnetic coupling term f_H in Eq. (6.11):

$$f_H = -\frac{1}{2\mu_0}\Delta\chi(\vec{H}\cdot\vec{n})^2 \tag{6.14}$$

where μ_0 is the permeability of free space. In an angle dependent NMR study, Eq. (6.14) can be rewritten using Eq. (6.1)

$$f_H = -\frac{1}{2\mu_0}\Delta\chi H^2(\frac{1}{2}\sin^2\theta\sin^2\theta_0\cos 2\phi + \frac{1}{2}\sin 2\theta\sin 2\theta_0\cos\phi) + f_{equil}(\theta,\theta_0) \tag{6.15}$$

where $f_{equil}(\theta,\theta_0)$, a function of θ and θ_0, contains the equilibrium terms under the CAA approximation. Here only the terms depending on ϕ angle, i.e. nonequilibrium terms, are of interest. Hence, the nonequilibrium (or phase-dependent) part of $f(\theta)$ for any θ w.r.t. $\vec{H}$ can be written as[11,21,28]

$$f = -\sin^2\theta_0\Lambda\frac{d\phi}{dz} + \sin^2\theta_0\frac{1}{2}K_{33}(\frac{d\phi}{dz})^2 - \frac{1}{2\mu_0}\Delta\chi H^2(\frac{1}{2}\sin^2\theta\sin^2\theta_0\cos 2\phi + \frac{1}{2}\sin 2\theta\sin 2\theta_0\cos\phi) \tag{6.16}$$

After the Euler-Lagrange minimization (Eq. (6.12)) of the f, the well-known Sine-Gordon equation for ϕ is derived:

$$2\xi'^2\frac{d^2\phi}{dz^2} + \frac{d}{d\phi}(\cos 2\phi + 4A\cos\phi) = 0 \tag{6.17}$$

where $A = \cot\theta\cot\theta_0$, $\xi' = \xi\frac{\sqrt{2}}{\sin\theta}$, and $\xi = \sqrt{\frac{\mu_0 K_{33}}{|\Delta\chi|H^2}}$ is the magnetic coherence length. By integrating the above equation over ϕ, one obtains the ϕ distribution

$$\frac{d\phi}{dz} = \frac{1}{\xi'\kappa}\sqrt{1 + 4A - \kappa^2(\cos 2\phi + 4A\cos\phi)} \tag{6.18}$$

where κ is a constant of integration which defines the degree of helical distortion ($0 \le \kappa \le 1$). When $\kappa = 1$, the helix is completely unwound, while $\kappa = 0$ means no distortion (see Eq. (6.18)). The pitch length of the soliton-like deformed helix is obtained by integrating Eq. (6.18)

$$p = \xi' k E_1(\kappa) = \xi'\kappa \int_0^{2\pi} \frac{d\phi}{\sqrt{(1+4A) - \kappa^2(\cos 2\phi + 4A\cos\phi)}} \tag{6.19}$$

where $E_1(\kappa)$ is defined by the above equation. To relate the κ value to H, we need to minimize F w.r.t. κ.[36] Although the minimization involves tedious calculations, the result is useful and is derived here. Substituting Eq. (6.18) into Eq. (6.16), the average free energy per unit volume (F) becomes

$$F = -\frac{\sin^2\theta_0 K_{33}}{2p} \int_0^p \left(\frac{4\pi}{p_0}\frac{d\phi}{dz} - \left(\frac{d\phi}{dz}\right)^2 + \frac{1}{\xi'^2}(\cos 2\phi + 4A\cos\phi)\right)dz$$

where p_0, given by Eq. (6.13), is a constant. By changing the integration variable from z to ϕ, F can be rewritten as

$$F = \frac{\sin^2\theta_0 K_{33}}{2}\frac{1}{p}\left[\int_0^{2\pi}\left(\frac{4\pi}{p_0} - \frac{d\phi}{dz}\right)d\phi + \int_0^{2\pi}\frac{1}{\xi'^2}(\cos 2\phi + 4A\cos\phi)\frac{dz}{d\phi}d\phi\right]. \tag{6.20}$$

The first integral I in the bracket of the above equation gives

$$I = \frac{8\pi^2}{p_0} - \frac{1}{\xi'\kappa}E_2(\kappa), \tag{6.21}$$

where $E_2(\kappa) = \int_0^{2\pi}\sqrt{(1+4A) - \kappa^2(\cos 2\phi + 4A\cos\phi})d\phi$, while the second integral II in the bracket is

$$\begin{aligned} II &= \frac{1}{\xi'\kappa}\int_0^{2\pi}\frac{\kappa^2(\cos 2\phi + 4A\cos\phi)}{\sqrt{(1+4A) - \kappa^2(\cos 2\phi + 4A\cos\phi)}}d\phi \\ &= \frac{1}{\xi'\kappa}[(1+4A)E_1(\kappa) - E_2(\kappa)]. \end{aligned} \tag{6.22}$$

Hence, the average free energy becomes

$$F = -\frac{\sin^2\theta_0 K_{33}}{2}\left\{\frac{8\pi^2}{p_0\xi'\kappa E_1(\kappa)} + \frac{1}{\xi'^2\kappa^2}\left[(1+4A) - 2\frac{E_2(\kappa)}{E_1(\kappa)}\right]\right\}. \tag{6.23}$$

To minimize F w.r.t κ, one also needs the relation:

$$\frac{d}{d\kappa}\left(\frac{E_2(\kappa)}{\kappa}\right) = -\frac{(1+4A)E_1(\kappa)}{\kappa^2}. \tag{6.24}$$

Using Eqs. (6.23)-(6.24), one has

$$\frac{dF}{d\kappa} = \left\{\frac{8\pi^2}{p_0\xi'} - \frac{2}{\xi'^2}\frac{E_2(\kappa)}{\kappa}\right\}\frac{d}{d\kappa}\left(\frac{1}{\kappa E_1(\kappa)}\right) \tag{6.25}$$

from which one obtains the following relation:

$$\frac{E_2(\kappa)}{\kappa} = \frac{4\pi^2\xi'}{p_0} = \frac{4\sqrt{2}H_c^0}{\sin\theta H} \tag{6.26}$$

where $H_c^0 = \frac{\pi\Lambda}{2K_{33}}\sqrt{\frac{K_{33}}{\mu_0|\Delta\chi|}}$ is the critical magnetic field to unwind the SmC* helix when $\vec{H}$ is applied perpendicular to the helical axis. Because our study focuses on small θ angles, the value of A is large. It is better to define

$$e(\kappa) = \frac{E_2(\kappa)}{2\sqrt{A}} = \int_0^{2\pi} \sqrt{(1+1/4A) - k^2(\cos 2\phi/4A + \cos\phi)}\,d\phi \tag{6.27}$$

to rewrite Eq. (6.26) as

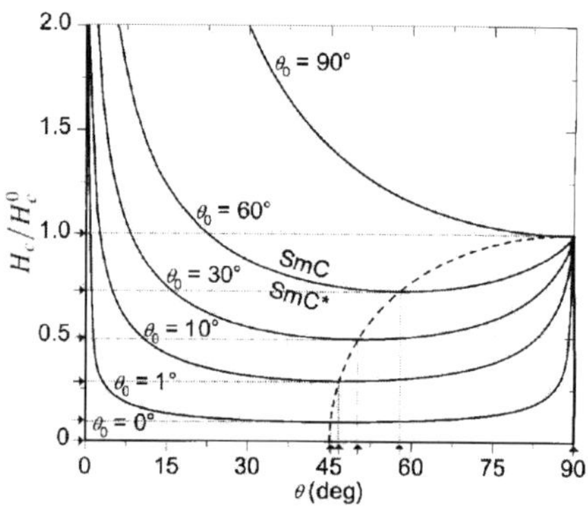

Fig. 6.5. The critical field H_c/H_c^0 as a function θ for some typical tilt angles θ_0. The dashed line shows the minimum H_c for a given θ_0. The angles θ and θ_0 are given in Fig. 6.3[21] (Reproduced with permission of APS).

$$\frac{\kappa}{e(\kappa)} = m\sqrt{\sin 2\theta} \tag{6.28}$$

where $m = H/(4\sqrt{\tan\theta_0}H_c^0)$. Now $e(1) = 2\sqrt{2}\sqrt{\tan\theta\tan\theta_0}[\sqrt{1+A} + A\ln\frac{1+\sqrt{1+A}}{\sqrt{A}}]$; the critical field H_c at $\theta > 0$ is[21]

$$\frac{H_c}{H_c^0} = \sin^{-1}\theta[\sqrt{1+A} + A\ln\frac{1+\sqrt{1+A}}{\sqrt{A}}]^{-1}. \tag{6.29}$$

Figure 6.5 reproduces a plot of H_c/H_c^0 versus θ for selected tilt angles θ_0. It is interesting to note that for a fixed θ_0, H_c/H_c^0 versus θ curve goes through a very shallow minimum. That is, the critical magnetic field H_c to unwind the SmC* helix depends on the $\vec{H}$ direction w.r.t. the helical axis. Except for $\theta_0 = 90°$, it is easier to unwind the helix when $\vec{H}$ is not perpendicular to the helical axes. The minimal critical field to unwind the helix is represented by a dashed line in Fig. 6.5. This provides a means to measure H_c^0 at relatively low magnetic fields. For the NMR field of 9.4T used in our laboratory, we have encountered two FLC which follow Eq. (6.28) to different extent w.r.t. θ. For a particular tilt angle θ_0 in our two samples, the NMR field in one case lies below the shallow minimum, so that the helical structure remains wound[47] always up to $\theta = 90°$, while in the other, the NMR field lies above the shallow minimum such that angular dependent spectra can only be achieved[22] for small rotation angles.

6.2.3. *Helical unwinding/winding transitions in a magnetic field*

To study the SmA-SmC* transition, f given in Eq. (6.10) is too simple to explain several experimental observables, such as the temperature behaviors of the pitch and the polarization. Over the past two decades, Eq. (6.10) has been modified by adding several phenomenological terms. These terms have to do with polarization on the smectic plane, $\vec{P} = (P_x, P_y)$. A first modification of the f for SmC* phase was made by Pikin and Indenbom,[48] who had included the flexoelectric parameter, μ, and a coupling term between θ_0 and the magnitude of polarization P_0 with C_0 being the piezoelectric coefficient:

$$
\begin{aligned}
f(z) = {} & \frac{1}{2}a_1(\xi_1^2 + \xi_2^2) + \frac{1}{4}b(\xi_1^2 + \xi_2^2)^2 - \Lambda\left[\xi_1\left(\frac{d\xi_2}{dz}\right) - \xi_2\left(\frac{d\xi_1}{dz}\right)\right] \\
& + \frac{1}{2}K_{33}\left[\left(\frac{d\xi_2}{dz}\right)^2 + \left(\frac{d\xi_1}{dz}\right)^2\right] + C_0(P_x\xi_2 - P_y\xi_1) \\
& + \frac{1}{2\chi_e}(P_x^2 + P_y^2) - \mu\left[P_x\left(\frac{d\xi_1}{dz}\right) + P_y\left(\frac{d\xi_2}{dz}\right)\right]
\end{aligned}
\tag{6.30}
$$

where χ_e is the dielectric susceptibility. The last three terms are new, and C_0 and μ are related to the chirality, that cause the symmetry breaking w.r.t the Z_M axis. Again a_1 is the only temperature dependent coefficient, viz. $a_1 = \alpha_1(T - T_0)$ with T_0 being the SmA-SmC transition temperature. By approximating the term $\sin\theta_0 \approx \theta_0$ and defining $\vec{q} = 2\pi/\vec{p}$ the pitch vector, then

$$
\begin{aligned}
\xi_1 &= \theta_0\cos(qz); & \xi_2 &= \theta_0\sin(qz); \\
P_x &= -P_0\sin(qz); & P_y &= P_0\cos(qz).
\end{aligned}
\tag{6.31}
$$

Finally f for the SmC* phase can be written as:

$$f_{C^*}(z) = \frac{1}{2}a_1\theta_0^2 + \frac{1}{4}b\theta_0^4 - \Lambda q\theta_0^2 + \frac{1}{2}K_{33}\theta_0^2 q^2 + \frac{1}{2\chi_e}P_0^2 - \mu P_0\theta_0 q - C_0 P_0\theta_0. \tag{6.32}$$

By minimizing the f w.r.t. θ_0, q, P_0, the SmA-SmC* transition temperature is found as:[38]

$$T_{c(SmA-SmC^*)} = T_0 + \frac{C_0^2 k\chi_e + \Lambda^2 + 2C_0\chi_e\Lambda\mu}{\alpha_1(k - \mu^2\chi_e)} = T_0 + \frac{\Delta}{\alpha_1} \tag{6.33}$$

where Δ is defined in Eq. (6.33). Also one can easily obtain θ_0^{min}, q^{min} and P_0^{min}, the values of tilt, wave vector and polarization, respectively at which f is minimum. Note that $f(z)$ is minimum in the SmA phase when θ_0, q, P_0 are zero, while $f(z)$ is null when values of θ_0, q and P_0 are different from zero in the SmC* phase. In the presence of a magnetic field, Eq. (6.32) has an additional magnetic contribution, i.e.

$$f_{C^*}(z) = \frac{1}{2}a_1\theta_0^2 + \frac{1}{4}b\theta_0^4 - \Lambda q\theta_0^2 + \frac{1}{2}K\theta_0^2 q^2 + \frac{1}{2\chi_e}P_0^2 - \mu P_0\theta_0 q - C_0 P_0\theta_0 + \frac{H^2\chi_a(\theta_0^2 - 1)}{2\mu_0}. \tag{6.34}$$

Note that only the magnetic term remains in the f of SmA phase. The magnetic contribution causes the SmA-SmC* transition temperature to decrease by $\Delta T \propto H^2$, viz.

$$T_{c(SmA-SmC^*)} = T_0 + \frac{C_0^2 k\chi_e + \Lambda^2 + 2C_0\chi_e\Lambda\mu}{\alpha_1(k - \mu^2\chi_e)} - \frac{H^2\chi_a}{\alpha_1\mu_0}. \tag{6.35}$$

Now a strong magnetic field does not simply shift the SmA-SmC* transition temperature, but may induce a new unwound phase, Sm$\bar{C}^*$(uSmC*) phase. Moreover, the experimental P/θ ratio decreases rapidly close to the transition, and the experimental pitch values do depend on the temperature. These could not be explained by the above calculations. To overcome these problems when approaching the transition, several authors[4,5] have proposed to add additional ad hoc terms, e.g. a biquadratic coupling between the tilt and polarization. The stability of the SmC* phase in the vicinity of the transition can also be increased by introducing a sixth power term of θ_0. The improved f can yield both more realistic polarization and pitch, though their theoretical expressions are more complex. However, the SmA-SmC* transition temperature is not influenced much by these new terms. For this reason, our numerical treatment of transition temperatures in Ref. 38 was based on the simpler f given in Eq. (6.30).

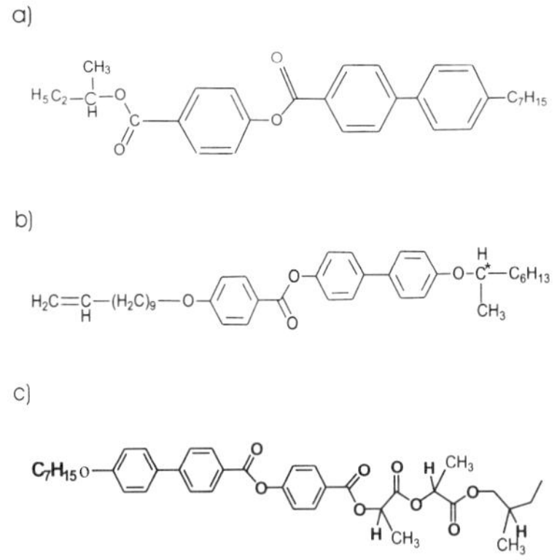

Fig. 6.6. Molecular structures of MBHB (a), 11EB1M7 (b), and ZLL 7/* (c).

In the literature, examples for unwinding the helix by $\vec{H}$ with a θ different from zero exist.[21] The unwinding of the ferroelectric helical axis at zero angle[37] has recently been investigated in three smectogens:[38] MBHB, 11EB1M7 and ZLL7/* (see Fig. 6.6), and their experimental findings have been rationalized by a simple phenomenological model. In particular, the effect of the H field on the phase transition temperatures among the different smectic phases has been discussed. In the uSmC* phase, chiral molecules are aligned on the average parallel to $\vec{H}$, while the smectic layers are tilted with molecules forming a tilt angle, x, (see Figure 6.7) with the layer normal. The angle x is basically the same as the tilt angle, θ_0, but a different symbol is used to distinguish the two phase structures in a magnetic field. The f of uSmC* is given by:

$$f_{uC^*}(z) = \frac{1}{2}a_2(\xi_1^2 + \xi_2^2) + \frac{1}{4}b(\xi_1^2 + \xi_2^2)^2 + C_0(P_x\xi_2 - P_y\xi_1) + \frac{1}{2\chi_e}(P_x^2 + P_y^2) - \frac{H^2\chi_a}{2\mu_0} \tag{6.36}$$

After substituting Eq. (6.31) (with θ_0 being replaced by x), one obtains

$$f_{uC^*}(z) = \frac{1}{2}a_2x^2 + \frac{1}{4}bx^4 + \frac{1}{2\chi_e}P_0^2 - C_0P_0x - \frac{H^2\chi_a}{2\mu_0} \tag{6.37}$$

To write down the above equation, several approximations are used in the uSmC* phase: 1. the temperature dependenct coefficient "a" in the Landau-de Gennes theory[49] is different in the SmC* and uSmC* phases, thus $a_1 \neq a_2$. The effect of decreasing temperature on the mesophase structure is indeed different for SmC* and uSmC* phases, because in the former

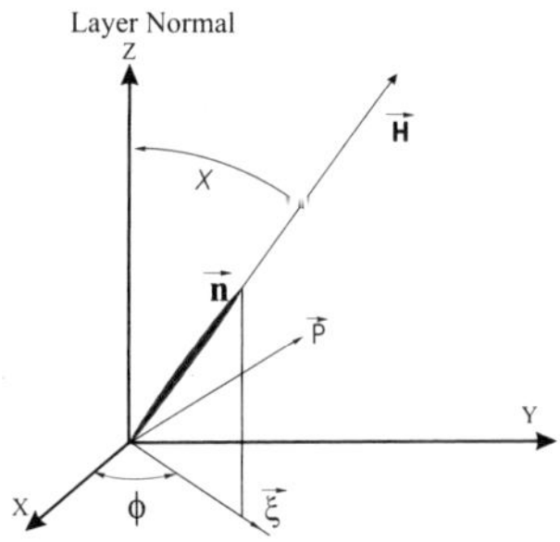

Fig. 6.7. A smectic plane of the uSmC* phase and the plane polarization, $\vec{P}$ are shown to lie on xy plane, together with the magnetic field, $\vec{H}$, the local director, $\vec{n}$, the characteristic *tilt*, x, and azimuthal, ϕ angles.

the change in the tilt orientation involves the molecules, while in the latter the smectic layer planes; 2. all other coefficients for uSmC* phase are taken same as in the SmC* phase, except the terms responsible for the helicoidal arrangement, i.e. those associated with K_{33}, Λ, and μ vanish since the derivatives in Eq. (6.30) are zero; 3. polarization P_0 is not averaged to zero, because of the smectic chirality. By minimizing the f_{uC^*} w.r.t. x and P_0, one obtains

$$T_{c(SmA-uSmC^*)} = T_0 + \frac{C_0^2\chi_e}{\alpha_2}. \tag{6.38}$$

Now the transition temperature between the uSmC∗ and SmC∗ phase is:

$$\begin{aligned} T_{c(uSmC^*-SmC^*)} &= T_0 + \frac{\Lambda^2 + 2C_0\chi_e\Lambda\mu + C_0^2\chi_e^2\mu^2}{(\alpha_1-\alpha_2)\cdot(k-\mu^2\chi_e\mu^2)} - \frac{H^2\chi_a}{(\alpha_1-\alpha_2)\cdot\mu_0} \\ &= T_0 + \frac{1}{(\alpha_1-\alpha_2)}\left[\Delta - \frac{H^2\chi_a}{\mu_0} - C_0^2\chi_e\right]. \end{aligned} \tag{6.39}$$

The transition between uSmC* and SmC* phases can occur only if the temperature expressed in Eq. (6.39) is less than the temperatures given in Eqs. (6.35) and (6.38). The three transition temperatures are summarized in Fig. 6.8. For the case of $H = 0$, one may note: 1. $T_{c(SmA-SmC^*)} > T_{c(SmA-uSmC^*)}$ means that the SmC* phase must appear at a temperature higher than that of the uSmC*, and not *vice versa*, thus leading to $\frac{\Delta}{\alpha_1} > \frac{C_0^2\chi_e}{\alpha_2}$, and 2. $T_{c(uSmC^*-SmC^*)} > T_{c(SmA-SmC^*)}$ means that the unwinding of the helical axis can not be followed by a rewinding by simply decreasing temperature, thus leading to $\alpha_1 - \alpha_2 > 0$.

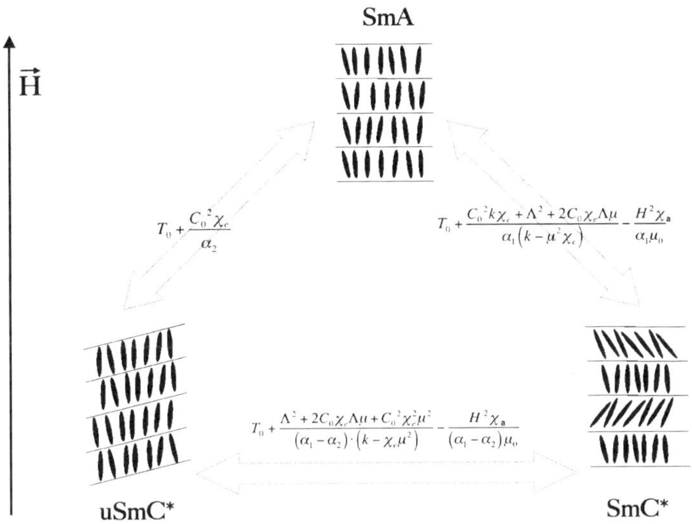

Fig. 6.8. Expressions of the transition temperatures among the three mesophases[38] in presence of the magnetic field. (Reproduced with permisson of Wiley)

For the case $H > 0$, one notices that 1. $T_{c(SmA-SmC^*)}$ depends on H^2 and decreases with increasing H, 2. $T_{c(SmA-SmC^*)} >$ (or $<$) $T_{c(SmA-uSmC^*)}$ depending on the fact that H is smaller (or larger) than the critical magnetic field H_c, where $H_c = \sqrt{\frac{\mu_0}{\chi_a}\left(\Delta - \frac{\alpha_1}{\alpha_2} C_0^2 \chi_a\right)}$, 3. at $H = H_C$ the three phase transition temperatures between SmA, SmC* and uSmC* are equal, yielding the LP point, and 4. $T_{c(uSmC^*-SmC^*)}$ depends on H^2 and decreases with increasing H, but this decrease is faster than for $T_{c(SmA-SmC^*)}$, because of the term $(\alpha_1 - \alpha_2)$. The above predictions will be discussed in light of the experimental findings presented below.

6.3. SmC* phase of 8BEF5

The sample (S)-[4-(2-methylbutyl)phenyl]-4'-n-noctylbiphenyl carboxylate (8BEF5) deuterated in the chiral chain and the phenyl moiety has been studied and reported previously by Catalano et al.[45] Its molecular structure and phase sequence are shown in Fig. 6.9(a). Solid echo sequence with the 90° pulse width equal to 3 μs was used to record the angle dependent spectra. The number of scans typically was 128 and the recycle time between scans was 1.5 s. Typical angular dependent spectra for the SmC*

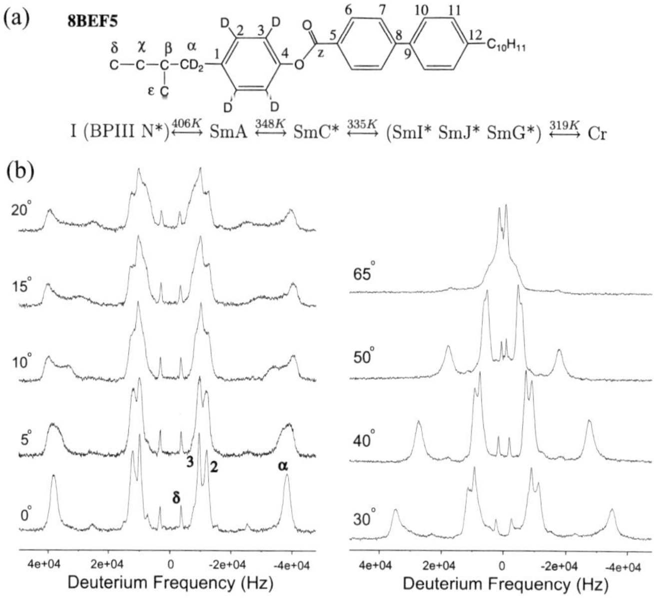

Fig. 6.9. (a) Molecular structure of 8BEF5 and its phase sequence. (b) Angular dependent ^{2}H NMR line shapes for some typical rotation angles in the SmC* phase of 8BEF5 at 340.4K.

phase of an aligned 8BEF5 sample are shown in Fig. 6.9(b).[22] The spectra can only be recorded up to $\theta = 65°$ at 340.4 K, because the H field causes orientational flows in the sample at larger θ angles. The orientational flow depends on the viscosity and $\Delta\chi$ of the sample, and the magnetic torque on the molecules is higher at large rotation angles. Our experiments show that there is no observable field-induced sample reorientation and layer destruction if θ angle is less than 30° in the studied temperature region. This has been checked experimentally by reproducing the spectrum after half an hour. To decrease the possibility of sample reorientation at larger rotation angles, the sample is first realigned at $\theta = 0$ before recording a spectrum for $30° < \theta < 65°$. One possible way to stabilize the sample is to place LC between thin glass plates to give a multiple-layer sandwich.[20] This has not been done here since small rotation angles ($< 25°$) suffice. Only the α deuteron signal has been studied in details (other deuterium peaks suffer severe overlaps), and its line shapes at small rotation angles bear some resemblance with those predicted in Fig. 6.4 for the reason given below. The observed intensities of $s_2^{\pm}$ singularities decrease with increasing θ angle, and disappear at a critical rotation angle θ_c. For the temperature shown in Fig. 6.9, θ_c is about 25°. The present case corresponds to the NMR field being above the shallow minimum shown in Fig. 6.5. At θ_c, the helix is completely unwound at 9.4 T field. Therefore, $s_1^\pm$ and $s_2^\pm$ singularities can only be observed when $\theta < \theta_c$. This phenomenon can be explained by distortions of the helical structure due to the magnetic field, and provides a method to measure the H_c^0 in the SmC* phase of 8BEF5.

The undistorted angular dependent DNMR spectrum can be calculated using Eq. (6.3), which is modified here to take into account helical distortions:

$$F(t) \propto \int_0^{2\pi} \int P(\theta_0' - \theta_0) \cos[2\pi\nu(\theta, \theta_0', \phi)t] d\theta_0' e^{-\sigma^2 t^2/2} (\frac{dz}{d\phi}) d\phi \quad (6.40)$$

where $P(\theta_0' - \theta_0)$, the distribution of the local $\vec{n}$, is assumed to be Gaussian[46]

$$P(\theta_0' - \theta_0) = \frac{1}{\varpi\sqrt{2\pi}} e^{-(\theta_0' - \theta_0)^2/2\varpi^2} \quad (6.41)$$

with θ_0' being the instantaneous tilt angle, and ϖ the full width at half maximum (FWHM) of the distribution. This is needed in our sample, as it is not too viscous and has some director fluctuations[19] to produce different broadening effects on singularities $s_1^\pm$ and $s_2^\pm$ (Eq. (6.5)). In the SmC* phase, the order parameter S is relatively high, making ϖ small. Letting $\Delta\theta_0 = \theta_0' - \theta_0$, one can simplify Eq. (6.40) using:

$$\begin{aligned}
&\int P(\theta_0' - \theta_0) \cos(2\pi\nu(\theta, \theta_0', \Phi)t) d\theta_0' \\
&\cong \int \frac{1}{\sqrt{2\pi}\varpi} e^{-(\Delta\theta_0)^2/2\varpi^2} \cos[2\pi(\nu + \frac{\partial\nu}{\partial\theta_0}\Delta\theta_0)t] d\Delta\theta_0 \\
&= \int \frac{1}{\sqrt{2\pi}\varpi} e^{-(\Delta\theta_0)^2/2\varpi^2} \cos(2\pi\nu t) \cos(2\pi\frac{\partial\nu}{\partial\theta_0}\Delta\theta_0 t) d\Delta\theta_0 \\
&= e^{-\frac{1}{2}\varpi^2(2\pi\frac{\partial\nu}{\partial\theta_0}t)^2} \cos(2\pi\nu t)
\end{aligned} \quad (6.42)$$

The line shape is simulated by numerically FT the following f.i.d.

$$F(t) \propto \int_0^{2\pi} \cos[2\pi\nu(\theta, \theta_0, \phi)t] R'(\phi, t) (\frac{dz}{d\phi}) d\phi \quad (6.43)$$

where $R'(\phi, t) = e^{-\sigma^2 t^2 - (2\pi\varpi\frac{\partial\nu}{\partial\theta_0})^2 t^2}$, and $dz/d\phi$ is given by Eq. (6.18). By simulating the line shapes, θ_0 and $\kappa(\theta)$ can be extracted. H_c^0 is then calculated from Eq. (6.28) by plotting $\kappa(\theta)$ versus $\sqrt{\sin 2\theta}$.

We have measured in the SmC* phase of 8BEF5 between 336K and 344K. Figure 6.10 shows a typical set of angular dependent spectral patterns

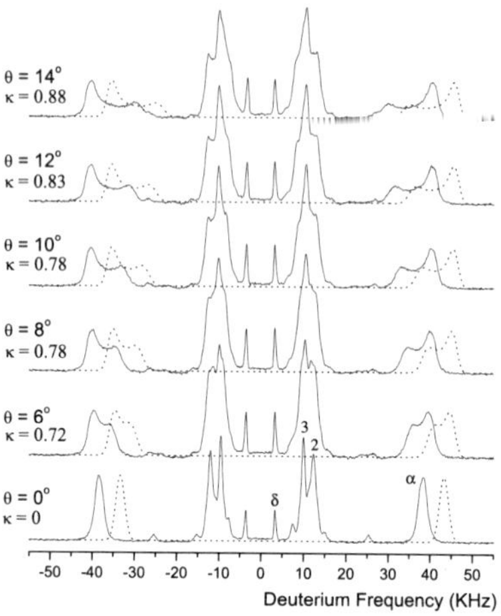

Fig. 6.10. Experimental (solid lines) and simulated (dashed lines) angular dependent ^{2}H NMR line shapes in the SmC* phase of 8BEF5 at 340.4 K. The simulated line shapes are shifted by 5 kHz for clarity, together with their best κ's.

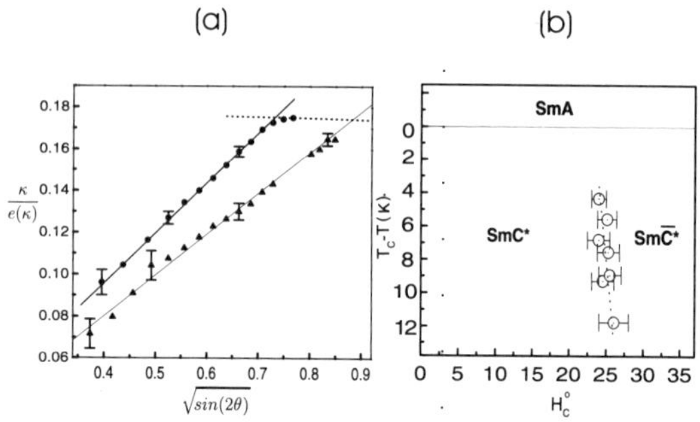

Fig. 6.11. (a) Plot of $\kappa/e(\kappa)$ versus $\sqrt{(\sin 2\theta)}$ for 8BEF5 at 338.7 K (triangle) and 343.7 K (circle). Dashed line indicates $\kappa = 1$ with $\theta_0 = 10°$, at which the helical structure is completely unwound. The solid lines are linear fitting lines given by Eq. (6.28). (b) The temperature dependent phase line between the SmC* and the Sm$\bar{C}$* phase.

observed at 340.4 K ($\theta_0 = 12.4°$) and at small rotation angles ($\theta < 25^o$). When simulating the angular dependent DNMR line shapes at a particular temperature, the tilt angle θ_0, time-averaged coupling constant $\bar{\nu}_Q$, line broadening parameters σ, and ϖ are set constant by trial and error, while the integration constant κ for a particular θ angle is varied to give the "best" simulated line shape as shown in Fig. 6.10. The κ values at three temperatures are listed in Table 6.1. We note that the θ_0 values are

Table 6.1. κ values at temperatures 342 K, 341 K and 336 K in the SmC* phase, starting from left.

θ(deg)	κ	θ(deg)	κ	θ(deg)	κ
6.5	0.62	6.0	0.65	4.5	0.55
7.5	0.66	7.0	0.72	6.0	0.62
9.0	0.70	8.0	0.75	7.0	0.67
10	0.76	9.0	0.78	8.0	0.69
11	0.72	10	0.82	9.0	0.74
12	0.77	11	0.85	10	0.76
13	0.80	12	0.88	11	0.79
14	0.82	13	0.90	12	0.82
15	0.89	14	0.93	13	0.84
16	0.92	15	0.95	14	0.85
18	0.94	16	0.96	17	0.90
21	0.95	19	0.996	21	0.96
23	0.98	—	—	23	0.98

about 2 degrees higher than those reported previously.[45] Within the fitting accuracy, $\bar{\nu}_Q$ increases slightly upon decreasing temperature as expected. Fig. 6.11(a) shows a plot of $\kappa/e(\kappa)$ versus $\sqrt{(\sin 2\theta)}$ for two temperatures, and the slope m gives the critical field at a particular temperature. We have studied the SmC* phase as a function of temperature to map out the $H_c(T)$ phase line between the SmC* and the Sm$\bar{C}^*$ phase. This is shown as a plot of the temperature versus the critical field in Fig. 6.11(b). It is clear that Landau theory is adequate in explaining the spectral behaviors in the SmC* phase of 8BEF5. However, near the unwound region Landau theory fails as the observed critical field is higher than that predicted by the theory (see Fig. 6.11(a) for deviation of data points from the solid line). The director distribution (Eq. (6.41)) appears to be important in reproducing the experimental spectral patterns.

In the SmC* phase of 8BEF5, the helical twist and the layer structure can prevent the molecules from following the field direction for some small θ angles at temperatures not close to the SmA-SmC* transition (348 K). At the same time, the tilt angle is large enough at temperatures far from the transition to measure the relative intensities of the two edge singularities. This is not the case as the temperature rises towards the phase transition

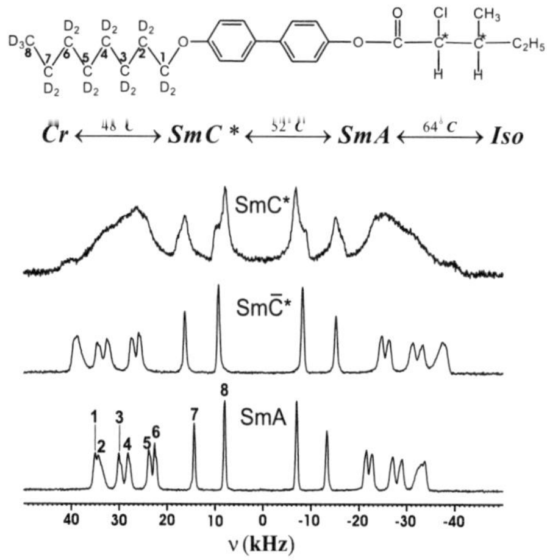

Fig. 6.12. Typical ^2H NMR spectra collected in different LC phases of BP8Cl. Peak assignments are labeled in the lower spectrum. (Reproduced with permission of APS)

where the observation of spectral patterns even at small θ becomes impossible. It is for this reason that $H_c(T)$ line can not be followed up to the SmC*-SmA transition. It is interesting to note from Fig. 6.11(b) that H_c^0 decreases slightly upon increasing the temperature in qualitative agreement with the theory (Fig. 6.1).[31] The magnitude of H_c^0 for 8BEF5 is on the order of 25 T.

6.4. SmC* phase of BP8Cl

The molecular structure of 4'-(octyloxy)-d$_{17}$ biphenyl-4-yl2-chloro-3-methylpentanoate (BP8Cl) with the deuteron labeled sites and its transition temperatures, together with typical deuterium 1D spectra obtained without rotating the sample[47] are shown in Fig. 6.12. Here, we focus on the methyl deuterons at the last C_8 carbon whose deuteron spin-lattice relaxation times are largest. Fig. 6.13(a) shows a typical set of angle dependent spectral patterns observed at 311.5 K . At this temperature, the sample shows no observable field-induced sample reorientation and layer destruction for θ up to 90°. It should be noted that the present sample behaves quite differently from 8BEF5, where the C_α deuteron(s) shows edge singularities only at small rotation angles ($\theta < 25°$). Hence, this sample can provide a better test of the Landau theory. When simulating the angular dependent DNMR line shapes, the tilt angle and time-averaged coupling

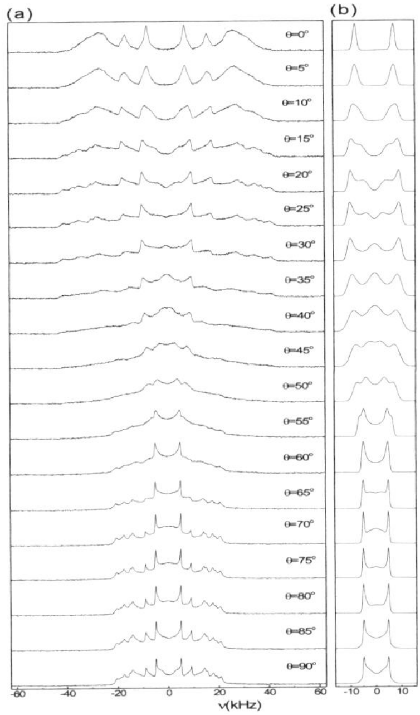

Fig. 6.13. (a) Experimental and (b) Simulated angular dependent ^{2}H NMR line shapes in the SmC* phase of BP8Cl at 311.5 K. (Reproduced with permission of APS)

constant are set at $\theta_0 = 23.3 \pm 0.2^0$ and $\bar{\nu}_Q = 10.0 \pm 0.5$kHz. For a particular angle θ, κ is again varied to give the "best" simulated line shape for C_8 deuterons as shown in Fig. 6.13(b). The line broadening parameters σ and ϖ are also varied to reflect the observed line widths. We note that κ is non-zero when the sample is rotated in the magnetic field, thereby indicating that the helical structure has suffered some soliton-like distortions. Although the line broadening parameter σ has been varied between $1.6 - 9$ at different θ angles, this variation only changes the line width without affecting the spectral features. Thus, κ has been uniquely determined. Figure 6.14 shows plots of $\kappa/e(\kappa)$ versus $\sqrt{(\sin 2\theta)}$ for θ between 0° and 90° at two different temperatures, and according to Eq. (6.28), their slopes determine the critical fields at these temperatures. As seen in Fig. 6.14, H_c^0 decreases slightly with increasing temperature in qualitative agreement with the theoretical prediction.[31] The magnitude of H_c^0 for BP8Cl is on the order of 30 T at low temperatures, which is about a factor of 3 higher than the NMR field. At higher temperatures, the NMR field in the direction of the pitch axes can in fact unwind the helices. Thus, the critical field must be lower

(~9 T) to give a Sm$\bar{C}$* (uSmC*) phase. At 9.4 T the SmC* phase only exists in a narrow temperature range, thereby excluding a detailed study of $H_c(T)$ in BP8Cl. Besides the sample viscosity and flow behaviors being rather sensitive to temperature, other factor(s) that is responsible for the uSmC*-SmC* transition will be discussed below.

In summary, spectral simulation of NMR spectra can give the tilt angles, and the critical field with the aid of Landau theory. Thus, the DNMR spectral patterns provide a powerful means to study soliton-like distortions in pure FLC, and to provide information on K_{33} and the Lifshitz term Λ in the Landau free energy expansion.

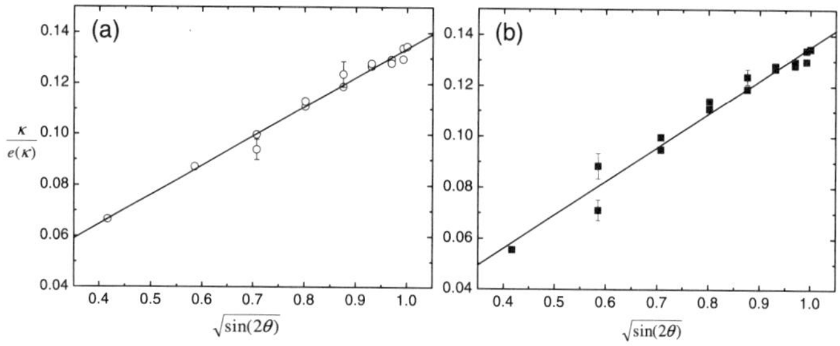

Fig. 6.14. Plot of $\kappa/e(\kappa)$ versus $\sqrt{(\sin 2\theta)}$ forBP8Cl at (a) 308.5 K(circle) and (b) 311.5 K(square). The solid lines are fitted lines according to Eq. (6.28). The H_c^0 is found to be 30±2T and 27±5T at 308.5K and 311.5 K, respectively. (Reproduced with permission of APS)

6.5. Anticlinic phases

Unwinding and field-induced distortions in AFLC are more complex due to the fact that chirality favors a twist as well as an anticlinic ordering. Hence, unwinding of the helices in AFLC involves also a transition from anticlinic to synclinic ordering. The first calculation of the antiferro- to ferro-electric transition in an electric field has revealed drastic changes.[50] Besides requiring a larger electric field than that for unwinding the ferroelectric helix, the larger electric field can noticeably change the tilt angle (the so called electroclinic effect), making the CAA approximation invalid. Calculations to account for the chirality and the change in the tilt angle in an electric field have recently appeared.[35] Because of the complexity of field-induced phenomena in AFLC, experimental investigations of soliton-like distortions are still lacking thus far. The first magnetic field effect on ferrielectric phases of 10B1M7 is now given.

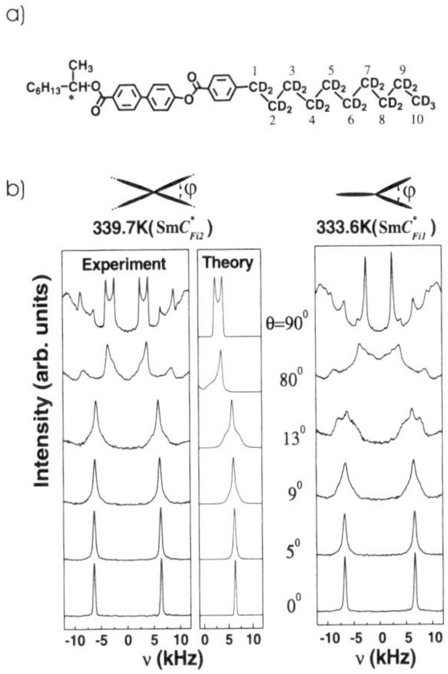

Fig. 6.15. (a) Molecular structure of 10B1M7 and its deuteron site labels. * indicates the chiral center. (b) Experimental angular dependent DNMR lineshapes (only emphasize C_{10} methyl signal) in the SmC^*_{Fi1}(334.6 K) and SmC^*_{Fi2}(339.7 K) phase of 10B1M7. Also shown are simulated lineshapes (positive halves only) for 339.7 K using Ω=5kHz, $\bar{\nu}_Q$=8.2kHz, $\theta_0 = 22.6°$, $\varphi = 20°$ and σ=1kHz.

6.5.1. SmC^*_{Fi} phases of 10B1M7

The chiral liquid crystal 1-methylheptyl 4'-(4-n-decyloxybenzoyloxy)-biphenyl-4-carboxylate (10B1M7) shown in Fig. 6.15(a) exhibits a sequence of mesophases at different ranges of temperature. The synthesis and its characterization have been reported elsewhere,[51,52] and the transition temperatures of 10B1M7-d_{21} determined by our ^{2}H NMR study are as follows:

$$\mathrm{Iso} \overset{397.5K}{\longleftrightarrow} \mathrm{Sm}A \overset{378.4K}{\longleftrightarrow} \mathrm{SmC}^* \overset{342K}{\longleftrightarrow} \mathrm{SmC}^*_{Fi2} \overset{335K}{\longleftrightarrow} \mathrm{SmC}^*_{Fi1} \overset{330.3K}{\longleftrightarrow} \mathrm{SmC}^*_A \overset{312.1K}{\longrightarrow} \mathrm{Sm}J^*$$

Only one ferri-phase was identified by an earlier optical study.[52] Our DNMR work has provided direct evidence of a phase transition between two ferrielectric phases in this compound,[53] viz. 3-layer (SmC^*_{Fi1}) and 4-layer (SmC^*_{Fi2}) superlattices. Because of the spectral overlap problem among most deuterium peaks, only the signal from methyl deuterons at C_{10} is emphasized for the study of ferrielectric phases.

Typical angular dependent spectra collected in the two ferrielectric phases are shown[53] in Fig. 6.15. In the small rotation angle region, their line shapes are quite similar to those observed in SmC^*_A phase, i.e. the two singularities[21,40] in the SmC^*_A phase were modulated by the interlayer diffusion of molecules to produce a lineshape showing a central peak together with two shoulders.[20] The spectra of the SmC^*_{Fi1} phase show obvious structural distortions at high rotation angles, since the intensity of one edge singularity is much higher than that of the other, especially at $\theta = 90°$. This means that the critical field H_c^0 to unwind the helical structure in SmC^*_{Fi1} phase is much smaller than that in SmC^*_{Fi2} and SmC^*_A. phase. This difference in H_c between the two ferri phases provides direct evidence of a phase transition between them in this compound. To simulate DNMR spectra in SmC^*_{Fi} phases, one can evaluate the evolution of the magnetization for one base unit (3- or 4-layer) based on Eq. (6.8). The observed line shape can be calculated by FT the f.i.d given by Eq. (6.9). The edge singularities are predicted to collapse to a single peak for the perfect "clock model"[53] at $\theta = 90°$. Hence only "asymmetric clock model" can explain the observed line shapes at this θ for the two ferri-phases. The simulated line shapes in SmC^*_{Fi2} (339.7K) with $dz/d\phi \approx$ constant are also plotted in Fig. 6.15 for direct comparison. From the spectra obtained at $\theta = 90°$, a critical field $H_c^0 > 40$ T is estimated in the SmC^*_{Fi2} phase of 10B1M7.

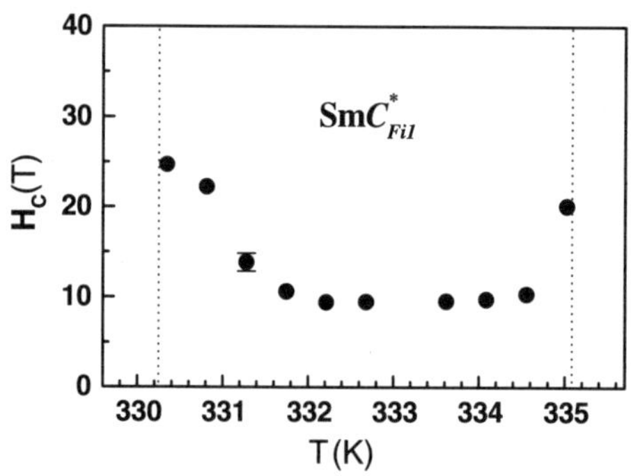

Fig. 6.16. Plot of critical field H_c versus the temperature in the SmC^*_{Fi1} phase of 10B1M7. (Reproduced with permission of APS)

Simulation of line shapes in the SmC^*_{Fi1} phase is harder due to its low H_c values and complex helicoidal distortions. This has prevented the determination of κ by spectral simualtions. However, κ has been obtained by fitting the decay curve of the echo intensity $I(\theta, 2\tau)$ for small θ angles.[53]

Table 6.2. The transition temperatures (°C) among the chiral smectic phases for MBHB and 11EB1M7, based on the DNMR measurements at different magnetic field strengths. A "X" is used when data are unavailable.

	H =2.35 T	H =4.70 T	H =7.05 T	H =8.89 T	H =9.39 T
MBHB					
SmA-SmC*	x	86.0	–	x	–
SmC*-uSmC*	x	–	86.0	x	86.0
uSmC*-SmC*	x	–	74.0	x	<55.0
11EB1M7					
SmA-SmC*	101	x	100.5	–	–
SmC*-uSmC*	–	x	–	105.5	105.5
uSmC*-SmC*	–	x	–	96.0	<74.0

In addition, to determine the critical field H_c, one can simulate the observed spectrum at $\theta = 90°$, since the soliton-like distortion of the helical structure at this rotation angle is simple and similar to those in SmC* and SmC$_A^*$ phases,[20,21] i.e.

$$\frac{dz}{d\phi}|_{\theta=90^\circ} = \frac{Constant}{\sqrt{1-\kappa^2 cos^2\phi}} \tag{6.44}$$

where κ is a fitting parameter, and related to H_c^0 by $H_c^0 = H \cdot E(\kappa)/\kappa$. Here $H = 9.4$ T is the NMR magnetic field and $E(\kappa)$ is the elliptical integral of the second kind. The critical field values in SmC$_{Fi1}^*$ phase, plotted in Fig. 6.16, are obtained by fitting the $\theta = 90°$ spectra taken at various temperatures. It can be seen that there are sudden jumps of the H_c field near the two transition temperatures from the SmC$_{Fi1}^*$ phase.

6.6. Magnetic field effect on the transition to SmC* phase

The unwinding of the helical axis by the NMR field applied along the pitch axes of smectogens BP8Cl,[47] MBHB, 11EB1M7 and ZLL7/* have recently been reported.[38] Several NMR spectrometers were used to record the DNMR spectra in the last 3 partially deuterated FLC. The temperature dependent phase structures under different NMR field strengths for MBHB and 11EB1M7, and their phase transition temperatures for winding/unwinding the helices among smectic phases are summarized in Table 6.2. The dependences of transition temperatures on the applied H field can in principle be predicted by using Eqs. (6.35), (6.38) and (6.39) provided that the molecular parameters K, α_1, b, C_o, χ_a, χ_e, μ, μ_0 and Λ,

Table 6.3. Some parameters from DOBAMBC.

K (N)	$1.0 \cdot 10^{-11}$	Λ (J/m^2)	$1.72 \cdot 10^{-5}$
b (J/m^3)	$3.0 \cdot 10^5$	χ_e (F/m)	$2.5 \cdot 10^{-11}$
C_0 (V/m)	$3.0 \cdot 10^6$	μ_0	$4\pi \cdot 10^{-7}$
μ (V)	$1.0 \cdot 10^{-1}$	α_1 $(J/(m^3K))$	$\approx 10^4$
χ_a	$4\pi \cdot 10^{-6}$		

are known from data for similar mesogens.[32] These values for DOBAMBC (p-(n-decyloxybenzyl)-p-amino-(2-methylbutyl)cinnamate) are given in Table 6.3 and used here. As an example, (H, T) phase diagram for MBHB is schematized in Fig. 6.17 on the basis of the treatment outlined in the theoretical section by setting $(\alpha_1 - \alpha_2) = 100$. The three dashed lines refer to the three H fields in which DNMR studies were done. It is noted that the SmA-SmC* transition temperature decreases by less than 0.5° with the H field increasing from 0 to 20 T when arbitrarily picking T_0= 100°C. This prediction is perhaps seen in 11EB1M7. The SmA-uSmC* transition temperature is, however, very much dependent on the values of H field, and $\alpha_1 - \alpha_2$[38] as seen in Eq. (6.39). When $H > H_c$, a transition from the SmA to uSmC* phase occurs. In this case, a second transition can occur at a low temperature from uSmC* to SmC* phase. Note that both these transition temperatures are strongly dependent on the $(\alpha_1 - \alpha_2)$ value. It is interesting to see that the phase diagram shown in Fig. 6.17 alreadly contains the essential features shown in the (H, T) diagram of DOBAMBC,[37] even at the present level of approximations in the Landau-de Gennes theory.

6.7. Conclusion

This chapter has reviewed how the angle dependent DNMR spectral patterns obtained in FLC samples can be used to study soliton-like distortions of the helical structure in chiral SmC* phases and/or subphases SmC^*_{Fi1} and SmC^*_{Fi2} phases. Spectral simulations can take helicoidal distortions into account via the Landau theory. Based on these simulations, one can obtain the critical field H_c and the jump diffusive rates[54] in the ferrielectric phases. As predicted by the theory, except with a tilt angle $\theta_0 = 90°$, the field strength to unwind the SmC* helices is not minimal with the H being applied perpendicular to the helical axis. In fact, the H_c field versus the rotation angle goes through a shallow minimum. We have chosen two FLC samples such that our NMR field of 9.4 T is higher than the minimum in

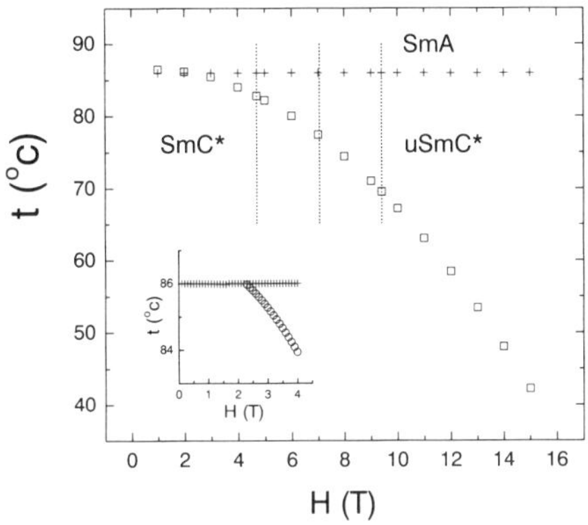

Fig. 6.17. Plot of (H, T) phase diagram of MBHB. All the transition temperatures reported were calculated using expressions summarized in fig. 6.8. The inset figure shows an expanded low field rgion.

one sample, while lower in the other for the particular tilt angle in each sample. Of course when the H_c field is too high for a particular FLC such as 10B1M7, it is then impossible to study soliton-like distortions in the SmC* phase. The H_c field in the chiral subphase SmC^*_{Fi1} is found lower than that in the SmC^*_{Fi2} and SmC^*_A phases of 10B1M7. As a consequence, two ferrielectric phases have been identified by NMR in this sample. The potential of using angle dependent DNMR spectra to extract structural and dynamic information in complex condensed phases is real. In addition, unwinding of helices in SmC* phases has been investigated by DNMR for the case when the NMR field is parallel to the helical axes. A simple Landau-de Gennes theory has been given and used to give the transition temperatures among the smectic phases. Finally, the necessity to deuterate the molecule of interest could, however, be difficult to meet, as deuteration process could be both time consuming and costly. However, the intramolecular nature of the nuclear quadrupole interaction of a deuteron spin has made the spectral interpretation particularly simple and attractive.

Acknowledgment

The financial support of Natural Sciences and Engineering Council of Canada is gratefully acknowledged. The author thanks Drs. A. Marini, J. Xu and J. Zhang for their works on some of materials given here.

References

1. R. B. Meyer, L. Liebert, L. Strzelecki, P. Keller, *J. Phys. Paris* **36**, L69 (1975).
2. J. P. F. Lagerwall, F. Giesselmann, *ChemPhysChem* **7**, 20 (2006).
3. A. D. L. Chandani, E. Gorecka, Y. Ouchi, H. Takezoe, A. Fukuda, *Jpn. J. Appl. Phys.* **28**, L1265 (1989).
4. C. C. Huang, S. Dumrongrattana, *Phys. Rev. A* **34**, 5020 (1986).
5. T. Carlsson, B. Zeks, A. Levstik, C. Filipic, I. Levstik, R. Blinc, *Phys. Rev. A* **36**, 1484 (1987).
6. A. Cady, J. A. Pitney, R. Pindak, L. S. Matkin, S. J. Watson, H. F. Gleeson, P. Cluzeau, P. Barois, A.-M. Levelut, W. Caliebe, J. W. Goodby, M. Hird, C. C. Huang, *Phys. Rev. E* **64**, 050702R (2001).
7. P. Mach, R. Pindak, A.-M. Levelut, P. Barois, H. T. Nguyen, C. C. Huang, L. Furenlid, *Phys. Rev. Lett.* **81**, 1015 (1998).
8. M. Gorkunov, S. Pikin, W. Haase, *JETP letters* **69**(3), 243–249 (1999).
9. L. S. Hirst, S. J. Watson, H. F. Gleeson, P. Cluzeau, P. Barois, R. Pindak, J. Pitney, A. Cady, P. Johnson, C. Huang, A. M. Levelut, G. Srajer, J. Pollmann, W. Caliebe, A. Seed, M. R. Herbert, J. W. Goodby, M. Hird, *Phys. Rev. E* **65**, 041705 (2002).
10. T. Akizuki, K. Miyachi, Y. Takanishi, K. Ishikawa, H. Takezoe, A. Fukuda, *Jpn. J. Appl. Phys.* **38**(8), 4832 (1999).
11. I. Muševič, M. Škarabot *Phys. Rev. E* **64**, 051706 (2003).
12. M. Čepič, E. Gorecka, D. Pociecha, B. Žekš, H. T. Nguyen, *J. Chem. Phys.* **117**(4), 1817 (2002).
13. P. M. Johnson, D. A. Olson, S. Pankratz, T. Nguyen, J. Goodby, M. Hird, C. C. Huang, *Phys. Rev. Lett.* **84**, 4870 (2000).
14. D. Konovalov, H. T. Nguyen, M. Čopič, S. Sprunt, *Phys. Rev. E* **64**, 010704R (2001).
15. P. Mach, R. Pindak, A.-M. Levelut, P. Barois, H. T. Nguyen, H. Baltes, M. Hird, K. Toyne, A. Seed, J. W. Goodby, C. C. Huang, L. Furenlid, *Phys. Rev. E* **60**, 6793 (1999).
16. M. Čepič, B. Žekš, *Mol. Cryst. Liq. Cryst.* **263**, 61 (1995).
17. V. L. Lorman, *Liq. Cryst.* **20**, 267 (1996).
18. A. Fukuda, Y. Takanishi, A Isozaki, K. Ishikawa, H. Takezoe, *J. Mater. Chem.* **4**, 997 (1994).
19. R. Y. Dong, *Nuclear Magnetic Resonance of Liquid Crystals*, Springer-Verlag, New York 1994.
20. B. Zalar, A. Gregorovic, R. Blinc, *Phys. Rev. E* **62**, R37–R40 (2000).
21. B. Zalar, A. Gregorovic, M. Simsic, A. Zidansek, R. Blinc, *Phys. Rev. Lett.* **80**, 4458–4461 (1998).
22. J. Xu, C. A. Veracini, R. Y. Dong, *Chem. Phys. Lett.* **416**, 47 (2005).
23. J. Xu, C. A. Veracini, R. Y. Dong, *Phys. Rev. E* **72**, 051703 (2005).
24. R. Seidin, D. Mukamel, D. W. Allender, *Phys. Rev. E* **56**, 1773–1783 (1997).
25. E. E. Burnell, C. A. de Lange, *NMR of ordered liquid crystal*, Kluwer Academic Publishers, 2003.

26. M. Yamashita, H. Kimura, H. Nakano, *Mol. Cryst. Liq. Cryst.* **68**, 79 (1981).
27. A. Michelson, *Phys. Rev. B* **16**, 577–584 (1977).
28. M. Skarabot, R. Blinc, I. Musevic, A. Rastegar, T. Rasing, *Phys. Rev. E* **61**, 3961–3968 (2000).
29. L. Lam, J. Prost, *Solitons in Liquid Crystals*, Springer-Verlag: New York, 1991.
30. M. Yamashita, *J. Phys. Soc. Jpn* **57**, 2051 (1988).
31. B. Kutnjak-Urbanc, B. Zeks, *Phys. Rev. E* **48**, 455–464 (1993).
32. I. Musevic, R. Blinc, B. Žekš, *The Physics of Ferroelectric and Antiferroelectric Liquid Crystals*, World Scientific, 2000.
33. L. Benguigui, A. E. Jacobs, *Phys. Rev. E* **49**, 4221–4227 (1994).
34. F. Ghoddoussi, M. A. Pantea, P. H. Keyes, R. Naik, P. P. Vaishnava, *Phys. Rev. E* **68**, 051706 (2003).
35. P. V. Dolganov, V. M. Zhilin, *Phys. Rev. E* **77**, 031703 (2008).
36. P. J. Collings, M. Hird, *Introduction to Liquid Crystals*, Taylor & Francis, 1997.
37. I. Musevic, B. Zeks, R. Blinc, T. Rasing, P. Wyder, *Phys. Rev. Lett.* **48**, 192–195 (1982).
38. V. Domenici, A. Marini, C. A. Veracini, J. Zhang, R. Y. Dong, *ChemPhysChem* **8** 2575 (2007).
39. J. W. Goodby, R. Blinc, N. A. Clark, S. T. Lagerwal, M. A. Osipov, S. A. Pikin, T. Sakurai, K. Yoshino, B. Žekš *Ferroelectric Liquid Crystals: Principles, Properties and Applications*, Gordon and Breach, 1991.
40. B. G. Wu, J. W. Doane, *J. Magn. Reson.* **75**, 39 (1987).
41. H. M. McConnell, *J. Chem. Phys.* **28**, 430 (1958).
42. J. H. Davis, K. R. Jeffrey, M. Bloom, M. I. Valic, T. P. Higgs, *Chem. Phys. Lett.* **42**, 390–394 (1976).
43. A. J. Vega, Z. Luz, *J. Chem. Phys.* **42**, 1615 (1965).
44. R. Blinc, B. Žekš, *Phy. Rev. A* **18**, 740 (1978).
45. D. Catalano, L. Chiezzi, V. Domenici, M. Geppi, C. A. Veracini, R. Y. Dong, K. Csorba, *Macromol. Chem. Phys.* **203**, 1594–1601 (2002).
46. A. Firouzi, D. J. Schaefer, S. H. Tolbert, G. D. Stucky, B. F. Chmelka, *J. Am. Chem. Soc.* **119**, 9466–9477 (1997).
47. J. Zhang, A. Ferraz, A. C. Ribeiro, P. J. Sebastiao, R. Y. Dong, *Phys. Rev. E* **74**, 061704 (2006).
48. V. L. Indenbom, S. A. Pikin, *Usp. Fiz. Nauk* **125**, 251 (1978).
49. P. G. de Gennes, *The Physics of Liquid Crystals*, Clarendon, Oxford 1975.
50. A. Roy, N. V. Madhusudana,it Eur. Phys. J. E **60**, 319 (2000).
51. D. Catalano, M. Cavazza, L. Chiezzi, M. Geppi, C. A. Veracini, *Liq. Cryst.* **27**(5), 621–627 (2000).
52. J. Goodby, J. Patel, E. Chin, *J. Mater. Chem.* **2**(2), 197–207 (1992).
53. J. Xu, C. A. Veracini, R. Y. Dong, *Phys. Rev. E* **73**, 011705 (2006).
54. R. Y. Dong, *Thin Solid Films*, **517**,1367 (2008).

Chapter 7

Deuteron NMR Study of the Effects of Random Quenched Disorder in 12CB Silica Dispersions

Daniele Finotello[1,2] and Vishal Pandya[1]

[1] *Department of Physics, Kent State University, Kent, OH 44242, USA*

[2] *Division of Materials Research, National Science Foundation, Arlington, VA 22230, USA*

We studied quenched disorder effects on the 12CB liquid crystal (LC) upon dispersion of hydrophilic silica spheres of $7nm$ diameter and $300m^2/g$ surface area, with hydroxyl groups covering the surface. The dispersions form a gel if the silica density exceeds the percolation threshold. For low densities and cooling from the isotropic phase, the LC director in the void volume is parallel to and follows the NMR field; a well defined and stable LC configuration forms. When a complete silica network forms and if the dispersion orientation in the field is changed, some silica links break and a few smectic layers realign. Mostly, the aerosil locks in the LC configuration which follows the $P_2(\cos\theta)$ dependence. Cooling outside the field, the smectic phase spectrum is a Pake powder pattern, consistent with a disordered structure, namely, an isotropic distribution of smectic domains. The field anneals the aerosil-induced random disorder up to a certain density beyond which, silica disordering effects dominate. The quenching of smectic 12CB at $\rho_S \approx 0.055g/cm^3$, is about one order of magnitude less than that for 8CB, a LC possessing a nematic phase. The aerosil induced static disorder, as quantified by the percent of LC molecules forming the powder pattern, is linearly dependent on the silica density.

7.1. Introduction

The study of random disorder effects on a variety of soft condensed matter systems has received considerable interest because of their fundamental and practical relevance. Some notable examples of systems widely investigated over the last few decades include superfluid helium in silica aerogels,[1] doped magnetic systems,[2] elastomers,[3] and liquid crystals (LC) imbedded in several different complex geometries.[4] More recently, LC-silica aerosil dispersions have been shown to be rich systems to study, as evidenced by many distinct experimental and theoretical efforts.[5–12] LC- aerosil systems are of great importance as they allow to access "soft" phases of continuous symmetry directly coupled to surfaces and external fields.[11] When nano-size random silica particles are imbedded in a LC, their presence destroys the LC long range orientational order and as a result there is an exponentially decaying local nematic order with a correlation length dependent on the disordering density of silica added to the LC.[13,14]

By varying the density ρ_S, of aerosil particles (spheres) dispersed in a LC, it is possible to access aerosil dispersions with different regimes of disorder. When only a few aerosil particles are added to the LC, lowest possible densities, the dispersions can be effectively treated, and the particles manifest their presence, in the same manner as when a small amount of impurities are introduced in a LC; the few silica particles introduced freely "float" in the LC material. If the silica density is increased, with ρ_S ranging from ~ 0.01 up to $0.1 g/cm^3$, because of the hydroxyl groups covering their spherical surface, the spheres are able to bond forming a network of strands of variable strength.[15] The network is known as a "soft-gel" since the bonding among the spheres is relatively weak; it may break under shear or in the presence of an external field, and the "soft-gel" network may rearrange itself on moderate time scales. At higher silica densities, the silica links are stronger, the network is thus more rigid, and the dispersion that is formed is known as a "stiff-gel"; at these densities, the dispersions behavior resembles that first found for LC imbedded in the more severe aerogel confinement.[16] In short, while the addition of aerosil introduces random disorder in the LC system, the richness of the so-formed dispersions is evident since the quantity of disorder can be controlled through the amount of silica added (the silica density of the dispersion), thus allowing investigations ranging from the no-gel to the soft-gel into the stiff-gel regime. No other LC confining system offers such a wide range where to probe disordering effect.

It was noted that aerosil dispersions exhibit similarities to another widely studied random disordering system: silica aerogel, a different type of fractal silica gel that has been used to confine LC.[11] Aerogels are formed by a reaction-limited aggregation process in which the basic silica units are chemically fused together. Aerosil and Aerogel have several similarities: fractal-like nature of the gel structure, surface composition and chemical properties, and eventually, silica density. The crucial difference between the two systems is their different response to elastic strains since aerogels have much greater shear modulus than aerosil gels and thus elastic strains cannot be easily overcome.[15] In *LC*-aerogel systems the silica network is fixed in space, and the elastic strains imposed by the random anchoring surfaces are fully quenched. The disorder introduced by even the lowest density aerogels is so severe that phase transition features are dramatically smeared.[11,16] In contrast, in aerosil gels, because of the weaker and better controlled random disorder that is introduced, elastic strains can be partially annealed thus reducing some of the silica-induced disorder effects. Also, aerosil allows the possibility to grow a network directly within the *LC* yielding structures of considerably higher porosity.[12] Ultimately, a tremendous advantage of aerosil systems is the ease of producing nearly arbitrary and widely ranging silica densities that permits the study of the competition between silica-induced disordering effects and external fields induced ordering effects.

Particularly noteworthy is to discuss the difference between aerogel and aerosil systems from the perspective of thermodynamic studies. For example, in octylcyanobiphenyl (8CB), the specific heat anomaly at the second order smectic-A to nematic (AN) transition, extremely prominent in bulk, is either absent or greatly broadened when confined to Aerogel, preventing the possibility of any critical behavior analysis.[13] In aerosil dispersions, phase transitions are affected differently depending on the silica density: for 8CB soft-gel dispersions, the specific heat at the AN transition is a sharp peak characterized by a critical exponent α that decreases monotonically with increasing ρ_S. The critical exponent changes from the pure bulk "effective" value towards the bulk $3D - XY$ value.[7,17,18] Effectively, the presence of silica particles weakens the coupling between the smectic and the nematic order parameter, eventually yielding the "true" critical exponent of the transition; this is identical to what occurs in homologous bulk systems with varying nematic range. At the weakly first order nematic to isotropic (NI) transition, and over a narrow density range, two closely spaced and sharp specific heat peaks were found; this could be related to molecular dynamics present in the system. At higher silica densities stiff-gel regime, the specific heat at both transitions consists of a single round peak suggesting an elastic strain smearing of the transitions. The random disorder introduced by the aerosil is fully quenched in this density regime. Quenched disorder refers to the fact that disorder variables are frozen in space on the time-scale of the experiments and thus do not anneal away. This is in contrast to what is seen in the soft-gel regime, where the random silica network and the elastic strains are partially annealed by the confined *LC* molecules through the breaking and rearranging of the aerosil strands, and disorder is not fully quenched. Theoretically, the effect of silica aerosil on the first-order nematic-isotropic is to shift the transition temperature, allowing the enthalpy jump to be phenomenologically described in the framework of the Landau–de Gennes theory. It has been shown that the aerosil network dispersed in the liquid crystal phase creates additional curvature perturbations, resulting in linear and nonlinear slopes both in the transition temperature shift and in the enthalpy jump.[19]

Investigation of dispersions has focused in the use and effects on smectic *LC*. X-ray scattering studies[20] were performed with 8CB confined by strained colloidal silica gels. The gels, comprised of aerosil particles, possess an anisotropic structure that stabilizes long-range nematic order in the *LC* while introducing random field effects that disrupt the smectic transition. The short-range smectic correlations that form within this environment are inconsistent with the presence of a topologically ordered state predicted for 3D random field XY systems; they are quantitatively like the correlations of smectics confined by isotropic gels. Detailed analysis reveals that the quenched disorder suppresses the anisotropic scaling of the smectic correlation lengths observed in the pure *LC*. It was further concluded that the observed smectic behavior was dictated by random fields coupling directly to the smectic order while fields coupling to the nematic director play a subordinate role. Also,[21] the short-range order which remains when the isotropic to smectic-A transition is perturbed by aerosil was studied using high-resolution synchrotron x-ray diffraction. The gels were created in situ in decylcyanobiphenyl (10CB), which has a strongly first-order isotropic to smectic-A transition and no nematic phase. The silica effects were determined by detailed analysis of the temperature and gel density dependence of the smectic structure factor. Compared to studies of the continuous nematic to smectic-A transition, the aerosil gel appeared to pin, at random, the phase of the smectic density modulation. For the isotropic to smectic-A transition the same gel perturbation yielded different results. The smectic correlation length decreased more slowly with increasing random-field variance in good quantitative agreement with the effect of a random pinning field at a transition from a uniform phase directly to a phase with one-dimensional translational order.

While most of the studies discussed above focused on how controlled disorder affects phase and configurational transitions, mostly using cyanobiphenyl *LC* dispersions, it is important to note that several groups have done extensive research on the optical applications of anisotropic aerosil gels, memory effects, and electric field-aided annealing.[22–24] In particular, Milette and coworkers[25] undertook a wide-line deuteron nuclear magnetic resonance (DNMR) study of both hydrophilic and hydrophobic silica dispersions formed in a Schiff-base-type ($EBBA$) LC; their investigation which employed a polar LC that consequently interacts with the silica surfaces in a different manner than nCB LC do, shows many similarities to the work discussed in this chapter, but more so with the original study of Tao and Finotello[8] as $EBBA$ possesses a nematic phase like 8CB.

In this chapter we present a systematic DNMR investigation of random disorder effects introduced on the orientational order of the dodecylcyanobiphenyl smectic liquid crystal (12CB), by dispersing in it hydrophilic type $A300$ aerosil particles.[26,27] After giving experimental details related to the sample preparation and the measuring technique, we show results with 12CB dispersions; these are presented according to the density regime. A short conclusions section ends the chapter.

7.2. Experimental Details

7.2.1. *Aerosil properties and sample preparation*

Hydrophilic silica particles type $A300$, are small silica spheres $7nm$ in diameter with a specific surface area $a = (300 \pm 30)m^2/g$; they have hydroxyl groups covering their surface consisting of about 5 free $OH\ groups/nm^2$.[22] When dispersed in an organic media, the hydroxyl groups induce hydrogen bonding between the silica spheres leading to the formation of aerosil strands. It is known[7] that aerosil particles form a thixotropic gel or network by diffusion-limited aggregation once the silica density ρ_S exceeds $0.01g/cm^3$, the so-called "gelation" or percolation threshold. In the specific case of 8CB-aerosil dispersions, with ρ_S ranging from ~ 0.01 up to $0.1g/cm^3$,[15] a network known as a "soft-gel" forms: links develop among the spheres but they are relatively weak, can easily break under shear and rearrange on moderate time scales. At higher silica densities, the network becomes more rigid and the dispersions are known as "stiff-gels".

For this investigation we used the dodecylcyanobiphenyl liquid crystal, 12CB, which in bulk exhibits a transition from the SmA to the isotropic phase at $T_{AI} = 331.6K$, crystallizing at $321K$. We thus studied 12CB-aerosil dispersions ranging from below the gelation threshold, or no-gel regime, to the soft-gel regime, to the stiff-gel regime, up to silica densities comparable to those in rigid aerogels. *Table* 7.1 lists several parameters characterizing the 12CB-aerosil dispersions[7,8,11] where ρ = grams of SiO_2 per cm^3 of the total volume; ρ_S = grams of SiO_2 per cm^3 of LC; $l_0 = 2/a\rho_S$ is the average void size in the aerosil network formed by $7nm$ particles with $a = 300m^2/g$ for $A300$ hydrophilic aerosil; $\phi = \rho/\rho_S$ is the porosity or volume fraction of "pores"; $p = l_b a \rho_S$ is the percentage of 12CB molecules anchored by the SiO_2 surface with $l_b = 3.2nm$ for the thickness of a 12CB boundary layer under the assumption of a bilayer by bilayer formation ($p*$ is also given for comparative purposes with 8CB and it was calculated assuming a typical $2nm$ thick boundary layer.[8]) However, it is important to note that the average void size definition is only applicable in the no-gel and soft-gel regimes since further addition of aerosil particles does not change the specific surface area given by $a = 300 - 103.8\rho_S$.[11] As the aerosil density is increased, the voids could be bound on all sides resulting in the condition $a_{void} \rightarrow \frac{a_{void}}{3}$ with l_0 also increasing by a factor of 3 to $l_0 = 6/a\rho_S$; the condition for a closed void is estimated to occur for densities $\rho_S = 1.97g/cm^3$.

Table 7.1: Properties of the 12CB silica dispersions

$\rho(g/cm^3)$	$\rho_S(g/cm^3)$	$l_0(nm)$	$3l_0(nm)$	ϕ	$p(\%)$	$p*(\%)$
0.003	0.003	2219	6658	0.9986	0.28	0.18
0.005	0.005	1330	3991	0.9977	0.48	0.30
0.010	0.010	664	1991	0.9955	0.96	0.60
0.0158	0.0159	419	1258	0.9928	1.53	0.95
0.0176	0.0187	376	1129	0.9920	1.70	1.06
0.0185	0.0187	357	1070	0.9916	1.79	1.12
0.020	0.020	330	991	0.9909	1.94	1.21
0.025	0.0253	264	791	0.9886	2.43	1.52
0.0556	0.057	117	351	0.9747	5.48	3.42
0.1	0.105	64	191	0.9545	10.1	6.29
0.219	0.243	27	82	0.9005	23.3	14.6
0.300	0.347	19	58	0.8636	33.3	20.8
0.452	0.569	12	35	0.7944	54.7	34.2
0.600	0.825	8	24	0.7272	79.2	49.5
0.900	1.523	4	13	0.5909	–	91.4

The 12CB-aerosil dispersions were prepared according to procedures originally described in Ref. [6] where: (a) The deuterated LC is heated 5 to $10K$ above the clearing temperature; (b) The proper amount of aerosil particles is introduced in the LC; (c) High purity acetone is added to the

12CB and aerosil mixture; (*d*) The solution is placed in a heated ultrasonic bath where it is kept agitating in excess of one day; (*e*) Finally, the mixture is placed in a vacuum oven, slowly heated to about $360K$ where it is kept for several hours to ensure complete evaporation of the solvent. In cases when the solvent is not fully removed or if the solvent is not of the highest purity, the dispersion may exhibit a yellowish appearance and would be discarded.

7.2.2. *Network types and relevant lengthscales*

By dispersing aerosil silica spheres in a LC, the LC is subjected to effects induced by a random field. Aside from introducing a random distribution of surfaces and distortions in the local LC directors, the additional LC-surface interaction results in two primary effects altering the orientational order (OO). First, the hydroxyl groups on the surface and the polar head of the LC likely yield a homeotropic LC alignment at the silica surfaces. Anchoring of LC molecules implies a surface-induced order, a local paranematic state whose order decays with the distance from the surface. The magnitude of this "field" effect is fixed by the intrinsic properties of the surface and specific LC; increasing the silica density increases its extent. This effect exists in both the isotropic phase and in the lower temperatures mesophases, and for instance, may shift phase transitions to higher temperatures since more thermal energy is necessary to destroy the additional surface-induced order. Second, in the mesophases, there are elastic strains (ES) arising from competing surface interactions and the distortion of the local OO of the LC, resulting in a random disorder whose magnitude, as it involves many surfaces, likely increases with silica density. The surface-induced order and the random disorder thus compete; the latter is more important at higher silica density, fully dominating beyond a critical density. The configuration in the aerosil gels can be thought of as random networks where the hosted LC molecules separate into many random domains (voids) having a fractal-like structure that is expected to exhibit a wide distribution of void length scales characterized by a mean void size, l_0.[7] The effects on a smectic fluid can be expected to be more severe when compared to those on a nematic fluid on simple grounds that the formation of smectic layer is more likely to be affected by the presence of disorder.

In contrast to a random network, in an anisotropic fluid such as a nematic or smectic LC that is fully aligned by a strong external field, the diffusion field is anisotropic yielding an anisotropic gel (anisotropic aerosil network, AAN). Therefore, to characterize the voids in an anisotropic network, it is necessary to define two length scales, $l_{||}$ and $l_{\perp}$, where $l_{||}$ is the mean void length and $l_{\perp}$ is the mean void width, with $l_{||} > l_0$ and $l_{\perp} < l_0$. A parameter $\sigma \equiv l_{||}/l_{\perp}$ can then be introduced to quantify the anisotropic extent of the network, with a larger σ indicating a more anisotropic network. In Fig. 7.1 we present an artist rendition of the difference between a silica formed random network with an anisotropic one.

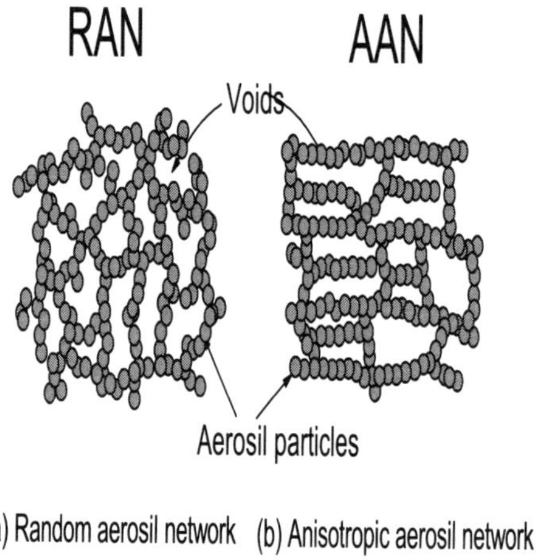

Fig. 7.1. Sketch illustrating the differences between a silica formed aerosil random network with an anisotropic aerosil network.

An anisotropic gel can also be formed through the annealing of a random network, as for instance, in the presence of external fields, which is termed field-aided annealing. Assuming a strong anchoring of the nematic director at the surfaces of aerosil particles forming the network, not only the surfaces act elastically on the director, but the director also acts on the surfaces and thus, on the network. Upon application of a NMR field, the confined LC molecules tend to align parallel to the field to minimize the free energy. This could lead to a rearrangement of the network through the breaking and re-linking of aerosil strands. If the aerosil network is not too rigid or if the mean void size is not too small, a random aerosil network (RAN) would anneal into an anisotropic one (AAN). Given a RAN, it could be expected that a stronger magnetic field would anneal it more easily resulting in an AAN of larger σ. Yet, with increasing aerosil density a RAN likely stiffens, the annealing becomes more difficult, and the anisotropy σ of the resulting annealed network decreases.

In an anisotropic network, if the magnetic field is to have an effect, it would be mostly determined by the smallest characteristic length of the voids: the mean void width $l_{\perp}$ and how it compares with the magnetic coherence length, ξ_M. In reality, in interconnected systems, the magnetic field affects the director configuration even when the mean void size is an order of magnitude smaller than ξ_M. This is to some extent a consequence of the wide distribution of void sizes, but more importantly, results from the interconnected nature of the gel voids. *LC* domains inside each void are not isolated from each other, rather, they are correlated to their neighboring domains; an independent (single) pore model[5] would not accurately predict the characteristics of the aerosil gels. For instance, in 8CB-aerogel systems, quasi-elastic light scattering results[28] demonstrated that nematic correlations occur over lengthscales comparable to 100 average void sizes. This is not surprising as order correlated to lengthscales greatly exceeding the confining size has also been found at the helium superfluid transition in Vycor and Aerogel glasses.[1,29]

Light scattering measurements in 6CB-aerosil dispersions by Bellini and coworkers[30] showed that the nematic director correlation length could grow to be one order of magnitude larger than the mean void size l_0. On the other hand, X-ray results for 8CB-aerosil[11] found that the smectic correlation length is comparable to l_0. These results suggest that when considering magnetic field ordering effects on aerosil dispersions, it is reasonable to assume that nematic domains effectively extend to lengthscales comparable to $10l_0$ or longer, while in the smectic phase each smectic domain does not exceed the void size. In this scenario, the magnetic field will have a much weaker influence on the gel structure in the smectic phase. A weaker field effect in the smectic phase can be understood since ξ_M, which in the smectic phase cannot be obtained from the single elastic constant approximation, is expected to be longer than in the nematic phase due to the larger elastic constants and higher viscosity in the SmA phase. Quenching will then occur at a lower silica density in smectic than in nematics; it will indeed be shown below that the density difference for the onset of quenched disorder is approximately one order of magnitude.

Finally, there is a significant difference between the *RAN* and the *AAN*. For the same aerosil density, the orientational order (and if appropriate, the translational order) of the *LC* in an *AAN* is likely of longer range than in a *RAN*. In an *AAN* the directors in neighboring domains are more alike than those in a *RAN*, and thus its nematic and smectic correlation lengths should be longer. Concomitant with the fact that there are fewer elastic strains in an AAN, a LC in an anisotropic network would be expected to possess a larger orientational order parameter than a LC in a random network.

7.2.3. *The Deuteron Nuclear Magnetic Resonance Technique for Confined Liquid Crystals*

DNMR spectra are obtained by a $\pi/2_x - \pi/2_y$ "solid echo" pulse sequence in a $B_0 = 9T$ magnet at the ^2H Larmor frequency ν_L =58.4MHz. The spectrum resulting from a domain of molecules with uniform director, namely, either a nematic or a smectic LC compound deuterated at a specific site, consists of two sharp absorption lines separated in frequency by:[8,16,31]

$$\delta\nu = \frac{1}{2}\delta\nu_0 Q\left[3\cos^2\theta - 1\right] \tag{7.1}$$

with $\delta\nu_0$ the maximum frequency splitting observable in a fully aligned bulk sample, Q the scalar orientational order parameter, and θ the angle between the director and the magnetic field. In bulk, $\theta = 0°$ due to the magnetic field-induced uniform alignment of the director along the field. In a confined LC, the quadrupolar splitting has a positional dependence $\delta\nu = \delta\nu(\overrightarrow{r})$ through a director structure $\theta = \theta(\overrightarrow{r})$ and an order parameter structure $Q(\overrightarrow{r})$. When such a positional dependence is taken into consideration, Eq. (7.1) becomes:

$$\delta\nu(\overrightarrow{r}) = \frac{1}{2}\delta\nu_0 Q(\overrightarrow{r})\left[3\cos^2\theta(\overrightarrow{r}) - 1\right] \tag{7.2}$$

For LC molecules oriented parallel to the direction of the applied field, $\theta(\overrightarrow{r}) = 0°$ and $\delta\nu(\overrightarrow{r}) = \delta\nu_0 Q$; if, on the other hand, all molecules are oriented perpendicular to the field, $\delta\nu(\overrightarrow{r}) = \frac{1}{2}\delta\nu_0 Q$ since $\theta(\overrightarrow{r}) = 90°$. Under confinement and thus in the presence of surfaces, the director structure induced by the surfaces would be affected by the external $9T$ magnetic field as long as the magnetic coherence length, $\xi_M = \sqrt{\frac{\mu_0 K}{B_0^2 \Delta\chi}} \simeq 520nm$, is less than the mean void size ($\xi_M < l_0$). In this single elastic constant approximation, K is the average Frank elastic constant and $\Delta\chi$ is a measure of the anisotropy of the magnetic susceptibility.[31]

DNMR spectra obtained in confined LC systems may also be affected by translational diffusion motion. The diffusion length measuring the average distance a molecule migrates over the time scale of a NMR measurement is estimated from[16]

$$x_0 = \sqrt{\frac{D}{\delta\nu_B}} \qquad (7.3)$$

with $\delta\nu_B$ the quadrupole frequency splitting of the bulk LC and D the average translational diffusion constant. Using typical values for cyanobiphenyls, $\delta\nu_B \simeq 40kHz$ and $D = 5 \times 10^{-12}m^2/s$, one finds $x_0 \simeq 150nm$ in the isotropic phase, decreasing by about an order of magnitude in the mesophases to $x_0 \simeq 15nm$.[16,32] If the mean void size is much larger than x_0, the $DNMR$ spectrum reflects the static director configuration and diffusion effects can be neglected. If the orientational order parameter and/or, the local director, varies considerably over a lengthscales comparable to x_0, there would be significant motional narrowing in the spectrum.

7.3. Experimental Results

7.3.1. *Low Silica Density Dispersions*

When the aerosil density is below the gelation threshold $\rho_S \simeq 0.01g/cm^3$, no thixotropic gel forms and the dispersions exhibit fluidity above the clearing point. No silica network forms, aerosil strands are isolated and float in the LC as impurities would. For the highest silica density dispersions within this regime, it might be possible that enough silica particles are present to nearly form a complete network of relatively large mean size voids ($l_0 \sim 1\mu m$). The links among silica strands would however be too weak to elastically strain the hosted LC and thus no thixotropic gel forms. In this regime the random disorder introduced is small thus the properties of these dispersions are not expected to differ much from those of the pure bulk LC. When 12CB -dispersions are cooled from the isotropic phase into the smectic phase, most but not all elastic strains (ES) are annealed at the transition through the rearrangement of aerosil particles. The dispersions are almost fully aligned by the magnetic field and the effect of random disorder on the orientational order while not substantial, it is still noticeable even for the lowest density studied. It will be stressed throughout the remainder of this chapter that the annealing process is not as effective in cooling from the isotropic phase directly into the smectic phase, as it is when the cooling is into the nematic phase.

The temperature dependence of the DNMR spectral patterns, except where noted, were taken in cooling from the isotropic phase in the presence of the $9T$ field. For the $\rho_S = 0.005\ g/cm^3$ dispersion (the units are dropped

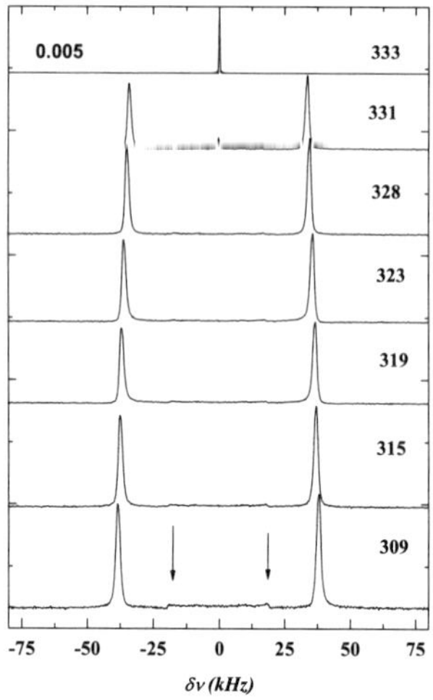

Fig. 7.2. Temperature dependence of the DNMR spectra for the 0.005 dispersion. Temperature listed are in Kelvin. The arrows for the spectrum at $309K$ are to indicate that disordering effects even at such a low silica density although weak are evident at the half-splitting frequency position.

hereafter), the temperature dependence of the spectral patterns is shown in Fig. 7.2. Spectra resemble those of a typical bulk 12CB sample: a single, narrow, and centered absorption peak in the isotropic phase evolving into a full-splitting doublet (the frequency separation between two absorption peaks) once in the smectic-A phase. At this low density the effect of random disorder on the spectral patterns and thus, on the LC director and the orientational order, does not appear to be significant. Yet, closer inspection reveals that disorder does play a weak role: the DNMR spectrum taken in the SmA phase at $T = 315K$ shows features beginning to appear between the two intense absorption peaks, becoming more pronounced as the sample is further cooled to $T = 309K$. Indeed, as indicated by the arrows, two weakly intense peaks that coincide with one-half of the full frequency splitting are noticeable. Their presence and weak intensity reflects that silica-induced disorder affects only a few smectic layers rather than the whole dispersion. Although not shown here, it is worth noting that a similar but less evident features were also observed for a 0.003 dispersion, the lowest density dispersion studied with 12CB.

Contrasting with comparable densities in 8CB dispersions, such disordering features were not seen; this is most likely because the transition is from the isotropic into the nematic phase. Disorder affects the trans-

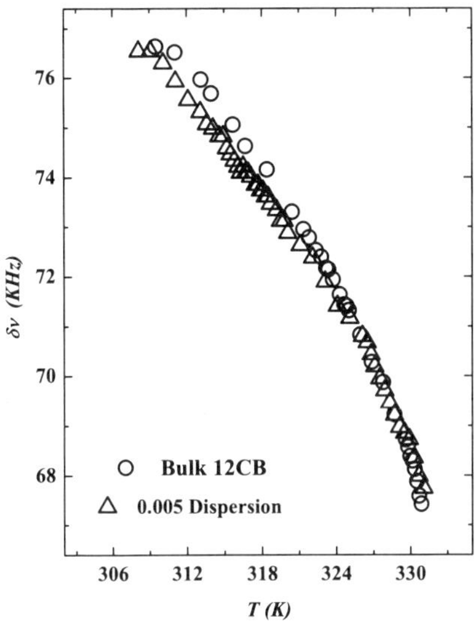

Fig. 7.3. Comparison of the temperature dependence of the full frequency splitting for bulk 12CB and the 0.005 dispersion. The slight reduced orientational order is consistent with an impurity effect.

lational order much sooner than orientational order: disordering effects anneal when cooling into the nematic phase, but they do not when cooling is into the smectic phase. For completeness, it must be noted that our 8CB studies took place in a $4.7T$ magnet which may have lacked the resolution to resolve such weak disordering effects. Reiterating, the addition of small amounts of silica affects the LC like impurities would. This is clearly reflected in Fig. 7.3 where we compare the temperature dependence of the full quadrupolar splitting (orientational order) for bulk 12CB and the 0.005 dispersion. Despite the large scale chosen, the reduction of orientational order in the dispersion as compared to bulk is not significant, with data overlapping over most of the temperature range.

Increasing the silica density to form dispersions in the neighborhood of the gelation threshold enhances the disordering effects on the LC smectic phase. For all densities in this regime, the spectrum in the isotropic phase is the typical bulk-like, single, narrow absorption peak. It is in the smectic phase where the disordering effects become more noticeable; this is emphasized in Fig. 7.4 where we present spectra at a temperature in the vicinity

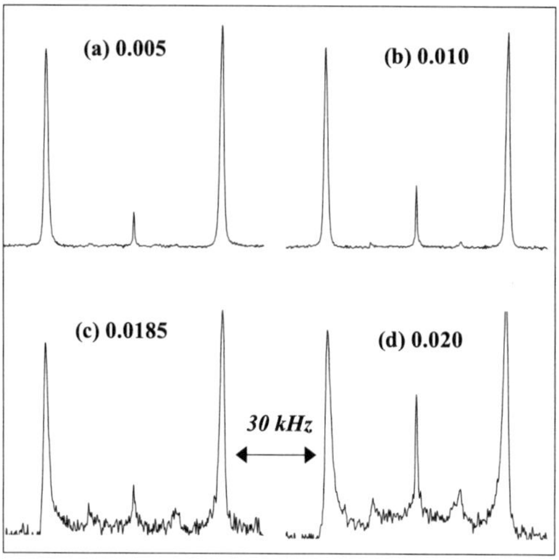

Fig. 7.4. Spectral patterns for four 12CB-silica dispersions in the smectic-A phase at a temperature in the vicinity of $329K$, selected to emphasize the growth of the half-splitting singularities with increasing silica density.

of $329K$ for four different dispersions, and in Fig. 7.5, where we show the overall temperature, and angular dependence at $329K$, of the spectra for the 0.010 (Figs. 7.5a and 7.5b) and for the 0.020 (Figs. 7.5c and 7.5d) dispersions. By angular dependence what is meant is that the dispersion is rotated within the magnetic field from its original orientation.

In the SmA phase, Fig. 7.4, and with increasing silica density, the singularities at the half-splitting position grow in intensity. The spectrum is mostly a full-splitting doublet, which, as shown in Fig. 7.5b for the 0.010 dispersion, exhibits the typical $P_2(\cos\theta)$ angular dependence. The fact that there is such an angular dependence in the presence of the large NMR magnetic field, indicates that the silica constraints 12CB layers in an anisotropic network and they do not re-orient if the dispersion orientation in the field is changed. As seen in the angular dependence studies for the 0.020 dispersion in Fig. 7.5d, the weakly intense peaks at the half-splitting position are present at all orientations in the magnetic field; if enlarged, they would be recognized to be the inner peaks of a small powder pattern. The area under this powder pattern is estimated to represent up to about 20% of the area

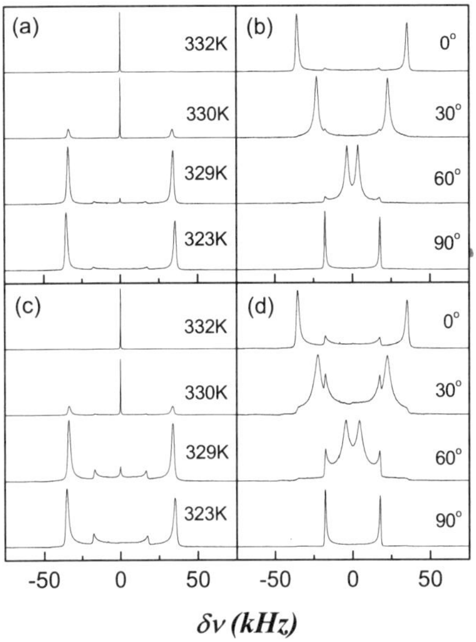

Fig. 7.5. Temperature and angular dependences of the DNMR spectra for the 0.01 ((a) + (b)) and 0.02 ((c) + (d)) 12CB-aerosil $A300$ gels in the smectic-A ($T = 323K$) phases. 0° is defined as the orientation of the sample when it is first placed in the magnetic field. The fact that there is an angular dependence indicates that some of the smectic layers are fixed by the presence of silica.

under the complete spectrum; recalling that the area under a spectrum is a measure of the number of molecules possessing a specific alignment, this effectively states that one fifth of the smectic layers are in random domains fixed by the silica and unaffected by the external magnetic field.

The previous results may be understood and summarized as follows. For the lowest density studied, the behavior is not unlike bulk and the aerosil behaves as if a small impurity amount was added to the LC. This was first noted in Fig. 7.3 and is further stressed in Fig. 7.6 below where we plot the temperature dependence of the smectic splitting for bulk 12CB and several dispersions; as silica is added, deviations from bulk, and thus reduction in orientational order become more apparent. Increasing the density to form the 0.010 dispersion, a random silica network begins to barely form; it is quite weak and it approaches the gelation threshold. When the sample is cooled in the field, the appearance of smectic order breaks a large number of silica links. As the layers align along the field, the silica strands rearrange to form a mostly anisotropic network where elastic strains are annealed. However, not all the silica links are broken. Some of the *LC* domains retain a random orientation because the silica links enclosing them are strong enough, and/or the domain sizes are small: smectic order cannot break the links and anneal the elastic strains. The silica configuration is a mixture of anisotropic and random networks, with the anisotropic component being much larger. In this 12CB dispersion regime, this picture is different from that in nematic 8CB soft-gel dispersions[8] where the elastic strains are almost completely annealed crossing into the nematic phase, yielding a pure anisotropic dispersion (silica network).

The field-aided annealing effect at the $A-I$ transition is weaker than that at the $N-I$ transition since the annealing of 12CB gels is not complete even for the weak soft-gel network that exists in the 0.010 dispersion. With further density increase to a 0.020 dispersion (beyond the gelation threshold), as seen in Figs. 7.4 and 7.5d, peaks at the half splitting position are visible at all orientations suggesting that parts of the dispersion consist of a random distribution of smectic domains while other parts of the sample are fully oriented by the external field.

The fact that there is an angular dependence in the SmA phase for the 0.01 dispersion may be due to the following mechanisms. First, while already more viscous (as compared to a nematic) and thus difficult to align, the viscosity of the 12CB-aerosil dispersions increases with increasing aerosil density. This "locks-in" the director configuration and the magnetic field is unable to re-align the smectic domains. Second, it is possible that the magnetic field enables a strongly anisotropic diffusion field removing most of the elastic strains and aiding the formation of a complete field-induced anisotropic silica network with large σ. In such a network, the width of all voids is smaller than the magnitude of the magnetic coherence length in the smectic phase, $l_{\perp}(0.005) < \xi_M(\text{SmA})$, and consequently the external magnetic field has no aligning effect. The *LC* configuration in dispersions up to the gelation threshold appear to be remarkably uniform.

To obtain additional insight on the dispersions behavior in this regime, and to better emphasize the competition between the ordering magnetic field and the disordering silica aerosil effects on the liquid crystal configuration, we have performed what we have termed *zero-field* studies and they are presented in Fig. 7.7. In such *zero-field* studies, the dispersions are cooled from the isotropic phase to room temperature outside the magnetic field; they are then inserted in the NMR magnet at room temperature (note

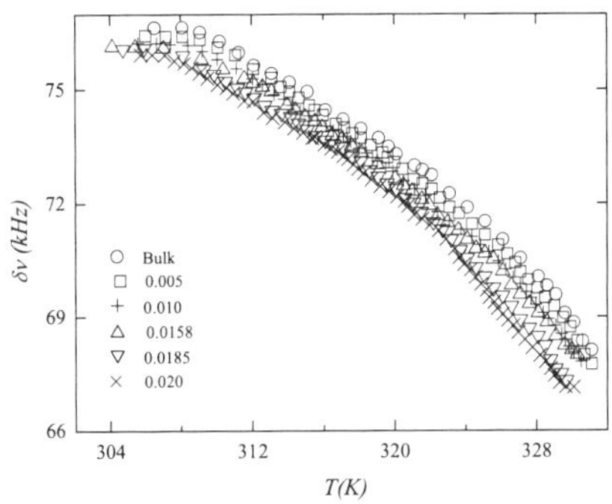

Fig. 7.6. Temperature dependence of the frequency splitting for bulk and several mixtures obtained in cooling from the isotropic phase. Notice that as a function of increasing aerosil density and, as measured by the reduced magnitude of the splitting, increasing the silica density decreases the orientational order.

that cooling and heating within the field as in the previously discussed results yields reproducible results, namely, no thermal history dependence). Under such *zero-field* cooling conditions, the *LC* configuration is established in the absence of an aligning field and it is solely induced (disordered) by the presence of aerosil.

When the *zero-field* cooled 0.005 dispersion is first placed in the magnet, one clearly sees a powder pattern, a consequence of cooling the sample in the presence of the disordering field: the liquid crystal configuration consists of randomly oriented domains confining smectic layers. As the temperature is increased, the external magnetic field begins to align a few of the domains; this is reflected by the decrease in intensity of the inner peaks. As the temperature is further increased and the transition into the isotropic phase is neared, the aligning magnetic field dominates, the dispersion is almost fully aligned and only weak aerosil introduced disorder remains. For this 0.005 dispersion, the transition from a disorder dominated configuration to a well-ordered configuration takes places rather smoothly. This same transition, also occurring in the more dense 0.0158 dispersion, by comparing the spectra at $328K$ and $329K$ in Fig. 7.7, appears to be more sudden. For dispersions with densities exceeding the gel threshold value, as in the case of the 0.0185 dispersion, the magnetic field is unable to create a fully aligned sample: even at $329K$ there is a visible remnant of a domain distribution, evolving into the isotropic peak at higher temperature.

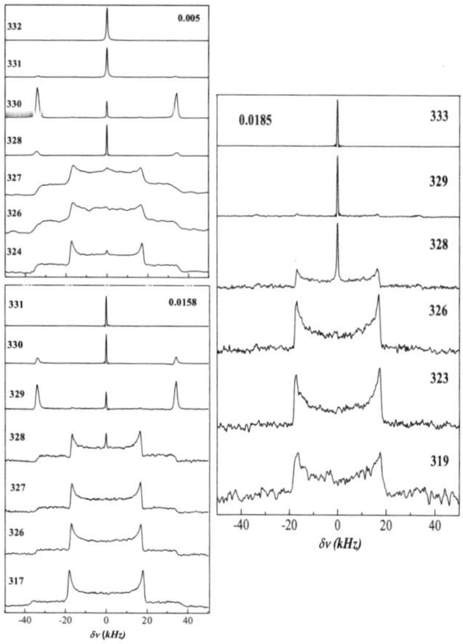

Fig. 7.7. Temperature dependence of the DNMR spectra obtained in heating from room temperature after "zero-field cooling" (cooling outside the magnetic field) for several dispersions. See text for details.

Comparing spectra after *zero-field* cooling at a temperature in the smectic phase near $328K$, the spectrum for the 0.005 dispersion is that of a sample which is mostly aligned; that found for the 0.0185 dispersion is mostly part a powder pattern. The intensity of the half-splitting peaks is greater than that of the full splitting peaks; this is opposite to what occurs in the 0.005 dispersion. Further, whether field ordering or silica disordering will be the dominant effect depends on the dispersion thermal history. If cooling is done in the magnetic field, the 0.005 (see Fig. 7.2) as well as the 0.0185 dispersions (see for example spectra in Fig. 7.4) exhibit a well-defined liquid crystal alignment and disordering effects only set in near room temperature. In short, the *zero-field* cooling results can be understood by noting that in the cooling process the diffusion field is random, the aerosil network that has formed is not complete, and only a few molecules are eventually aligned by the field. An alternative view one could conceive is that the RAN is complete but $l_0(0.005) < \xi_M(\mathrm{SmA})$ is not satisfied for all voids; LC molecules in the larger voids become aligned by the magnetic field once the sample is placed in the NMR field.

It is appropriate at this time to compare results for low densities aerosil dispersions with theoretical predictions. A theoretical study of quenched random disorder (QRD) effects on smectic ordering in liquid crystals by Radzihovsky and Toner[33] concluded that arbitrarily small amounts of QRD would destroy even quasi long range order (LRO) and consequently a true smectic phase could not exist given that the smectic correlation length would remain finite at all temperatures and for all disordering strengths. This theoretical conclusion has been confirmed through a recent x-ray study with 8CB aerosil dispersions by Park and co-workers.[12] This theoretical expectation for our NMR work, given the evidence of a weakly intense powder pattern found even at the lowest densities spectra may be interpreted as follows: (a) complete field annealing cannot occur even at the lowest densities studies; and (b) elastic distortions are present at all temperatures thus preventing the formation of a true smectic phase. Indeed, for all spectral patterns obtained close and below the T_{AI} transition, the presence of a central absorption Lorentzian peak reflects a co-existence of phases; yet, the very intense absorption peaks at the full splitting position suggest that most of the dispersion is forming a smectic phase. Nonetheless, for all aerosil densities, absorption peaks of varied intensity strength that increases with increasing amount of silica are apparent at the half-splitting position (becoming more evident with decreasing temperature); this indicates that randomly oriented smectic domains due to the aerosil induced disorder are present. Consequently, even an arbitrarily small amount of QRD begins to destroy true smectic phase.

7.3.2. *Higher Silica Density Dispersions*

As previously discussed, increasing the aerosil density in the soft-gel regime, increases disordering effects and introduces a larger reduction in orientational order as compared to bulk and as measured by a decreased frequency splitting. However, since some field aided annealing may be possible in this regime, the reduction in orientational order is not drastically lower than bulk. The disordering effects are however enhanced by adding more silica particles.

When a dispersion with a density of 0.0556 is studied cooling from the isotropic phase, as originally, in the presence of the NMR magnetic field, several interesting features are uncovered. Inspection of the temperature dependence of the spectral patterns shown in Fig. 7.8 reveals that in cooling from the isotropic into the smectic phase, the now broader isotropic

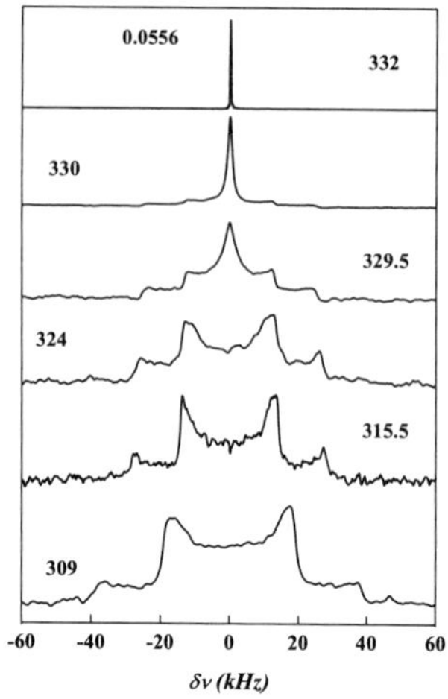

Fig. 7.8. Temperature dependence of the DNMR spectral patterns for the 0.0556 aerosil dispersion. Notice the Pake powder pattern emergence with decreasing temperature.

peak evolves into prominent peaks at the half splitting position (see for instance the spectra at $330K$ and $329.5K$) more intense than those at the full splitting position. If the temperature is lowered further, the isotropic peak disappears and a clear Pake powder pattern forms. Again, a Pake powder is the manifestation that orientationally ordered LC domains are present but the director is randomized throughout the dispersion. It is important to note that the silica network at this density is not expected to be completely rigid, rather, it is a candidate for thixotropic bonds as known from other network studies.[11] However, the smectic domains are constrained at random orientations. In contrast, when similar densities dispersions were studied for $8CB$[8], they showed milder disordering effects in the nematic phase when compared to the severe disordering effects found here with the smectic 12CB. It is easier for disorder to affect translational order as compared to orientational order.

The results for the 0.0556 dispersion may be understood by considering that the mean void length ($l_0 = 117nm$) is smaller than the magnetic coherence length, which is the case for both 8CB and 12CB. Yet, the silica induced disorder effects on nematic 8CB are not as severe as in smectic 12CB at the same silica density due to the more fluid, less viscous nature of the nematic. Correlations between smectic domains may extend up to $\approx 1l_0$, whereas correlations of nematic domains may reach lenghts comparable to $100l_0$.[34] Thus, the magnetic coherence length for the nematic phase becomes smaller than the effective ($100l_0$) domain length, while the magnetic coherence length for the smectic phase is smaller than the effective ($1l_0$) domain length. As a result, field aided annealing is possible at larger aerosil densities in cooling through the $N - I$ phase transition, but not likely in cooling through the $A - I$ phase transition. This explains why quenching in the smectic phase occurs at a silica density nearly one order of magnitude lower than quenching in the nematic phase.

It has been predicted[11] that ρ_S may be linearly related to the QRD strength for low to moderate density values. Since the no-gel and soft-gel dispersions regimes correspond to low and moderate values of ρ_S, it is possible to verify this prediction using the results obtained so far. The area under the absorption peaks is directly proportional to the number of LC molecules aligned at that particular orientation. Thus, calculating the ratio of the area under the peaks at the half splitting position to the total area under the complete spectral pattern yields information on the percent of LC molecules constrained by the aerosil network; this can be viewed as a measure of the QRD strength of the network. As shown in Fig. 7.9 below, we calculated the percent of LC molecules constrained by the random network and indeed found that the QRD strength is linearly dependent on ρ_S. In addition, at a fixed silica density, it is almost constant independent of temperature. The % of smectic layer composing the powder pattern ranges from 8% for a 0.005 dispersion to $\sim$ 95% for a 0.06 dispersion.

A recurring feature is that disorder effects onset for smectics much sooner than for nematics; it is easier to prevent a layered formation than a molecular orientation. In particular for the 0.0556 dispersion, approximately 95% of the LC molecules are constrained by the aerosil network, as evidenced by the powder spectra and the results of Fig. 7.9. Further, inspection of the spectra at $329.5K$ in Fig. 7.8, shows the presence of the isotropic phase (center absorption peak), co-existing with the full splitting shoulders as well as singularities emerging at the half splitting position. This result is again suggestive that smectic layers constrained (pinned by the silica particles) by the aerosil network compose the majority of the dispersion; the NMR ordering field has a minimal effect in aligning the smectic domains.

It was reported[35] that for LC molecules in a micropore ($< 0.7\mu m$) and ($1\mu m$) glassy matrix, the surface anisotropic energy is of the order of $10^{-3} J/m^2$. Although this estimate may not hold for aerosil networks, it could be reasonable to assume that the pinning energy per unit area for

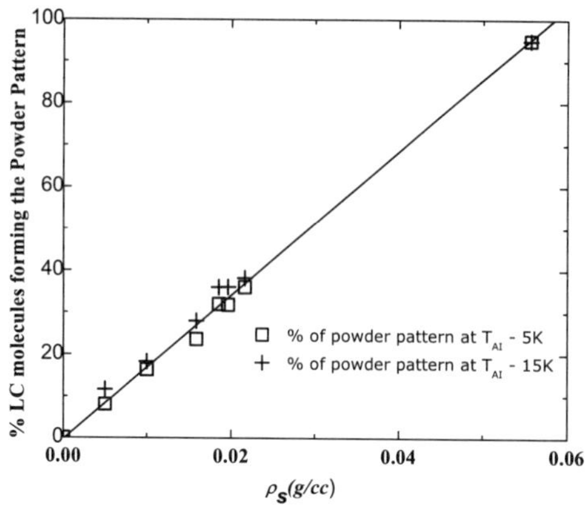

Fig. 7.9. Percent of LC molecules in a powder pattern as a function of aerosil density at two different temperatures below the $A-I$ transition.

the aerosil network assumes a constant value. It was also reported[36] that this energy saturates for $n = 9$ for the nCB homologous series. Thus, if we assume that the anisotropic surface interaction energy is of same order of magnitude for 8CB and 12CB, then, the drastic difference: mild disorder for 8CB while severe disorder for 12CB, can be attributed to the fact that 12CB forms the smectic phase while 8CB forms the nematic phase in cooling from the isotropic phase. Disordering effects are more severe because of the additional broken symmetry in the smectic phase. For similar strength disordering effect to occur in a nematic, the silica density must be considerably higher. For 8CB dispersions this takes place at a density in the vicinity of 0.35, compared to approximately 0.06 for 12CB.

In short, quenching of the smectic phase occurs in dispersions with a silica density which is at least seven times lower than for nematics.

7.3.3. *Stiff-gel dispersions*

At the highest densities, the so-called stiff gel regime where $\rho_S \succeq 0.1$, the aerosil density is large enough that a rigid isotropic network is formed. Spectral patterns obtained for all samples studied in this regime exhibit a similar behavior, thus, we will present results for a single dispersion. Specifically, spectral patterns determined for a 0.4523 density are shown in Fig. 7.10. The spectra are very interesting, since the initially measured isotropic peak, already much broader than a bulk isotropic peak, becomes

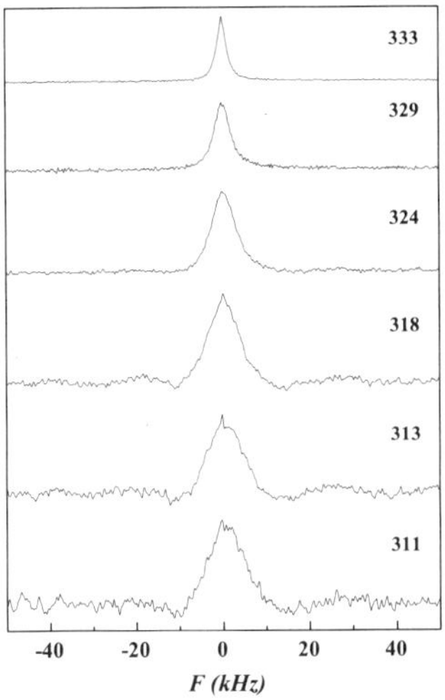

Fig. 7.10. Temperature dependence of the spectral pattern for the 0.4523 density dispersion showing the continuous increase of the width of the absorption peak. The broadening of the peak reflects a continuous evolution of orientational order.

even broader cooling into the smectic phase. This continuous broadening of the peak should not be interpreted as reflecting the absence of the smectic phase. Rather, it means that smectic domains are present, yet, with a wide (and rather smooth) orientational order angular distribution. The full width at half maximum of the absorption peak changes from 3.3kHz in the isotropic phase (one order of magnitude broader than bulk) to nearly 10kHz at the lowest tempearture shown of $311K$. It should be stressed that the reason for the large width of the isotropic peak is that, although we have repeatedly stated that silica introduces disorder, this is indeed the case in the LC mesophases; in the isotropic phase, the presence of the silica network surfaces induces orientational order via a wetting mechanisms.

Results for stiff-gel dispersions may be understood as follows. At these densities, and in particular for the 0.4523 being discussed, the order parameter assumes a lower value as compared to that in the no-gel or soft-gel dispersion regimes. Also, for this particular dispersion, the mean void size l_0 is approximately $12nm$, whereas, the distance traveled by a LC molecule on the NMR experimental time scale, namely, the diffusion length x_0, is estimated from Eq. (7.3) to be of order of $15nm$. Since x_0 and l_0 are the relevant lengthscales for these dispersions and are of the same order of magnitude, one may expect a motionally narrowed absorption spectrum. Indeed, as shown in our earlier aerogel studies[16] quadrupolar absorption spectra would be considerably motionally affected if:

$$\varepsilon \cong \frac{\delta\nu l_0^2}{6D} \leq 10 \tag{7.4}$$

a condition that is met for these silica density dispersions with $\rho_S \succeq 0.1$. Thus, the *LC* molecules 'see' various orientations while diffusing on the NMR experimental time scale. Hence, the Pake powder pattern which was found in lesser dense dispersions collapses into a broad isotropic peak. We conclude by stating that all dispersions in this regime (we reiterate that they are not shown because of their similarities with the 0.4523 one) with densities ranging from 0.1 to 0.9, all spectra are motionally averaged.

7.4. Conclusions

When hydrophilic aerosil particles are dispersed in a *LC* material, and depending on the silica density, they form a random interconnected network. The disordering effects, controlled through the amount of silica particles added to the *LC*, are manifest through the distortion of the local director and the suppression of the long range orientational and/or translational order of a *LC* phase. Depending on the aerosil silica density, the behavior in zero external field shows that there are different dispersions regimes:

(1) no-gel, where the amount of introduced disorder is minimal or impurity-like;

(2) soft-gel, in which the dispersions are in the presence of disorder that is eventually fully quenched; and

(3) fully quenched stiff-gel dispersions where the disorder is severe, strongly affecting the translational order.

As silica is added, disorder increases and the liquid crystal configuration is affected until smectic domains, while present, are randomly oriented throughout the dispersion. With increasing density the DNMR spectra evolve from a well-established configuration, to a Pake powder pattern, to an extremely broad isotropic-like peak, indicating a continuous evolution of orientational order.

From our current investigations of 12CB silica dispersions and those with 8CB, combined with the considerable body of work found in the literature, it can be concluded that aerosil dispersions are incredibly rich and broad systems to study. Specifically:

- The addition of aerosil affects the smectic phase more strongly than the nematic phase.
- Disorder distorts the local director; it suppresses the long range translational and orientational order.
- Disorder distorts even quasi *LRO* since the smectic correlation length is finite for all arbitrarily low aerosil densities and for all temperatures; this is revealed by the Pake powder pattern appearing upon going through the isotropic to smectic transition. The powder powder is more pronounced at lower temperatures and higher silica densities.
- As expected, the *QRD* strength is linearly related to the aerosil density.
- Due to the short smectic correlation length, the field is unable to anneal elastic strains.
- Increasing aerosil density induces downshift in the isotropic to smectic transition temperatures, and the isotropic/smectic coexistence persists over a broad temperature range.
- The powder pattern intensity increases with increasing silica density. At the highest densities studied, almost all molecules are in smectic domains randomly orientationally distributed through the sample.
- Diffusion effects become important in the stiff-gel regime for smectics, unlike the situation with nematics. This is likely the consequence of the void sizes being comparable to the translational diffusion length of the *LC* molecule on the NMR experimental time scale. Correlations among smectic domains is approximately two orders of magnitude smaller than in the case of nematics.

Acknowledgments

This work was supported by the NSF-INT USA-Slovenia 03-06851 and by a Research Challenge from the State of Ohio OBOR B-8709 grants; it was also supported by the Kent State Physics Department, the College of Arts and Sciences, and NSF-DMR through the IR/D program. This chapter while based on the dissertation work of Vishal Pandya, has greatly benefitted from work performed by previous graduate and postdoctoral students G.S. Iannacchione, T. Jin, H. Zeng at Kent State, and B. Zalar at the University of Ljubljana, Slovenia. Any opinions, findings, and conclusions or recommendations expressed in this publication are those of the authors and do not necessarily reflect the views of the National Science Foundation.

References

1. M. Chan, N. Mulders, and J. Reppy, *Phys. Today* **49** (8), 30 (1996); and references therein.
2. Q. J. Harris, Q. Feng, Y. S. Lee, R. J. Birgeneau, and A. Ito, *Phys. Rev. Lett.* **78**, 346 (1997); and references therein.
3. S. M. Clarke and E. M. Terentjev, *Phys. Rev. Lett.* **81**, 4436 (1998).
4. see for instance "*Liquid Crystals in Complex Geometries Formed by Polymer and Porous Networks*", edited by G.P. Crawford and S. Zumer, (Taylor & Francis, London, 1996).
5. T. Bellini, C. Chiccoli, P. Pasini, and C. Zannoni, *Phys. Rev. E* **54**, 2647 (1996).
6. B. Zhou, G. S. Iannacchione, C. W. Garland, and T. Bellini, *Phys. Rev. E* **55**, 2962 (1997).
7. G. S. Iannacchione, C. W. Garland, J. T. Mang, and T. P. Rieker, *Phys. Rev. E* **58**, 5966 (1998).
8. T. Jin and D. Finotello, *Phys. Rev. Lett.* **86**, 818 (2001); *ibid*, *Phys. Rev. E* **69**, 041704 (2004).
9. M. Marinelli, A. K. Ghosh, and F. Mercuri, *Phys. Rev. E* **63**, 061713 (2001); F. Mercuri, A. K. Ghosh, and M. Marinelli, *Phys. Rev. E* **60**, R6309 (1999).
10. T. Bellini, L. Radzihovsky, J. Toner, and N. A. Clark, *Science* **294**, 1074 (2001).
11. G. S. Iannacchione, S. Park, C. W. Garland, R. J. Birgeneau, and R. L. Leheny, *Phys. Rev. E* **67**, 011709 (2003).
12. S. Park, R.L. Leheny, R.J. Birgeneau, J.L. Gallani, C.W. Garland, and G. S. Iannacchione, *Phys. Rev. E* **65**, 050703 (2002)
13. T. Bellini, N. A. Clark, C. D. Muzny, L. Wu, C.W. Garland, D.W. Schaefer, and B.J. Olivier, *Phys. Rev. Lett.* **69**, 788 (1992).
14. T. Bellini, M. Buscaglia, C. Chiccoli, F. Mantegazza, P. Pasini, and C. Zannoni, *Phys. Rev. Lett.* **85**, 1008 (2000).
15. H. Sonntag and K. Strenge, "*Coagulation and Structure Formation*", (Plenum, New York, 1987), pp. 134-145 and pp. 172-177.
16. H. Zeng, B. Zalar, G. S. Iannacchione, and D. Finotello, *Phys. Rev. E* **60**, 5607 (1999).
17. S. Qian, G.S. Iannacchione, and D. Finotello, *Phys. Rev. E* **53**, R4291 (1996).
18. H. Zeng and D. Finotello, *Phys. Rev. Lett.* **81**, 2703 (1998).
19. A.V. Zakharov and J. Thoen, *Phys. Rev. E* **69**, 011704 (2004).
20. D. Liang, M.A. Borthwick, and R.L. Leheny, *J. Phys.: Condens. Matter* **16**, S1989–S2002 (2004).
21. M.K. Ramazanoglu, P.S. Clegg, R.J. Birgeneau, C.W. Garland, M.E. Neubert, and J.M. Kim, *Phys. Rev. E* **69**, 061706 (2004).
22. I. V. Kreuzer, T. Tschudi, W. H. de Jeu, and F. L. Eidenschink, *Appl. Phys. Lett.* **62**, 1712 (1993).
23. A. Glushchenko, H. Kresse, V. Reshetnyak, Y.U. Reznikov, and O. Yaroshchuk, *Liq. Cryst.* **23**, 241 (1997).
24. N.J. Diorio Jr., M.R. Fisch, and J.L. West, *Liq. Cryst.* **29**, 589 (2002).

25. J. Milette, C. T. Yim, and Linda Reven, *J. Phys. Chem. B* **112**, 3322 (2008).
26. Degussa Corporation, Silica Division, Aerosils, Frankfurt, Germany.
27. A. A. Chuiko, and Y.I. Gorlov, "*Chemistry of a Silica Surface*", Naukova Dumka, Kyiv, Ukraine 1992; G. Michael, H. Ferch, "*Basic Characteristics of AEROSIL*", Technical Bulletin Pigments No. 11, Degussa: Frankfurt, Germany (1993).
28. A. Mertelj and M. Copic, *Phys. Rev. E* **55**, 504 (1997).
29. D. Finotello, K. A. Gillis, A. Wong, and M. H. W. Chan, *Phys. Rev. Lett.* **61**, 1954 (1988).
30. T. Bellini, N. A. Clark, V. Degiorgio, F. Mantegazza, and G. Natale, *Phys. Rev. E* **57**, 2996 (1998).
31. M. Vilfan and N. Vrbancic-Kopac, in chapter 7 of Ref. 4.
32. G. P. Crawford and J. W. Doane, *Mod. Phys. Lett. B* **28**, 1785 (1993).
33. L. Radzihovsky and J. Toner, *Phys. Rev.B* **60**, 206 (1999).
34. R. L. Leheny, S. Park, R. J. Birgeneau, J. L. Gallani, C. W. Garland and G. S. Iannacchione, *Phys. Rev. E* **67**, 011708(2003); and references therein.
35. S. Kralj, G. Lahajnar, A. Zidansek, N. Vrbancic-Kopac, M. Vilfan, R. Blinc, and M. Kosec, *Phys. Rev. E* **48**, 340 (1993).
36. T. Jin, B. Zalar, A. Lebar, M. Vilfan, S. Zumer, and D. Finotello, *Eur. Phys. J. E* **16,** 159 (2005).

Chapter 8

Dynamics of Liquid Crystals by means of Deuterium NMR Relaxation

Carlo Alberto Veracini* and Valentina Domenici

Dipartimento di Chimica e Chimica Industriale, University of Pisa, via Risorgimento 35, 56126 Pisa, Italy
**email: verax@dcci.unipi.it*

In the recent years deuterium NMR spectroscopy has been widely applied to the study of molecular dynamics of liquid crystalline materials such as the thermotropic (rod-like, banana-shaped and dendrimers) and the lyotropic (micellar, columnar and lamellar) systems. The development of suitable molecular models and the advances in computational tools for the analysis of relaxation data, as well as the enlargement of the examined cases and the comparison with other experimental techniques have made deuterium NMR Relaxation a solid and self-consistent approach for the characterization of overall molecular reorientations, such as the tumbling and spinning motions, the internal and the collective motions.

8.1. Introduction

Molecular motions occurring both in thermotropic and lyotropic mesophases deeply influence many of their microscopic and macroscopic liquid crystalline properties and because of their relevance many techniques are used to study dynamics in liquid crystals (LC). Among them, NMR is a very important technique, since information on different motional processes can be extracted from the many nuclear properties typical of NMR spectroscopy.[1,2] Spin-lattice relaxation times of ^{1}H, ^{2}H and ^{13}C nuclei, and spectral features, such as chemical shifts and linewidths, are the most widely used. The power of NMR in the study of molecular dynamics mainly relies on its versatility. In fact, different types of relaxation times, linewidths and exchange experiments, can provide information over a very wide range of frequencies, thus potentially allowing several types of molecular motion occurring on very different time scales to be investigated. The different behaviors of nuclei that are present or can be quite easily inserted in the liquid crystalline molecules, such as ^{1}H, ^{13}C and ^{2}H, allows one to explore different dynamic aspects with different levels of details.

For instance, from ^{2}H spin-lattice relaxation times it is possible to get individual information on "fast" motional processes (internal inter-conformational reorientations, molecular spinning and tumbling), exploiting the intra-molecular character of the quadrupolar interaction and the selective labeling of particular molecular moieties.[3-6] In particular, ^{2}H NMR relaxation is unique in getting insight in the internal motions affecting labeled moieties. ^{2}H NMR spin-spin relaxation methods have been mostly applied to uniaxial phases, such as nematics, both in low and high molecular weight liquid crystals. This method put in evidence dynamic processes occurring at low frequencies, typically collective motions.[1] In contrast, the extraction of quantitative information from ^{1}H spin-lattice relaxation times is more difficult, since the inter-molecular contributions are non-negligible and the information is not site-specific. However, field-cycling NMR techniques allow one to extend the measurements down to frequencies of a few hundreds of Hz, thus also rendering accessible the investigation of slow motions, such as order director fluctuations in nematic phases and layer undulations in smectic phases. [7,8] (see Chap. 5)

Solid-state NMR techniques, such as cross polarization (CP), high-power ^{1}H-decoupling and magic angle spinning (MAS), substantially improve the resolution of the ^{13}C spectra allowing the exploitation of ^{13}C nuclei. In fact, ^{13}C MAS NMR techniques can be used without making recourse to sample isotopic enrichment.[9,10,11] However, the extraction of information in terms of dynamic features is not straightforward due to complexity of spectral features and partial overlapping of ^{13}C signals, particularly true in large molecules, such as liquid crystals. ^{13}C T_1 relaxation measurements have been mainly applied to the study of nematic and uniaxial smectic phases, but also, more recently, to tilted chiral phases.[12,13] Several efforts have been devoted to the investigation of the change of dynamic properties at the SmA-SmC* phase transition, as well as of the different motional behaviors in ferroelectric and anti-ferroelectric phases.[14] However, these studies are limited to the qualitative interpretation of the T_1 trends.

The ^{2}H NMR spectra, in contrast with those of ^{13}C and ^{1}H NMR, are much simpler, site dependent and not influenced by intermolecular interactions. Moreover, the ^{2}H relaxation is nearly completely driven by motions modulating the electric field gradient around ^{2}H nuclei. The only disadvantage to study deuterated mesogens is the need to specifically label the samples under investigation. Natural abundance ^{2}H NMR spectra of mesogens can be obtained but the sensitivity is at the moment too poor to study relaxation. Additional specific NMR techniques, such as pulsed-field-gradient NMR[15,16,17] or exchange deuterium NMR techniques,[18,19,20] allow one to study either intra- or inter-smectic layers translational self-diffusion in smectic LC phases or more complex diffusive motions in frustrated LC phases, such as twist grain boundary and cholesteric phases.

8.2. Molecular Motions in Liquid Crystals

Molecular motions occurring in liquid crystalline mesophases can be studied by different NMR techniques. For instance, translational diffusion is a fundamental dynamic process for characterizing mesophases. The influence of translational diffusion in ^{1}H NMR Relaxometry will be presented later, while the use of NMR Field Gradient methods for self diffusion studies is the subject of chapter 9. In

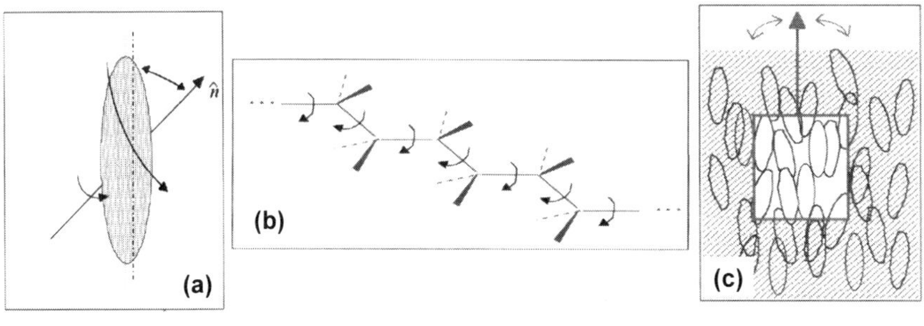

Fig. 8.1. Scheme of the main motions typical of Liquid Crystals: (a) overall molecular, (b) internal and (c) collective motions.

the present chapter we focus our attention to deuterium NMR Relaxation times (spin-lattice and spin-spin ones) and to the motions accessible from these nuclear properties.

For this purpose, it is convenient, invoking time rate order of magnitude arguments,[21] to separate non-translational molecular motions into three types (see Fig. 8.1). However, this separation is far from being safe and this is a weak point in the theories developed to extract dynamic properties from NMR relaxation. These three types of motions are:

1) Overall rotational diffusion:

Various models have been proposed to describe overall molecular rotational diffusion. In the first one, proposed by Nordio,[22] the diffusion tensor is assumed to be diagonal in a molecular reference axes system with two principal components: the "spinning", corresponding to the motion around the principal molecular axis, and the "tumbling", the orthogonal motion having a greater inertia moment. Other models have been put forward and they will be briefly mentioned later on.

2) Internal motions:

Molecular internal rotations around single or double bonds and *trans-gauche* isomerizations of alkyl chains are examples of internal motions.

3) Collective motions:

Collective motions (sometime called cooperative or cluster motions) are low frequency motions.[1] These motions are due to viscoelastic properties typical of LC phases, and referred to as director fluctuations. In chiral smectics other cooperative motions such as the tilt angle fluctuations (soft modes) and rotations of helical axis (Goldstone modes) can be present.

8.3. Dynamic Motions and NMR Techniques

We will here make an attempt to classify NMR techniques with respect to measurements of various dynamic phenomena occurring in liquid crystalline mesophases.[21] One of the possible ways is to order dynamic processes with respect to the Larmor frequency, ω_0: in this respect molecular motions can be *fast, intermediate, slow or very slow.* In all these cases, assuming a sole process as responsible of the relaxation, a correlation time τ_c characteristic of the motion can be defined. In case of fast motions, having frequencies of the order of the Larmor frequency, ($\tau_c \approx 2\pi/\omega_0$, usually $\tau_c \approx 10^{-7}$-10^{-11} s), the ^{2}H NMR spectra are determined by averaged Hamiltonian operator elements and the dynamic information can be accessible only by measuring longitudinal relaxation times. Within the slow regime ($\tau_c \geq 10^{-3}$ s) bidimensional ^{2}H NMR techniques with magnetization transfer, first introduced by Spiess,[18,19] are available and they can be used to investigate slow exchange processes in polymers. In the intermediate regime the exchange between states (or conformations) occurs with frequencies comparable with interactions determining the spectral features ($\tau_c \approx 10^{-4}$-10^{-5} s). In such case, the dynamic processes can be studied fitting the ^{2}H NMR line-shape by suitable models. In the case Solid State NMR techniques, whenever applied to liquid crystals, are used to observe the ^{13}C nucleus (such as MAS and high-power ^{1}H decoupling)[18] spectral line-shape widening can occur when the dynamic exchange between different sites matches the MAS frequencies (typically 10^3-10^5 Hz). Moreover, the conformational exchanges ranging between 10^2 and 10^3 Hz may affect the chemical shifts for non equivalent sites. On the other hand, the ^{13}C T_1 relaxation times are sensitive to slow dynamic processes, such as the Goldstone mode typical of ferroelectric phases.

Fast switching methods are employed also in the ^{1}H NMR field-cycling experiments, but, in this case the preparation field is lowered to very low values corresponding to low frequencies (of the order of few KHz), where the collective motions influence substantially relaxation. For this reason, this technique is particularly helpful for studying specific motional mechanisms typical of liquid crystals in the slow and very slow motional regime. Among other dynamic phenomena, diffusion along helical axes or across synclinic or anticlinic layers in chiral smectics (ferroelectric, ferri or antiferroelectric) can affect spectral linewidths (see Chap. 6) and can be investigated by suitable NMR methods. This aspect has been recently studied although at a deeper level.[23]

8.3.1. *Relaxation times*

^{2}H NMR lineshapes and relaxation times depend principally on the interaction between the ^{2}H quadupolar moment and the electric field gradient tensor surrounding the ^{2}H nucleus in a molecular site. The electric field gradient tensor is dramatically affected by averaging from different types of molecular motions and this effect is observed in spectral lineshapes as well as in the measured relaxation times.[24] However, in order to extract dynamic information through relaxation times measurements we need to model the relevant dynamic processes occurring in the liquid crystal mesophases and to build a consistent link between these models and the various measured quantities.

For the spin-lattice relaxation times, a fast motion regime is usually assumed: in this case, the Redfield theory is a good approximation.[25] If this approximation is not valid, as it is usually encountered for spin-spin relaxation times, the solution of the Stochastic Liouville equation is needed.[26] Within Redfield approximation the interactions between spins and surroundings are treated as random perturbations within the formalism of density matrix. Analytical expressions of relaxation times can be obtained as a function of spectral densities $J_m(\omega)$, defined as Fourier transforms of suitable autocorrelation functions $C_m(\tau)$:[24]

$$J_m(\omega) = \int_0^{\infty} C_m(\tau) e^{-i\omega\tau} d\tau. \tag{8.1}$$

Here the $C_m(\tau)$ are defined as:

$$C_m(\tau) = \overline{F_m(t) F_m(t+\tau)} \tag{8.2}$$

where $F_m(t)$ is a general random time function reporting on the time dependence of Hamiltonian operator. In the case of quadrupolar interaction, the dominating interaction for deuterium nuclei, Redfield theory gives the following expressions for the spin-lattice relaxation times:

$$\frac{1}{T_{1Z}} = K_Q \left[J_1(\omega_0) + 4J_2(2\omega_0) \right], \tag{8.3}$$

$$\frac{1}{T_{1Q}} = 3K_Q J_1(\omega_0). \tag{8.4}$$

T_{1Z} and T_{1Q} are "Zeeman" and "Quadrupolar" spin-lattice relaxation times describing the processes of recovery to equilibrium of Zeeman and quadrupolar magnetization orders,[2] respectively. $K_Q = (3\pi^2/2)(e^2qQ/h)^2$, where $e^2qQ/h(=\nu_q)$ is the quadrupolar coupling constant.

Different from the T_1, the spin-spin relaxation is more complex. The decay of the transverse magnetization component (xy plane), referred to as T_2*, depends not only on the dynamics of the specific spin system but also on external conditions (e.g. spatial heterogeneity of the magnetic field). On the other hand the transverse decay as measured normally by echo techniques, (i.e. by refocusing inhomogeneities), is referred to as T_2. Several types of T_2, however, can be defined as constant decays of the different spin coherences.

A commonly used expression for the experimental T_2^* is that related to the NMR spectral linewidth, according to:

$$T_2^* = \frac{1}{\pi \cdot \Delta\nu_{h/2}} \tag{8.5}$$

with $\Delta\nu_{h/2}$ being the spectral linewidth at half height, which is measurable in simple cases, such as that of highly oriented systems having few not overlapped deuterium signals. In different cases, including those with different averaged quadrupolar tensor orientations, it is most useful to evaluate the T_2^* from the second moment of the spectrum, M_2, defined as:

$$M_2 = \frac{\int_{-\infty}^{+\infty} (\omega - \langle\omega\rangle)^2 f(\omega)\, d\omega}{\int_{-\infty}^{+\infty} f(\omega)\, d\omega}. \tag{8.6}$$

For instance, in the case of Gaussian distribution of the frequencies $f(\omega)$, T_2^* is equal to $\sqrt{2/M_2}$. Beside the differences between T_2^* and T_2, the extraction of dynamic parameters from the measurement of spin-spin relaxation times is less straightforward than for spin-lattice relaxation times.

For an isolated deuterium nucleus (I=1) three independent T_2 can be defined, each of them are related to the time-evolution of different coherences. Several theoretical approaches have been developed based on the time evolution of the density matrix with different levels of approximation. For instance, if the Redfield approximation is valid, the following three spin-spin relaxation times can be defined as a function of the spectral densities:[24]

$$\frac{1}{T_{2a}} = \frac{K_Q}{2} \cdot \left(3J_0(0) + 5J_1(\omega_0) + 2J_2(2\omega_0)\right), \tag{8.7}$$

$$\frac{1}{T_{2b}} = \frac{K_Q}{2} \cdot \left(3J_0(0) + J_1(\omega_0) + 2J_2(2\omega_0)\right), \tag{8.8}$$

$$\frac{1}{T_{2D}} = K_Q \cdot \left(J_1(\omega_0) + 2J_2(2\omega_0)\right). \tag{8.9}$$

The first two relaxation times can be measured separately or as a sum of two contributions, depending on the NMR pulse sequences but also on the nature of the sample (if isotropic or anisotropic system). This aspect is discussed in Sec. 8.3.1.2. Moreover, it should be pointed out that the "real" cases usually do not refer to isolated single nuclei, with the direct consequence that proton-deuteron or deuteron-deuteron coupling may influence the spin-spin relaxation giving rise to the impossibility of measuring pure T_2.

8.3.1.1. *^{2}H NMR spin-lattice relaxation times*

Translational motions do not influence either quadrupolar interactions or ^{2}H NMR spin-lattice relaxation, which are essentially dependent on reorientation of molecular field gradients. As stated before, ^{2}H spin-lattice relaxation in liquid crystals is usually sensitive to three types of molecular motions: rotational diffusion of mesogens, internal motion of flexible groups (alkyl chains, ring rotations, conformational changes) and collective motions. By time scale arguments these processes are assumed to contribute separately and their contribution is simply added to get the total spectral densities. The difference in time scale can be considered valid for cooperative motions which are effective toward low frequencies (low with respect to Larmor frequency). On the contrary, this assumption has to be made with care for internal motions which can have a time scale not far from the overall rotational diffusion times. On the other hand, the NMR methods and in particular ^{2}H NMR relaxation are unique with respect to other methods used to investigate dynamic properties since ^{2}H NMR is unique in giving information on internal motions.

Zeeman and Quadrupolar relaxation times, T_{1Z} and T_{1Q}, can be measured either by Jeener-Broekaert pulse sequence,[27,28] or by the modification of J-B sequence made by Wimperis[29] (called broadband version of J-B), which allows one to measure longitudinal relaxation times for deuterons having large quadrupolar splitting. This last sequence ($90_x - \tau_1 - 67.5_y - \tau_1 - 45_{-x} - (\tau_1/2) - 45_{-x} - \tau_2 - 45_{-x}$ – Acq) has two critical delays, τ_1 and τ_2. The delay τ_1 must be calibrated experimentally to produce the best anti-phase quadrupolar doublet when τ_2 vanishes. The delay τ_2 is the variable delay whose values range from very low ones (until 10μs) to larger ones (up to 100 ms) allowing the recovery and decay of the Zeeman and Quadrupolar order, respectively. The power of the J-B and broadband J-B sequences lays the possibility to measure simultaneously T_{1Z} and T_{1Q}. The sum (M_+) and the difference (M_-) of the integrals of the component of the quadrupolar doublet depend on the variable delay through the following relationships:

$$M_+(\tau_2) = c1\left(1 - c2\exp\left(-\frac{\tau_2}{T_{1Z}}\right)\right), \tag{8.10}$$

$$M_-(\tau_2) = c3\exp\left(-\frac{\tau_2}{T_{1Q}}\right) + c4 \tag{8.11}$$

with the parameters c1, c2, c3, c4 allowed to float in order to compensate deviations from the theoretical behavior, that is, a complete magnetization inversion when τ_2=0, c2=2 and c4=0. The fits to the sum and difference between the experimental integrals according to Eqs. (8.10) and (8.11) give the spin-lattice relaxation times, T_{1Z} and T_{1Q}.

8.3.1.2. *^{2}H NMR spin-spin relaxation times*

In the case of deuterium (I=1) three different spin-spin relaxation times T_2 can be defined, as reported previously (Eqs. (8.7)-(8.9)). Typically two sequences are used in order measure the spin-spin relaxation times, namely the quadrupolar echo[30] (QE), which is made by two 90 degree pulses separated by a variable delay τ, and the quadrupolar version of the Carr-Purcell-Meiboom-Gill (CPMG) sequence,[31,32] which can be seen as a quadrupolar echo followed by (n-1) elementary sequences in the form

$\tau - 90 - \tau$. When the QE sequence is applied to ordered phases, such as the nematic one, it is easy to show that $1/T_2^{QE}$, the exponential coefficient of the FID decay as a function of τ (see Eq. (8.12)), is equal to $1/T_{2a}+1/T_{2b}$.

$$I(2\tau) = I_0 \cdot e^{(-2\tau/T_2^{QE})} . \tag{8.12}$$

The T_2 measured by the CPMG sequence, according to the following relationship,

$$I(2n\tau) = I_0 \cdot e^{(-2n\tau/T_2)} \tag{8.13}$$

can be written as:

$$\frac{1}{T_2} = \frac{1}{T_{2a}} + 2J_1(\omega_0) \cdot \left(\frac{\sin(2\omega_q \tau)}{2\omega_q \tau} \right) . \tag{8.14}$$

Two limiting situation may arise: if $\omega_q\tau << 1$, then $1/T_2=(1/T_{2a}+1/T_{2b})$, if $\omega_q\tau >> 1$, the T_2 measured by the CPMG sequence is exactly T_{2a}.

The discussion of the effect of the CPMG sequence in ordered phases is far from being the subject of this chapter and for an exhaustive treatment we refer to Frezzato *et al.*[33,34,35] The main problem of the extraction of dynamic information from T_2 is the lack of validity of the Redfield theory. This approximation is valid for treating the faster molecular rotations, but less useful for collective motions characterized by much slower rates. In these cases a slow-motional approach based on the solution of the stochastic Liouville equation is needed.

8.3.2. *Modeling of different motions in liquid crystals*

8.3.2.1. *Rotational diffusion in ordered phases*

Overall molecular motions may be described assuming that the mesogenic molecule behaves like a rotational ellipsoid. This ellipsoid should correspond to a sort of averaged shape over internal and conformational motions. The motion of this ellipsoid is treated within a small step diffusional model in the presence of the anisotropic orientational potential. In addition to this very simple model that treats mesogenic molecules as uniaxial rotors diffusing in uniaxial phases,[1] several theories have been provided for the rotational diffusion of biaxial molecules in uniaxial phases as well as for uniaxial molecules in biaxial phases.[36,37] However, these theories have not yet found so many applications.[1-4] Here, we illustrate the simplest theory which assumes for the molecular diffusion a model of axial rotors in uniaxial phases: this is the case of uniaxial nematics and uniaxial smectic phases.

Correlation functions $C_m(t)$, previously introduced in Eq. (8.2), can be expressed in terms of the auto-correlation functions $C_{m,k,k'}(t)$:

$$C_m(t) = \sum_{k,k'} d_{0,k}(\beta) d_{0,k'}(\beta) C_{m,k,k'}(t) \tag{8.15}$$

where $d_{0,k}(\beta)$ is the Wigner reduced matrices and β is the second Euler angle connecting the principal molecular axis to the PAS of interaction tensor. In the case of quadrupolar interaction this angle is, in a good approximation, the angle between the molecular principal axis and the direction of the C-D bond.

$$C_{m,k,k'}(t) = \left\langle \left\{ D^2_{m,k}[\Omega(0)] - \overline{D^2_{m,k}} \right\} \left\{ D^{2*}_{m,k'}[\Omega(t)] - \overline{D^{2*}_{m,k'}} \right\} \right\rangle. \tag{8.16}$$

$D_{m,k}[\Omega]$ are second rank Wigner matrices, and $\Omega(t)$ is the angle between lab reference frame and molecular one. Depending on the model assumed to describe molecular motions, $C_{m,k,k'}(t)$ can take different forms. If a small step rotational diffusion of a symmetric rotor is taken as a model (with $D_{//}$ the diffusion constant of the molecular main axis, while: $D_{xx} = D_{yy} = D_\perp$ for the orthogonal ones), the autocorrelation functions from solving the rotational diffusion equation take the form of either single[22] or multi decreasing exponential[38] expressions.

In the latter case, $C_{m,k,k'}(t) = g_{m_L,m_M,m_K}(t)$:

$$g_{m_L,m_M,m_K}(t) = g_{m_L,m_M}(t) = \delta_{m_M,m_K} c_{m_L,m_M} \sum_j a^{(j)}_{m_L,m_M} \exp\left(\frac{t}{\tau^{(j)}_{m_L,m_M}} \right). \tag{8.17}$$

Here, c_{m_L,m_M} is the starting value of the function, $a^{(j)}_{m_L,m_M}$ are the weights of exponential decays, and $\tau^{(j)}_{m_L,m_M}$ the correlation times, which contain the information about the molecular rotational diffusion.

By Fourier transforming the correlation functions the following expression for the spectral densities, $J_{mL}(m_L\omega_0)$ is obtained:

$$J_{m_L}(m_L\omega_0)=\frac{3\pi^2}{2}(v_q)^2\sum_{m_M=-2}^{2}c_{m_Lm_M}[d^2_{m_M0}(\beta_{M,Q_0})]^2\sum_{j=1}^{3}a^{(j)}_{m_Lm_M}\frac{(\tau^{(j)}_{m_Lm_M})^{-1}}{(m_L\omega_0)^2+(\tau^{(j)}_{m_Lm_M})^{-2}} \tag{8.18}$$

v_q is the quadrupolar coupling constant of the particular deuterium nucleus, and β_{M,Q_0} is the angle between the principal molecular axis and the C-D bond direction. The correlation times $\tau^{(j)}_{m_L,m_M}$ are defined in terms of the diffusion coefficients as follows:

$$\frac{1}{\tau^{(j)}_{m_Lm_M}}=\frac{6D_\perp}{b^{(j)}_{m_Lm_M}}+m_M^2(D_\parallel-D_\perp) \tag{8.19}$$

in the case of Nordio theory,[22,39] and

$$\frac{1}{\tau^{(j)}_{m_Lm_M}}=\frac{6D_\beta}{b^{(j)}_{m_Lm_M}}+m_L^2(D_\alpha-D_\beta) \tag{8.20}$$

for the anisotropic viscosity model.[40,41] In the Nordio model, $D_{//}$ and $D_\perp$ are the principal components of the diffusion tensor in the molecular reference frame describing the "spinning" and "tumbling" motions, respectively. In the "anisotropic viscosity" model, D_α and D_β are the principal components of the diffusional tensor diagonal in the laboratory reference frame describing the precession and "tumbling" motions of the principal molecular axis with respect to the director.

In the model called "*third rate anisotropic viscosity model*", which is an extension of the anisotropic viscosity model, the rotation of the molecule around the longest molecular axis (γ *motion*) is simply added and assumed to be uncorrelated with α and β motions. The total correlation function describing the overall motion is written as product between the correlation function of the γ motion around the principal (longest) axis ($\Gamma_{mM}(t)$) and that of the α and β motions ($g_{m_L,m_M}(t)$):

$$C_{m_Lm_M}(t)=\Gamma_{m_M}(t)g_{m_Lm_M}(t). \tag{8.21}$$

The γ motion can be described by strong collision model[42] or small step diffusion[43] around the principal axis:

$$\Gamma_{m_M}(t)=e^{-\frac{t}{\tau_{m_M}}} \tag{8.22}$$

where

$$\tau_{m_M} = \frac{1}{k_{m_M}}, \; k_1 = D_\gamma, \; k_2 = (3p+1)D_\gamma \tag{8.23}$$

$p = 0$ if the strong collision model is valid, while $p = 1$ for the small step diffusion.

In the *"third rate anisotropic viscosity"*, the correlation time can be written as:

$$\frac{1}{\tau^{(j)}_{m_L m_M}} = \frac{6D_\beta}{b^{(j)}_{m_L m_M}} + m_L^2(D_\alpha - D_\beta) + \xi(m_m)D_\gamma \tag{8.24}$$

$\zeta(m_M)$ is either $(1-\delta_{m_M})$, in the case of strong collision model,[42] or m_M^2, if we use the small step diffusional model.[43]

8.3.2.2. *Internal motions*

In the case that the deuterium nuclei are in a molecular site (fragment) undergoing either internal rotations or large conformational motions, the internal dynamics becomes important for relaxation. To take in due account the internal motions, two possibilities arise for the evaluation of spectral densities: the "superimposed rotations" model[42] and the model referred as "decoupled model".[44] In both cases, the independence of internal with respect to overall motions is assumed. While in the first model the motions are considered to be independent from each other, this is not the case for the decoupled model. Because of this the first model is used mainly for rotations of single isolated groups while the second model can deal with conformational changes occurring for instance in aliphatic chains. Within the model called "superimposed rotations" the internal motion is treated as the γ motion in the "third rate anisotropic viscosity model". The diffusion constant D_{int} is introduced to characterize the γ motion and this contribution results in a term to be added in the spectral density expression:

$$J_{m_L}(m_L\omega_0) = \frac{3\pi^2}{2}(v_q)^2 \sum_{m_M=-2}^{2} \sum_{m_R=-2}^{2} c_{m_L m_M} [d^2_{m_R 0}(\beta_{i,Q_i})]^2 [d^2_{m_M m_R}(\beta_{M,i})]^2$$
$$\times \sum_j a^{(j)}_{m_L m_M} \frac{(\tau^{(j)}_{m_L m_M})^{-1} + \xi(m_R)D_{\text{int}}}{(m_L\omega_0)^2 + [(\tau^{(j)}_{m_L m_M})^{-1} + \xi(m_R)D_{\text{int}}]^2} \tag{8.25}$$

where β_{i,Q_i} is the angle between C-D bond and internal rotation axis, and $\beta_{M,i}$ is the angle between this and the molecular long axis.

In the "decoupled model" the internal motions of aliphatic chains are considered as transitions between different conformational states. One bond, two bond and three bond transitions are described by three rate constants: k_1, k_2 and k_3. Here the rotational diffusion tensor is assumed to be unaffected by internal motions and related to an "averaged" conformer or molecular structure. For each carbon atom of the chain the spectral density expression becomes:

$$J_{mL}\left(m_L\omega_0\right)=\frac{3\pi^2}{2}\left(v_q\right)^2\sum_{m_M=-2}^{2}\sum_{n=1}^{N}c_{m_Lm_M}\left[\sum_{k=1}^{N}d^2_{m_M0}(\beta^k_{M,Q})\cdot\exp\left[-im_M\alpha^k_{M,Q}\right]\cdot x_k^{(1)}\cdot x_k^{(n)}\right]^2$$

$$\times\sum_j a^{(j)}_{m_Lm_M}\frac{\left(\tau^{(j)}_{m_Lm_M}\right)^{-1}+|\lambda_n|}{\left(m_L\omega_0\right)^2+\left[\left(\tau^{(j)}_{m_Lm_M}\right)^{-1}+|\lambda_n|\right]^2} \tag{8.26}$$

where n and k indicate a particular conformer and the sum is over the N conformers, $\beta^{\kappa}{}_{M,Q}$ and $\alpha^{\kappa}{}_{M,Q}$ are the polar angles of a C-D bond of the k conformer in the molecular reference frame, fixed on the molecular core. The terms λ_n and $x^{(n)}$ are the eigenvalues and eigenvectors, respectively, associated with the "master equation" describing chain motion obtained by diagonalization of **R** matrix, which contains rate constants of the conformational transitions. An improvement of this approach has been introduced by substituting the three bond rate constant k_3, which requires a too high activation energy, with the so-called "type II" motions, introduced by Helfand,[45] and consisting either the production of *gauche* pairs or the so-called *gauche* migration. This method is different only in the building up of the **R** matrix, although the formula of the spectral density remains the same.

8.3.2.3. *Collective motions*

Collective (cooperative) motions include undulation modes in smectic phases, both rotations and translations in cholesterics and director fluctuations in nematics and smectics. The possibility that fluctuating director may serve as a spin relaxation mechanism has attracted much attention and theories[46,47] have been provided for contributions from these motions to spectral densities. Usually in the applications, these motions have been considered as decoupled both from reorientational and internal dynamics. The theories predict a vanishing or small contribution to $J_2(2\omega_0)$, and a typical $\omega_0^{-1/2}$ dependence to $J_1(\omega_0)$ in nematic phases, while for smectics a dependence of either $\omega_0^{-1/2}$ and/or ω_0^{-1} is expected. All these contributions are considered additive to the total $J_1(\omega_0)$ of a single deuterium site. For instance, the contribution to $J_1(\omega_0)$ of the order director fluctuations (ODF) according to Pincus and Blinc's models[46,47] is:

$$J_1^{DF}(\omega_0) = \frac{3\pi^2}{2}(v_q)^2[d_{00}^2(\beta_{M,Q_0})]^2(S_{ZZ})^2 T \frac{a_{DF}}{(\omega_0)^{\zeta}} \qquad (8.27)$$

where the ζ is 0.5 or 1 for a nematic or a smectic behavior, respectively. S_{ZZ} is the order parameter for the assumed axis of cylindrical symmetry and the factor a_{DF} depends on macroscopic parameters such as the average Frank elastic constant, the viscosity coefficient and the self-diffusion translational constant.

More complex treatment concerns the contribution of the collective motions to the $J_0(0)$ spectral density which is accessible only by measuring T_2.[2,3]

8.4. Selection of Experimental Cases

Several reviews[2,3,5,6] have been recently published presenting some aspects of the dynamic investigation of liquid crystals by means of ^{2}H NMR spectroscopy. This section is mainly devoted to the description of the recent state of the art developments[5,6,48] of the ^{2}H NMR relaxation studies of liquid crystals done by our group, ranging from the uniaxial and biaxial smectic phases formed by rod-like LC to the nematic and smectic phases of more complex LC, such as dendrimers, bent-core and polymer LCs. Several examples will be presented in order to give a complete view of the possibilities arising from the application of spin-lattice and spin-spin NMR relaxation methods to the study of dynamic properties in liquid crystals.

8.4.1. *Determining the diffusion coefficients for the overall molecular and internal motions from T_1 analysis*

The first studies based on ^{2}H NMR T_1 relaxation times were performed on uniaxial phases of relatively small LC molecules[2,3] by using the "decoupled model"[44] to get information on both internal and overall molecular motions. These first studies, however, were limited to very restrict sets of data and the diffusion coefficients as well as the jump rates were determined temperature by temperature. A Global Target approach was then introduced by Dong[49] in order to get more reliable analysis and to extend the set of experimental data to be analyzed. Within this approach, the diffusion coefficients are assumed to have an Arrehnius trend (Eq. (8.28)) as a function of temperature, thus rendering possible to fit simultaneously larger sets of experimental spectral densities recorded in the temperature range of stability of the LC mesophase.

$$D(T) = D^{\infty} \cdot e^{-\Delta E_a / RT}. \tag{8.28}$$

Each diffusion process is described by two parameters: the pre-exponential factor, D^{∞}, and the activation energy, ΔE_a. The versatility of the global analysis, which allows one to simultaneously analyze also spectral data recorded for different deuterium labeled sites and, in principle, at several Larmor frequencies, was then applied to several nematics and smectic A LC systems giving good results, namely reasonable values of diffusion coefficients with satisfactory reproduction of experimental spectral densities, in particular for the aromatic core of LCs. However, the development of a new mathematic tool, namely the CAGE software,[52] and a critical analysis over the previous published works revealed a large correlation among fitting parameters mainly due to the complexity of the models used for the analysis and to the number of parameters to be fixed (see for instance Eqs. (8.25) and (8.26)). It was then clear that the quantitative modeling of spin-lattice relaxation times in terms of dynamic parameters requires a careful control of the input parameters of the global target fitting, following a preliminary exploration of the reliable values of the diffusion and jump rate coefficients describing the dynamic processes active in the LC under investigation. Several works[48,50,51,52] based on careful and critical analysis of T_{1Z} and T_{1Q} relaxation times put in evidence that the determination of the diffusion coefficient for the tumbling motion ($D_{\perp}$) is not reliable in case of small data set.

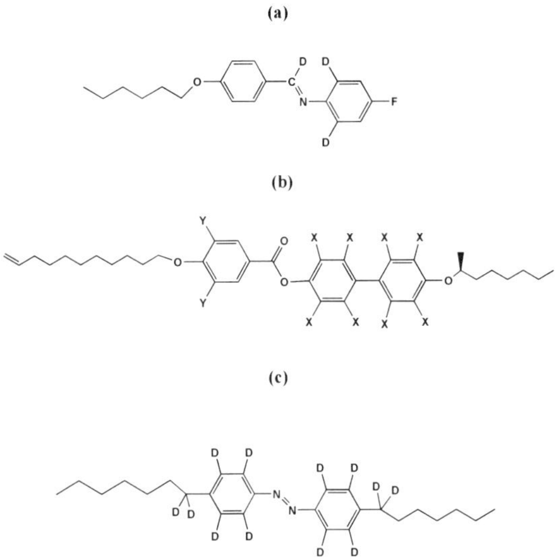

Fig. 8.2. Molecular structure of the smectogens: FAB-OC6-d_3 (a); 11EB1M7-d_2 (Y=D, X=H) and 11EB1M7-d_8 (Y=H, X=D) (b); HAB-d_{12} (c).

A significant example is that of the sample FAB-OC6-d_3 whose molecular structure is showed in Fig. 8.2(a). This liquid crystalline mesogen, having three mesophases, namely nematic, SmA and SmB ones, was selectively labeled in the aromatic core, as indicated in Fig. 8.2(a). The analysis of the experimental spectral densities $J_1(\omega_0)$ and $J_2(2\omega_0)$, derived from the measured relaxation times T_{1Q} and T_{1Z} through Eqs. (8.3) and (8.4), was performed[48,51] in the three phases according to Eq. (8.25) by using the Nordio model[22] as well as the third anisotropy model[38] for the spinning and tumbling motions and either the small step diffusion[42] or the strong collision[43] model to describe the internal motion, which in this case is the reorientation of the deuterated aromatic ring around its *para* axis.

A critical analysis of the experimental data[48] recorded at a single Larmor frequency revealed that two different sets of dynamic parameters give good and comparable reproduction of experimental spectral densities. In Figs. 8.3(a) and 8.3(b) the two sets of diffusion coefficients for the three motions, namely the spinning ($D_{//}$), the tumbling ($D_{\perp}$) and the internal reorientation of the phenyl ring (D_R), obtained by using the Nordio model and small step diffusion for the overall and internal motions, respectively, are reported for the nematic and SmA mesophases. As it is shown in Fig. 8.3(c), in the smectic B phase, a complete indetermination of the diffusion coefficient for the tumbling motion is observed. This means that all values of $D_{\perp}$ lower than $10^5 s^{-1}$ give a satisfactory reproduction of the experimental data.

The solution to the indetermination of $D_{\perp}$ was the subject of intense works mainly reported in V. Domenici's PhD thesis.[48] A first observation

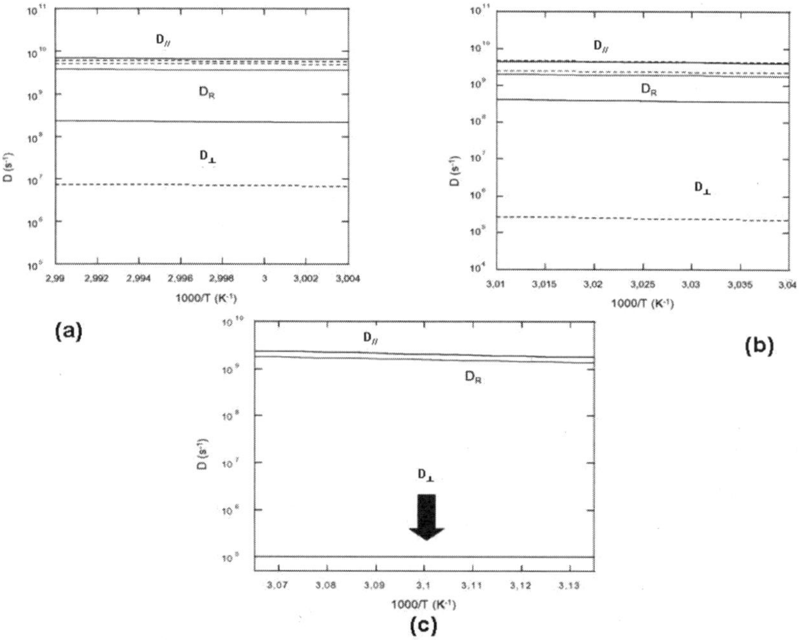

Fig. 8.3. Diffusion coefficients (s-1) vs 1000/T (K) as obtained from the global fitting [50] of the experimental spectral densities of the FAB-OC6-d3 in the nematic (a), smectic A (b) and smectic B (c). Solid and dashed lines refer to different sets of best fitting parameters corresponding to similar good reproductions of the experimental spectral densities. In the SmB phase a total indetermination is found for the tumbling motion. $D_{//}$, $D_{\perp}$ and D_R refer to spinning, tumbling and internal reorientation diffusion coefficients. These graphics have been modified from V. Domenici's PhD thesis.

coming from these works is that one critical issue in the analysis of ^2H spin-lattice relaxation times is the dimension of the experimental data set. In particular, limiting the global fitting analysis on spectral densities of a single deuterium type recorded at a single Larmor frequency in a restricted range of temperatures is very critical and normally results in an indetermination of dynamic parameters, mostly of the tumbling motion.

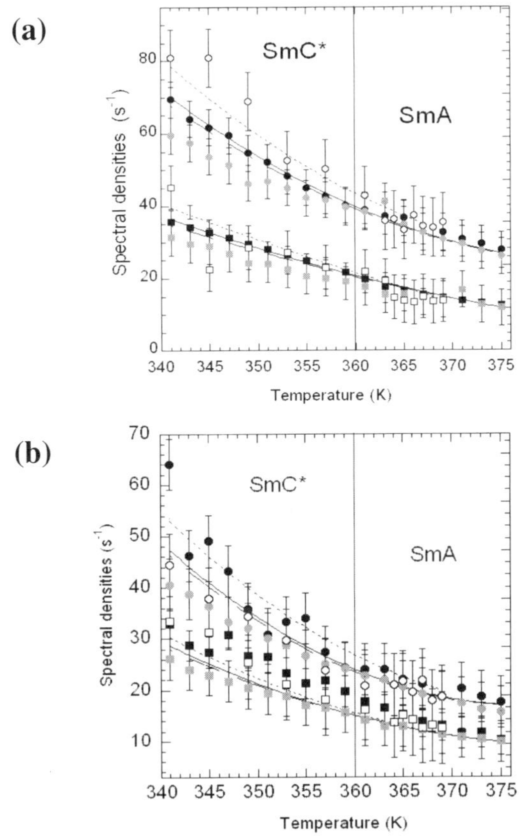

Fig. 8.4. Spectral densities (s^{-1}) vs Temperature (K) of the sample 11EB1M7 as obtained from a multifrequency study applied to the SmA and SmC* phases of two isotopomers, 11EB1M7-d_2 (a) and 11EB1M7-d_8 (b). Black, grey and empty symbols refer to data recorded at 58, 46 and 15 MHz (deuterium Larmor frequencies). Circles and squares refer to $J_1(\omega_0)$ and $J_2(2\omega_0)$, respectively. Dotted, solid and dashed curves refer to the calculated spectral densities at 58, 46 and 15 MHz, as obtained from the global fitting analysis. These graphics are modified from ref. [51].

A first step toward the overcoming of this indetermination consists in fixing some of the fitting parameters according to available theoretical models. For instance, a geometric model developed by Perrin[54] was used in few cases to fix the ratio $D_{//}/D_{\perp}$ which can be evaluated by assuming the LC molecule as a rigid ellipsoid and calculating the above ratio as a function of the semi-major and semi-minor axes. Within this model, and by assuming that this ratio is temperature independent, we could fit the experimental spectral densities for several ferroelectric mesogens.[48,53,55] However, this approach cannot be considered satisfying due to the "artificial" fixing of some parameters. The basic idea to overcome this problem is the enlargement of the experimental data set through different procedures such as: 1) angular-dependent ^{2}H NMR T_1 relaxation measurements[56] (however, it should be noticed that this approach can be applied only to LC systems which do not reorient with respect to the external magnetic field during the experiment); 2) ^{2}H NMR T_1 relaxation measurements for different isotopomers labeled on different molecular sites which allow one to analyze all relaxation data in a global fitting;[53,55] 3) ^{2}H NMR T_1 relaxation measurements recorded at different Larmor frequencies (so-called "multi-frequency" study).[57-59]

An example of the combined effect of points 2 and 3 is that reported in refs. [57] and [58] concerning the 11EB1M7 ferroelectric mesogen, whose molecular structure is shown in Fig. 8.2(b). The molecular dynamic of this liquid crystal was first explored in the uniaxial SmA phase by analyzing T_1 relaxation times recorded at one Larmor frequency[53] for two isotopomers as labeled in Fig. 8.2(b). In that work[53] we fixed the ratio $D_{//}/D_{\perp}$ to 11 according to Perrin's model.[54] In a subsequent study,[57] we added new measurements at two more Larmor frequencies for both isotopomers: we could analyze all these data in a single global target fitting without fixing any parameters. The results for the two isotopomers are shown in Figs. 8.4(a) and 8.4(b), where the experimental and calculated spectral densities at the three Larmor frequencies are in good agreement. The best fitting of these data in the SmA phase was obtained by modeling the overall molecular motions, namely tumbling and spinning ones, by the Nordio model. Moreover, two internal motions, namely the reorientation of the phenyl and biphenyl moieties around their *para* axes, were accessible thanks to the selective deuteration (see Fig. 8.2(b)).

These two internal motions were modeled either by the small step diffusion or by the strong collision model, however, only the second one gave reliable results with a good reproduction of the experimental spectral densities (see Fig. 8.4). Surprisingly this is not the only case showing selectivity between the two models describing internal motions. In particular, aromatic reorientations in ferroelectric liquid crystals are well described only by the strong collision model.[53]

In the specific case of the 11EB1M7 smectogen, the best fitting diffusion parameters for the four motions, namely the tumbling ($D_{\perp}$), spinning ($D_{//}$), reorientation of the phenyl ($D_{R\ (phen)}$) and reorientation of the biphenyl ($D_{R\ (biph)}$) fragments are reported in Fig. 8.5 and in Table 8.1. In this case, among the main results, the multi-frequency study allowed us a self-consistent analysis getting information on both overall and internal motions. Moreover, this study confirmed that the ratio between the spinning and tumbling diffusion coefficients is almost temperature

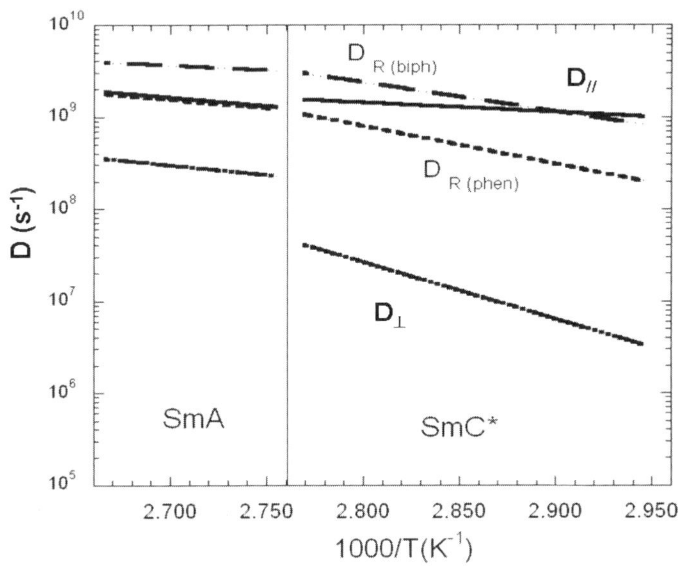

Fig. 8.5. Diffusion coefficients (s^{-1}) vs 1000/T (K^{-1}) of the sample 11EB1M7 as obtained from a multifrequency study applied to the SmA and SmC* phases of two isotopomers, 11EB1M7-d_2 (a) and 11EB1M7-d_8 (b). This graphic is modified from ref. [51].

independent and it has the same order of magnitude as that calculated by using the Perrin's model, thus validating the use of this method when the available data set is poor.

In Table 8.1 the values of the activation energies and diffusion coefficients, taken in the middle of the range of existence of the SmA phase, for the spinning, tumbling and internal motions are reported for several LC samples investigated within this approach. All the cases reported in Table 1 have been obtained in a self consistent way having at our disposal very large sets of experimental spectral densities (with more than one isotopomer and/or at more than one Larmor frequency). These data show that tumbling motions are usually one or two orders of magnitude slower than spinning and internal reorientations. The latter two motions, on the other hand, have often the same order of magnitude. The activation energy of the tumbling motion is always higher or at least equal to that of the spinning motion. Moreover, the activation energy of the ring reorientations was never found higher than the tumbling motions, except for the 8BEF5 sample.[61]

Table 8.1. Rotational diffusion coefficients (s^{-1}) and activation energies (kJ/mol) for the spinning, tumbling and internal motions, determined from ^2H NMR data analysis for several rod-like smectogens in the SmA phase. The values of D are taken at a temperature T in the middle of the temperature range of stability ΔT of the SmA phase.

Mesogens	Diffusion coefficient D (s^{-1})			Activation energy ΔEa (kJ/mol)		
	$D_{//}$	$D_{\perp}$	D_R	(//)	(⊥)	(R)
11EB1M7-phen-D_2 [57]	1.5×10^9	1.3×10^8	1.5×10^9	35.0	40.0	35.0
11EB1M7-biph-D_2 [57]	1.5×10^9	1.3×10^8	3.5×10^9	35.0	40.0	20.0
HAB-phen-α-D_4 [60]	1.3×10^9	1.5×10^7	1.1×10^9	25.5	39.2	35.0
HAB-phen-β-D_4 [60]	1.3×10^9	1.5×10^7	2.1×10^9	25.5	39.2	35.0
8BEF5-phen-D_4 [61]	7.9×10^8	1.1×10^8	1.9×10^{10}	29.6	29.6	33.3
MBHB-biph-D_8 [62]	2.5×10^9	5.0×10^8	3.2×10^9	32.1	32.1	29.1
10B1M7-phen-D_2 [63]	1.3×10^9	3.0×10^8	1.3×10^9	40.0	40.0	35.0
10B1M7-biph-D_2 [63]	1.3×10^9	3.0×10^8	3.5×10^9	40.0	40.0	40.0
ZLL/7*-phen-D_2 [64]	2.0×10^9	4.7×10^7	1.6×10^8	45.9	54.4	50.5
ZLL/7*-biph-D_2 [64]	2.0×10^9	4.7×10^7	1.1×10^9	45.9	54.4	44.2

8.4.2. ^2H NMR T_{1Z} and T_{1Q} analysis applied to ferroelectric (biaxial) smectic phases

The quantitative analysis of relaxation times in case of biaxial mesophases is much more complicated due to several difficulties in linking the measured relaxation times to the dynamic parameters.[5] The complications are both theoretical and experimental ones partially due to the phase biaxiality, which as a consequence introduces many parameters, and the dependence of the relaxation times on the polar angles defining the position of the local phase director with respect to the external magnetic field. In fact, on one side theoretical models taking into account phase biaxiality are quite complex,[65,66] on the other side such a model would require an unavailable large set of experimental data. The ferroelectric phase, namely the SmC*, is one of the most studied biaxial phases and the possibility to investigate this phase from the dynamic point of view was a great stimulus for us in developing a method[61] to analyze T_{1Q} and T_{1Z} relaxation times.

This method bases on the assumption of uniaxiality of the helical supramolecular structure, thus discarding the effect of the azhimutal angle, but taking into account the effect of the tilt angle on the spectral densities. Details of this approach are reported elsewhere.[58,61,62] So far it has been applied to few ferroelectric smectogens as reported in Table 8.2. From these works we can conclude some general remarks. First, the activation energies of the overall and internal molecular motions in the SmC* phase are usually different from those in the SmA phase. This fact determines a different slope in the trend of the diffusion coefficients as a function of temperature (see for instance Fig. 8.5) when passing from the SmA to the SmC* phases. In few cases, and in particular for the tumbling motion, in addition to a different value of the activation energy (higher in the SmC* phase than in the SmA phase) we observed also a discontinuity, or a jump, at the SmA-SmC* transition. An example of this behavior is that of the 11EB1M7 smectogen (see Fig. 8.5).[58]

8.4.3. ^2H NMR T_{1Z} and T_{1Q} studies vs other experimental techniques

Among different experimental techniques, dielectric relaxation[67] has been widely used to study the molecular dynamics of nematic and smectic liquid crystals. The advantage of this spectroscopy is the possibility to investigate dynamic properties in a wide range of frequencies, from few KHz to few GHz. Several theoretical models for the calculation of dielectric properties from NMR quantities or vice versa have been developed by Dong and Zakharov.[68,69] The purpose of these theoretical efforts was to link relaxation times measured either by NMR or dielectric relaxation to diffusion coefficients of the overall molecular motions. In the recent years, this theory has been applied to several experimental cases,[48] such as the rod-like mesogen HAB (see Fig. 8.2(c)).[60,70] The basic idea of this theory consists in the calculation of the dielectric susceptibility and permittivity quantities, which are first rank tensors, by modeling the autocorrelation functions as a sum of exponential functions. Similarly to the models used to calculate NMR properties such as the spectral densities, also within this approach[68,69] the exponential coefficients of the autocorrelation functions, namely the correlation times, are related to the diffusion coefficients describing the tumbling ($D_\perp$) and spinning ($D_{//}$) motions and to the molecular principal order parameter S_2. The following relationships can be indeed derived[68,69] for some of the correlation times, in the single-exponential approximation:

Table 8.2. Rotational diffusion coefficients (s^{-1}) and activation energies (kJ/mol) for the spinning, tumbling and internal motions, determined from ^2H NMR data analysis for several rod-like smectogens in the SmA phase. The values of D are taken at a temperature T in the middle of the temperature range of stability ΔT of the SmC* phase.

Mesogens	Diffusion coefficient D (s^{-1})			Activation energy ΔEa (kJ/mol)		
	$D_{//}$	$D_\perp$	D_R	(//)	(⊥)	(R)
11EB1M7-phen-D_2 [58]	1.2×10^9	1.1×10^7	4.1×10^8	20.0	119.0	78.0
11EB1M7-biph-D_2 [58]	1.2×10^9	1.1×10^7	1.5×10^9	20.0	119.0	61.0
8BEF5-phen-D_4 [61]	1.0×10^9	1.1×10^8	7.2×10^9	70.0	70.0	19.0
MBHB-biph-D_8 [62]	1.2×10^9	2.3×10^8	1.1×10^9	52.0	52.0	36.0
ZLL/7*-phen-D_2 [64]	7.8×10^8	2.7×10^7	4.6×10^8	59.6	50.5	45.6
ZLL/7*-biph-D_2 [64]	7.8×10^8	2.7×10^7	8.5×10^8	59.6	50.5	41.9

$$\tau^{1}_{00} = \left(D_{\perp} \frac{2-2S_2}{1+2S_2} \right)^{-1}, \tag{8.29}$$

$$\tau^{1}_{10} = \left(D_{\perp} \frac{2+2S_2}{1-S_2} \right)^{-1}, \tag{8.30}$$

$$\tau^{1}_{01} = \left(D_{//} + D_{\perp} \frac{1+2S_2}{1-S_2} \right)^{-1}. \tag{8.31}$$

From dielectric relaxation it is possible to measure parallel and perpendicular components of the complex permittivity as a function of frequency, which can be analyzed[67] to obtain high frequency and low frequency relaxation times, $\tau_{h.f.}$ and $\tau_{l.f.}$. The relationship between these experimental relaxation times and the correlation times $\tau^{1}_{m,n}$ is not straightforward. However, recent studies, both theoretical and experimental ones,[70,71,72,73] show that $\tau_{l.f.}$ corresponds to $\tau^{1}_{0,0}$ and $\tau_{h.f.}$ is mainly related to $\tau^{1}_{0,1}$, even though for the high frequency process several conditions are to be fulfilled in order to exclude the influence of $\tau^{1}_{1,1}$.[72] The validity of this theory has been tested in the case of the HAB sample[60,70] for which self-consistent analyses could be performed by means of both NMR and dielectric relaxation. The main result is displayed in Fig. 8.6 in which the experimental dielectric $\tau_{l.f.}$ (Fig. 8.6(a)) and $\tau_{h.f.}$ (Fig. 8.6(b)) relaxation times are compared with the $\tau^{1}_{0,0}$ and $\tau^{1}_{0,1}$ calculated from the NMR data (namely the orientational order $S_{zz}=S_2$ and diffusion coefficients $D_{//}$ and $D_{\perp}$ evaluated from the T_{1Q} and T_{1Z} analysis). For high frequency processes (Fig. 8.6(b)) $\tau^{1}_{0,1}$ was calculated by using the $D_{//}$ coefficient obtained from NMR for the spinning motion and internal reorientations of the two phenyl rings. The good agreement found in the case of HAB is very promising: in principle, the relationships reported in Eqs. (8.29)-(8.31) can be used when the self-consistency of the NMR analysis is not guaranteed by the small NMR data set and if dielectric relaxation data are available.

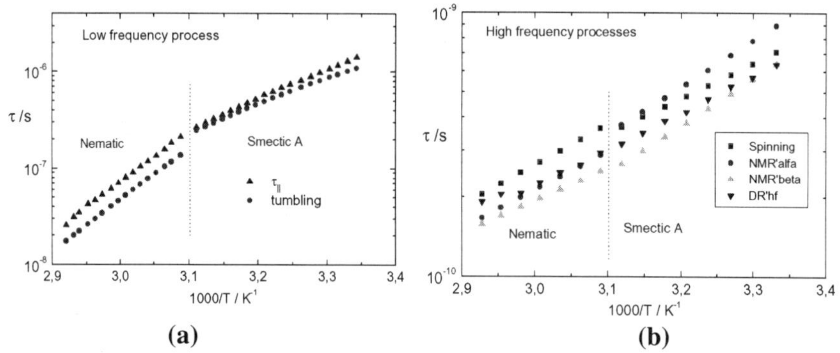

Fig. 8.6. (a) Comparison between the low frequency relaxation times (s) vs 1000/K (s^{-1}) of the HAB sample, as obtained from Dielectric Relaxation (▲) and NMR Relaxation (●), according to Eq. (8.29); (b) Comparison among the high frequency relaxation times (s) vs 1000/K (s^{-1}) of the HAB sample, as obtained from Dielectric Relaxation (▼) and NMR Relaxation according to Eq. (8.31) by using the diffusion coefficients for the spinning (■), internal phenyl α (●) and phenyl β (▲) motions. [modified from refs. [48] and [60].

The comparison between ^{2}H NMR relaxation and different experimental techniques has been the object of several works mainly with the purpose to have a more complete description of the molecular dynamics of liquid crystals. Recent efforts have been spent in the comparison between ^{2}H NMR relaxation and ^{13}C CP-MAS NMR spectroscopy[74] as well as neutron scattering techniques.[74] Very recently, ^{2}H NMR and ^{1}H NMR relaxation were also compared in the case of a bent-core nematic LC.[75] However, this topic is now in progress and in the near future it would be important to extend these comparative studies to other liquid crystalline molecules.

8.4.4. *Applications of ^{2}H NMR T_2 relaxation to liquid crystals*

Deuterium NMR T_2 relaxation[76] has been applied to several experimental cases and a specific theoretical model was developed by Frezzato *et al.*[33-35] to extract information about collective motions, such as order director fluctuations. This approach, whose validity has been demonstrated for nematic phases formed by rod-like LC mesogens as well as LC polymers, and for lyotropic systems such as lamellar phases and bilayered membranes,[77-79] should also be applicable to smectic phases. However, at the moment several difficulties, mainly experimental ones, prevented the investigation of dynamic features in smectic LC systems by means of T_2 analysis. As stated in the previous part of the chapter, T_2 relaxation times are sensitive to dynamic processes in the so-called "slow motional regime" and this is the reason why T_2 studies are helpful in characterizing collective motions.

However, ^2H spin-spin relaxation has been also applied to the study of single molecular motions, such as overall molecular reorientation diffusions, in some complex thermotropic liquid crystals. Recent studies include bent-core liquid crystals (BLCs)[80-84] and dendrimers[85] forming smectic phases. The case of BLCs deserves to be mentioned since it is a representative example of the potential of T_2 studies applied to liquid crystals. Two bent-core mesogens, whose molecular structure is reported in Fig. 8.7(a), have been investigated by means of ^2H NMR.[48] Interestingly their ^2H NMR spectra recorded by using the QE sequence[30] in the isotropic phase, even though far above the isotropic-mesophase transition, presented an unusual line-broadening, not ascribable to pre-transitional effects. The measurement of spin-spin relaxation times by the QE sequence was indeed very helpful both at a qualitative and quantitative levels. In fact, the comparison between the measured T_2 and the experimental T_2^* (according to Eq. (8.5)) gave a clear indication of the dynamic nature of the observed line-broadening. In order to identify the kind of motion responsible of such broadening a first rough model was adopted and it consisted in assuming a single molecular motion, characterized by a correlation time τ_c, in the Redfield approximation.[80] The relaxation times T_2 can be indeed expressed as reported in Eq. (8.7) and the spectral densities can be modeled as follows:

$$J(\omega) = 2\tau_c / \left(1 + \omega^2 \tau_c^2\right). \tag{8.32}$$

By assuming an Arrenhius trend for the correlation time as a function of temperature,

$$\tau_c(T) = \tau^{\infty} \cdot e^{-\Delta E_a / RT} \tag{8.33}$$

(a)

(b)

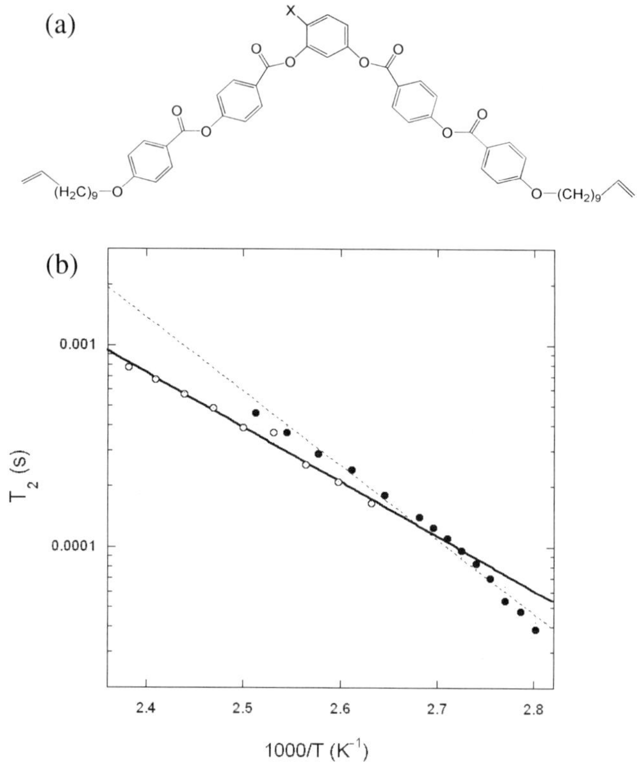

Fig. 8.7. (a) Molecular structure of the banana-shaped liquid crystalline mesogens Pbis11BB (X=H) and ClPbis11BB (X=Cl); (b) Spin-spin (T_2) relaxation times (s) vs 1000/T (K^{-1}) recorded by using the QE sequence in the isotropic phase of the Pbis11BB (○) and ClPbis11BB (•) samples. Solid and dotted curves result from the best fitting of the T_2 trend as described in the text. Figure modified from refs. [48] and [80].

we could fit the measured T_2 in the isotropic phase of the two BLC mesogens (see Fig. 8.7(b)).

Even though this approach may be considered too simplistic, it serves to put in evidence the occurrence of an unusual slow dynamic process in the isotropic phase of BLC systems, which has later been confirmed by other experimental techniques on similar samples.[6] A different approach was applied to the investigation of dynamic properties of one of the BLC samples shown in Fig. 8.7(a), namely the ClPbis11BB compound, in its nematic phase. For this sample, a drastic line-broadening was observed not only in the isotropic, but also in the nematic phase, in which the ^{2}H NMR spectra were characterized by quadrupolar doublets having a Lorentzian shape.[81,86]

As in the isotropic phase, also in the nematic phase the main contribution to the line-width is dynamic one, as confirmed by the comparison between T_2 and T_2^*. However, in the nematic phase the Redfield approximation is not acceptable since the values of the T_2 (below 30μs) clearly indicate a very slow dynamics. The measured T_2 were indeed analyzed by analytically solving the Liouville equation[82] with a model developed for this particular BLC system which allowed us to obtain information about the overall molecular diffusion reorientations, namely the spinning and tumbling motions. This study shows that T_2 relaxation times can be analyzed not only to investigate collective or cooperative motions, but also single molecule dynamic processes.

8.5. Conclusions and Future Developments

At the moment ^{2}H NMR spin-lattice relaxation applied to the study of fast dynamic processes in liquid crystals is probably one of the most sound techniques, since lots of experimental cases have been investigated in the last two decades, showing that it is possible to get reliable results both in uniaxial phases, such as nematic and SmA ones, but also in biaxial phases, such as the SmC* one. The recent experimental and theoretical efforts, including the comparison between ^{2}H NMR relaxation and other NMR methods,[74,75,85] such as ^{1}H NMR relaxometry and ^{13}C CP-MAS NMR linewidth, give a strong indication of how well this method is in describing molecular motions in the fast regime, in particular the overall molecular and internal motions. However, some improvements could be performed in the development of new analytical tools to analyze in a single global fitting of ^{2}H NMR spin-lattice and spin-spin relaxation times as well as ^{13}C NMR spin-lattice relaxation times. On the experimental side future developments are related to the measurement of T_2 in the presence of dipolar coupling and the application of T_2 studies to complex phases such as chiral and frustrated LC phases.

References

1. G. R. Luckhurst and C. A. Veracini, Eds., *The Molecular Dynamics of Liquid Crystals*, NATO ASI Ser. A, Vol. 431 (Reidel, Dordrecht, 1989).
2. R. Y. Dong, *Progr. Nucl. Magn. Reson. Spectr.* **41**, 115 (2002).
3. R. Y. Dong, *Annual Reports on NMR Spectroscopy* **53**, 67 (2004).
4. R. Y. Dong, Nuclear Magnetic Resonance of Liquid Crystals (Springer, New York, 1997).
5. V. Domenici, M. Geppi and C. A. Veracini, *Prog. Nucl. Magn. Reson. Spectr.* **50**, 1 (2007).
6. V. Domenici, *Pure Appl. Chem.*, **79**, 21 (2007).
7. F. Noack, *Prog. Nucl. Mag. Res. Spectr.* **18**, 171 (1986).
8. R. Kimmich and E. Anoardo, *Prog. Nucl. Mag. Res. Spectr.* **44**, 257 (2004).
9. H. Bender, F. Noack, M. Vilfan and R. Blinc, *Liq. Cryst.* **5**, 1233 (1989).
10. A. Yoshizawa, H. Kikuzaki, T. Hirai and M. Yamane, *Jpn. J. Appl. Phys. Part 2 Lett.* **29**, 1153 (1990).
11. H. Satoh, K. Hiraoka and Y. Uematsu, *Ferroelectrics* **212**, 99 (1998).
12. K. Miyayama, Y. Shikayama, A. Iwahori, T. Nose and K. Hiraoka, *Jpn. J. Appl. Phys. A.* **44**, 3130 (2005).
13. T. Seki, K. Miyayama, H. Satoh, M. Kanekiyo, K. Hiraoka and Y. Uematsu, *Jpn. J. Appl. Phys. Part 1* **42**, 3535 (2003).
14. H. Fujimori, JP. Bayle and S. Miyajima, *J. Phys. Soc. Jpn.* **69**, 3090 (2000).
15. R. Blinc, D. L. Hogenboom, D. E. O'Reilly and E. M. Peterson *Phys. Rev. Lett.* **23**, 969 (1969).
16. R. Blinc, J. Pirš and I. Zupančič, *Phys. Rev. Lett.* **30**, 546 (1973).
17. S. V. Dvinskikh and I. Furó *J. Chem. Phys.* **115**, 1946 (2001).
18. K. Schmidt-Rohr and H. W. Spiess, *Multidimensional Solid-state NMR and Polymers*, (Academic Press Inc., London, 1994).
19. C. Schmidt, B. Wefing, B. Blumich and H. W. Spiess, *Chem. Phys. Lett.* **130**, 84 (1986).
20. C. Boeffel, Z. Luz, R. Poupko and A. J. Vega, *Isr. J. Chem.* **28**, 283 (1988).
21. G. L. Hoatson and Y. K. Levine, in *The Molecular Dynamics of Liquid Crystals*, Eds. G. R. Luckhurst and C. A. Veracini, NATO ASI Ser. A, Vol. 431, Chapter 1, (Reidel, Dordrecht, 1989).
22. P. L. Nordio and P. Busolin, *J. Chem. Phys.* **55**, 5485 (1971).
23. I. Musevic, R. Blinc, and B. Zeks, in *The Physic of Ferroelectric and Antiferroelectric Liquid Crystals*, World Scientific, Singapore, **2000**.
24. R. L. Vold and R. R. Vold, in *The Molecular Dynamics of Liquid Crystals*, Eds. G. R. Luckhurst and C. A. Veracini, NATO ASI Ser. A, Vol. 431, Chapter 7 (Reidel, Dordrecht, 1989).
25. A.G. Redfield, *Adv. Magn. Reson.* **1**, 1 (1965).
26. C. Polnaszek, G.V. Bruno and J.H. Freed, *J. Chem. Phys.* **58**, 3185 (1973).

27. J. Jeener and P. Broekaert, *Phys. Rev.* **157**, 232 (1967).
28. R. L. Vold, W.H. Dickerson and R. R. Vold, *J. Magn. Reson.* **43**, 213 (1981).
29. S. Wimperis, *J. Magn. Reson.* **86**, 46 (1990).
30. Z. Luz and S. Meiboom, *J. Chem. Phys.* **39**, 36, (1963).
31. H. Y. Carr and E. M. Purcell, *Phys. Rev.* **94**, 630 (1954).
32. S. Meiboom and D. Gill, *Rev. Sci. Instr.* **29**, 688 (1958).
33. D. Frezzato, G. Kothe and G. Moro, *J. Phys. Chem. B* **105**, 1281 (2001).
34. D. Frezzato, G. Kothe and G. Moro, *J. Phys. Chem. B* **119**, 6931 (2003).
35. D. Frezzato, G. Kothe and G. Moro, *J. Phys. Chem. B* **119**, 6946 (2003).
36. E. Berggren, R. Tarroni and C. Zannoni, *J. Chem. Phys.* **99**, 6180 (1993).
37. E. Berggren and C. Zannoni, *Mol. Phys.* **85**, 299 (1995).
38. R.R. Vold, R.L. Vold, *J. Chem. Phys.* **88**, 1443 (1988).
39. P. L. Nordio, G. Rigatti and U. Segre, *J. Chem. Phys.* **56**, 2117 (1972).
40. C. F. Polnaszek, G. V. Bruno and J. H. Freed, *J. Chem. Phys.* **58**, 3185 (1975).
41. C. F. Polnaszek and J. H. Freed, *J. Phys. Chem.* **79**, 2283 (1975).
42. P. A. Beckmann, J. W. Emsley, G. R. Luckurst and D. L. Turner, *Mol. Phys.* **59**, 97 (1986).
43. D. E. Woessner, *J. Chem. Phys.* **36**, 1 (1962).
44. R. Y. Dong and G. M. Richards, *Chem. Phys. Lett.* **171**, 389 (1990).
45. E. Helfand, *J. Chem. Phys.* **54**, 4651 (1971).
46. P. Pincus, *Solid St. Commun.* **7**, 415 (1969).
47. R. Blinc, D. Hogenboom, D. O'Reilly and E. Peterson, *Phys. Rev. Lett.* **23**, 969 (1969).
48. V. Domenici, in Ph. D. thesis: Structure, orientational order and dynamics of rod-like and banana-shaped liquid crystals by means of deuterium NMR: new developments, University of Pisa, Pisa, 2005.
49. R. Y. Dong, *Mol Phys.* **88**, 979 (1996).
50. L. Calucci and M. Geppi, *J. Chem. Inf. Comput. Sci.* **41**, 1006 (2001).
51. D. Catalano, M. Cifelli, M. Geppi and C. A. Veracini, *J. Phys. Chem. A* **105**, 34 (2001).
52. S. Urban, B. Gestblom, C. Gandolfo, C. A. Veracini, *Zeitsch. Natur. A*, 57, 819 (2002).
53. D. Catalano, L. Chiezzi, V. Domenici, M. Geppi and C.A. Veracini, *J. Phys. Chem. B*, **107**, 10104 (2003).
54. F. Perrin, *J. Phys. Radium* **5**, 497 (1934).
55. L. Chiezzi, V. Domenici, M. Geppi, C. A. Veracini and R. Y. Dong, *Chem. Phys. Lett.* **358**, 257 (2002).
56. M. Cifelli, C. Forte, M. Geppi and C. A. Veracini, *Mol. Cryst. Liq. Cryst.* **372**, 81 (2001).
57. V. Domenici, M. Geppi, C. A. Veracini, R. Blinc, A. Lebar and B. Zalar, *ChemPhysChem* **5**, 559 (2004).

58. V. Domenici, M. Geppi, C.A. Veracini and A.V. Zakharov, *J. Phys. Chem. B* **109**, 18369 (2005).
59. V. Domenici, A. Marini, R. Menicagli, C. A. Veracini, A. Bubnov, M. Glogarova, in *Proceeding of SPIE - International Conference of Optics and Opto-electronics*, Vol. 6587, pp. F1-F13 (SPIE publishing, Washington 2007).
60. V. Domenici, J. Czub, M. Geppi, B. Gestblom, S. Urban and C.A. Veracini, *Liq. Cryst.* **31**, 91 (2004).
61. V. Domenici, M. Geppi and C.A. Veracini, *Chem. Phys. Lett.* **382**, 518 (2003).
62. V. Domenici, M. Geppi, C. A. Veracini and R. Y. Dong, *Liq. Cryst.* **33**, 479 (2006).
63. R. Y. Dong, J. Zhang, C. A. Veracini, *Solid State NMR*, **28**, 173 (2005).
64. V. Domenici, A. Marini, C. A. Veracini, R. Menicagli, C. Malanga, *J. Phys. Chem. B*, submitted.
65. E. Berggren, C. Zannoni, *J. Chem. Phys.* (1993) **99**, 6180.
66. R. Tarroni, C. Zannoni, *J. Chem. Phys.* (1991) **95**, 4550.
67. G. Williams, in *The Molecular Dynamics of Liquid Crystals*, Eds. G. R. Luckhurst and C. A. Veracini, NATO ASI Ser. A, Vol. 431, Chapter 17, (Reidel, Dordrecht, 1989).
68. A.V. Zakharov and R.Y. Dong, *Phys. Rev. E* **63**, 011704 (2000).
69. A.V. Zakharov and R.Y. Dong, *Phys. Rev. E* **64**, 031701 (2001).
70. S. Urban, V. Domenici, J. Czub, M. Geppi, B. Gestblom and C. A. Veracini, in *Neutron Scattering and Complementary Methods in Investigations of Condensed Phase*, Ed. Janusz Chrusciel, Vol. 1 (University of Podlasie Publishing House, Siedlce, 2003) pg. 121-128.
71. W. T. Kozak, J. Mosciski, G. Williams, *Mol. Cryst. Liq. Cryst.* **201**, 1 (1991).
72. W. T. Coffrey, Yu P. Kalmykov, *Adv. Chem. Phys.* **113**, 487 (2000).
73. J. Jadzyn, C. Legrand, G. Czechowski, D. Bauman, *Liq. Cryst.* **24**, 689 (1998).
74. W. Zajac, S. Urban, V. Domenici, M. Geppi, C. A. Veracini, M. Telling and B. J. Gabrys, *Phys. Rev. E* **73**, 051704 (2006).
75. V. Domenici, T. Apih and C. A. Veracini, *Thin Solid Films*, **517**, 1402, (2008).
76. G. Khote and K. Muller, in *The Molecular Dynamics of Liquid Crystals*, Eds. G. R. Luckhurst and C. A. Veracini, NATO ASI Ser. A, Vol. 431, Chapter 19, (Reidel, Dordrecht, 1989).
77. J. C. Blackburn and P. K. Kilpatrick, *Langmuir* **8**, 1679 (1992).
78. C. R. Rodriguez, D. J. Pusiol, A. M. F. Neto, C. A. Martin, *Phys. Rev. E* **69**, 041708 (2004).
79. P. Hakansson, P. O. Westlund, *Phys. Chem. Chem. Phys.* **17**, 4321 (2004).
80. V. Domenici, M. Geppi, C. A. Veracini, A. Lebar, B. Zalar and R. Blinc, *J. Phys. Chem. B* **109**, 769, (2005).
81. V. Domenici, K. Fodor-Csorba, D. Frezzato, G. Moro and C. A. Veracini, *Ferroelectrics* **344**, 19, (2006).
82. V. Domenici, D. Frezzato and C. A. Veracini, *J. Phys. Chem. B*, **110**, 24884 (2006).

83. J. Xu, R. Y. Dong, V. Domenici, K. Fodor-Csorba and C. A. Veracini, *J. Phys. Chem. B*, **110**, 9434 (2006).
84. M. Cifelli and V. Domenici, *Phys. Chem. Chem. Phys.* **9**, 1202 (2007).
85. V. Domenici, M. Cifelli, C. A. Veracini, V. E. Agina, N. V. Boiko, V. P. Shibaev, *J. Phys. Chem. B*, **112**, 14718 (2008).
86. V. Domenici, C. A. Veracini, K. Fodor-Csorba, G. Prampolini, I. Cacelli, A. Lebar, B. Zalar, *ChemPhysChem* **8**, 2321 (2007).

Chapter 9

Translational Self-diffusion Measurements in Thermotropics by means of Static Field Gradients NMR Diffusometry

Mario Cifelli

Dipartimento di Chimica, Pisa University,
Via Risorgimento 35, 56126, Pisa, Italy
email: Mario@dcci.unipi.it

Nuclear Magnetic Resonance Diffusometry (NMR Diffusometry) is a convenient and non invasive way to investigate translational spin motions. Hahn echo based experiments performed in the presence of pulsed as well as static magnetic field gradients allow us to measure the displacement of the observed spins. This methodology has also been successfully applied to liquid crystals as testified by a recently published review in this field. In this context, we will present and discuss in details the implementation and the exploitation of a Static Field Gradient approach to the study of molecular translational self diffusion in calamitic liquid crystals.

9.1. Introduction: Why Translational Diffusion in Liquid Crystals?

Nuclear Magnetic Resonance (NMR) Diffusometry is nowadays widely considered as a powerful and versatile means to study mass transport phenomena in fluids materials[1] as well as in biological sciences.[2] The exploitation of specific spatial dependent magnetic field gradients, either pulsed or constant throughout the experiment, and the unnecessary need of using probes such as radio tracers or spin-probes provide a full non invasive way of studying molecular diffusion processes.[3,4,5] Among these processes, one may say that the most basic and ubiquitous one is indeed the so-called "self-diffusion" that stems from the intrinsic kinetic energy of molecules. Due to this energy molecules in fact scatter around in homogeneous fluids at the chemical and thermal equilibrium. These random molecular motions can be described by a distribution of the probability density, $p(\mathbf{r},t)$ that a molecule is displaced of a distance **r** in the time t.[5] That is, the probability to find the displaced molecule in a region defined as $x+dx$, $y+dy$, $z+dz$ is $p(\mathbf{r},t)dxdydz$. In analogy to Fick's law for diffusion, for free isotropic diffusion the time evolution of $p(\mathbf{r},t)$ can be expressed as:

$$\frac{\partial p(\mathbf{r},t)}{\partial t} = D\nabla^2 p(\mathbf{r},t) \tag{9.1}$$

where $p(\mathbf{r},t)$, is normally referred in this context as "diffusion propagator" and the constant D, which quantifies the proportionality between the two terms of the equation, is the so-called *diffusion coefficient*. This coefficient expresses the natural tendency of particles to move, for the same spatial variation of $p(\mathbf{r},t)$; particles with a higher D will scatter faster while slower ones will be characterized by a lower D. This is mathematically expressed by the Stoke-Einstein equation:

$$D = \frac{kT}{f} \tag{9.2}$$

with k being the Boltzmann constant, T the temperature and f the frictional factor. It is easy to understand that "kT" expresses the kinetic energy of the diffusing particle, while f is a kind of "fingerprint", as it depends on the viscosity η of the medium and also on the "shape" of the molecule. In the simple case of a spherical particle, for example:

$$f = 6\pi\eta R \tag{9.3}$$

with R being the radius of the particle. Surely the reader has noticed that the determination of D can also be useful to evaluate both viscosity as well as the shape of a diffusing particle; this is indeed true, but as this subject is beyond the scope of this introductory section the interested readers are referred to specialized monographs.[4,5] Now, to solve Eq. (9.1) one can consider, the initial condition $p(\mathbf{r},0)=\delta(\mathbf{r})$, the boundary condition $p(\mathbf{r},t)\rightarrow 0$, as $\mathbf{r}\rightarrow\infty$, and the normalization condition, to obtain:

$$p(\mathbf{r},t)=\frac{1}{(4\pi Dt)^{\frac{3}{2}}}\exp\left(-\frac{r^2}{4Dt}\right). \tag{9.4}$$

Equation (9.4) will be useful in the next Sec. where how NMR can be used to determine the diffusion coefficient will be recalled. Now, let us go back to the question posed in the title of this introduction:

"Why do we want to measure translational diffusion in thermotropic liquid crystals (LC)?"

Citing F. Noack, the answer might be: "Diffusion, in particular self-diffusion, belongs to the properties of liquid crystals which are only poorly understood ...."[6] and then "The unsatisfactory situation is primarily due to experimental limitations and related to theoretical problems arising from the evaluation of the measurements ...".[6] Of course the situation has improved a little from the 80's and recently it has been shown that the accuracy of measurements is now sufficient to exploit experimental data for testing theoretical models and for phase behaviour discriminations.[7] However, it must also be pointed out that the occurrence of novel phases, such as chiral smectic ones,[8] as well as the mesophase formed by even more exotic systems, such as V-shaped molecules,[9] LC dendrimers,[10] and mesogens tailored to stabilize biaxial nematic phases,[11] poses new questions to people (like the author) working in this field. Moreover, under a more practical point of view, it is noted that self-diffusion is intimately related to properties with potential implications in LC's applications, such as viscosity, for example. While at a molecular level, diffusivity (and its anisotropy) is also strictly related to the order and supramolecular structure of LC phases. In other words, self-diffusion can also be a valuable means to investigate structural properties in liquid crystal phases. An example of this exploitation will be presented in Sec. 9.5.1, where how positional order parameters in smectic phases can be evaluated from diffusion data is discussed. Hence, it is hoped that the reader will agree that there are still good reasons to study and to improve our understanding of translation diffusion processes in liquid crystals.

9.2. The Basic Principles of NMR Diffusometry: From the Echo Decay to the Self-Diffusion Coefficient

As anticipated in the introduction, NMR diffusometry has proved to be a valuable tool to study mass transport in fluids and the technique has been extensively reviewed[3] as well as several monographs have been devoted on the subject.[4,5] Assuming that the reader has familiarity with the basic principles of NMR diffusometry, here how the exploitation of magnetic field gradients makes Hahn echoes sensitive to spin displacements is recalled. In order to do this one considers a simple pulsed gradient Hahn spin echo, depicted in Fig. 9.1, applied to three equivalent spins in three different positions along the gradient direction (the z axis) in the following three motional conditions:

(a) Non moving spins

(b) Spins flowing in one direction

(c) Spins scatter around randomly under Brownian motions

In this experiment the first pulsed gradient creates a spatial dependent static field that produces a z dependent phase shift for the spins:

$$\Delta\varphi(z) = \gamma G_z \delta \times z$$

where γ is the nuclear gyromagnetic ratio, G_z the gradient strength and δ the duration of the pulsed gradient, considered as rectangular in shape. The 180° hard radio frequency (RF) pulse reverses the transverse spin magnetization, that is the sign of $\Delta\phi(z)$ and the second pulse gradient does the same job as the first one. After the second gradient pulse, the echo amplitude depends on if and how the spins have moved during the time Δ. It is easy to understand that case (a), corresponding to static spins, is equivalent to performing the experiment in a homogenous field. The dephasing due to the first gradient pulse is completely compensated by the second gradient with no loss of signal. Case (b) corresponds to a collective coherent displacement of spins, namely a flow. As shown in the figure, this kind of displacement produces a phase shift on the echo but no intensity reduction. On the other hand, case (c) represents a random spin displacement (i.e. spins exchange positions) that compromises the phase shift refocusing by the second gradient, eventually leading to a signal loss.

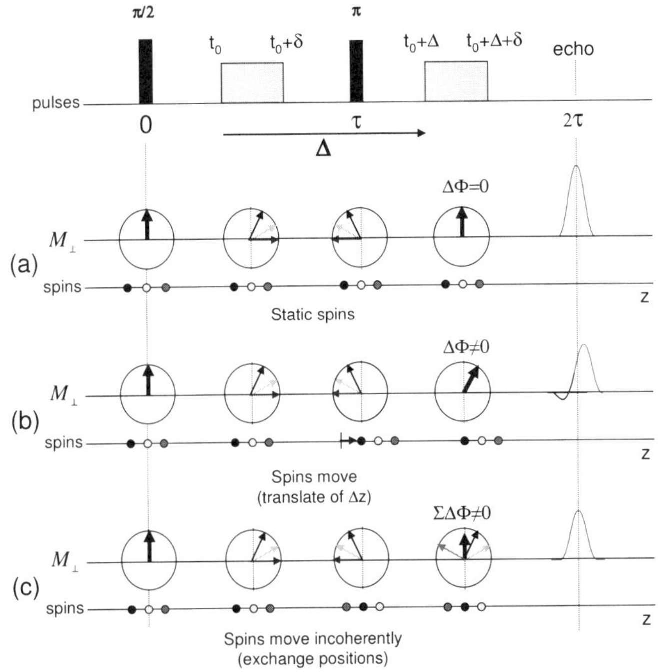

Fig. 9.1. A simple Spin Echo experiment with pulsed field gradient in the case of (a) static spins, (b) spins flowing along the z direction, (c) spins scattering randomly along the z direction.

Now, what one has to do is to relate this signal loss to the diffusion coefficient D. An easy and direct way to do this is writing down the transverse magnetization $M_\perp$ at the echo time 2τ as follows:

$$M_\perp = \iint \exp\left[i\left(\gamma G_z \delta \times \left(z_0 - z_f\right)\right)\right] p\left(z_0\right) P\left(z_0, z_f, \Delta\right) dz_0 dz_f \,. \quad (9.5)$$

Here, the term $\left[i\left(\gamma G_z \delta \times \left(z_0 - z_f\right)\right)\right]$ represents the phase shift difference gained by a spin that is at the starting position z_0 during the first pulsed gradient and at the final position z_f during the second one, while $p(z_0)$ is the spins distribution at the beginning of the experiment. Again one has the conditional probability $P\left(z_0, z_f, \Delta\right)$ describing the probability of finding a particle at position z_f after a time Δ, starting at position z_0. For free diffusion along the gradient direction, it can be expressed by Eq. (9.4). In case of an homogenous fluid $p(z_0)$ is actually the density of fluid and can be considered constant. Then, after the integration, in the range $(-\infty,+\infty)$ Eq. (9.5) gives the echo intensity expression:

$$A(\delta,\Delta,G) = A_0 \exp\left[-(\gamma G\delta)^2 D\left(\Delta-\frac{\delta}{3}\right)\right]. \tag{9.6}$$

That is, the echo intensity coming from the first 90° pulse A_0 decays exponentially as a function of the gradient strength G, as well as the time delays δ and Δ. The slope of the decay depends on the diffusion coefficient D along the gradient direction.

9.2.1. *Spin echo versus stimulated echo*

In the above, it is seen how the echo intensity can be used to evaluate the diffusion coefficient D via its attenuation by varying k (standing for δ or G)* or Δ. However, the diffusion effect on the echo intensities adds to those from other processes, such as relaxation, and a more complete expression for the echo decay should be:

$$A_{SE} = A_0 \exp\left\{-k^2 D\left(\Delta-\frac{\delta}{3}\right)\right\}\exp\left\{-\frac{2\tau}{T_2}\right\} \tag{9.7}$$

where the last term stands for spin-spin relaxation occurring during the evolution time 2τ. Now, it is easy to notice that the shorter the T_2, the shorter is the time window available for the gradient pulsing ($\delta \leq \tau$); that is, fast transverse relaxation is a limitation for the gradient application.

---

*The definition $k=\gamma Gt$, with t being the time the gradient is applied, is an important concept in MRI. It has the dimension of the inverse of a length as it represents the observable in the reciprocal space. In this context it is considered as a compact way to define the effect of the gradient on the echo decay. That is, the higher the k, the fastest the decay.

It is learnt from another chapter in this volume that as in solid systems, transverse spin relaxation times in liquid crystal phases are quite short, while longitudinal relaxation time can be relatively much longer. For example, for protons, one can say that if T_2 is estimated around 1 *ms*, T_1 can be as long as hundreds of milliseconds, that is $T_1 >> T_2$. This condition is indeed quite unfavorable for a direct echo experiment, as depicted in Fig. 9.1, but can be circumvented to extend the diffusion time Δ if one considers the so-called Stimulated Echo (STE), reported in Fig. 9.2.[3-5] In this experiment, the basic concept is that

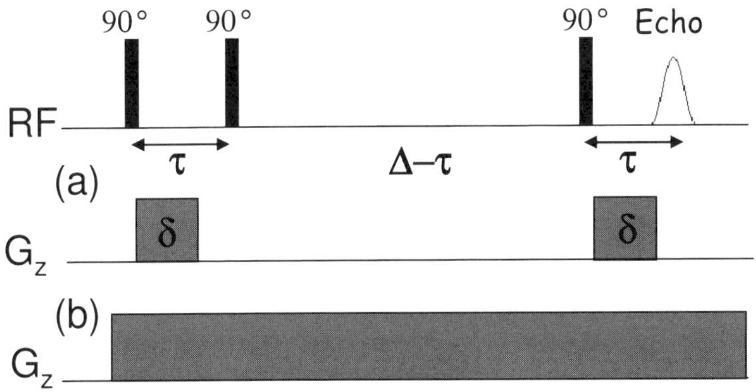

Fig. 9.2. Stimulated Echo Experiment: (a) with Pulsed Field Gradient along the z axis and (b) in presence of a constant gradient along the z axis. During the time (Δ–τ) magnetization evolves under longitudinal relaxation.

after the first "phase encoding" of spin position, the magnetization is "stored" along the z axis and during the diffusion time Δ, it evolves mostly under the longitudinal relaxation, leading to an echo that is refocused at time $t = \Delta + \tau$ with the following expression for the echo decay:

$$A_{STE} = \frac{A_0}{2} \exp\left\{ -k^2 D \left(\Delta - \frac{\delta}{3} \right) \right\} \exp\left\{ -\frac{2\tau}{T_2} \right\} \exp\left\{ -\frac{\Delta - \tau}{T_1} \right\}. \quad (9.8)$$

It's easy to understand that this experiment, even if it refocuses half of the magnetization comparing to the direct echo, allows much longer times for diffusion.

9.2.1.1. *Echo attenuation in a Static Field Gradient (SFG)*

So far pulsed field gradients are considered, that is, field gradients that can be turned on and off during the experiment. As a methodology will be discussed in Sec. 9.4 that employs a constant static field gradient, it is noted that the same expressions for the echo decay can essentially be obtained in this case. In particular for the stimulated echo, one can simply consider that $\delta = \tau$ for the gradient being on during the whole τ delays.[4]

$$A_{STE} = \exp\left\{ -k^2 D\left(\Delta - \frac{\tau}{3} \right) \right\} \exp\left\{ -\frac{2\tau}{T_2} \right\} \exp\left\{ -\frac{(\Delta - \tau)}{T_1} \right\} \quad (9.9)$$

where $k = \gamma G\tau$. The commonly operative condition of $\Delta >> \tau$ simplifies the echo decay to:[4]

$$A_{STE} = \exp\{-k^2 D\Delta\} \exp\left\{ -\frac{2\tau}{T_2} \right\} \exp\left\{ -\frac{\Delta}{T_1} \right\}. \quad (9.10)$$

9.2.2. A couple of words on the dynamic range of the measurements: which nucleus and which gradient?

Since the Stimulated Echo is the best experiment to measure diffusion in LC phases, it is now worthy to spend a few words about the dynamic range one expects to achieve in such experiments, in terms of *the smallest diffusion coefficient one can measure in the presence of specific relaxation rates*. As seen from Eq. (9.10), spin relaxation is a competing factor with diffusion attenuation. This is quite clear by considering the following expression:[6]

$$G_{\max} = \frac{4}{(\gamma T_2) T_1 D}. \quad (9.11)$$

This expression gives the maximum gradient strength required to reduce the initial echo intensity A_0 of a factor ($1/e$) in the presence of a diffusion coefficient D, while time delays are kept in the range of the relaxation times, that is $2\tau = T_2$ and $\Delta = T_1$. Of course, the gradient strength required depends on the relaxation times as well as on the diffusion coefficient, but also, on the *gyromagnetic ratio* of the nucleus exploited. Considering for example proton, that has the *highest* γ, typical relaxation times of $T_1 = 1s$ and $T_2 = 0.1ms$ and a typical diffusion coefficient $D = 10^{-11}\, m^2/s$ for LC phases,[7] as will be seen in Sec. 9.5, the maximum gradient strength needed should be $G \approx 4000 G/cm$, that is indeed quite high. It is anticipated that the situation gets worse as the relaxation becomes faster and the diffusion slower, as it normally happens when a liquid crystal is cooled down.

As proton presents so short T_2, one would be tempted to consider other nuclei. In this case one has to consider that what one may (and let us underline "may") gain for example in extending time delays (i.e., because of longer relaxation times, for example), but lose in signal intensity as well as in gradient efficiency. Just to give an example, comparing proton to carbon-13, even in the case of enriched samples, the loss is about one order of magnitude in sensitivity while for the gradient the factor is about 1/16, as $k^2 = (\gamma G \tau)^2$. This is because of the rescaling of the *gyromagnetic* ratio. Possibly a good alternative to proton might be fluorine, if present, only for the case of favorable relaxation conditions. This should be carefully checked, with suitable relaxation experiments. In short, proton seems to be the best choice and this chapter thus focuses on proton NMR diffusometry.

9.3. NMR Diffusometry in Thermotropic Liquid Crystals: Measuring Diffusion in Anisotropic Systems

People working in LC science surely know that the most fundamental property of these systems is the presence of a certain degree of long range molecular order, either orientational or positional, in a fluid context. This molecular order induces a certain degree of anisotropy in LC physical properties and molecular translational diffusion is no exception; it is in fact easy to imagine that in the nematic phase of calamitic LC, molecules diffuse faster along the phase director than across it. As translational self diffusion is orientation dependent, it has to be represented by a second rank tensor. Thus Eq. (9.10), expressing the echo decay for diffusion measured in a generic Laboratory reference frame (X,Y,Z) along a generic gradient $\vec{\mathbf{G}} = (G_X, G_Y, G_Z)$, changes into:

$$A = \frac{A_0}{2}\exp\left[-\gamma^2\tau^2\Delta\left(\vec{\mathbf{G}}\cdot\tilde{\mathbf{D}}\cdot\vec{\mathbf{G}}\right)\right]\exp\left\{-\frac{2\tau}{T_2}\right\}\exp\left\{-\frac{\Delta}{T_1}\right\} \quad (9.12)$$

where $\tilde{\mathbf{D}}$ is the symmetric second rank self-diffusion tensor expressed in the Laboratory (gradient) frame (LAB):

$$\tilde{D}^{LAB} = \begin{pmatrix} D_{XX} & D_{XY} & D_{XZ} \\ D_{YX} & D_{YY} & D_{YZ} \\ D_{ZY} & D_{ZY} & D_{ZZ} \end{pmatrix}.$$

Hence, in the echo decay expression, the gradient orientation also becomes important as:

$$\vec{G}\cdot D^{LAB}\cdot\vec{G} = D_{XX}^{LAB}G_X^2 + D_{YY}^{LAB}G_Y^2 + D_{ZZ}^{LAB}G_Z^2$$
$$+2D_{XZ}^{LAB}G_XG_Z + 2D_{XY}^{LAB}G_XG_Y + 2D_{YZ}^{LAB}G_YG_Z.$$

In general, in order to determine the full information on self-diffusion one should determine five different terms in the matrix, measuring diffusion (at least) along five different gradient directions.[3] This quite daunting task can be strongly simplified by considering the diffusion tensor in its principal axis frame (PAS).

$$D = \begin{pmatrix} D_{XX} & D_{XY} & D_{XZ} \\ D_{YX} & D_{YY} & D_{YZ} \\ D_{ZY} & D_{ZY} & D_{ZZ} \end{pmatrix} \xrightarrow[\textit{Diagonalization}]{} \begin{pmatrix} D_{xx} & 0 & 0 \\ 0 & D_{yy} & 0 \\ 0 & 0 & D_{zz} \end{pmatrix}$$

In this special reference frame, one has one axis parallel to the phase director for the case of nematic and smectic A phases, and only three diagonal terms of the tensor should be determined. Moreover, for phases with a uniaxial symmetry, the tensor further reduces by the conditions $D_{//} = D_{zz}$ and $D_{\perp} = D_{xx} = D_{yy}$. As far as calamitic $\Delta\chi > 0$ LCs are concerned, this condition can be easily achieved by noticing that the nematic, as well as the smectic A phases, align with the phase director parallel to the static magnetic field direction, as shown in Fig. 9.3, and the PAS frame is equivalent to the LAB frame. That is, if self diffusion is measured in the PAS frame, one can just determine the diffusion coefficient along ($D_{//}$) and across ($D_{\perp}$) the phase director, in order to determine the full tensor $\tilde{\mathbf{D}}$. This can be achieved measuring diffusion along the z direction and along x (or y) direction of the LAB frame.

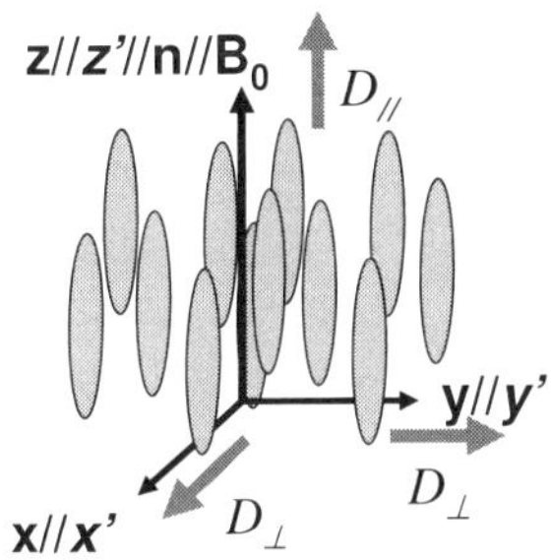

$$\mathbf{D}_{LAB} = \mathbf{D}_{PAS} \equiv \begin{pmatrix} D_{\perp} & 0 & 0 \\ 0 & D_{\perp} & 0 \\ 0 & 0 & D_{//} \end{pmatrix}$$

LAB = PAS

Fig. 9.3. The alignment of a uniaxial phase formed by $\Delta\chi>0$ mesogens in the magnetic field. The phase director is parallel to the static field direction and the laboratory frame and the PAS frame of the diffusion tensor coincide.

In normal conditions, that means having available equipment with only a gradient along the z direction, as it is the case of commercial diffusion probes, only the parallel component of the diffusion tensor can be measured. In order to measure also one normal component one can imagine two possibilities: (i) applying gradients along a direction normal to the phase director, (ii) rotating the sample in order to let the phase director forming an angle $\theta \neq 0$ with the (fixed) gradient direction. The former approach can be pursued if, for example, a micro-imaging probe with a x, y, z pulsed gradient set is available, while the latter depends on the possibility of rotating the sample in the magnetic field, for example, by means of a goniometric probe. Of course, in this case the sample should not realign in the magnetic field after rotation. In Sec. 9.5, the case of uniformly oriented smectic A phases are addressed, where the static field strength is not high enough to induce sample reorientation.

9.3.1. *Further "anisotropic complications" of residual couplings*

It is well known that LC from an NMR point of view present "solid-like" behaviors. In particular, even if molecular dynamics can be fast in the NMR timescale the so-called motional averaging is still incomplete due to the anisotropy of molecular motions (see Chap. 8). For example looking at proton NMR, spectra are strongly affected by residual dipolar couplings. The presence of these couplings that can be as high as tens of KHz in frequency, also affects diffusometry experiments. The next paragraphs will discuss these effects in details.

9.3.1.1. *Fast transverse relaxation in liquid crystals*

As seen in the previous section, NMR diffusion experiments give a measure of the attenuation of the echo signal due to the increasing of the *blurring* of the phase encoding during the gradient application caused by the spin displacements as k increases (either by increasing G or the gradient time, τ or δ). In the case of LC, k values are limited by the fact that T_2 is quite short; as already discussed in Sec. 9.2.2, for protons it can be less than $100\ \mu s$. Just to be more specific, recalling Eq. (9.11) even to determine diffusion coefficient of $10^{-10}\ m^2/s$, a gradient of more than $1000\ G/cm$ is needed. Such high pulsed gradients (PFG) can be achieved only on dedicated diffusion probes with high current generators, but are for example out of range for microimaging as well as for standard NMR diffusometry set-up.[¥] This limitation can be tackled basically in two ways:

1. Applying line-narrowing techniques, such as the one used in solid state NMR, where the line broadening needs to be alleviated.
2. Exploiting very high static gradients in order to make the encoding-decoding times very short. This approach is the subject of this chapter and will be discussed in details in Sec. 9.4.

[¥] Nowadays, diffusion dedicated probes can reach PFG as high as 3000 G/cm, limiting the gradient to the z direction, while microimaging systems can provide three-dimensional gradients up to about 300 G/cm only.

9.3.1.2. *Dipolar correlation effect (DCE): slow modulation of residual dipolar couplings*

The residual coupling previously mentioned derives from the incomplete motional averaging due to fast anisotropic motions, where fast is referred to the NMR time-scale. In LC, the "dynamic landscape" is quite various, as different kinds of motions are present with different correlation times. Let us now consider the stimulated echo experiment (STE):

$$\left(\frac{\pi}{2}\right)_x - \tau - \left(\frac{\pi}{2}\right)_x - (\Delta - \tau) - \left(\frac{\pi}{2}\right)_x - \tau - echo \qquad (9.13)$$

and take into account motions with correlation times of the same order of magnitude as the time delays $\tau_c \approx \tau, \Delta$. These motions can modulate the dipolar coupling during the experiment when the time delays are changed, as shown in Fig. 9.4. This is the so-called dipolar correlation effect (DCE), first used by Callaghan and co-workers to investigate slow motions in polymer melts.[12]

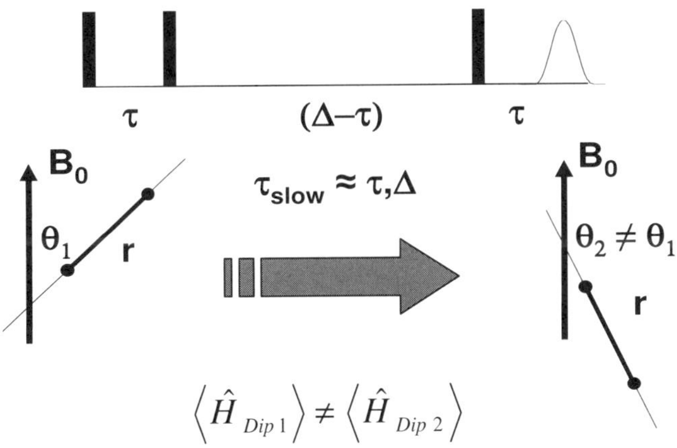

Fig. 9.4. A pictorial representation of DCE for the dipolar interaction between two protons. Assuming that at the beginning of the experiment the angle between $\boldsymbol{r}$ and the static field is θ_1, corresponding to a dipolar interaction expressed by $\langle \hat{H}_{Dip1} \rangle$. As time passes, the orientation of $\boldsymbol{r}$, modulated by dynamics with correlation time $\tau_c \approx \tau, \Delta$, also changes leading to a different dipolar interaction $\langle \hat{H}_{Dip2} \rangle$ at the echo time. The correlation between the initial and final orientation of the dipole can be exploited to investigate the dynamics involved in the process.

As far as LCs are concerned, motions that can induce DCE are present, as for example the director fluctuation in the nematic phase,[13] as well as layer undulations in the smectic A phase. These motions in fact are characterized by correlation times of the same order of magnitude as the delays in the STE (tens of microseconds to milliseconds). This effect of course, is not a real problem in diffusometry experiment, when it is performed keeping the delays constant and changing the gradient strength, but can be a complication if experiments are performed in a Static Field Gradient (SFG). In this case time delays have to be varied, and DCE has to be taken into account.

9.4. A SFG Approach to Measure Translational Self-diffusion

As seen in the previous section, measuring self-diffusion in LC presents peculiar complications related to their unique anisotropic nature. Since the very beginning of NMR studies of LC, different approaches to circumvent these complications have been proposed and exploited. A few of them requires specially designed hardware, such as field cycling coupled with pulsed field gradients,[14] specially designed magnets and NMR probes,[15] while other are based on combining homonuclear decoupling to STE experiments in order to extend the transverse relaxation times.[7] A recent review on this subject has been published by Dvinskikh and Furó.[7] Detailed information on these methodologies can be found in this review and in the references therein. In this chapter, an alternative methodology that exploits the very high steady gradient in the fringe field of commercial superconducting magnets is presented. This approach, despite presenting some drawback in comparison with pulsed field gradient techniques, also has the relevant advantage of making an experimental set-up equivalent to a commercial diffusion dedicated probe with essentially no hardware upgrade requirements.

9.4.1. *Stray field NMR diffusometry: Exploiting the SFG present in the fringe field of the magnet*

As already discussed in Sec. 9.3.1.1, an alternative way to overcome fast transverse relaxation could be through exploiting a gradient high enough to fulfill the conditions discussed in Sec. 9.2.2. It has already been noticed that a gradient strength of about $G = 4000 G/cm$ could be sufficient to measure a diffusion coefficient as low as $10^{-11} m^2/s$, in systems with a transverse relaxation time of about 100 μs. A static field of this feature is always present in the fringe field of commercial superconducting magnets. As depicted in Fig. 9.5, the magnet is assembled to have a very high static and homogenous field B_0 in the isocenter and at the same time to let the field decrease sharply as one moves farther from that region. Historically, the first exploitation of this static gradient that can be as high as 10^4 G/cm was presented by Samoilenko in the context of NMR microimaging of solids.[16] Later, Kimmich and Fischer applied the methodology to measure translational diffusion in polymer melts.[17] In that case, diffusion coefficients as low as $10^{-14} m^2/s$ have been measured by exploiting a gradient of about

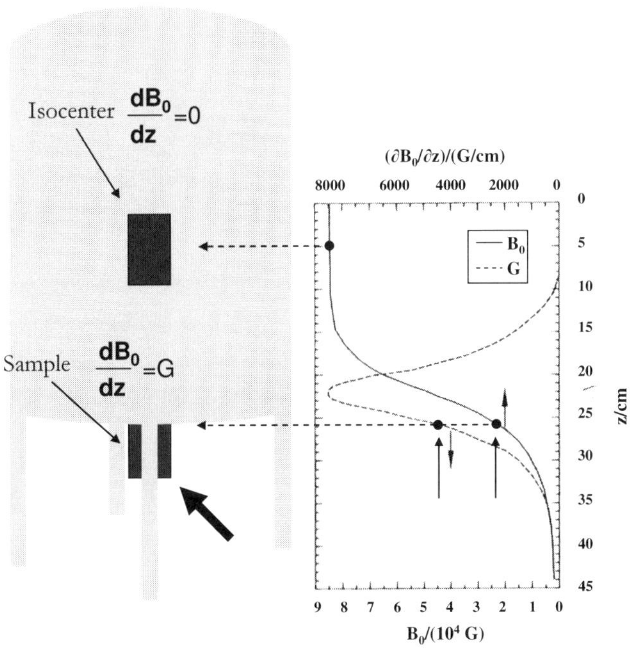

Fig. 9.5. A schematic of static field in a superconducting magnet and its field gradient profile along the z axis. The field has its maximum strength in the isocenter, where the gradient is practically zero (also thanks to a good shimming), However, the field strength goes down steeply and a huge gradient is present down along the z axis.

$G = 5000G / cm$.[17] This technique has also been recently reviewed by Geil.[18] In this excellent review however, the emphasis is on applications devoted to measure translational self-diffusion in slow diffusing systems, such as polymer melts and supercooled fluids. Here the challenge is the presence of very slow molecular translation, more than fast relaxation. This chapter can hence be seen also as a *compendium* to that review, as SFG NMR diffusometry is applied here to a class of compounds characterized by relative fast translational motions and very fast transverse relaxation (besides strong anisotropy). In this discussion some aspects relevant for a correct experimental set-up, such as RF pulses, gradient strength calibration and sample positioning are highlighted.

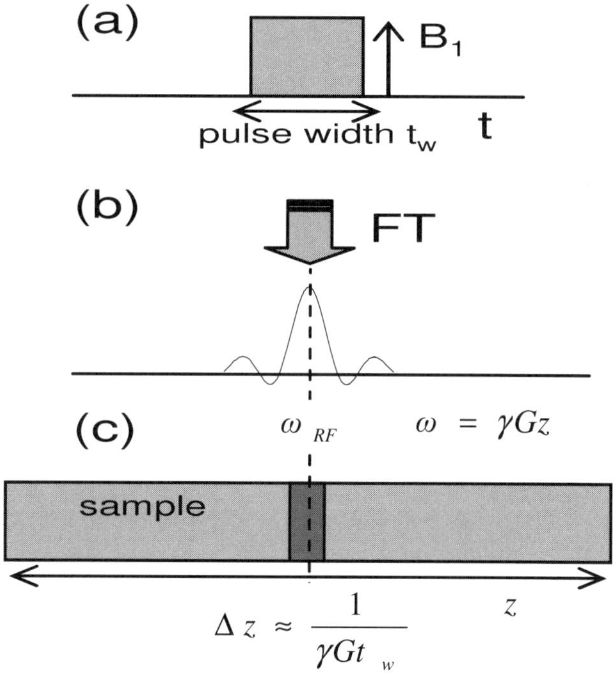

Fig. 9.6. (a) An RF pulse in a high field gradient of strength G behaves like a soft pulse with a frequency profile sketched in (b). It excites spins only in a limited spatial region depicted as Δz (the dark slice in (c)) that depends on the pulse width as well as on the gradient strength.

9.4.1.1. *Pulses in the SFG and their calibrations*

Acquiring NMR experiments in the presence of steady gradients (between $G = 3000 \div 10000 G/cm$) in the fringe field of superconducting magnets, implies a few differences when comparing to conventional NMR and is addressed here. First of all, it is noted that such a gradient produces a spread of resonant frequencies (for ^{1}H) of tens of MHz/cm. Hence, even a square pulse of about 1μs, normally considered hard in conventional NMR has to be considered soft here, as it excites a thin slice of thickness of about $0.1 \div 1mm$. Moreover, the strong inhomogeneous field leads to very short signal decay (T_2^* of less than a few μs) that is well below the typical spectrometer dead time. This leads to two consequences:

1. No FID can be acquired under the condition,
2. Any RF pulse produces a (practically inhomogeneous) local excitation in a sample such as shown in Fig. 9.6.

The former limitation can be easily overcome by taking advantage of Hahn echo that yields signal well clear of the spectrometer dead time in the form of a sharp echo. Regarding the latter, it is noticed that NMR information can be extracted on a limited portion of the sample (about 10-100 μm). This peculiarity is the basis of the STRAFI microscopy technique,[19] and it can also be suitably exploited to produce a profile along the gradient direction of the excited region with a resolution of a few microns.[16] Alternatively the echo itself despite it completely lacks spectral resolution, can be used as a single point of a profile: other points along the gradient direction can be then acquired moving the sample in the field as discussed in Ref. [19]. As far as self-diffusion measurements are concerned, the slice thickness, as represented in Fig. 9.6, should not represent a problem, besides reducing the effective sample being inspected. In fact if the sample is homogeneous, the attenuation of STE can be well expressed by Eq. (9.10), i.e.

$$A(k,\Delta) = \exp\{-k^2 D\Delta\}\exp\left\{-\frac{2\tau}{T_2}\right\}\exp\left\{-\frac{\Delta}{T_1}\right\} \qquad (9.14)$$

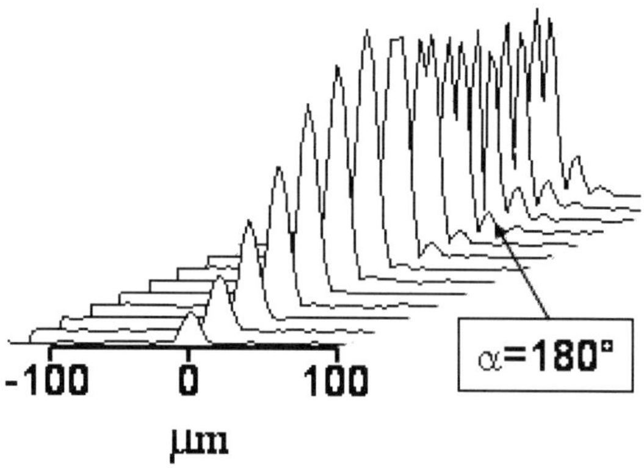

Fig. 9.7. Pulse calibration in a static fringe of G=5300 G/cm performed on a rubber sample of about 5 mm thickness along the gradient direction. The τ delay was kept 40 µs and pulse width was fixed to 10 µs while its amplitude was increased gradually; in this condition the slice excited was kept constant to about 20 µm as shown. The 180° pulse can be easily identified from the Fourier transformed echo that presents a "hole in the middle" corresponding to no transverse magnetization on resonance.

as in the PFG experiment. The most important difference consists of the fact that using the stimulated echo acquired in the presence of a strong field gradient, no spectral resolution can be exploited to distinguish multi-component systems. However, other features can be used, such as, obvious difference in diffusion coefficients, as well as different relaxation times.[18] This concept has also been exploited to develop an exotic way to perform DOSY via SFG diffusometry.[20]

So far it is seen that RF pulses refocus the signal in a form of sharp echoes in the SFG. However, the stimulated echo experiment should require 90° pulses, in order to work properly. While it is quite straightforward to calibrate pulses in homogeneous fields, for example operating on resonance, one can imagine that in the SFG things may be different as in fact it is. So the concept of "pulse calibration" should be redefined and specific methods to calibrate pulse lengths in the SFG have been proposed.[21-23] Of course, the calibration refers to the "on resonance" position. Furthermore, for carrying out stimulated echo experiment one also requires for example 180° pulses (see Sec. 9.4.2.1). Here a SFG pulse calibration method is based on the behavior of the echo in the following experiment:[23]

$$\alpha_x - \tau - \alpha_y - \tau - echo\,. \tag{9.15}$$

In this experiment the pulse angle α, can be calibrated by checking the shape of the Fourier transformed echo as the pulse changes. In general, it is preferable to keep the pulse length t_w constant, as this preserves the excited slice dimension constant (see Fig. 9.6), and the pulse is changed by increasing its amplitude. The experiment is performed by choosing a pulse length, a suitable echo delay τ and then the amplitude, starting from a low value, is gradually increased as shown in Fig. 9.7.

When the pulse amplitude is so that the pulse angle $\alpha = 180°$, the FT echo has a sharp hole in the center. Once the 180° pulse is identified, the 90° can be easily settled by halving the amplitude.[23] Other methods are, in the author experience, equally effective in the isotropic phase. But it seems that a single echo approach is more suitable for calibrations in the LC phases, where residual dipolar couplings may strongly affect the echo intensities in the pulse trains,[21,22] that at least can be substantially reduced by the transverse relaxation times.

9.4.2. *Pulse sequences for measuring diffusion with SFG*

It is now very clear that STE experiments in the SFG can be performed by only changing the time delays in the pulse sequence, as the gradient strength (and direction) is fixed. Of course, if one changes the delays the effects of relaxation and dipolar correlations on the echo are felt. These effects must be taken into account in order to correctly determine the diffusion coefficient. Here two methodologies are proposed that produce echo decays where the effects of spin relaxation and residual dipolar couplings are kept constant and only diffusion attenuation is effective.

9.4.2.1. *Two stimulated echo ratio*

The first approach exploits the combination of the following stimulated echo experiments:

$$\left(\frac{\pi}{2}\right)_x - \tau - \left(\frac{\pi}{2}\right)_x - (\Delta - \tau) - \left(\frac{\pi}{2}\right)_x - \tau - echo \tag{9.16}$$

and

$$\left(\frac{\pi}{2}\right)_x - \left(\frac{\tau+\delta}{2}\right) - \pi_y - \left(\frac{\tau-\delta}{2}\right) - \left(\frac{\pi}{2}\right)_y - (\Delta-\tau) - \left(\frac{\pi}{2}\right)_y - \left(\frac{\tau-\delta}{2}\right)$$
$$-\pi_y - \left(\frac{\tau+\delta}{2}\right) - \left(\frac{\pi}{2}\right)_y - echo \tag{9.17}$$

with respective echo decays expressed as:

$$A_{STE} = \exp\left\{-k^2 D\left(\Delta - \frac{\tau}{3}\right)\right\}\exp\left\{-\frac{2\tau}{T_2}\right\}\exp\left\{-\frac{(\Delta-\tau)}{T_1}\right\}A^{DCE}(\Delta,\tau) \tag{9.18}$$

with $k=\gamma G\tau$ and

$$A_{5p}^{STE} = \exp\left\{-\frac{1}{6}\gamma^2 G^2 D\left(\tau^3 - 3\tau^2\delta + 3(\tau+2\Delta)\delta^2 + 3\delta^3\right)\right\}$$
$$\times\exp\left\{-\frac{\Delta}{T_1}\right\}\exp\left\{-\frac{2\tau}{T_2}\right\}A^{DCE}(\Delta,\tau) \tag{9.19}$$

where the term $A^{DCE}(\Delta,\tau)$ represents the DCE contribution to the echo decay as time delays Δ and τ are changed.

Inspecting the expressions of respective echo decays it is noticed that, if the diffusion time Δ is changed, the contributions of relaxation and DCE to the two decays are the same, as both echoes refocus at the same time ($\tau+\Delta$) and the 180° pulses (generated doubling the amplitude of the pulse but keeping the same width as the 90° pulse) in the five pulses STE do not affect dipolar couplings during the time delays τ, as already discussed in Ref. [12]. Hence the ratio of the echo decays gives an expression that depends only on diffusion attenuation, i.e.

$$\frac{A^{STE}}{A_{5p}^{STE}} = \exp\left\{-(\gamma G)^2 D\Delta\left(\tau^2 - \delta^2\right)\right\} \tag{9.20}$$

Note the ratio has been calculated for the condition $\Delta >> \tau > \delta$, normally used for this kind of experiments as $T_1 >> T_2$. This approach has been successfully used to measure translational diffusion in different thermotropics.[24-27]

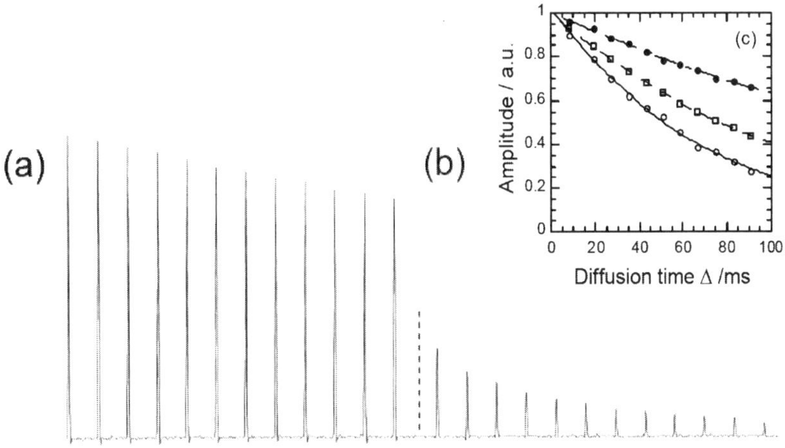

Fig. 9.8. Proton stimulated echoes sampled as a function of the diffusion time Δ in the smectic A phase of the liquid crystal 8OCB (T=335 K) with pulse sequences STE5p (a) and STE (b) in a SFG of G=5800 G/cm with ω_0=235 MHZ. Experimental delays have been chosen as $\tau = 35\ \mu s$, $\delta = 10\ \mu s$ and Δ varied from 5 to 93 ms in steps of 8 ms and echoes have been collected with an acquisition time of 40 μs and accumulating 40 transients, in order to obtain a good signal to noise ratio. Notice the relevant difference of diffusion damping on the echo intensity passing from STE to the STE with 180° pulses in the τ delays. In the insert the normalized echo decays are plotted: STE (○), STE5p (•) and the ratio (□).

As an example, Fig. 9.8 reports the two echo decays used to calculate the diffusion coefficient along the phase director in the SmA phase of 4-n-octyloxy-4'-cyanobiphenyl (8OCB) at T=335 K. As can be seen from the comparison of the two STE decays, acquired by increasing the diffusion time Δ while keeping the other delays constant, the attenuation is mostly due to translational diffusion, as shown by the similarity between the ratio and the STE decays in the insert of Fig. 9.8.[24]

9.4.2.2. *Constant time double stimulated echo (CTSTE)*

Recently, Kimmich and Rainer showed that a secondary stimulated echo, performed in constant total time conditions, is also efficient in removing the effects of DCE, and spin relaxation on the echo decay[28] by using

$$\left(\frac{\pi}{2}\right)-(\tau_{1,\min}+\Theta)-\left(\frac{\pi}{2}\right)-\tau_2-\left(\frac{\pi}{2}\right)-(\tau_{1,\min}+\tau_{1,\max})-\left(\frac{\pi}{2}\right)$$
$$-\tau_3-\left(\frac{\pi}{2}\right)-(\tau_{1,\max}-\Theta)-echo. \tag{9.21}$$

The echo attenuation of the secondary stimulated echo is:

$$\begin{aligned} A^{CTSTE} &= \frac{A_0}{4} A_{diff}\left(\Theta,\tau_{1,\min},\tau_{1,\max},\tau_2,\tau_3\right) \\ &\times \exp\left\{-\frac{(\tau_2+\tau_3)}{T_1}\right\} \exp\left\{-\frac{2\left(\tau_{1,\min}+\tau_{1,\max}\right)}{T_2}\right\} A^{DCE} \end{aligned} \tag{9.22}$$

with A^{DCE} again being the contribution to the echo intensity coming from dipolar correlation. As the experiment is performed as a function of Θ keeping all the other delay constant, relaxation does not affect the echo decay as it is kept constant throughout the experiment.[29] The explanation concerning DCE is not so straightforward, as also discussed in Ref. [28]. Simply speaking, it is seen that, by looking at Fig. 9.4, the modulation due to DCE is constant as the total time is constant, and does not affect the echo decay. The experiment has been so far successfully tested to measure translational self-diffusion in polymer melts,[28] where DCE effects have already been pointed out to be effective in modulating the STE echo decay.[12] Hence the only decay as Θ increases derives only from diffusion and can be expressed as:

$$\begin{aligned} A_{diff} = \exp\Big\{-(\gamma G)^2 D\Big[\Theta^2(\tau_2+\tau_3)-2\Theta\left(\tau_{1,\max}\tau_3-\tau_{1,\min}\tau_2\right) \\ +2\Theta\vartheta(\Theta-\rho)\Big]\Big\} \end{aligned} \tag{9.23}$$

with $\vartheta=\tau_{1,\min}+\tau_{1,\max}$ and $\rho=\tau_{1,\max}-\tau_{1,\min}$. Any overlap coming from other secondary echoes that can be refocused by the pulses is avoided by operating with $\tau_2 >> \tau_{1,\min}+\Theta$ and $\tau_3 >> \tau_{1,\max}-\Theta$.[28]

9.4.2.3. *Which method is preferable?*

Comparing the two approaches, indeed CTSTE should be preferable as it reduces the measurement to a single experiment. However, one must notice that the necessity of refocusing a second stimulated echo reduces the echo intensity to one half of a simple stimulated echo ($\frac{A_0}{4}$ versus $\frac{A_0}{2}$), with a consequent loss of signal. A further loss of intensity, in comparison with the STE comes from relaxation. If one considers for example, the simple case with $\tau_{1,\max} = \tau_{1,\min} = \tau$ and $\tau_2 = \tau_3 = \Delta/2$, with $\Delta >> \tau$, the CTSTE decay simplifies to:

$$A^{CTSTE} = \exp\left\{(\gamma G)^2 D\left[\Theta^2 \times \Delta\right]\right\} \times \exp\left\{-\frac{\Delta}{T_1}\right\} \exp\left\{-\frac{4\tau}{T_2}\right\} \qquad (9.24)$$

where Θ can span from 0 to τ. A STE experiment with the same τ and Δ, is instead affected by just half of the attenuation for transverse relaxation, i.e. $\exp\{-2\tau/T_2\}$. When relaxation is quite fast (short T_2), as it is the case for LCs, the limitation on $\tau_{1,\min}$ (and $\tau_{1,\max}$) cannot allow a sufficient dynamic range for the echo decay that depends on Θ. This aspect should be carefully considered when the experimental procedure is chosen.

One case in which the CTSTE should surely be preferred is for diffusion measurements in the isotropic phase; as in this case fast transverse relaxation is not a limitation. Another favorable case is in the presence of thermal convections steaming from temperature gradients in the sample. Without going into details on this subject, it is noticed that CTSTE decay is not affected by flow, considered as a coherent spin displacement, while in the STE the effect can be present and may introduce errors in the evaluation of the diffusion coefficient.[4] However, one also has to bear in mind that thermal convection, being due to thermal gradient across the sample, can also be alleviated by reducing the dimension of the sample itself. For example for diffusion measurements in the fringe field of the magnet, where the sensitive region normal to the z direction is about 1 mm, limiting sample dimension along this direction should help, without introducing a reduction in the signal intensity.

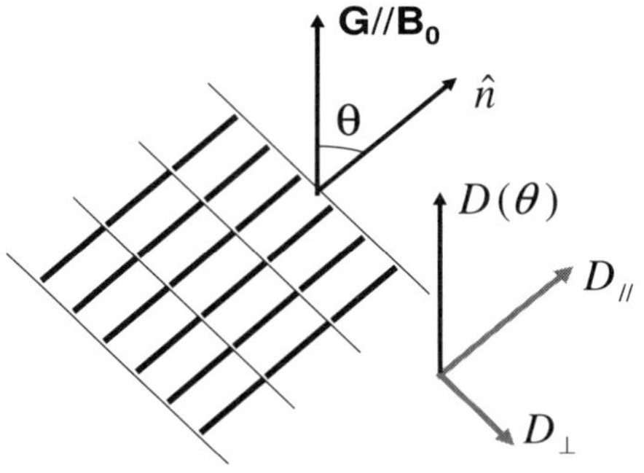

$$D(\theta) = D_{//} \cos^2 \theta + D_{\perp} \sin^2 \theta$$

Fig. 9.9. Sketch of diffusion tensor measurement in a fixed static gradient $G//B_0$ by exploiting sample rotation. The effective diffusion coefficient *D(θ)*, in the case of a uniaxial phase, such as the smectic A, is a linear combination of the two principal diffusion components and can be expressed as in Eq. (9.25).

9.4.3. *Sample rotation for full tensor diffusion measurements*

So far we have not explicitly considered the tensorial nature of diffusion in LC phases. As discussed in Sec. 9.3, thermotropic LC with positive $\Delta\chi$ align with the phase director parallel to the magnetic field direction. Now, it is well known that nematic phases of calamitic LC easily align in quite weak static field ($B_0 \approx 0.5T$).[§] Hence, working in a SFG with field strengths bigger that 1 T, for example, should assure a homogeneous nematic phase alignment with the director parallel to the magnetic field (for $\Delta\chi > 0$). Of course, as the gradient direction is fixed, this sample geometry limits the access only to the measurement of $D_{//}$, which is the case also for commercial diffusion probes. In principle, in case that the sample, once aligned, could be rotated in the field, it would be possible to measure the full diffusion tensor as the phase director could span an arbitrary angle θ with the gradient direction as shown in Fig. 9.9. In this situation access to the whole tensor is possible by changing the angle θ and measuring the diffusion coefficient as a function of that angle. In case of a uniaxial phase, this angular dependence can be expressed as:

[§] "quite weak" in terms of magnetic fields normally used in NMR spectroscopy.

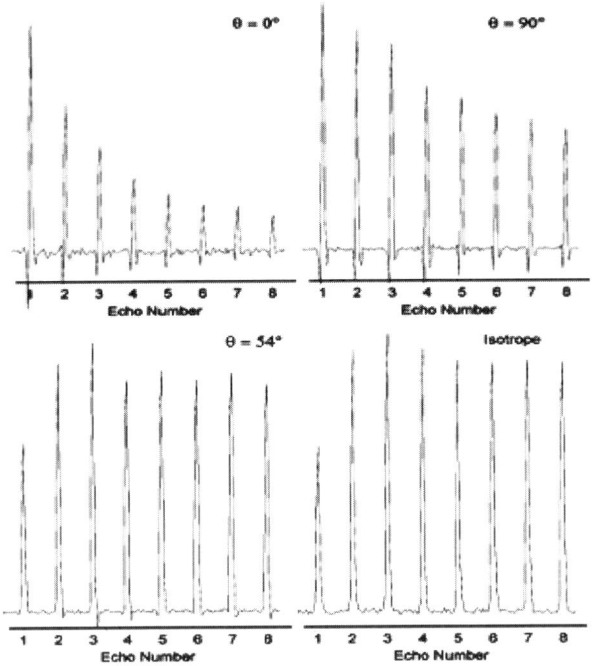

Fig. 9.10. Echo trains acquired as a function of the angle between the phase director and the magnetic field.

$$D(\theta) = D_{//} \cos^2 \theta + D_{\perp} \sin^2 \theta \tag{9.25}$$

Of course, sample rotation is not feasible in the nematic phase, as it realigns quickly (in a few ms) to the $\theta = 0$, $\hat{n} // B_0$ orientation. In case of a smectic A phase, the situation is different as it has been shown that for field $B_0 \leq 2.4T$, once the director has been oriented in the nematic phase, rotation in the smectic A is feasible, using a goniometric probe.[25-27] Of course, once rotated, sample alignment and stability should be checked in the SFG; that is, we have to be sure that the director does not flow back parallel to the magnetic field. As no spectra can be inspected as in conventional NMR, alternative methods should be employed. A possibility is, for example, to check the behaviour of an echo train such as:[25]

$$\alpha_x - \tau - (\alpha_y - \tau - echo)_n \tag{9.26}$$

where n denotes the nth echo refocused by the pulse train. The intensities of the echoes should depend on the angle θ, as residual dipolar couplings strength, $\langle H_{DD} \rangle$, can affect transverse relaxation and hence the decay rate depends on the θ angle , between the phase director and the static magnetic field according to:

$$\langle H_{DD} \rangle \propto (1 - 3\cos^2 \theta) \tag{9.27}$$

As an example, Fig. 9.10 gives the echo train behaviors acquired in the smectic A phase of the azoxy-benzene 4-4'-octyloxy-azoxybenzene (OAB) at T=328K by rotating the sample in a field of 2.4 T. As expected, the echoes in the trains show a slower decay moving from θ=0° to θ=90°. In particular, one sees that with θ=54° (close to the magic angle), the echo train behavior is similar to the one found in the isotropic phase (reported in the figure for comparison) as dipolar coupling are almost completely removed ($\langle H_{DD} \rangle \approx 0$).

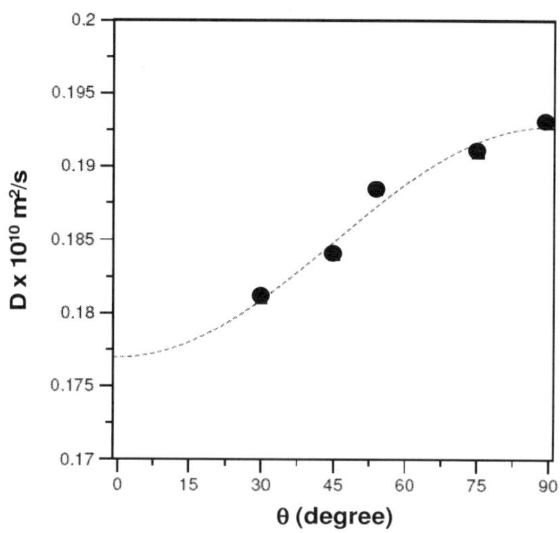

Fig. 9.11. Translational self diffusion coefficients measured as a function of the θ angle in the smectic A phase of HAB in a static field gradient of 3600 G/cm at a on resonance frequency of 100 MHz for proton.

Under these conditions then, diffusion can be measured at different angles using one of the methods previously described (see 9.4.2) and the behavior fitted with Eq. (9.24) to determine the principal diffusion coefficients $D_{//}$ and $D_{\perp}$.[25-27] This approach has also been used to measure inter-layer diffusion in phospholipidic bilayers. As an example, Fig. 9.11 reports the diffusion coefficient measured in SFG as a function of the angle θ in the smectic A phase of 4-4'-heptyloxy-azoxybenzene (HAB), previously aligned in the nematic phase. The experimental data have been fitted with Eq. (9.24) to yield the optimized principal diffusion coefficients. This has allowed the determination of the full diffusion tensor in the smectic A phase, and will further be discussed in Sec. 9.5 In the case of phospholipids, the bilayers have been previously aligned outside the magnet in thin films and a sample of a stack of these films was used.[30]

9.4.4. *Where should one place the probe to do the measurements?*

So far we have discussed how to exploit a static field gradient to measure diffusion in fluids in general and in particular in LC, but a relevant point still remains:

"Where does one have to place the probe in the fringe field of a superconducting magnet?"

At a first glance, there is a region as seen in Fig. 9.5 where the gradient reaches a maximum, and this presents the best place to perform the measurements. However, in order to find this "best position" the mapping of the fringe field should be performed, for example by measuring the resonant frequencies and gradient strengths along the z axis of the magnet. Different methodologies have been proposed in the literature to give, in the context of imaging,[19] a detailed mapping of the fringe field.[31] In practice, in the author's experience, unless one wants to build from scratches a probe tuned to the "best position" frequency, what can be done is just using a commercial multinuclear probe. The heteronuclear X channel normally spans from the ^{31}P frequency down to very low γ nuclei, such as ^{15}N. A good start can then be to exploit for example the ^{13}C frequency that is about a factor of 4 smaller than proton in the isocenter. Operationally, to tune the X channel to the chosen frequency with a suitable phantom, for example a piece of rubber (high proton signal and long transverse relaxation times), the probe is moved down along the z axis till one finds the place in the fringe field where the protons resonate. In the case of a 9.4 T Oxford magnet used by the author to perform the measurements reported in Sec. 9.5, protons resonate at ω = 100MHz 308 mm below the isocenter. The position was simply measured by a ruler and found by moving the probe down in one millimeter steps by acquiring an echo with a short pulse (about $1\mu s$) and a short τ delay ($20\mu s$) in each step. The right position is easily identified, where protons in the sample resonate by the sudden appearance of a sharp echo in the noise. In general one can imagine that, as the sensitive slice can be as small as 10-100 micrometers a high spatial precision is needed to fix the position. Again, for diffusometry purposes this requirement can be confidently relaxed, as long as the used sample is sufficiently long. To clarify this concept, consider for example a homogenous sample of 5 mm thickness along the z gradient direction. As shown in Fig. 9.12, a soft pulse that covers about 1 mm needs a "spatial resolution" of about $\pm 1mm$ to detect the position of interest.

Once the sample enters in the sensitive region (Fig. 9.12(a)), an optimization can be performed by just looking at the intensity of the NMR signal that, as the sample is homogenous, corresponds to the amount of sample in the region covered by the soft pulse. One would expect that by moving the sample down, the signal should increase but then fade away again as it moves out of the sensitive region (Fig. 9.12(c)). Of course, it's important to bear in mind that in case of shorter sample, the accuracy should be increased accordingly.

When the optimal position is found the second step is to calibrate the gradient strength needed for the diffusion measurements. An easy way to do this is just measuring the diffusion coefficient of a well known sample. In this case the unknown is the gradient strength that can then be determined as a parameter in fitting the diffusion decay from a suitable experiment. A good candidate as diffusion standard can be, for example distilled water, as its self diffusion coefficients have been reported for this purpose.[32] The calibration that the author has carried out in the position considered using a phantom of distilled water at T=20°C, that has a diffusion coefficient D=2.1x10^9 m^2/s. is shown in Fig. 9.13, and the echo decay of a CTSTE has been used to evaluate the gradient strength.

The chosen experimental delays were $\tau_{1,\min} = \tau_{1,\max} = 20\mu s$, $\Theta = 5\mu s$, and $\tau_2 + \tau_3 = 2.11ms$. In this case, the echo decay has been built up changing τ_2 and τ_3 but leaving their sum constant. The concept of CTSTE is preserved and the fitting result gave a gradient strength $G = 3600 \pm 30G/cm$. Of course, other positions can be found, exploiting the multi-frequency probe, repeating the same procedure for a different frequency. But, even more important, other nuclei can also be investigated at the same position found for proton just by re-tuning the probe to a suitable frequency via rescaling the frequency with the gyromagnetic ratios γ_X / γ_H , where X is the nucleus of interest.

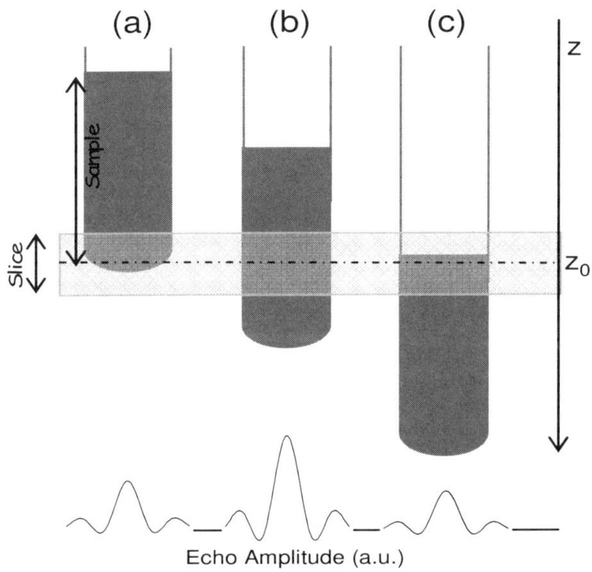

Fig. 9.12. Sample positioning can be checked by looking at the signal intensity of a simple Hahn echo experiment. Let z_0 be the centre of the slice irradiated by the "soft" pulse, the maximum intensity is achieved when the best filling of the excited slice is reached, as in case (b) If the sample does not completely fill the excited slice on the other hand, the signal decreases accordingly (case (a) and (c). Moreover in case that the (homogeneous) sample is wider than the excited slice, positioning is not so critical as a relatively broad region of the sample is available.

Now, there is still one point that should be addressed: is the gradient along z actually linear and what happens in the radial directions? To answer the first part of the question, one should notice that as a very narrow portion of the field is employed one can consider that in that region the gradient is linear, that is $\partial B_z/\partial z = G$. Concerning radial homogeneity, it is well known that for imaging purposes the best choice is a position where $\partial^2 B_z/\partial R^2 = 0$, in order to avoid distortions in the excited slice.[19,23] In our case, we can expect that where $\partial B_z/\partial z = G$ is very high (that is what we want), the condition $G \gg \partial^2 B_z/\partial R^2$ holds[19] and curvature effect due to radial gradients can be safely neglected.

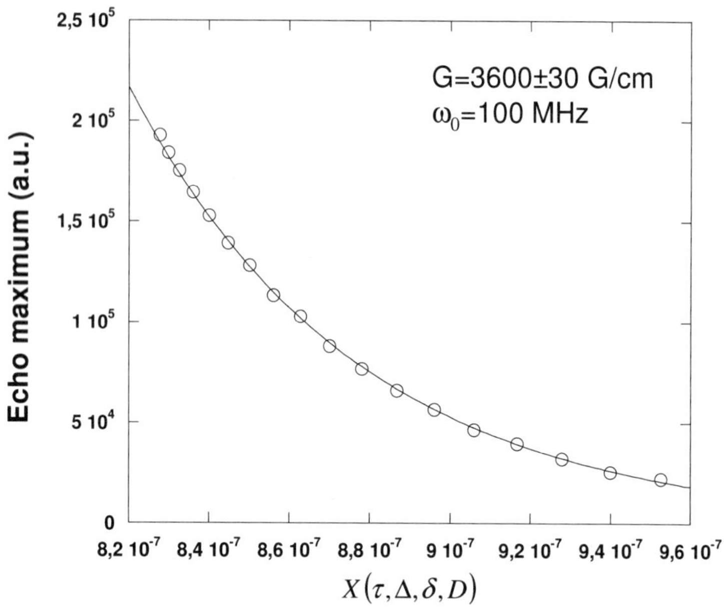

Fig. 9.13. STE decay from Eq. (9.18) measured in the fringe field on a sample of pure water at T=20°C. The chosen experimental delays $\Delta = 5\ ms$, $\tau = 100\ \mu s$ were kept constant and δ was varied from *6* to *40* μs. The gradient strength G was determined by fitting the decay with Eq. (9.18) with G being the only parameter allowed floating.

One other consideration that might be useful is that an estimation of the 90° pulse length in the fringe field can be performed taking into account the value obtained in the isocenter. If, for example, the 90° pulse for ^{13}C in the isocenter of a 9.4 T magnet (resonant frequency 100 MHz) is about 5 μs, we can estimate that in the fringe field where we look for protons at 100 MHz, the 90° pulse, at the same amplitude should be about 1.25 μs. This is based on that on resonance the pulse angle is calculated by:

$$\theta = \gamma_i B_1 t_w \tag{9.28}$$

and the ratio $\gamma_H / \gamma_C \approx 4$. In other words our multinuclear probe can provide (just by chance, let us say) quite short pulses in the fringe field when looking at protons. Now, short pulses can be important in increasing the sensitivity, as a bigger amount of sample is covered by the "soft pulse".

9.5. Application of the Static Field Gradient Approach to Nematic and Smectic A phase in HAB and OAB

After describing the different aspects of NMR diffusometry SFG approach to measure translational self diffusion coefficients in LC phases, it would be useful to conclude by showing a couple of data sets for such determinations. Translational self diffusion coefficients of two azoxybenzenes, 4-4'-heptyloxy-azoxybenzenes (HAB)[26] and 4-4'-octyloxy-azoxybenzenes (OAB)[25] are shown in Fig. 9.14. Experiments have been performed in the SFG of $G = 3600 G/cm$ using a CTSTE

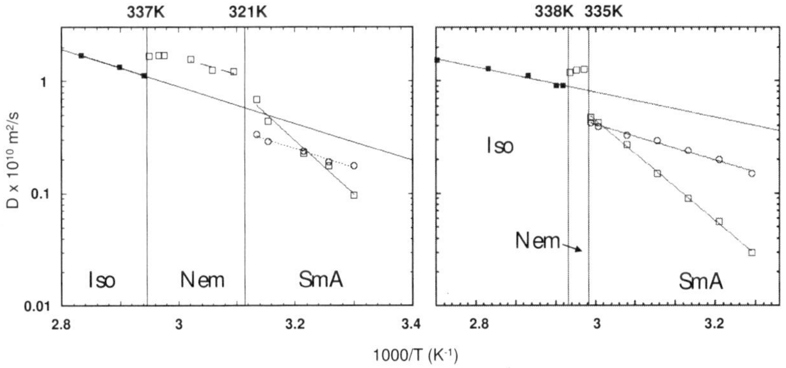

Fig. 9.14. Diffusion coefficients D_{iso} (■), $D_{//}$ (□) and $D_{\perp}$ (○) measured in the isotropic nematic and smectic A phase of the two azoxybenzenes HAB (on the left) and OAB (on the right). The lines represent Arrhenius fittings of the diffusion coeffcients.

experiment in the isotropic phase and the double STE approach (see Sec. 9.4.2.1) in the LC phases.[25,26] In particular, the full diffusion tensor was determined in the smectic A phase by exploiting sample rotation starting with the sample aligned at $\theta = 0$ through cooling from the nematic phase. The two principal diffusion coefficients $D_{//}$ and $D_{\perp}$ have been determined as the best fitting parameters in Eq. (9.24), as seen in Fig. 9.11 for HAB.

As already discussed, this approach does not allow a direct measurement of diffusion normal to the phase director; anyway the isotropic nematic phase transition can be easily identified from the sudden increase of the diffusion coefficient that changes from D_{iso} to $D_{//}$. The nematic to smectic A phase again is well marked by a sudden change in the diffusion coefficient. Moreover in this case $D_{\perp}$ was also measured in the smectic A phase where the lines in the figure represent Arrhenius fitting of the diffusion coefficients by

$$D_i = D_i^{\infty} \exp\left(-\frac{E_a^i}{RT} \right) \tag{9.29}$$

with $i = //, \perp$. It is quite evident that $E_a^{//} > E_a^{\perp}$. This is a typical behavior, confirmed in several cases.[24-27,33] The layer structure in the phase hinders diffusion along the phase director while less affects the diffusion perpendicular to it.

The determination of the full tensor also allows one to inspect and compare the diffusion anisotropy $D_{//}/D_{\perp}$ for the two mesogens. First of all, it is noticed that at a certain temperature (let's call it T_{inv}) a change from $D_{//}/D_{\perp} > 1$ to $D_{//}/D_{\perp} < 1$ occurs. Now, if one defines a reduced temperature $T^R = T/T_{N-SmA}$, the temperature rescaled by the nematic to smectic A transition temperature, one finds:

$$T_{inv}^R(OAB) > T_{inv}^R(HAB) \tag{9.30}$$

This behavior can be rationalized by looking at the McMillan theory for the nematic to smectic A phase transition.[34] Simply speaking, the theory indicates that the "strength" of the transition depends on the relative extension of the nematic and smectic A phase; the shorter the nematic range, the stronger the change in physical properties moving from one phase to the other. This is the case for translation self diffusion as it is clear from Fig. 9.14. Without going into much detail, it is worthy to notice that this picture can be further complicated by the fact that there are different ways in which a smectic A phase can assemble its layered structure.[35] For example, cyanobiphenyls are well known to form interdigitated smectic layers.[35] Now, comparing the LC behavior of 8CB with HAB, their nematic phase have roughly the same range (about 15 K), however, 8CB shows a nematic like diffusion anisotropy ($D_{//}/D_{\perp} > 1$) through the whole smectic A phase, as shown in Ref. [33]. The same behavior has been reported in other cases where interdigitated smectic A phase have been considered.[36] This can be explained by the different layer structures between smectic A and interdigitated smectic A phase, formed by cyanobiphenyl dimers interacting through their cyano groups. An interesting experiment that points out this conclusion has been reported by Oishi and Miyajima[37]. The authors presented a case where an inversion of the diffusion anisotropy from $D_{//}/D_{\perp} > 1$ to $D_{//}/D_{\perp} < 1$ was obtained changing the cyano group with a CF_3O group in the same mesogen.[37]

9.5.1. *Translational order from diffusion data*

In the previous section, it has been pointed out that translation diffusion, in particular the parallel component is affected by the smectic layer supermolecular structure. This behavior was also modellized by Moro and coworkers[38] and Volino and Dianoux[39] that reached basically the same conclusions. These can be briefly summarized as follows. Translational diffusion is modelled assuming the presence of a periodic potential *V* (not dependent on the temperature) along the smectic phase director:

$$V(z) = -(V_1/2)\cos(2\pi z/d) \tag{9.31}$$

where z is the layer normal direction, V_1 is a constant and *d* is the layer spacing in the smectic A phase. The periodic potential leaves unaffected the in-layer diffusion (namely $D_{\perp}$) but creates a barrier for molecules crossing among the layers. This is expressed mathematically by the following equations.[38,39]

$$D_{//}^{smectic} = D_{//}^{nematic}\left[I_0\left(V_1/2RT\right)\right]^{-2} \tag{9.32}$$

and

$$D_{\perp}^{smectic} = D_{\perp}^{nematic} \tag{9.33}$$

where I_0 is the zeroth order modified Bessel function of the first kind and $D_i^{nematic}$ ($i = //, \perp$), the nematic diffusion behavior, is also extended into the smectic A phase region. That is, model predicts a discontinuity at the nematic – smectic A transition for diffusion across the layer but no change for diffusion in the layers. In this model the kth positional order parameter can be defined as.[38,39]

$$\tau_k = \langle \cos k2\pi z/d \rangle = \frac{I_k\left(V_1/2RT\right)}{I_0\left(V_1/2RT\right)} \tag{9.34}$$

Looking at Eq. (9.34) one is tempted to exploit it to evaluate the positional order parameters τ_k if the potential V_1 can be determined from the measured diffusion coefficient. This approach has been tested both theoretically by means of simulations and experimentally on the data sets reported in Fig. 9.14. In particular, a detailed analysis including also the exploitation of theoretical models for the nematic phase is presented in Ref. [26]. Here, a simpler approach, analogous to the one reported in reference 24 is proposed, where only the parallel component of the diffusion tensor is considered. An example of the procedure, based on the HAB data set, is reported in Fig. 9.15. As shown in the figure, the temperature behavior of the diffusion coefficient measured in the nematic and smectic A phase have been fitted by means of the equation:

$$D_{//}(T) = D_\infty \exp\left(\frac{-E_{//}^{nem}}{RT}\right)\left[I_0\left(Step[T_{N-SmA}]V_1/2RT\right)\right]^{-2} \tag{9.35}$$

where $E_{//}^{nem}$ is the Arrhenius activation energy and D_∞ is the extrapolated diffusion coefficient at $T \to \infty$ from diffusion coefficients determined in the nematic phase. *Step[$T_{N\text{-}SmA}$]* is a step function such that *Step[$T_{N\text{-}SmA}$]* = 0 in the nematic phase and *Step[$T_{N\text{-}SmA}$]* = 1 in the smectic phase. As shown in the figure, two different potentials V_1 are tested. A temperature independent one, as described in the original model, and a temperature dependent one:[33]

$$V_1(T) = V_0(1 - T/T_0)^p \qquad (9.36)$$

where V_0, T_0 and p are phenomenological parameters to be determined. The best parameters obtained are $E_a = 35\ KJ/mol$, $V_1 = 6.9KJ/mol$ for the temperature independent potential. For the temperature dependent one the best parameters are instead $E_a = 26KJ/mol$, $V_0 = 22KJ/mol$ and $p = 0.25$. It is quite clear that when the temperature dependent potential is considered the fitting improves substantially, as already pointed out in Refs. [33] and [24], leading to the conclusion that a positional order (see τ_1 in Fig. 9.15), gradually consolidates as the temperature decreases,

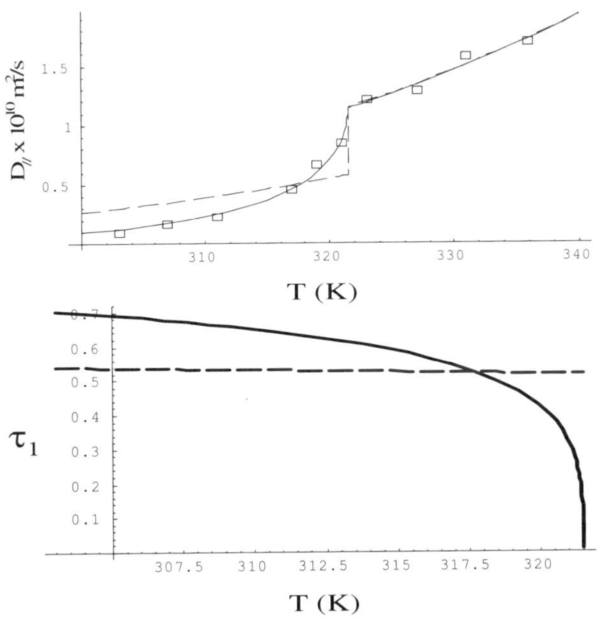

Fig. 9.15. Top: Fitting results of the experimental diffusion coefficients parallel to the phase director in the nematic and smectic A phase of HAB, as obtained using Eq. (9.35), with orginal V_1 (broken line) and temperature dependent $V_1(T)$ (continuous line); the best fitting parameters are reported in the text. Bottom: position order parameter τ_1 determined by the two potentials V_1 by using Eq. (9.34).

thereby giving a more realistic picture of the smectic phase. As already discussed in Ref. [26], this approach can be quite valuable, to the author's knowledge, as only X ray diffraction have been used so far to determine positional order in smectic phases.[40] However, it is noted that the method, assuming an accurate experimental data, is also strongly dependent on the theoretical model used. At the moment, it seems that the present theoretical models can fairly reproduce the nematic to smectic A transition behavior but may still need some refinements.[26]

For example, it has been shown that the assumption of a continuous behavior of $D_{\perp}$ at the nematic-smectic A phase transition is valid for cyanobiphenyls that present an interdigitated smectic A phase where the relation $D_{//}/D_{\perp} > 1$ holds, but it may fail in systems with more ordered smectic A phases, such as for example azoxybenzenes.[23] Thus, it is reasonable to conclude that though the method has to be refined, it can lead to information on the layer structure in smectic phases. In particular, it could represent a good start to study positional order and layer structure of more complex systems such as smectic phases formed by chiral mesogens.

9.6. Conclusions

In this chapter, a methodology is described to measure translational self diffusion in liquid crystals by exploiting the ubiquitous static gradient present in the fringe field of superconducting magnets. Of course it is important to mention that, where available, pulsed field gradient approaches[7] are indeed preferable, even though they require specifically designed hardware to reach high enough gradient strength. In particular on diffusion along the x,y direction, commercial probes, exhibiting x,y,z gradients, are normally realized in NMR microimaging systems even if limited in strength. However, it is also worthy to bear in mind that in general, the stray field diffusometry approach gives access to an extremely high gradient diffusion probe simply by using a wideline multinuclear probe at essentially no cost for hardware upgrade and can be a valuable means to perform translational diffusion measurements on systems other than liquid crystals.[18]

References

1. E. Hawlicka, *Chem. Soc. Rev.*, 34, 13743 (1995).
2. K. Nicolay, K.P.J. Braun, R.A. de Graaf, *et al.*, *NMR in Biomedicine*, 14(2), 94 (2001).
3. W.S. Price, *Concept Magnetic. Res.*, Part 1, 9, 299 (1997); (b) W.S. Price, Concept Magnetic. Res., Part 1, 10, 197 (1998).
4. R. Kimmich, *NMR Tomography Diffusometry Relaxometry*, Springer-Verlag (1997).
5. Callaghan, *Principles of Nuclear Magnetic Resonance Microscopy*, Oxford University Press, USA (1994).
6. F. Noack, *Mol. Cryst. Liq. Cryst.*, 113, 247 (1984).
7. S.V. Dvinskikh and I. Furó, *Russ. Chem. Rev.*, 75, 497 (2006).
8. I. Musevic, R. Blinc and B. Zeks, *The Physics of Ferroelectric and Antiferroelectric Liquid Crystals*, World Scientific Publishing Company (2000).
9. H. Takezoe and Y. Takanishi, *Jpn. J. Appl. Phys.*, Part 1, 45, 597 (2006).
10. M. Marcos, R. Martin-Rapun, A. Omenat, J.L. Serrano, *Chem. Soc. Rev.*, 36(12), 1889 (2007).
11. G.R. Luckhurst, *Thin Solid Films*, 393, 40 (2001).
12. R. Kimmich, E. Fischer, P. Callaghan and N. Fatkullin, *J. Magn. Reson. A*, 117, 53 (1995).
13. F. Grinberg and R. Kimmich, *J. Chem. Phys.*, 103, 365 (1995).
14. Noack, St. Becker and J. Struppe, *Annu. Rep. NMR Spectrosc.*, 33, 1 (1997).
15. O. Oishi and S. Miyajima, *J. Magn. Reson. Ser. A*, 123, 64 (1996).
16. A.A. Samoilenko, D.U. Artemov and L.A. Sibel'dina, *JETP Lett.*, 47, 417 (1988).
17. R. Kimmich, W. Unrath, *et. al.*, *J. Magn. Reson.*, 91, 136 (1991).
18. Geil B., *Concept Magnetic. Res.*, 10(5), 299 (1998).
19. P.J. McDonald, *Prog. Nucl. Magn. Reson. Spectrosc.*, 30, 69 (1997).
20. D.H. Wu and C.S.E. Johnson, *J. Magn. Reson A.*, 116(2), 270 (1995).
21. W. Randall and D.G. Gillies, *J. Magn. Reson A.*, 121, 411 (1996).
22. P. Kinchesh, S.A. Smith, A.R. Preston and E.W. Randall, *J. Magn. Reson.*, 154(2) 252 (2002).
23. J. Godward, L. Ciampi. M. Cifelli and P.J. McDonald, *J. Magn. Reson.*, 155(1), 92 (2002).
24. M. Cifelli, P.J. McDonald, C.A. Veracini, *Phys. Chem. Chem. Phys.*, 6(19), 4701 (2004).
25. M. Cifelli and C.A. Veracini, *Phys. Chem. Chem. Phys.*, 7(19), 3412 (2005).
26. M. Cifelli, L. De Gaetani and G. Cinacchi, *J. Chem. Phys.*, 125(19), 164912 (2006).
27. M. Cifelli, *Mol. Cryst. Liq. Cryst.*, 465, 309 (2007).
28. E. Fischer and R. Kimmich, *J. Magn. Reson.*, 166(2), 273 (2004).
29. D.H. Wu, *J. Magn. Reson.*, 106, 229 (1994).
30. P. Karakatsanis and T.M. Bayerl, *Phys. Rev. E*, 54(2), 1785 (1996).
31. P. Jeglic, A. Lebar, T. Apih and J. Dolinsek, *J. Magn. Reson.*, 190(1), 39 (2001).

32. R. Mills, *J. Phys. Chem.*, 77, 685 (1973).
33. S.V. Dvinskikh, I. Furó, H. Zimmermann and A. Maliniak, *Phys. Rev. E*, 65, 061701 (2001).
34. W.L. McMillan, *Phys. Rev. A*, 4, 1238 (1971).
35. P. DeGennes and J. Prost, *The Physics of Liquid Crystals*, Oxford University Press, London (1993).
36. S. Miyajima, A.F. McDowell and R.M. Cotts, *Chem. Phys. Lett.*, 212, 277 (1993).
37. O. Oishi and S. Miyajima, *J. Magn. Reson.*, 160, 74 (2003).
38. G. Moro, P.L. Nordio and U. Segre, *Chem. Phys. Lett.*, 105, 440 (1984).
39. F. Volino and A.J. Dianoux, *Mol. Phys.*, 36, 389 (1978).
40. W.L. McMillan, *Phys Rev. A*, 6(3), 936 (1972).

Chapter 10

Deuterium NMR Studies of Static and Dynamic Director Alignment for Low Molar Mass Nematics

Akihiko Sugimura[1] and Geoffrey R. Luckhurst[2]

[1]*Department of Information Systems Engineering, Osaka Sangyo University, 3-1-1 Nakagaito, Daito, Osaka 574-8530, Japan*
email: sugimura@ise.osaka-sandai.ac.jp

[2]*School of Chemistry, University of Southampton, Highfield, Southampton, SO17 1BJ, United Kingdom*
email: G.R.Luckhurst@soton.ac.uk

There have been many investigations of the alignment of nematic liquid crystals by either a magnetic and/or an electric field. The basic features of these hydrodynamic processes have been characterized for the systems in their equilibrium and non-equilibrium states. To complement the experiments theoretical models, based on continuum theory, have been developed which successfully describe the static and dynamic phenomena. Such macroscopic behaviour has been investigated with deuterium nuclear magnetic resonance (NMR) spectroscopy. This technique has proved to be especially important for the investigation of liquid crystals. Since the quadrupolar splitting for deuterons observed in the liquid crystal phase is determined by the angle between the director and the magnetic field, time-resolved deuterium NMR spectroscopy has been employed to investigate the dynamic director alignment process in a thin nematic film following the application or removal of an electric field. In this Chapter, we describe some of our studies of the field-induced director dynamics for low molar mass nematics using deuterium NMR but with an electric as well as a magnetic field to orient the director.

10.1. Introduction

Nuclear magnetic resonance spectroscopy (NMR) is widely used in the study of liquid crystals. Deuterium NMR has proved to be especially important for the investigation of liquid crystals because the spectra of specifically or fully deuteriated materials are rather simple compared to the corresponding proton NMR spectra.[1-5] The quadrupolar splitting for deuterons observed in the liquid crystal phase is related to the second rank orientational order parameter for the C-D bond direction and so deuterium NMR has been widely used for the study of the orientational order of liquid crystals as well as their phase transitions.[2] The quadrupolar splitting is also determined by the angle between the director and the magnetic field. In consequence, deuterium NMR spectroscopy has proved to be a powerful method with which to investigate the director orientation and its distribution, as well as the director dynamics in nematic[6-23] and smectic liquid crystals.[24-26] In more recent years deuterium NMR spectroscopy, combined with continuum theory, has been applied successfully to investigate the static director distribution in thin nematic liquid crystal cells, with different film thicknesses and different surface anchoring strengths, subject to both magnetic and electric fields.[11-13] Deuterium NMR spectroscopy has also been employed to investigate the dynamic director alignment process in a thin nematic film following the application or removal of an electric field.[14-23] This technique has the added advantage that the presence of the magnetic field of the spectrometer ensures that during the electric field-induced alignment the director rotates as a monodomain,[13] which facilitates the analysis of the results, except for the case when the angle made by the magnetic field with electric field is close to right angle.

In this Chapter, we describe some of our studies of the static director distribution in thin nematic liquid crystal cells using a combination of deuterium NMR spectroscopy and continuum theory. The nematic liquid crystal, 4-pentyl-d_2-4'-cyanobiphenyl (5CB-d_2) deuteriated in the α-position of the pentyl chain, was confined between two glass plates with a weak anchoring condition; the anchoring strengths were measured using a saturation voltage method.[27] A series of deuterium NMR spectra was acquired as a function of the applied electric field, which can be used to explore the director deformation. We also describe the application of deuterium NMR spectroscopy to investigate the director dynamics in the same nematic liquid crystal (5CB-d_2) confined between two glass plates and subject to magnetic, **B**, and AC electric, **E**, fields.

The cell was set in the NMR probe with the electric field, whose direction is normal to the substrate surface, making an angle of about 45°, with the magnetic field. This experimental geometry allows a unique director motion in the alignment process; recent studies suggest that this is also true for larger angles. In the absence of the electric field the director for 5CB aligns parallel to the magnetic field because the diamagnetic anisotropy, $\Delta\tilde{\chi}$, is positive. When an electric field, which is strong enough to overcome the magnetic torque, is applied then the director will tend to be parallel to the electric field since the dielectric anisotropy, $\Delta\tilde{\varepsilon}$, is also positive. After the electric field is switched off, the director will then move from being inclined to the magnetic field to being parallel to it. The dynamics of the director relaxation can be followed by monitoring the NMR spectrum during this alignment process, as a function of time. That is, the time dependence of the director orientation during and after the application of an electric field is studied. This is possible because the NMR spectrum for a monodomain sample with one group of equivalent deuterons, having a negligible dipolar interaction, contains a simple quadrupolar doublet whose separation is determined by, among other things, the angle, θ, made by the director, $\mathbf{n}$, with the magnetic field.

Studies of the field-induced alignment of the nematic director for low molar mass materials using vibrational spectroscopy have revealed some fascinatingly unexpected results. Several studies have found that the relaxation time for the director alignment is dependent on the group in the molecule used to monitor the director orientation. Seemingly this undermines the basic concept of the Leslie-Ericksen hydrodynamic theory of nematics. Here we describe complementary studies using deuterium NMR spectroscopy of perdeuteriated 4-pentyl-4′-cyanobiphenyl (5CB-d_{19}) where the director is aligned by a magnetic or an electric field. This particular technique was chosen because the spectral peaks associated with each rigid group in the molecule are clearly resolved and of comparable intensity. We have investigated the director alignment using different temperatures, electric field strengths and angles between the magnetic and electric fields, each of which influences the director relaxation time. For all of the experiments we find that the relaxation times are independent of the group used to determine the director orientation during the alignment process.

The layout of this Chapter is as follows. In the next section we give the theoretical background to both the NMR experiment and the continuum theory for the director orientation. The NMR experiments are described in Sec. 10.3. The results for the static behaviour of the director are given in Sec. 10.4. The results for the director dynamics are described in Sec. 10.5 where they are discussed on the basis of the hydrodynamic theory analysis. This allows the ratio $\Delta\tilde{\varepsilon}/\Delta\tilde{\chi}$, the diamagnetic anisotropy $\Delta\tilde{\chi}$, the field-induced relaxation times and the rotational viscosity coefficient to be determined provided $\Delta\tilde{\chi}$ is known independently. In Sec. 10.6 studies using deuterium NMR spectroscopy of perdeuteriated 5CB-d_{19} to explore the site-dependence of the dynamics of director alignment are shown. Our conclusions are in Sec. 10.7.

10.2. Theoretical Background

Deuterium has a nuclear spin of one and so possesses a quadrupole moment, which interacts with the electric field gradient at the nucleus, to give a quadrupolar interaction tensor. This does not influence the number of lines in the deuterium NMR spectrum for a normal liquid because the rapid and random rotational molecular motion averages the quadrupolar interaction to zero. The NMR spectrum of a single deuteron, therefore, contains a single line composed of a pair of degenerate transitions as indicated schematically in Fig. 10.1(a). In a liquid crystal phase this degeneracy is removed because of the intrinsic long range orientational order combined with the quadrupolar interaction of the deuterium nuclei.

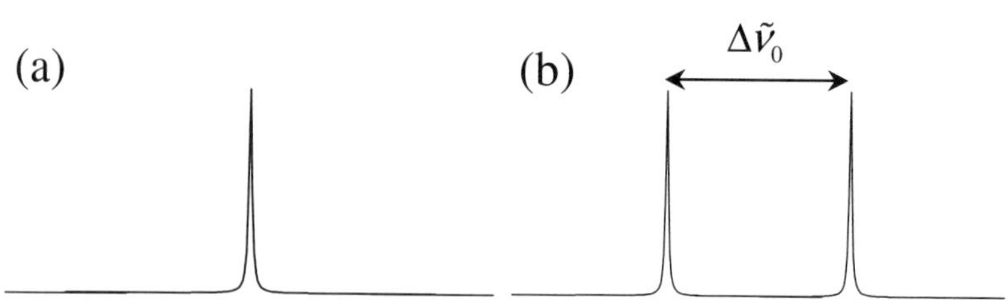

Fig. 10.1. A typical deuterium NMR spectrum of a single deuteron in an isotropic phase (a) and a partially ordered system (b).

For a monodomain sample, in which the director is uniformly aligned with respect to **B,** the NMR spectrum consists of a single doublet (see Fig. 10.1(b)) this is also observed for sets of equivalent deuterons provided the dipolar interaction is negligible in comparison with the linewidth. The separation, $\Delta\tilde{\nu}$, between the quadrupolar split lines is related to the components of the Saupe ordering matrix, S, and the quadrupolar tensor, $\mathbf{q}$. However, for aliphatic deuterons the quadrupolar tensor is, to a good approximation, cylindrically symmetric about the C-D bond direction, and then the quadrupolar splitting, partially averaged by the molecular rotation, is given by[2]

$$\Delta\tilde{\nu}_0 = \frac{3}{2} q_{CD} S_{CD}, \tag{10.1}$$

where S_{CD} is the second rank orientational order parameter for the C-D axis and q_{CD} is the quadrupolar coupling constant. In this analysis the director is taken to be parallel to the magnetic field. However, when the angle the director makes with the magnetic field changes, the splitting will also change in a well-defined way. Thus if the director makes an angle, θ, with the magnetic field of the NMR spectrometer the quadrupolar splitting is given by[7]

$$\Delta\tilde{\nu}(\theta) = \Delta\tilde{\nu}_0 P_2(\cos\theta), \tag{10.2}$$

where $\Delta\tilde{\nu}_0$ is the splitting when the director is parallel to the magnetic field (see Eq. (10.1)) and $P_2(\cos\theta)$ is the second Legendre polynomial. This result follows from the uniaxial symmetry of the nematic phase and the fast molecular rotational motion; it holds irrespective of the molecular symmetry or that of the quadrupolar tensor. As the director moves away from being parallel to the field the splitting is predicted and observed to decrease, pass through zero at the magic angle ($\theta = 54.74°$) and then to increase to one half of the original splitting, $\Delta\tilde{\nu}_0$, when the director is orthogonal to the magnetic field. Strictly the quadrupolar splitting changes sign at the magic angle but the sign of the splitting is not directly available from the spectrum. This would create a difficulty when the splitting is less than or equal to one half of $\Delta\tilde{\nu}_0$; for example if $|\Delta\tilde{\nu}(\theta)/\Delta\nu_0|$ was exactly 1/2 then the angle could be either 38.26° or 90°. Fortunately, this is not a problem in our experiments for, as we shall

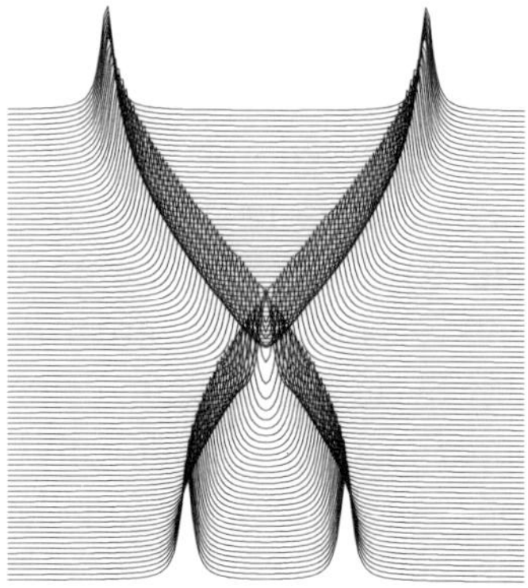

Fig. 10.2. The simulated dependence of the deuterium NMR spectrum on θ, the angle between the director and the magnetic field. The value of θ is increased from 0 to 90° in steps of 1°.

see, the quadrupolar splitting varies continuously from $\Delta\tilde{\nu}_0$ and so its sign or rather its sign relative to $\Delta\tilde{\nu}_0$ is not in doubt. Thus, the value of the angle θ can be determined directly from Eq. (10.2) by measuring the quadrupolar splittings, $\Delta\tilde{\nu}(\theta)$ and $\Delta\tilde{\nu}_0$. The angular variation of the deuterium NMR spectrum predicted from Eq. (10.2) is illustrated in Fig. 10.2.

As we have noted one of the prime advantages of NMR spectroscopy in studying the director orientation with respect to the magnetic field is that the form of the spectrum is also influenced by deviations of the distribution from a monodomain that is when the director is uniformly distributed with respect to the magnetic field. Provided the functional form of the director distribution, $P(\theta)$, that is the probability density for finding the director at an angle, θ, to the magnetic field is known then the observed spectrum can be simulated from

$$I(\nu) = \int L\left(\nu, \tilde{\nu}_{\pm}(\theta), T_2^{-1}(\theta)\right) P(\theta) \sin\theta d\theta. \tag{10.3}$$

Here $L(\nu, \tilde{\nu}_{\pm}(\theta), T_2^{-1})$ denotes the normalised shape of a spectral line centered at either $\tilde{\nu}_+(\theta)$ or $\tilde{\nu}_-(\theta)$ and with a linewidth related to $T_2^{-1}(\theta)$. We shall take the lineshape to be a Lorentzian, that is

$$L(\nu, \tilde{\nu}_{\pm}(\theta), T_2^{-1}) = \pi^{-1} T_2^{-1} / \left[\left(T_2^{-1}\right)^2 + \left(\nu - \tilde{\nu}_{\pm}(\theta)\right)^2 \right], \tag{10.4}$$

where the linewidth $T_2^{-1}(\theta)$ is half the distance between the points of maximum slope of the line. The angle dependent resonance frequency is

$$\tilde{\nu}_+(\theta) = \nu_0 + (\Delta\tilde{\nu}_0/2)P_2(\cos\theta), \qquad (10.5)$$

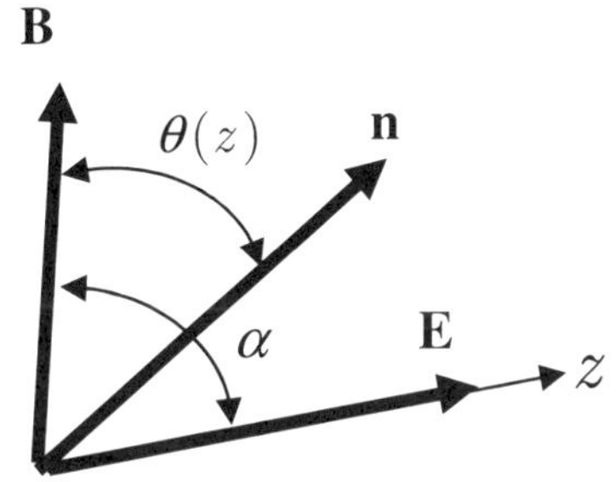

Fig. 10.3. The geometry used for the experiment and theoretical analysis. The z-axis is taken as being parallel to the electric field. The magnetic field (**B**), electric field (**E**) and director (**n**) are in the *xz*-plane. The director and electric field make the angles θ and α, respectively, with the magnetic field.

with an analogous expression for the resonance frequency, $\tilde{\nu}_-(\theta)$, of the other component of the quadrupolar doublet. In the spectral simulations it is convenient to set the central frequency, ν_0, coming from the Zeeman splitting to zero.

Deuterium NMR provides a very good method for investigating the director distribution in a thin nematic cell subject to an electric field because of the sensitivity of the spectrum to the director orientation. By varying the strength of the electric field, the total balance of the torques originating from the magnetic field, the electric field, the surface anchoring and the elasticity can be controlled. In other words the application of an electric field makes it possible to investigate the variation of the director distribution in a thin nematic cell with the field strength even in the presence of the strong magnetic field of the NMR spectrometer. However, to help interpret the deuterium NMR spectra, especially when the director is uniformly aligned it is valuable to have some theoretical guidance as to the form of the director relaxation. For a monodomain director subject to both electric and magnetic fields, ignoring surface anchoring effects, which is a reasonable assumption for a thick cell with untreated glass surfaces, the hydrodynamic theory gives the form of the director relaxation. We shall, therefore, now consider the Leslie-Ericksen based theory[28] for the time dependence of the nematic director subject to both magnetic and electric fields.

Essentially all of the deuterium NMR spectra in our study are dominated by a single quadrupolar doublet which allows us to determine the director orientation associated with a monodomain sample. In this analysis, we therefore treat the director as being uniformly aligned. As shown in Fig. 10.3, the electric field makes an angle α with the magnetic field. The rate of change of the director orientation is given, for the turn-on process, by the torque-balance equation[29] which for a monodomain nematic[14] is

$$\gamma_1 \frac{d\theta(t)}{dt} = -\frac{\Delta\tilde{\chi}}{2\mu_0} B^2 \sin 2\theta(t) + \frac{\varepsilon_0 \Delta\tilde{\varepsilon}}{2} E^2 \sin 2(\alpha - \theta(t)), \qquad (10.6)$$

where γ_1 is the rotational viscosity coefficient, μ_0 is the magnetic permeability and ε_0 is the dielectric permittivity of a vacuum. The solution of Eq. (10.6) is obtained analytically as[14]

$$\theta(t) = \theta_\infty + \tan^{-1}\left[\tan(\theta_0 - \theta_\infty)\exp(-t/\tau)\right], \qquad (10.7)$$

where θ_∞ is the limiting value of $\theta(t)$ when t tends to infinity, τ is the relaxation time, and θ_0 is the initial angle between **n** and **B**. The value of θ_∞ is given by

$$\cos 2\theta_\infty = \frac{1 + \rho\cos 2\alpha}{\left(1 + 2\rho\cos 2\alpha + \rho^2\right)^{\frac{1}{2}}}, \qquad (10.8)$$

where

$$\rho = \frac{U_{\mathrm{E}}}{U_{\mathrm{M}}} = \mu_0\varepsilon_0 \left(\frac{E}{B}\right)^2 \left(\frac{\Delta\tilde{\varepsilon}}{\Delta\tilde{\chi}}\right), \quad U_{\mathrm{M}} = -\frac{\Delta\tilde{\chi}}{2\mu_0} B^2, \quad U_E = -\frac{\varepsilon_0 \Delta\tilde{\varepsilon}}{2} E^2. \qquad (10.9)$$

The parameter, ρ, is the ratio of the magnetic and electric anisotropic energies, U_{M} and U_{E}, respectively. The limiting value of the director orientation depends, therefore, on the material property, $\Delta\tilde{\varepsilon}/\Delta\tilde{\chi}$, and the known experimental parameters, E/B and α. The predicted relationship between the limiting or equilibrium director orientation, θ_∞, and ρ is shown in Fig. 10.4 for various values of α. It is clear from the results in this figure that the limiting angle is not exactly the same as the angle between the two fields because the value of ρ is insufficient to align the director parallel to the electric field. What is more apparent is that there is a strong dependence of the variation of θ_∞ with ρ on the angle between the two fields as this tends to 90°. Indeed, when the fields are orthogonal the director orientation is predicted to be 0° when $\rho < 1$ and 90° when $\rho > 1$ with a discontinuous change when ρ is unity, that is the two field energies are identical. Given the relationship between the quadrupolar splitting and the director orientation (see Eq. (10.2)) then, $\Delta\tilde{\nu}$, will also show an analogous variation with ρ, and hence with the applied voltage.

We now turn to the dynamic behaviour of the director. When an electric field, which is strong enough to overcome the magnetic torque, is applied then the director will be aligned with respect to the electric field (turn-on process). After the electric field is switched off, the director then moves from being approximately parallel to the electric field to being parallel to the magnetic field (turn-off process). The relaxation times for the turn-on (τ_{ON}) and turn-off (τ_{OFF}) processes are

$$\tau_{\mathrm{ON}} = \left(\tau_{\mathrm{M}}^{-2} + 2\tau_{\mathrm{M}}^{-1}\tau_{\mathrm{E}}^{-1}\cos 2\alpha + \tau_{\mathrm{E}}^{-2}\right)^{-\frac{1}{2}} = \frac{\tau_{\mathrm{M}}}{\left(1+2\rho\cos 2\alpha+\rho^2\right)^{\frac{1}{2}}} \qquad (10.10)$$

and

$$\tau_{\mathrm{OFF}}(\equiv \tau_{\mathrm{M}}) = \gamma_1/(2U_{\mathrm{M}}), \qquad (10.11)$$

$$\tau_{\mathrm{E}} = \gamma_1/(2U_{\mathrm{E}}), \qquad (10.12)$$

where τ_{E} is the electric field-induced relaxation time for the director. It is apparent from the equation for τ_{ON} that the theory predicts an interesting variation with ρ, which depends on α. Thus for α equal to 45° $\tau_{\mathrm{ON}} = \tau_{\mathrm{M}}/(1+\rho^2)^{1/2}$ and so the relaxation time decreases with increasing ρ. In contrast as α tends to 90° so τ_{ON} tends to $\tau_{\mathrm{M}}/(1-\rho)$ so that τ_{ON} now increases with ρ.[5] Equations (10.8)-(10.12) give the following simple relationships used in the analysis of our results,

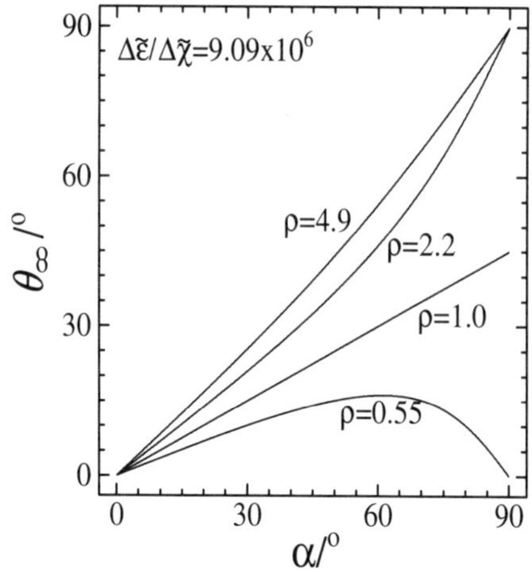

Fig. 10.4. The predicted relationship between the equilibrium angle, θ_∞, and α calculated for various values of ρ. The equilibrium angle is seen not to be exactly the same as α because the value of ρ is insufficient to align the director parallel to the electric field.[30]

$$\sin^2 2\alpha = \frac{R^2 \sin^2 2\theta_\infty}{R^2 - 2R\cos 2\theta_\infty + 1}, \tag{10.13}$$

where

$$R = \frac{\tau_{\mathrm{OFF}}}{\tau_{\mathrm{ON}}} = \sqrt{\rho^2 + 2\rho\cos 2\theta_\infty + 1}. \tag{10.14}$$

It is evident from Eqs. (10.13) and (10.14) that α can be determined using θ_∞ measured from the quadrupolar splitting. We can see from Eq. (10.8) that the static experiments give ρ from which the material property $\Delta\tilde{\varepsilon}/\Delta\tilde{\chi}$ can be determined. It is also evident from Eqs. (10.10)-(10.12) that the dynamic experiments provide the two pure relaxation times, τ_E and τ_M, for the alignment of the director by electric and magnetic fields, respectively. Since the field strengths are known the relaxation times are related to the material properties $\gamma_1/\Delta\tilde{\varepsilon}$ and $\gamma_1/\Delta\tilde{\chi}$, respectively. The ratio of these is just $\Delta\tilde{\varepsilon}/\Delta\tilde{\chi}$ which permits a check with the static experiments. Both relaxation times, τ_{ON} and τ_{OFF}, are related to the rotational viscosity coefficient but there are advantages to determining τ_{ON} since it is possible to bring this into a measurable range via changes in α and E. To obtain the rotational viscosity coefficient it is necessary to know $\Delta\tilde{\chi}$ or $\Delta\tilde{\varepsilon}$ from separate experiments. In practice it proves to be easier to determine accurate values for $\Delta\tilde{\varepsilon}$ than for $\Delta\tilde{\chi}$.

As we have just seen the angle, α, between the electric and magnetic fields is one of the important experimental parameters needed to investigate the static and dynamic director orientation. However, it is difficult to determine this angle precisely and directly because the electric field strength used in our measurements is not quite high enough to align the director parallel to the electric field. As reported previously,[19] Eqs. (10.2) and (10.8) show that the quadrupolar splitting ratio for a certain applied voltage, V, gives the limiting angle, θ_∞. This was achieved by rearranging Eq. (10.8) and combining it with Eq. (10.2) to give

$$g(\Delta\tilde{\nu}/\Delta\tilde{\nu}_0) = \frac{1}{a\sin 2\alpha}\cdot\frac{1}{V^2} + \cot 2\alpha, \tag{10.15}$$

where

$$g(\Delta\tilde{\nu}/\Delta\tilde{\nu}_0) = \frac{4(\Delta\tilde{\nu}/\Delta\tilde{\nu}_0) - 1}{2\sqrt{\left[2 - 2(\Delta\tilde{\nu}/\Delta\tilde{\nu}_0)\right]\left[1 + 2(\Delta\tilde{\nu}/\Delta\tilde{\nu}_0)\right]}},$$
$$a = \frac{\mu_0\varepsilon_0}{B^2 d^2}\left(\frac{\Delta\tilde{\varepsilon}}{\Delta\tilde{\chi}}\right), \tag{10.16}$$

and d is the cell thickness. The function $g(\Delta\tilde{\nu}/\Delta\tilde{\nu}_0)$ is seen to be linear in $1/V^2$, accordingly α and a, that is $\Delta\tilde{\varepsilon}/\Delta\tilde{\chi}$, can be found from a linear least squares fit of the experimental NMR data for the voltage dependence of the quadrupolar splittings to Eq. (10.15).[22]

10.3. Experimental

The nematogen used for our detailed studies of the static director distribution as well as for its dynamics was 5CB-d_2, which had been specifically deuteriated in the α-position of the pentyl chain. This was prepared using a procedure described elsewhere[31] but with the reduction of the ketyl group performed using lithium aluminium deuteride rather than the hydride. The nematic was contained in a sandwich cell with the glass plates coated with transparent In_2O_3 to act as electrodes. The cell was held together by a special glue which is stable in the presence of the cyanobiphenyls and which can be cured using UV radiation for a few minutes. The saturation voltage method[27] was employed to measure the surface anchoring strength, A, at the interface of 5CB with the substrate surface. All of the measurements were made at different temperatures in the nematic phase. The spectra were recorded using a JEOL Lambda 300 spectrometer, which has a magnetic flux density, B, of 7.05T. The spectra were obtained using a quadrupolar echo sequence, with a 90° pulse of $7.7\,\mu\text{s}$ and an interpulse delay of $40\,\mu\text{s}$. The repetition rate was 0.3s. The number of free induction decays (FIDs) used to produce spectra with good signal-to-noise varied from 2,000 to 20,000 depending on the sharpness of the spectral lines observed during the alignment process. The nematic cell was held in the NMR probe head so that the electric field, whose direction is normal to the substrate surface, is inclined to the magnetic field. An amplifier and a function generator were used to provide a $1-10\text{kHz}$ sinusoidal AC electric field to the cell. This frequency is sufficiently high to overcome the alignment effects of ionic conduction. By changing the strength of the electric field, the director orientation rotates in a plane defined by **B** and **E**, since the diamagnetic anisotropy and the dielectric anisotropy of 5CB are both positive.

10.4. Static Director Orientation

The nematic cell was $195\,\mu\text{m}$ thick and so the glass surfaces were not treated meaning that the effective anchoring strength was essentially zero. The final adjustment of the orientation of the cell, to ensure that the electric field makes the desired angle with the magnetic field, was carried out by switching on a high electric potential ($200V_{rms}$) and rotating the cell by a few degrees using an ultrasonic stepping motor until the appropriate quadrupolar splitting was obtained (see Eq. (10.2)).

The deuterium NMR spectra, measured as a function of the applied electric potential, are given in Fig. 10.5. The voltage dependence of the spectra shows that with increasing electric field strength the quadrupolar

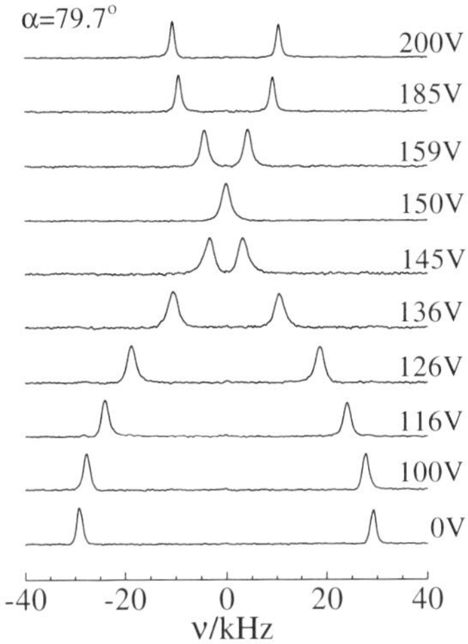

Fig. 10.5. The deuterium NMR spectra measured as a function of the applied electric potential at 288K for the 195 μm nematic cell with untreated glass surfaces.

splitting is reduced, passes through zero and then increases again to less than half its original value. It would seem that the director orientation changes more or less continuously from being parallel to the magnetic field to being almost parallel to the electric field, as its strength grows. This change of the quadrupolar splitting follows the pattern illustrated in Fig. 10.2 in accord with Eq. (10.2). It is of interest to note that the spectra are essentially, but not exactly, the same for the same electric field strengths irrespective of the cell thickness, as expected from theory. Also as the electric field is increased, the lines appear to broaden necessarily leading to a decrease in the signal-to-noise ratio. For higher electric fields, however, the lines sharpen again resulting in an improved signal-to-noise ratio. Thus the experimental results indicate a continuous change in the director orientation with increasing electric field and suggest a slight broadening of the director distribution. This broadening is especially apparent in the center of the range when the director is at angle of 45° to the magnetic field, as expected theoretically[32] and shown experimentally.[30] However, the differences in the linewidths shown in the initial ($0V_{rms}$) and final spectrum ($200V_{rms}$) can be associated with the unresolved D-H dipolar couplings.[30]

10.5. Director Dynamics

A thin nematic sandwich cell ($d = 56.1\,\mu$m) was prepared. The transparent electrodes were not treated in any way. The surface anchoring strength was found to be 1.0×10^{-7} J/m^2 corresponding to weak anchoring at the untreated electrodes. The nematic cell was held in the NMR probe head so that the electric field, whose direction is normal to the substrate surface, makes an angle, $\alpha \approx 45°$, with the magnetic field (see Fig. 10.3). Figure 10.6 shows the spectrum recorded without an electric field at 295K; the quadrupolar splitting, $\Delta\tilde{\nu}_0$, is 54.2kHz. The weak feature in the centre of the spectrum originates from a small amount of the isotropic phase of 5CB-d_2 possibly dispersed in the glue holding the cell together. The final adjustment of the cell alignment to ensure that the electric field makes an angle of essentially 45° with the magnetic field was carried out by switching on a large electric potential (100V_{rms}) and rotating the cell until a doublet splitting was obtained which is 1/4 of the splitting without the electric field (this condition would mean that the substrate surface makes an angle about 45° with the magnetic field). However, the electric field will not be at exactly 45° with respect to the magnetic field because 100V_{rms} is insufficient to align the director parallel to the electric field. An amplifier and a function generator were used to provide a 10kHz sinusoidal AC electric potential (50V_{rms}) to the cell.

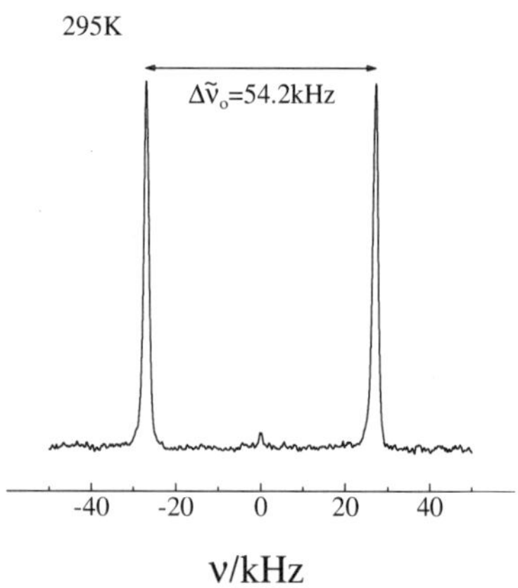

Fig. 10.6. The deuterium NMR spectrum of 5CB-d_2 recorded without an electric field at 295K, the quadrupolar splitting, $\Delta\tilde{\nu}_0$, is 54.2kHz and the effective spin-spin relaxation time, T_2, is 1ms.[19]

Deuterium NMR spectra obtained during (a) the turn-on and (b) the turn-off processes at 295K are shown in Figs. 10.7(a) and (b), respectively. For the turn-on process the director relaxation was monitored at several values of t_{ON} between 0 and 25ms following the application of the electric potential ($50V_{rms}$). For the turn-off process, the director relaxation was measured at several values of t_{OFF} in the range 0 to 25ms. An electric potential of $50V_{rms}$ was applied for $t_a = 50$ms to obtain the initial director alignment for the turn-off experiments.

In the turn-on process, the quadrupolar splitting decreases and then saturates with time after about 3ms. In the turn-off process, the quadrupolar splitting increases because the director moves from being at approximately 30° to the magnetic field to being parallel to it. The spectra recorded in the fast time region for the turn-on and turn-off processes contain weak oscillatory spectral features associated with the director rotation during the acquisition for the free induction decay. We

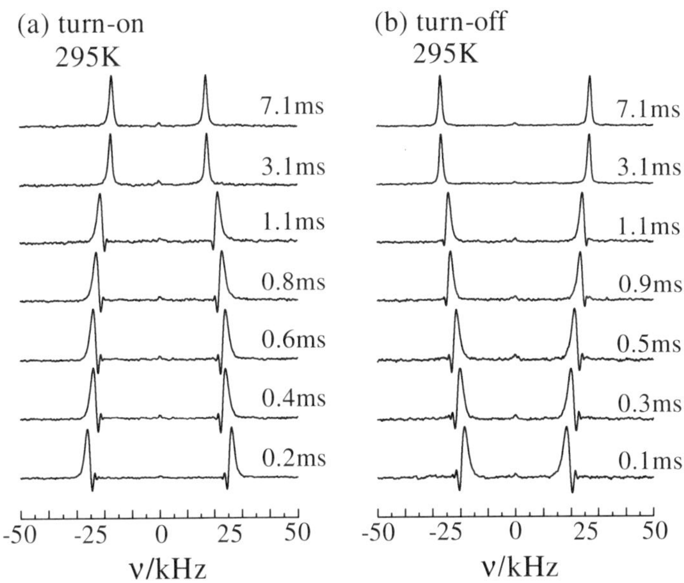

Fig. 10.7. The deuterium NMR spectra for (a) turn-on and (b) turn-off processes recorded at 295 K for 5CB-d_2. In the turn-on process, the spectra were measured by changing the application time for the electric potential (t_a). In the turn-off process, the spectra were measured at a time t_r after the electric potential was turned off.[19]

describe this as the fast time region since it corresponds to the variation of the quadrupolar splitting during the acquisition of the FID. The origin of the oscillatory spectral features is well understood[33] but they are not of importance for this investigation. However, it is of interest to note that the oscillations appear on the inside of the quadrupolar doublet during the turn-on process but on the outside of this doublet for the turn-off process. They give, therefore, the sense of the director motion. The time resolved director orientation can be easily determined from results because of the simple relation between the quadrupolar splitting and the director orientation, as we shall now discuss. If we neglect the non-uniform director distribution that causes the slight broadening of the spectral lines[19] in the hydrodynamic theory then an appropriate form of the time dependence of the director orientation can be determined relatively easily. This is a good assumption since there is no significant elastic deformation in the nematic slab used in our experiments. The observed quadrupolar splitting can then be used together with Eq. (10.2) and the value of the splitting when the director is parallel to the magnetic field to determine the director orientation.

According to continuum theory,[34] we should consider the one-dimensional distortion of the director across the cell. However, all of the deuterium NMR spectra in our study appear to be dominated by a single doublet which allows us to determine the director orientation associated with an essentially monodomain sample. In this analysis, we therefore treat the director as being uniformly aligned. Figure 10.8 shows the temporal variation in the ratio of the quadrupolar splittings, $\Delta\tilde{\nu}(t)/\Delta\tilde{\nu}_0$, determined from the time-resolved deuterium NMR spectra shown in Figs. 10.7(a) and (b) for the turn-on and turn-off processes. We can see in Fig. 10.8 that for the turn-on process the director rotates from the initial angle $\theta_0 = 0°$ for $\Delta\tilde{\nu}_0 = 54.2\,\text{kHz}$ and then aligns at the limiting angle, θ_∞, of 29.8° for $\Delta\tilde{\nu}_\infty = 34.1\,\text{kHz}$ (see Eq. (10.2)), the limiting value of $\Delta\tilde{\nu}(t)$ as t tends to infinity. The limiting value of θ_∞ was obtained from $\Delta\tilde{\nu}_\infty$ and Eq. (10.2). In the turn-off process, the time dependence of the director orientation was obtained in the same way and is also shown in Fig. 10.8. The director rotates back parallel to the magnetic field and the time taken

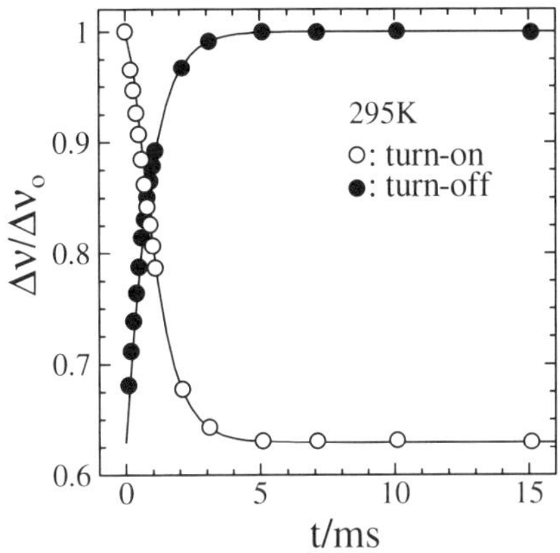

Fig. 10.8. The time dependence of the ratio, $\Delta\tilde{\nu}/\Delta\tilde{\nu}_0$, for (a) the turn-on and (b) the turn-off processes determined from the deuterium NMR spectra of 5CB-d_2 shown in Figs. 4.7(a) and (b) at 295K.[19]

for the alignment process is slower in the turn-off process than in the turn-on process. This is clearly apparent from the results listed in Table 10.1 but is not so obvious from the different time dependences of the quadrupolar splittings shown in Fig.10.8, because this shows the time dependence of $P_2(\cos\theta(t))$ and not $\tan(\theta(t)-\theta_\infty)$ for the turn-on process and not $\tan\theta(t)$ for the turn-off process. The large difference in the relaxation times for the turn-on and turn-off processes, essentially a factor of two, follows from the theory. Thus the relaxation time, τ_{OFF}, for the turn-off process is independent of the electric field (see Eq. (10.11) whereas τ_{ON} for the turn-on process depends on both the electric field strength and its orientation, α, with respect to the magnetic field (see Eq. (10.10)). In our experiments α is approximately 45° so that the cross term in the denominator is negligible, accordingly the quadratic term in τ_E simply adds to that in τ_M thus increasing the denominator and so decreasing the relaxation time τ_{ON} in comparison with τ_{OFF}, as we find experimentally. From the experimental values of $\Delta\tilde{\nu}_0$ and $\Delta\tilde{\nu}_\infty$ combined with Eqs. (10.2) and (10.7), the values of the two relaxation times, τ_{ON} and τ_{OFF}, were obtained by fitting the ratio of the quadrupolar splittings as a function of time for the turn-on and turn-off processes at each temperature. The solid lines in Fig. 10.8 show the best-fit lines giving the values of τ_{ON} and τ_{OFF} to be 0.766ms and 1.54ms, respectively. We have chosen to fit the time dependence of the quadrupolar splittings rather than that for the director orientation calculable from it because the splitting constitutes the primary experimental data and the absolute error associated with each point is more or less constant. The values for these relaxation times at other temperatures were obtained in the same manner and are listed in Table 10.1. An accurate value for the angle α between the magnetic and electric fields was calculated by substituting the values for θ_∞, τ_{ON} and τ_{OFF} into Eq. (10.13); they are listed in Table 10.1.

Our NMR results allow us to calculate the ratio, $\Delta\tilde{\chi}/\Delta\tilde{\varepsilon}$, via Eq. (10.9), the results for this important quantity are shown in Table 10.2 and plotted against the reduced temperature, T/T_{NI}, in Fig. 10.9, where the nematic-isotropic transition temperature for 5CB-d_2 is taken as 306K. It is immediately apparent that the ratio is independent of temperature within experimental error. Similar results have been reported in Refs. [5] and [34], however, Bunning *et al.*[35] have claimed a small but significant variation in this ratio with temperature. This may have resulted from the fact that $\Delta\tilde{\varepsilon}$ and $\Delta\tilde{\chi}$ were measured in independent experiments unlike our results which are determined directly as the ratio from the NMR experiment. Although the results reported by Bunning *et al.*[35] are almost certainly in error, it is of interest to consider their explanation for the temperature dependence. They noted that mesogenic molecules are not

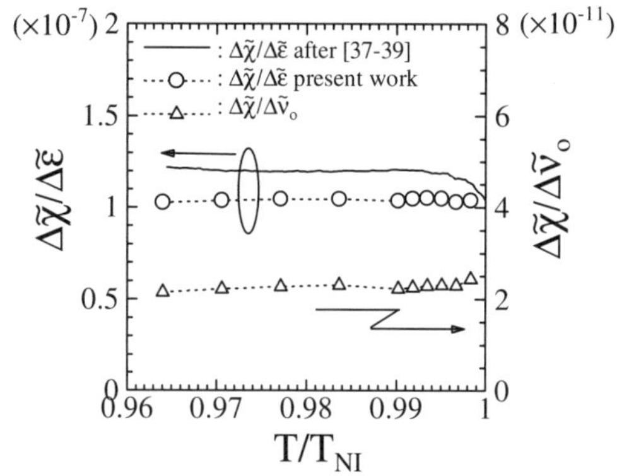

Fig. 10.9. The ratios $\Delta\tilde{\chi}/\Delta\tilde{\varepsilon}$ and $\Delta\tilde{\chi}/\Delta\tilde{\nu}_0$ for 5CB-d_2 as a function of the reduced temperature, T/T_{NI}, where T_{NI} denotes the nematic-isotropic transition temperature which is 306K for 5CB-d_2.[19]

cylindrically symmetric so that the contribution of the biaxial ordering of the molecule to the anisotropic susceptibilities must be included. In general the anisotropy $\Delta\tilde{A}(=\tilde{A}_{\parallel}-\tilde{A}_{\perp})$ of some partially averaged second rank tensorial property, $\tilde{\mathbf{A}}$, is given by

$$\Delta\tilde{A}=(3/2)S_{zz}A'_{zz}+(3/4)(S_{xx}-S_{yy})(A'_{xx}-A'_{yy}). \tag{10.17}$$

Table 10.1. The temperature dependence of the quadrupolar splitting, $\Delta\tilde{\nu}_0$, the ratio, $\Delta\tilde{\nu}_\infty/\Delta\tilde{\nu}_0$, of the quadrupolar splittings with and without the electric field, the turn-on, τ_{ON}, and turn-off, τ_{OFF}, relaxation times and the angle, α, between the electric and magnetic fields.[19]

T/K	$\Delta\tilde{\nu}_0$/kHz	$\Delta\tilde{\nu}_\infty/\Delta\tilde{\nu}_0$	τ_{ON} / ms	τ_{OFF} / ms	α / °
295	54.2	0.630	0.766	1.54	44.7
297	52.1	0.632	0.672	1.34	44.7
299	49.6	0.636	0.594	1.18	44.6
301	46.4	0.636	0.514	1.02	44.6
303	44.6	0.627	0.443	0.88	44.7
303.5	42.6	0.636	0.424	0.84	44.7
304	41.2	0.637	0.404	0.80	44.5
304.5	39.5	0.636	0.384	0.76	44.7
305	37.6	0.634	0.348	0.70	44.7
305.5	34.7	0.623	0.314	0.62	44.7

Here $\mathbf{A}'$ is the traceless tensor set in a molecular frame and $\mathbf{S}$ is the Saupe ordering matrix given in the same frame. Here, either the principal axes of $\mathbf{S}$ or $\mathbf{A}$ are used to define the molecular frame, *xyz*, and for convenience this is usually taken to be the principal axis system for the ordering matrix. The z-axis is taken as that which approximates closest to the molecular symmetry axis, accordingly S_{zz} is the major order parameter and $(S_{xx}-S_{yy})$ is the biaxial order parameter and measures the difference in the ordering of the axes orthogonal to *z*. The ratio, $\Delta\tilde{\chi}/\Delta\tilde{\varepsilon}$, is then given by

$$\frac{\Delta\tilde{\chi}}{\Delta\tilde{\varepsilon}}=\frac{S_{zz}\chi'_{zz}+(1/2)(S_{xx}-S_{yy})(\chi'_{xx}-\chi'_{yy})}{S_{zz}\varepsilon'_{zz}+(1/2)(S_{xx}-S_{yy})(\varepsilon'_{xx}-\varepsilon'_{yy})}, \tag{10.18}$$

where the influence of internal electric fields have been ignored and $\boldsymbol{\varepsilon}'$ is in essence an effective traceless dielectric tensor for a perfectly oriented system. Since the major and the biaxial order parameters have

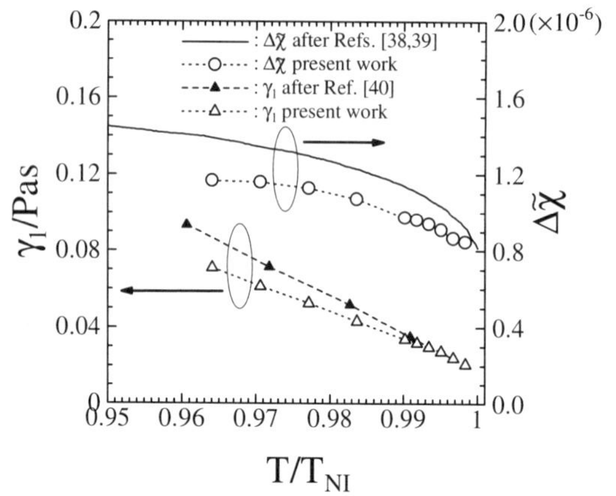

Fig. 10.10. The rotational viscosity coefficient, γ_1, and the diamagnetic anisotropy, $\Delta\tilde{\chi}$, for 5CB-d_2 as a function of the reduced temperature.[19]

different temperature dependences it is to be expected that the ratio, $\Delta\tilde{\chi}/\Delta\tilde{\varepsilon}$, will, in general, be temperature dependent even though χ' and ε' are usually temperature independent. The simplest way to ensure that $\Delta\tilde{\chi}/\Delta\tilde{\varepsilon}$ is temperature independent is if the biaxial terms are small in comparison with the major terms. Certainly for 5CB $(S_{xx} - S_{yy})$ is extremely small in comparison with S_{zz} [36] and so provided the biaxialities of the associated molecular quantities are not abnormally large in the same axis system the biaxial contributions to $\Delta\tilde{\chi}/\Delta\tilde{\varepsilon}$ (see Eq. (10.18)) can be neglected leaving the major order parameter which will cancel and the ratio will be independent of temperature as we find. This independence would also be expected if the biaxial order parameter was linear in S_{zz} which can occur for temperatures close to the nematic-isotropic transition which is the region in which our measurements are made. It seems likely that both factors play a role in the temperature independence which we have observed.

The angle α between the electric and magnetic fields can also be determined from our results, without making any further assumptions, from Eq. (10.13). The results for α are listed in Table 10.1 and are equal, on average, to 44.7° and, as we can see, this angle is independent of temperature, as required. This temperature independence provides an internal check on our experiments and analysis for once the sample was positioned in the probe head of the NMR spectrometer this was not changed for the studies of the sample at different temperatures. Accordingly α should not change with temperature which is in accord with our results.

In order to calculate the values of $\Delta\tilde{\chi}$ and γ_1 from our results we need the value of $\Delta\tilde{\varepsilon}$ at the same temperature. For this we have used the dielectric anisotropies determined by Dunmur and Miller.[37] The resultant

Table 10.2. The temperature dependence of the dielectric anisotropy, $\Delta\tilde{\varepsilon}$ [37], the diamagnetic anisotropy, $\Delta\tilde{\chi}$, the rotational viscosity coefficient, γ_1, the ratio of the anisotropic susceptibilities, $\Delta\tilde{\chi}/\Delta\tilde{\varepsilon}$, and the ratio of the diamagnetic anisotropy to the quadrupolar splitting, $\Delta\tilde{\chi}/\Delta\tilde{\nu}_0$.[19]

T/K	$\Delta\tilde{\varepsilon}$ [37]	$\Delta\tilde{\chi}\times10^6$	γ_1 / Pa s	$(\Delta\tilde{\chi}/\Delta\tilde{\varepsilon})\times10^7$	$(\Delta\tilde{\chi}/\Delta\tilde{\nu}_0)\times10^{11}$ / kHz
295	11.4	1.17	0.0710	1.03	2.15
297	11.2	1.16	0.0614	1.04	2.22
299	10.8	1.13	0.0526	1.04	2.27
301	10.2	1.07	0.0432	1.05	2.31
303	9.40	0.973	0.0339	1.04	2.23
303.5	9.19	0.962	0.0319	1.05	2.26
304	8.95	0.940	0.0297	1.05	2.29
304.5	8.68	0.909	0.0273	1.05	2.30
305	8.40	0.865	0.0239	1.03	2.30
305.5	8.15	0.846	0.0207	1.04	2.44

temperature dependences of $\Delta\tilde{\chi}$ and γ_1 are listed in Table 10.2 where $\Delta\tilde{\varepsilon}$ used in the calculation is given. Fig. 10.10 shows the dependence of $\Delta\tilde{\chi}$ and γ_1 on the reduced temperature. The open circles and solid line indicate the temperature dependence of $\Delta\tilde{\chi}$ from the present work and the literature,[38,39] respectively. The open and closed triangles indicate the temperature dependence of γ_1 from the present work and Ref. [40], respectively. In Fig. 10.10 the dashed lines for the present work serve only as a guide for the eye. We note that our values for $\Delta\tilde{\chi}$ and γ_1 are in relatively good agreement with those obtained by other studies. Of course, analogous information concerning the director dynamics could be obtained by other methods, such as capacitance and transmitted light measurements. However, it is difficult to determine both the diamagnetic anisotropy and the rotational viscosity at the same time. On the other hand, using deuterium NMR measurements, the diamagnetic anisotropy and the rotational viscosity are both determined using the simple relation between the relaxation times for the turn-on and turn-off processes. In addition, deuterium NMR spectroscopy presents the possibility of being able to explore the spatial distribution of the director in nematic cells as a function of time during the turn-on and turn-off processes. This is of special importance because it allows us to see if the director is uniformly distributed which is required if the hydrodynamic theory which we have used is to be strictly applicable.

10.6. Molecular Site Dependence of Director Dynamics

The macroscopic behaviour of a nematic liquid crystal can be described at the static level in terms of the Zocher-Oseen-Frank continuum theory which gives the director distribution resulting from a set of competing constraints.[41] Here, the director is identified with the symmetry axis of this uniaxial phase and hence with the symmetry axes of such second rank tensorial properties as the dielectric and magnetic susceptibilities. The rate at which the director changes from a non-equilibrium to an equilibrium state is described by the Leslie-Ericksen hydrodynamic theory.[41] The key quantities occurring in this are the viscosity coefficients related both to director dynamics and to flow. Of particular interest and technological importance are the director viscosity coefficients that determine the rate at which the director orientation changes. These coefficients can be measured in a variety of ways.[42] One of which is to apply an external field to a uniformly aligned nematic so as to create a non-equilibrium state and then to monitor the change in the director orientation as the system returns to equilibrium, as we have seen in the previous section. Experimentally this time dependent director orientation can be determined by measuring a component of a particular macroscopic property along an axis set in the laboratory. However, with the development of spectroscopic techniques for the study of liquid crystals it has become possible to determine the director orientation using tensorial properties associated with specific fragments of the mesogenic molecules. For example, for a nematogen, such as 4-pentyl-4′-cyanobiphenyl, vibrational spectroscopy can be used to measure the director orientation via peaks associated with the aromatic core or the pentyl chain attached to it. Indeed early studies using FTIR of the electric field-induced director orientation as a function of time for a thin film of 5CB revealed that the director associated with the chain moved faster than that associated with the aromatic core.[43] This surprising result was confirmed independently using time-resolved IR spectroscopy[44] and Raman spectroscopy[45]. The result is of particular importance because the implicit observation of different viscosity coefficients depending on which molecular group is used to determine the director orientation during its alignment would undermine the macroscopic hydrodynamic theory of director dynamics. The result is also surprising because the rates at which both parts of the low molar mass mesogenic molecule move are considerably faster than the rate at which the director orientation changes. This significant timescale separation of the molecular and macroscopic dynamic processes should ensure that they are uncorrelated; in other words the director orientation measured as a function of time should be independent of the molecular group used to determine it.

Given the potential importance of these early observations there have been many attempts to confirm or refute them with a bewildering array of results. For example, an investigation, also of 5CB and also using FTIR to determine the director orientation while this was being changed by an electric field, failed to reveal any difference between the motion monitored by vibrations in the chain or the core.[46] Similarly a study, again of 5CB, using extremely large electric fields resulting in relaxation times for the director which were up to 1000 times faster than those achieved in the earlier studies could not detect any difference in relaxation rates based on observation of the chain or the core.[47] This result contrasts with that found in an investigation of the field-induced director dynamics in the nematic phase of 4-pentyl-(4′-cyanophenyl) cyclohexane which is closely related structurally to 5CB.[48] Here it was discovered, using two-dimensional Raman spectroscopy, that whereas the director associated with the pentyl chain started to move immediately on the application of the electric field that for the phenylcyclohexane core exhibited an induction period. However, the subsequent relaxation times for the core and the chain were found to be essentially the same. It is of interest that although the data shown in an analogous time resolved Raman study of 5CB reveals an induction period for both core and chain, that for the core is longer than for the chain.[49]

Other spectroscopic techniques can also provide molecular site specific information and one of the most important of these for liquid crystals is NMR spectroscopy. This technique has been used to monitor the director alignment in 5CB by observing the change in chemical shifts of carbon-13 nuclei in both the chain and the core.[50] In this experiment the director alignment is controlled by the magnetic field of the spectrometer and the non-equilibrium orientation of the director was created by spinning the sample rapidly (1-2 kHz) at an angle of 46° to the magnetic field. When the spinning is stopped there is an induction period for both groups in the molecule. Then the pentyl chain is observed to be realigned at a slightly slower rate than for the core, in contrast to the early studies based on vibrational spectroscopy. Deuterium NMR has also been used to study the dynamics of the director alignment by a magnetic field for 5CB-d_{15} but now the initial non-equilibrium orientation was produced by slow spinning ($\leq$ 5 Hz) of the sample about an axis orthogonal to the field.[51] Again on stopping the rotation there is an induction period where the director orientation does not change. Then the relaxation time for the alignment of the director parallel to the magnetic field is found to be the same for the CD_3 and all of the CD_2 groups in the pentyl chain as well as the deuteriated phenyl ring in the core. In contrast to the Raman studies the induction period is observed to be the same for the biphenyl core and pentyl chain segments.

In view of the differences in the results obtained with the NMR experiments and because of the continuing controversy concerning the site dependence of field-induced director dynamics we undertook a new study using deuterium NMR spectroscopy. The nematogen investigated was also 5CB in order to facilitate comparison with previous work for this compound which is clearly extensive. The experiment differs from the two earlier NMR investigations in that an electric field was used to create the non-equilibrium state. This avoids the need to spin the sample which necessarily induces flow that could be responsible for the induction period. In addition, the director dynamics during alignment by the electric field was investigated since this makes comparison with the

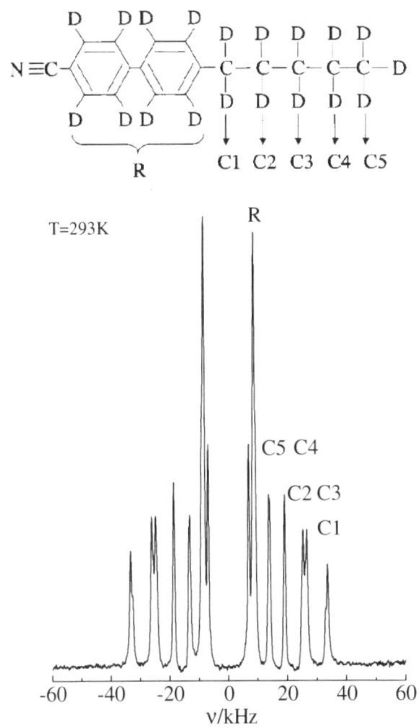

Fig. 10.11. The chemical structure of perdeuteriated 4-pentyl-4′-cyanobi-phenyl together with its deuterium NMR spectrum in the nematic phase at 293K with the director parallel to the magnetic field.[18]

experiments using vibrational spectroscopy more direct. There is the added advantage that the relaxation time can be altered by varying the strength of the electric field. This is important because the relaxation times for the core and chain have been found to be equal for certain field strengths.[45]

The sample used in our experiments was perdeuteriated 4-pentyl-4′-cyanobiphenyl; this had been prepared using the route described in Ref. [52] but with perdeuteriated precursors. The nematic-isotropic transition temperature for 5CB-d_{19} was 306.7 K, in agreement with the literature value of 308.0 K. The sample was contained in a thin cell made of In_2O_3 coated glass plates. The cell thickness was determined optically and found to be uniform and equal to 55.8 μm; thus ensuring that the electric field is also uniform. The surfaces were untreated so that the surface anchoring strength was weak which ensures that the director is uniformly aligned within the cell by an external field. An AC electric field with a frequency of 10 kHz was used in the experiments. Our primary concern for these was to vary the director relaxation time and to see how this influences its site dependence. According to Eqs. (10.10)-(10.12) this variation in τ_{ON} can be achieved by changing several variables.

We begin with the deuterium NMR spectrum of 5CB-d_{19} with the director aligned parallel to the magnetic field, here the quadrupolar splittings are the largest (see Eq. (10.2)). The spectrum shown in Fig. 10.11 was measured for the nematic phase at 293K; it is clearly composed of a series of quadrupolar doublets centred essentially on a common frequency. An additional small splitting is apparent on the outer pair of lines which originates from a dipolar interaction between the deuterons.[53] The seven quadrupolar splittings observed have been assigned to the groups of deuterons in 5CB-d_{19} [54] and this assignment is shown on the peaks according to the labelling indicated on the chemical structure, also given in Fig. 10.11. The pentyl chain contains five groups of equivalent deuterons, four methylene and one methyl, and the quadrupolar doublets associated with all of these are readily apparent in the spectrum. The eight deuterons in the two phenyl rings are only strictly equivalent in pairs but because the C-D bonds make essentially the same angle with the para-axis of the phenyl rings the quadrupolar splittings are expected to be very similar. In practice three of the pairs of deuterons have the same splitting while the remaining pair has a slightly smaller quadrupolar splitting. It is apparent that the peaks associated with the core and the chain are well separated in the spectrum and so it is an easy task to use NMR spectroscopy to follow the dynamics for the director associated with these two fragments. Indeed, the peaks originating from the five groups in the pentyl chain are also well-resolved and so it is possible to monitor the dynamics not just for the chain but for the individual rigid groups within it. This is an advantage over the use of vibrational spectroscopy which is normally unable to separate the peaks coming from the different groups in the chain. The ability to study them individually is important because the methyl group is most likely to behave differently from the other groups in the chain as they approach the attachment to the core.

We now consider our results for the turn-on and turn-off experiments to follow the field-induced alignment of the director. In view of the conflicting results obtained with site-dependent director relaxation studies for 5CB we consider it of special importance to show some, at least, of the primary experimental data. Typical deuterium NMR spectra

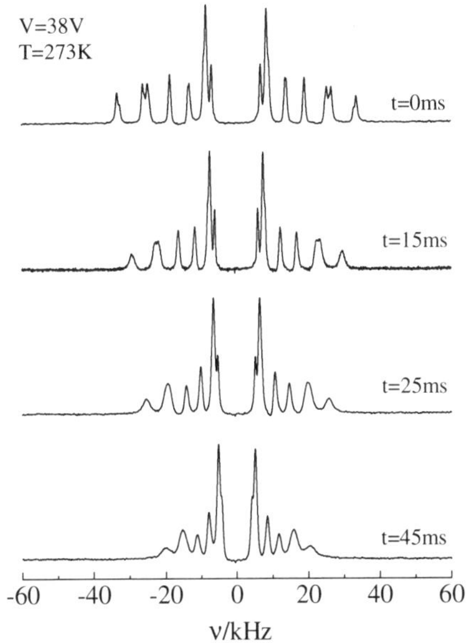

Fig. 10.12. The time dependence of the deuterium NMR spectrum of 5CB-d_{19} following the application of the electric field. For this turn-on process the electric field was 0.681 MV m^{-1}, it was inclined at an angle of ~55° to the magnetic field and the sample temperature was 273K.[18]

for the turn-on process are shown in Fig. 10.12 for the sample at 273K. In this process the director starts parallel to the magnetic field and so following the application of the electric field (0.681 MV m^{-1}) the quadrupolar splittings decrease with increasing time. They reach their minimum values when the director reaches the limiting angle of θ_∞ with respect to the magnetic field (see Eq. (10.8)). These anticipated changes are readily apparent in the spectra shown in Fig. 10.12. After 15 ms the quadrupolar splittings have decreased as the director moves away from being parallel to the magnetic field. The spectral lines remain relatively sharp showing that the director is still uniformly aligned as its orientation changes. There is, however, a small broadening of the spectral lines which is seen to increase with the quadrupolar splitting. Since the methyl and aromatic deuterons have similar splittings which are the smallest the broadening of these spectral lines is also similar as well as small; it will not, therefore, affect the accuracy with which the director orientation associated with these two extreme groups in the molecule can be determined. After 25ms the quadrupolar splittings have decreased still further (see Fig. 10.12) showing that the director has continued to move away from the magnetic field. In keeping with this increase in θ is an increase in the spectral linewidths, especially for those associated with the largest quadrupolar splitting. This is thought to be caused by a small deviation, $\delta\theta$, of the director from a uniform alignment and the resultant broadening is given by $\delta\Delta\tilde{\nu}(\theta) = (3/2)\Delta\tilde{\nu}_0 \sin 2\theta\delta\theta$ [30,32] and so grows with $\Delta\tilde{\nu}_0$. The increase in the linewidths means that the difference in the quadrupolar splittings for the deuterons at positions 2 and 3 in the pentyl chain of 5CB is not resolved. Nonetheless those for the other deuterons are clearly separated which allows the director orientation to be determined for the other positions in the molecule, in particular for the aromatic core and the methyl group at the end of the chain. The continuing motion of the director away from parallel to the magnetic field and towards the electric field is clearly apparent from the spectrum recorded after 45ms (see Fig. 10.12). Again there is an increase in the linewidth associated with the increase in θ and $\delta\theta$. Despite this increase in the width of the spectral lines the resolution is clearly sufficient to determine the quadrupolar splittings for the aromatic core and for three of the positions in the pentyl chain needed to determine the director orientation for these different groups in the molecule.

As we saw in Sec.10.2 the ratio of a quadrupolar splitting, $\Delta\tilde{\nu}(\theta)$, to its value when the director is parallel to the magnetic field, $\Delta\tilde{\nu}_0$, is dependent only on the orientation of the director with respect to the magnetic field via $P_2(\cos\theta)$. Accordingly if the director associated with different groups within 5CB align at different rates then a plot of the time dependence of the ratios, $\Delta\tilde{\nu}(\theta)/\Delta\tilde{\nu}_0$, where $\Delta\tilde{\nu}(\theta)$ is the quadrupolar splitting at time, t, should result in a series of curves. Conversely if the field-induced relaxation of the director is site independent then the points for the various molecular sites will fall on a common curve. The results are shown in Fig. 10.13 for the group of six biphenyl deuterons in the core and for positions 1, 4 and 5 in the pentyl chain. It is immediately apparent that the points fall on a common curve and that there is no site dependence within the experimental error. This result is in keeping with an earlier deuterium NMR study[51] as well as some investigations using vibrational spectroscopy[46,47] but contrasts with other investigations using this technique.[43-45] As it has been suggested that such difference in behaviour found for the director dynamics might be associated with the timescale for the director alignment process we have determined the

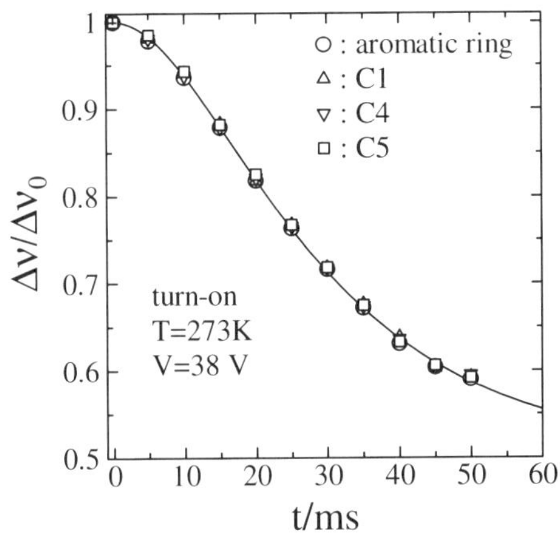

Fig. 10.13. The time dependence of the ratio of the quadrupolar splittings, $\Delta\tilde{\nu}(\theta)/\Delta\tilde{\nu}_0$, for the six biphenyl deuterons (○) and for the deuterons in the pentyl chain at positions 1 (○), 4(□) and 5(□) following the application of an electric field (0.681 MV m^{-1}) at an angle of ~75° to the magnetic field.[18]

relaxation time by fitting our results for the time dependence of $P_2(\cos\theta)$ to the Leslie-Ericksen prediction contained in Eq. (10.7). The best fit is shown as the solid line in Fig. 10.13 and is clearly in good agreement with experiment. The optimum value for the relaxation time, τ_{ON}, is found to be 20.4 ms which is comparable to the value determined, using Raman spectroscopy, for the chain at which the difference between that for the core is the greatest.[45]

It is also apparent from the experimental results in Fig.10.13 that the director orientation starts to change immediately the electric field is applied; that is there is no induction period. This contrasts with the behaviour found in some studies using vibrational spectroscopy.[44,45,48]

However, this difference is to be expected simply from the predictions of the Leslie-Ericksen theory (see Eq. (10.7)). As we have seen from the fit in Fig. 10.13 no induction period is predicted for our experiments, however as θ_∞ the final director orientation grows so the rate at which the director orientation changes falls. Thus for θ_∞ close to 90° the theory

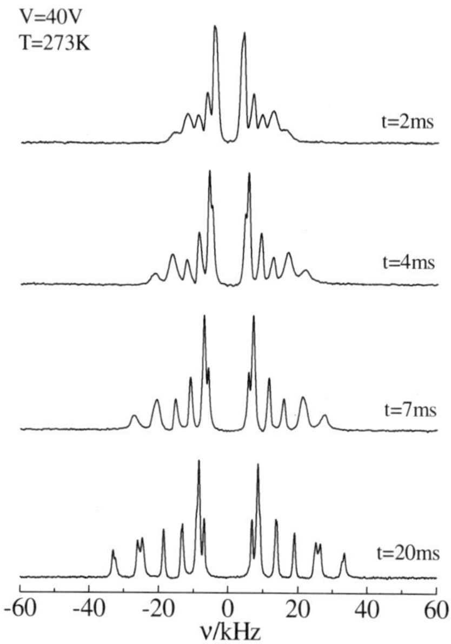

Fig. 10.14. The time dependence of the deuterium NMR spectrum of 5CB-d_{19} following the removal of the electric field. For this turn-off process the electric field was 0.717 MV m^{-1}, it was inclined at an angle of ~75° to the magnetic field and the sample temperature was 273K.[18]

predicts an initial period where the change in θ is very small followed by a region where the director orientation changes rapidly.[55] This initial region could certainly be mistaken for an induction period and so is to be expected for those experiments where the electric field is both large and orthogonal to the initial director orientation. In addition, it is of interest to note that the experiments based on vibrational spectroscopy reveal a site dependence for the induction period.[44,45,48] This cannot occur, of course, for the results which we have obtained using NMR spectroscopy since they do not exhibit an induction period.

We now come to the turn-off experiment in which the director is first aligned at an angle to the magnetic field, the electric field is removed and the director is aligned by the magnetic field. A selection of the deuterium NMR spectra recorded after the removal of the electric field of 0.717 MV m^{-1} is shown in Fig. 10.14. The first spectrum was recorded just 2ms after the removal of the electric field and the quadrupolar splittings are small showing that the director is at an angle to the magnetic field; from the splittings this is estimated to be about 36°. As might be expected at this orientation the spectral lines are somewhat broad which reduces the resolution and hence the accuracy with which the quadrupolar splittings can be determined, although not to any significant extent. After 4ms the spectral resolution has improved considerably largely because the quadrupolar splittings have increased although in addition the linewidths have become slightly smaller. Both of the changes are expected as the director alignment approaches that of the magnetic field. The trend continues as the spectrum recorded 7ms after the removal of the electric field shows. The changes in the spectra are rather modest and it is necessary to wait 20ms until the director is almost parallel to the magnetic field. The spectrum recorded at this time is characterised by narrow lines and large quadrupolar splittings (see Fig. 10.14); it is essentially identical to that shown in Fig. 10.12 for the sample with the director parallel to the magnetic field. Although the spectral lines are broadened to a decreasing extent in the turn-off process they are comparable to those observed during the turn-on process which indicates that the breadth, $\delta\theta$, of the director distribution is also small being of the order of 1°. In other words, the director is aligned by the magnetic field essentially as a monodomain.

The ratios of the quadrupolar splittings, $\Delta\tilde{\nu}(\theta(t))/\Delta\tilde{\nu}_0$, for the turn-off process are plotted as a function of time in Fig. 10.15 for the six deuterons in the biphenyl core and for those at positions 1, 4 and 5 in the pentyl chain. It is immediately apparent that all of the points fall on a common curve. This shows that there is no molecular site dependence for the magnetic field-induced director relaxation, as for the electric field-induced alignment. It is also clear that the relaxation process is faster than for the turn-on process and fitting Eq. (10.7) to our results gives a relaxation time τ_{OFF} of ~7.9ms. This difference in behaviour

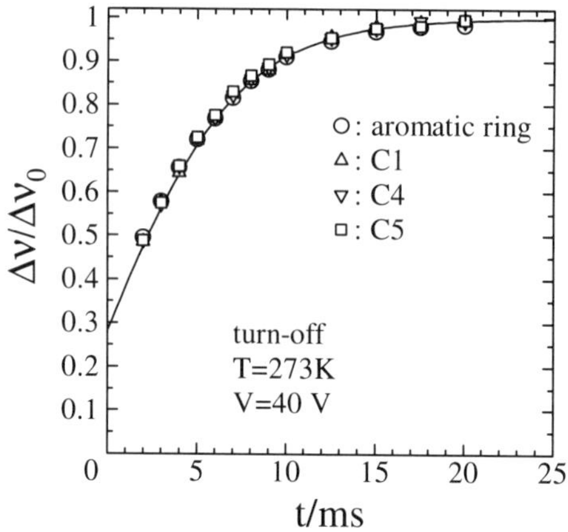

Fig. 10.15. The time dependence of the ratio of the quadrupolar splittings, $\Delta\tilde{\nu}(\theta(t))/\Delta\tilde{\nu}_0$, observed during the turn-off process for the six biphenyl deuterons (○) and for the deuterons in the pentyl chains at positions 1 (Δ), 4(∇) and 5(□). An electric field of 0.717 MV m^{-1} applied at an angle of ~75° to the magnetic field was used to provide the initial director orientation.[18]

follows from Eqs. (10.10)-(10.12) and the large angle between the two fields in the turn-on experiment. We see, therefore, that changing the relaxation time by a factor of about 2.5 does not alter the site independence of the field induced director alignment. The other feature of interest in these results is the absence of an induction period following the removal of the electric field. This contrasts with the earlier deuterium NMR investigation[51] where a viscous torque held the director at an angle to the magnetic field. The fact that the director did not move immediately when sample spinning was stopped presumably shows that the flow of the sample is responsible for the induction period.

We have performed similar studies at 273K with different electric field strengths and experimental geometries. In all cases the time dependence of the ratios of the quadrupolar splittings exhibits a single behaviour thus confirming the result that there is no molecular site dependence for the field-induced director relaxation. The relaxation times determined from these results are given in Table 10.3 and agree to within experimental error. We have also made measurements at the higher temperature of 293K where γ_1 is considerably reduced and so the relaxation times are much faster than at 273K (see Eq. (10.10)). This increase in the relaxation rate presents a challenge to capture the NMR spectra without distortion of the spectral lines because the quadrupolar splitting varies during the acquisition of the FIDs.[56] Consequently only the turn-on experiment was performed using a range of electric field strengths. In all of these studies the director relaxation was found to be independent of the site of the deuterons used to monitor the director orientation. The relaxation times determined from these results are listed in Table 10.3 and range from 0.84 to 1.61ms. They are clearly much smaller than that found for 5CB at 273K (20.2ms) but this variation does not have any effect on their site independence.

The analysis of the spectra used to demonstrate the site-independence of the field-induced director relaxation times requires the measurement of the quadrupolar splittings and their scaling with the values when the director is parallel to the magnetic field. However, there is a unique situation when this is not necessary and the common orientation of the director for the different molecular sites at the same time is immediately

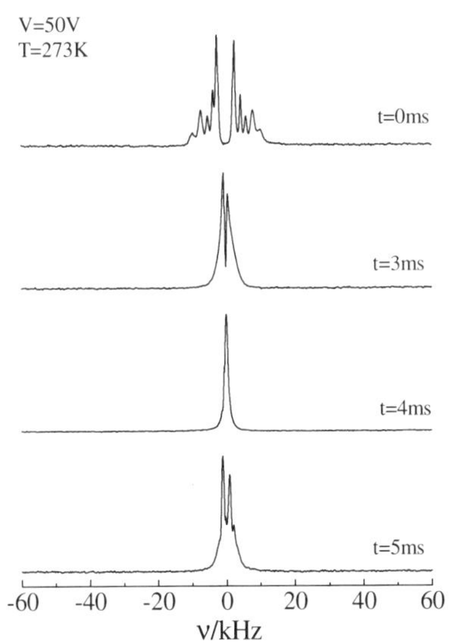

Fig. 10.16. The time dependence of the deuterium NMR spectra recorded during the turn-off process for an electric field of 0.896 MV m^{-1} at an angle of ~75° to the magnetic field. The temperature of the 5CB-d_{19} sample was 273K.[18]

apparent from the spectrum. This occurs when the angle between the director and magnetic field is $\cos^{-1} 1/\sqrt{3}$ or 54.74°, that is the so-called magic angle, for then $P_2(\cos\theta)$ vanishes and all of the quadrupolar splittings are zero irrespective of $\Delta\tilde{\nu}_o$ (see Eq. (10.2)). A series of spectra recorded during a turn-off experiment with an electric field of 0.896 MV m^{-1} showing this novel effect are given in Fig.10.16. The quadrupolar splittings are small in the spectrum recorded prior to the removal of the electric field. It is interesting to note that the spectrum is not quite symmetric because the centre for the quadrupolar doublets originating from the aromatic deuterons differs slightly from that for the aliphatic deuterons. This difference is caused by the small difference in their scalar chemical shifts; it also influences the spectra in Figs. 10.11, 10.12 and 10.14 although because the quadrupolar splittings are larger the resultant spectral asymmetry is not so apparent. The spectrum obtained 3ms after the electric field (see Fig. 10.16) was removed is essentially a single doublet with a small splitting and an unusual lineshape. It shows clearly a major reduction in the quadrupolar splittings

Table 10.3. The experimental conditions used to study the field-induced director relaxation of perdeuteriated 4-pentyl-4′-cyanobiphenyl. In all of the experiments the magnetic flux density, B, was 7.05T.[18]

Process	T/K	E/MV m^{-1}	α/°	θ_0/°	θ_∞/°	τ/ms
turn-on	273	0.448	~50	0	12.0	1.61
turn-on	293	0.538	~50	0	17.5	1.49
turn-on	293	0.717	~50	0	23.5	1.12
turn-on	293	0.896	~50	0	30.0	0.84
turn-on	273	0.645	~75	0	28.1	20.2
turn-on	273	0.681	~75	0	36.5	20.4
turn-off	273	0.717	~75	45.5	0	7.90
turn-off	273	0.896	~75	68.4	0	7.63
turn-off	273	0.896	~55	36.6	0	7.64

as the director approaches the magic angle with respect to the magnetic field. After 4ms the spectrum has collapsed to a single line with a hint of a shoulder at its base. This spectrum shows immediately that the directors associated with the different deuteriated groups in 5CB-d_{19} all make the same angle, namely the magic angle, with respect to the magnetic field. Then after 5ms the quadrupolar splittings have reappeared although they are small and so poorly resolved. Another consequence of the smallness of the quadrupolar splittings is that the influence of the difference in the chemical shift is more apparent in the marked asymmetry of the spectrum.

10.7. Conclusions

Deuterium NMR spectroscopy has been employed to investigate the static director orientation in a thin nematic film. The dependence of the quadrupolar splitting on the electric field strength has been measured for various cells. For one with weak anchoring, we find that the director orientation changes continuously from being parallel to the magnetic field to being orthogonal to it. The electric field dependence of the quadrupolar splitting can be understood from the continuum theory predictions for the director distribution. Our experimental results and their comparison with theory show that deuterium NMR provides a valuable technique with which to investigate the director distribution in thin liquid crystal samples.

Deuterium NMR spectroscopy has also been used to investigate the director dynamics in a nematic slab of 5CB with weak anchoring. A series of deuterium NMR spectra, obtained using a quadrupolar echo sequence, was acquired as a function of time. When the electric field, whose strength is controlled so that the director is inclined with respect to the magnetic field is applied to the nematic film, the director moves from being parallel to the magnetic field to being at an angle θ to it. After the electric field is switched off, the director relaxes back to being parallel to the magnetic field. In this way a non-equilibrium director orientation with respect to the field can be created without causing flow to occur within the nematic sample. Deuterium NMR spectra were recorded during the turn-on and the turn-off alignment processes as a function of time. We have studied the time dependence of the director orientation for the turn-on and turn-off processes at different temperatures in the nematic phase. From the relaxation times measured, we have determined the rotational viscosity coefficient and the diamagnetic anisotropy of 5CB at different temperatures. Both of these quantities are in reasonably good agreement with values reported in the literature. The ratio of the anisotropic susceptibilities, $\Delta\tilde{\chi}/\Delta\tilde{\varepsilon}$, was found to be independent of temperature in agreement with other but not all experimental studies; it can be understood in terms of the effective cylindrical symmetry of the Saupe ordering matrix for the 5CB molecules. The time variation of the complete director dynamics was predicted by a hydrodynamic theory and found to be in good agreement with our experimental results as well as the predictions of the simple theory in which elastic deformations and surface anchoring are ignored. The present results indicate that deuterium NMR spectroscopy provides a valuable technique with which to investigate the turn-on and the turn-off alignment processes as a function of time in thin nematic slabs.

Deuterium NMR spectroscopy has also been used to investigate the molecular site dependence of the field-induced director relaxation for 5CB using deuterium NMR spectroscopy. These studies have revealed that, over a wide range of conditions, the director dynamics are independent of the molecular site of the deuterons used to determine the director orientation. This independence is particularly striking because, unlike vibrational spectroscopy, the site dependence of the NMR technique allows the study of the biphenyl core and the methyl group at the end of the chain for which the difference in dynamic behaviour is expected to be maximal. The absence of the site dependence in the director alignment is found to hold whether the director is aligned by a magnetic field or an electric field commonly used in studies based on vibrational spectroscopy. In addition, changing the rate of director alignment over a significant range by using turn-on or turn-off experiments or by changing the electric field strength or by altering the temperature has no influence on the site independence of the director dynamics. It is also revealed that there is no induction period for the alignment process for both turn-on and turn-off experiments in accord with the predictions of the Leslie-Ericksen theory, unlike the unusual results sometimes observed using Raman spectroscopy. In addition the induction periods found in the vibrational spectroscopy experiments reveal a site dependence which is necessarily absent from the results of our NMR experiments. There is, however, no clear explanation for the curious site dependence occasionally found with vibrational spectroscopy. What is clear is that deuterium NMR spectroscopy is unable to detect any such site dependence over a wide range of experimental conditions comparable to those used in vibrational spectroscopic studies. This site independence is in keeping with the large difference molecular and director rotational relaxation times for low molar mass nematics.

Acknowledgments

This work was supported by a Scientific Grant-in Aid from the Japan Society for the Promotion of Science (JSPS) and the HRC project of the Ministry of Education, Culture, Sports, Science and Technology of Japan, and was carried out as an Anglo-Japanese joint research project of the International Exchange program supported by The Royal Society and JSPS.

References

1. J. C. Rowell, W. D. Phillips, L. R. Melby and M. Panar, J. Chem. Phys. **43**, 3442 (1965).
2. G. R. Luckhurst, J. Chem. Soc., Faraday Trans. 2, **84**, 961 (1988).
3. A. F. Martins, P. Esnault and F. Volino, Phys. Rev. Lett., **57**, 1745 (1986).
4. M. Winkler, D. Geschke and P. Holstein, Liq. Cryst., **17**, 283 (1994).
5. M. Bender, P. Holstein, and D. Geschke, J. Chem. Phys., **113**, 2430 (2000).
6. R. Stannarius, G. P. Crawford, L. C. Chien and J. W. Doane, J. Appl. Phys., **70**, 135 (1991).
7. S. M. Fan, G. R. Luckhurst and S. J. Picken, J. Chem. Phys., **101**, 3255 (1994).
8. J. R. Hughes, G. Kothe, G. R. Luckhurst, J. Malthete, M. E. Neubert, I. Shenouda, B. A. Timimi and M. Tittelbach, J. Chem. Phys., **107**, 9252 (1997).
9. E. Ciampi, J. W. Emsley, G. R. Luckhurst and B. A. Timimi, J. Chem. Phys., **107**, 5907 (1997).
10. P. Esnault, J. P. Casquilho, F. Volino, A. F. Martins, and A. Blumstein, Liq. Cryst., **7**, 607 (1990).
11. A. Sugimura, K. Nakamura, T. Miyamoto, P. J. Le Masurier, B. A. Timimi, T. H. Payne, and G. R. Luckhurst, Proc. of The 4th Int. Display Workshops, 65 (1997).
12. G. R. Luckhurst A. Sugimura and B. A. Timimi, Mol. Cryst. Liq. Cryst., **347**, 297 (2000).
13. G. R. Luckhurst, T. Miyamoto, A. Sugimura, T. Takashiro and B. A. Timimi, J. Chem. Phys., **114**, 10493 (2001).

14. C. J. Dunn, G. R. Luckhurst, T. Miyamoto, A. Sugimura and B. A. Timimi, Mol. Cryst. Liq. Cryst., **347**, 167 (2000).
15. G. R. Luckhurst, T. Miyamoto, A. Sugimura, and B. A. Timimi, Thin Solid Films, **393**, 399 (2001).
16. G. R. Luckhurst, T. Miyamoto, A. Sugimura and B. A. Timimi, Mol. Cryst. Liq. Cryst., **347**, 167 (2000).
17. G. R. Luckhurst, T. Miyamoto, A. Sugimura and B. A. Timimi, Mol. Cryst. Liq. Cryst., **402**, 103 (2003).
18. G. R. Luckhurst, A. Sugimura, and B. A. Timimi, J. Chem. Phys., **116**, 5099 (2002).
19. G. R. Luckhurst, T. Miyamoto, A. Sugimura, and B. A. Timimi, J. Chem. Phys., **117**, 5899 (2002).
20. G. R. Luckhurst, T. Miyamoto, A. Sugimura, B. A. Timimi, and H. Zimmermann, J. Chem. Phys., **121**, 1928 (2004).
21. D. Kamada, G. R. Luckhurst, K. Okumoto, A. Sugimura, B. A. Timimi, and H. Zimmermann, Mol. Cryst. Liq. Cryst., **441**, 131 (2005).
22. G. R. Luckhurst, A; Sugimura, B. A. Timimi, and H. Zimmermann, Liq. Cryst., **32**, 1389 (2005).
23. G. R. Luckhurst, A. Sugimura, and B. A. Timimi, Liq. Cryst., **32**, 1449 (2005).
24. G. R. Luckhurst, Mol. Cryst. Liq. Cryst., **347**, 121 (2000).
25. G. R. Luckhurst, T. Miyamoto, A. Sugimura and B. A. Timimi, Mol. Cryst. Liq. Cryst., **347**, 147 (2000).
26. G. R. Luckhurst, T. Miyamoto, A. Sugimura and B. A. Timimi, Mol. Cryst. Liq. Cryst., **394**, 74 (2003).
27. A. Sugimura, T. Miyamoto, M. Tsuji, and M. Kuze, Appl. Phys. Lett., **72**, 329 (1998).
28. See, for example, I. W. Stewart, *The Static and Dynamic Continuum Theory of Liquid Crystals* (Taylor & Francis, 2004), p133.
29. G. Labrunie and J. Robert, J. Appl. Phys., **44**, 4869 (1973).
30. D. Hamasuna, G. R. Luckhurst, A. Sugimura, B. A. Timimi, K. Usami, and H. Zimmermann, Thin Solid Films, **517**, 1394 (2008).
31. G. W. Gray and A. Mosley, Mol. Cryst. Liq. Cryst., **35**, 71 (1976).
32. S. A. Brooks, G. R. Luckhurst and G. F. Pedulli, Chem. Phys. Lett., **11**, 159 (1971).
33. A. M. Kantola, G. R. Luckhurst, A. Sugimura, and B. A. Timimi, Mol. Cryst. Liq. Cryst., **402**, 117 (2003).
34. Hp. Schad, G. Baur and G. Meier, J. Chem. Phys, **71**, 3174 (1979).
35. J. D. Bunning, D. A. Crellin and T. E. Faber, Liq. Cryst., **1**, 37 (1986).
36. J. W. Emsley, G. R. Luckhurst, and C. P. Stockley, Mol. Phys., **44**, 565 (1981).
37. D. A. Dunmur and W. H. Miller, J. Physique, **40**, 361 (1979).
38. B. Bahadur, *Liquid Crystals- Applications and Uses Vol.1* (World Scientific, 1990), p.156.
39. D. A. Dunmur, M. R. Manterfield, W. H. Miller, and J. K. Dunleavy, Mol. Cryst. Liq. Cryst., **45**, 127 (1978).

40. K. Skarp, S. T. Lagerwall and B. Stebler, Mol. Cryst. Liq. Cryst., **60**, 215 (1980).
41. P. G. de Gennes and J. Prost, *The Physics of Liquid Crystals*, (2nd ed. Oxford University Press, Oxford, 1993).
42. W. H. de Jeu, *Physical Properties of Liquid Crystalline Materials*, (Gordon and Breach, New York, 1980).
43. V. G. Gregoriou, J. L. Chao, H. Toriumi and R. A. Palmer, Chem. Phys. Lett., **179**, 491 (1991).
44. T. Urano and H. Hamaguchi, Chem. Phys. Lett., **195**, 287 (1992).
45. K. M. Booth and H. J. Coles, Liq. Cryst., **13**, 677 (1993).
46. J. de Bleijser, L. H. Leyte-Zuiderweg, J. C. Leyte, P. C. M. van Woerkom and S. J. Picken, App. Spect., **50**, 167 (1996).
47. W. G. Jang and N. A. Clark, Phys. Rev. E, **63**, 031707 (2001).
48. K. Huang and G. G. Fuller, Liq. Cryst., **25**, 745 (1998).
49. J. M. W. Hanwer and H. J. Coles, Mol. Cryst. Liq. Cryst., **262**, 235 (1995).
50. M. L. Magnuson and B. M. Fung, J. Chem. Phys., **100**, 1470 (1994).
51. E. Ciampi, J. W. Emsley, G. R. Luckhurst, B. A. Timimi, G. Kothe and M. Tittlebach, J. Chem. Phys., **107**, 5907 (1997).
52. R. Dabrowski, Z. Witkiewicz and K. Kenig, Mol. Cryst. Liq. Cryst., **58**, 251 (1980).
53. See, for example, *NMR of Liquid Crystals*, edited by J. W. Emsley (Reidel, Dordrecht, 1985).
54. J. W. Emsley and D. L. Turner, J. Chem. Soc. Faraday Trans. II, **77**, 1493 (1981).
55. A. Kantola, G. R. Luckhurst, A. Sugimura and B. A. Timimi, in preparation.

Chapter 11

Viscoelastic Properties of Liquid Crystals: Statistical-mechanical Approaches and Molecular Dynamics Simulations

Alex V. Zakharov

Saint Petersburg Institute for Machine Sciences,
The Russian Academy of Sciences,
Saint Petersburg 199178, Russia

The problem of predicting structural and dynamical behaviors associated with liquid crystalline materials remains a very interesting one in the physics of condensed matter. A useful route in studies of macro- and microscopic properties of liquid crystals (LCs) is provided by a combination of statistical-mechanical theories, computer simulation techniques, and NMR spectroscopy. We believe that this chapter shows some useful routes not only for further examining the validity of theoretical description of the viscous properties of real LCs, but also for analyzing their elastic and flexoelectric properties.

11.1. Introduction

The problem of predicting the physical properties of liquid crystalline compounds, based upon information on molecular shape and intermolecular interactions remains a most fundamental one in liquid crystal (LC) physics. In recent years, the Gay-Berne (GB) model [1] and its generalised versions[2] have been frequently used for describing interactions between anisotropic molecules. Since mesogenic molecules often possess strong polar groups, the dipolar GB potential[3–6] with an axial dipole at the center or near the end of the molecule is more appropriate for the description of such systems. In the case of polar LCs, such as n-*cyanobiphenyls*, the angle between the para axis and the long molecular axis is small, and this leads to a large axial value of dipole moment, for example, $\sim$ 5 D for $4-$*pentyl*$-4'-$*cyanobiphenyl* (5CB). Among the most important properties of such compounds, viscosity coefficients and Frank elastic constants play a crucial role in several applications such as twisted and super twisted nematic displays.[7] The elastic constants

(ECs), being material properties, reflect the underlying microscopic interactions and correlations. In order to elucidate the role of these correlations in the calculations of bulk ECs, several microscopic theories have been proposed,[4–6,8–13] where the ECs are related to structural quantities such as the single-particle orientational distribution function (SODF), and the direct correlation function (DCF) of a nematic system. The central difficulty of these theories is the calculation of the DCF in the nematic phase. The DCF for such anisotropic system as the nematic phase can be calculated in the framework of two independent approaches: an equilibrium statistical-mechanical theory,[14] taking intercell pair correlations[5,15] into account, and molecular dynamics (MD) simulation applied to the description of nematic LCs.[16–19]

The hydrodynamic properties of liquid crystalline substances have been investigated from the points of view of both basic knowledge and application.[20] The formulation of Ericksen-Leslie theory has allowed the shear stress tensor (ST) to be expressed in terms of five independent viscosity coefficients, and experimental effort has directed to measuring all five Leslie coefficients by two independent groups at approximately the same time.[21,22] Further studies have examined the temperature and order parameter dependence of viscosity coefficients,[23] and extended the microscopic theory of these coefficients.[24–30]

On the other hand, the theoretical treatment of both structural and dynamical properties of flexible molecules in anisotropic medium is not an easy task, and requires a certain number of simplifying assumptions which may only be justified by comparison between model predictions and experiment. Thus the combination of NMR and MD techniques with theoretical treatments provides a powerful tool for the investigation of both structural and dynamical properties of mesogenic compounds containing flexible moieties, especially hydrocarbon chains. This chapter aims to show some useful routes not only for further examining the validity of theoretical description of the viscous properties of real LCs, but also for analyzing their elastic and flexoelectric properties.

11.2. Structure and elastic properties of nematic LCs

There are several methods to investigate the coupling between macroscopic LC properties, such as the elastic coefficients, and molecular structure. A usual route in these studies is provided by a combination of statistical-mechanical theories[4,6,8–13,31] and computer simulation techniques.[16–19]

11.2.1. *Elastic properties*

In an ideal (infinitely extended) nematic LC (NLC) sample, the director field $\hat{\mathbf{n}}(\mathbf{r})$ is strictly constant throughout space;[20] on the other hand, the presence of thermal fluctuations or confining surfaces perturbs this ideal configuration. Distortion of the orientational order occurs and, in general, the director field changes in space, in a more or less smooth way. The distortion of the director field $\hat{\mathbf{n}}(\mathbf{r})$ can be determined by minimizing the free energy functional, having the form

$$F = \int dV f_F + \int dS \, (f_{13} + f_{24} + f_s), \tag{11.1}$$

where

$$f_F = \frac{1}{2}\left[K_1 (\nabla\hat{\mathbf{n}})^2 + K_2 (\hat{\mathbf{n}}\nabla \times \hat{\mathbf{n}})^2 + K_3 (\hat{\mathbf{n}} \times \nabla \times \hat{\mathbf{n}})^2\right]; \tag{11.2}$$

$f_{13} = k_{13}\hat{\mathbf{k}} \cdot \hat{\mathbf{n}}(\nabla \cdot \hat{\mathbf{n}})$, $f_{24} = -(K_2 + k_{24})\hat{\mathbf{k}} \cdot [\hat{\mathbf{n}}\nabla \cdot \hat{\mathbf{n}} + \hat{\mathbf{n}} \times \nabla \times \hat{\mathbf{n}}]$, and f_s is the anchoring energy at the interface, and $\hat{\mathbf{k}}$ denotes a unit vector normal to the surface S confining the NLC. Here k_{13} and k_{24} are surface-like bulk elastic terms and can be omitted in an ordinary curvature elastic treatment with a fixed boundary condition, because these constants do not enter the Euler-Lagrange equation for bulk equilibria.[32] The first term in Eq. (11.1) describes the energy associated with three basic types of deformation, i.e. the bulk Frank elastic constants (ECs) K_1, K_2, K_3, for splay, twist, and bend distortions,[20] respectively. Over the years, quite a few microscopic theories for the bulk ECs K_i ($i = 1, 2, 3$) have been developed.[8–13] Using Mayer cluster expansion method, the expression for the bulk ECs may be written as[8]

$$K_1 = \frac{k_B T}{2} \int r_x^2 C(\mathbf{r}, \hat{\mathbf{e}}_i, \hat{\mathbf{e}}_j) f_0'(\cos\beta_i) f_0'(\cos\beta_j) e_{i,x} e_{j,x} r^2 dr d\Omega d\Omega_i d\Omega_j, \tag{11.3}$$

$$K_2 = \frac{k_B T}{2} \int r_x^2 C(\mathbf{r}, \hat{\mathbf{e}}_i, \hat{\mathbf{e}}_j) f_0'(\cos\beta_i) f_0'(\cos\beta_j) e_{i,y} e_{j,y} r^2 dr d\Omega d\Omega_i d\Omega_j, \tag{11.4}$$

$$K_3 = \frac{k_B T}{2} \int r_x^2 C(\mathbf{r}, \hat{\mathbf{e}}_i, \hat{\mathbf{e}}_j) f_0'(\cos\beta_i) f_0'(\cos\beta_j) e_{i,z} e_{j,z} r^2 dr d\Omega d\Omega_i d\Omega_j, \tag{11.5}$$

where $C(\mathbf{r}, \hat{\mathbf{e}}_i, \hat{\mathbf{e}}_j)$ is the DCF, β_i denotes the polar angle, i.e., the angle between the long axis of the molecule i and the nematic director $\hat{\mathbf{n}}$, taken to coincide with the z axis, $d\Omega = \sin\beta_{ij} d\beta_{ij} d\psi_{ij}$, β_{ij} and ψ_{ij} are the polar and azimuthal angles of the intermolecular unit vector $\hat{\mathbf{e}} = \mathbf{r}/|\mathbf{r}|$, related to the molecular separation vector $\mathbf{r} = \mathbf{r}_i - \mathbf{r}_j$, where $\mathbf{r}_i$ and $\mathbf{r}_j$ are the center-of-mass coordinates of molecules i and j, respectively. Furthermore, $\hat{\mathbf{e}}_i$ and $\hat{\mathbf{e}}_j$ denote the unit vectors defining their orientations, $d\Omega_i = \sin\beta_i d\beta_i d\psi_i$ and $e_{i,\alpha}$ and r_α are the projections of the corresponding vectors $\hat{\mathbf{e}}_i$ and $\mathbf{r}$ on the axes $\alpha = x, y, z$, respectively. Here k_B is the Boltzmann constant, and T is the temperature. Finally, $f_0'(\cos\beta_i)$ is the derivative of the SODF $f_0(\cos\beta_i)$ with respect to $\cos\beta_i$. As already mentioned, the main difficulty with Eqs. (11.3)-(11.5) is their dependence on the DCF $C(\mathbf{r}, \hat{\mathbf{e}}_i, \hat{\mathbf{e}}_j)$ of a reference state. Progress in the evaluation of the ECs is connected to a simplified approach in finding the DCF. Within the framework of the theory,[12] based on a density-functional approach, the bulk ECs of a nematic depend on the DCF of the corresponding isotropic liquid, treated as a reference state. A drawback of that approach is that the original nematic phase has been replaced by an "effective" hard-sphere system interacting through an "effective" potential. In the case of Ref. [16], this reference state is for an unoriented liquid crystal, and achieved by averaging the DCF over all possible orientations of the director and representing the remaining part by a spherical series. A few years ago, based also on a density-functional approach and using the model of perfect local orientational order, explicit expressions for the Frank ECs have been obtained by Osipov and Hess (OH).[11] In the framework of the named approach, a fluid of perfectly oriented ellipsoidal particles is substituted by a reference fluid of spherical particles at the same density; the DCF $C(r)$ was calculated using an approximate closure coupled to the Ornstein-Zernike-type equation. In such an approximation, the bulk ECs may be written as[11]

$$K_1/K = 1 + \lambda_1 + 9\lambda_1 z_1 + \left(6 + \frac{39}{11}\lambda_1\right) z_1^2, \tag{11.6}$$

$$K_2/K = 1 - 2\lambda_1 - 3\lambda_1 z_1 + \left(6 - \frac{141}{11}\lambda_1\right) z_1^2, \tag{11.7}$$

$$K_3/K = 1 + \lambda_1 + 12\lambda_1 z_1 + \left(6 + \frac{102}{11}\lambda_1\right) z_1^2, \tag{11.8}$$

where $\lambda_1 = \frac{2}{7}\omega$, and ω is a geometric factor for the molecule, usually taken as $\omega = (\gamma^2 - 1)/(\gamma^2 + 1)$, $\gamma = \sigma_{\parallel}/\sigma_{\perp}$ is the length-to-breadth ratio of the molecule, $z_1 = \overline{P}_4/\overline{P}_2$, and $\overline{P}_{2L}$ $(L = 1, 2, 3)$ are the corresponding order parameters (OPs) of even rank. The quantity K is an arithmetic mean of splay, twist, and bend distortion modes and can be expressed as

$$K = \frac{1}{3}(K_1 + K_2 + K_3) = \frac{1}{9}M_2 \frac{\gamma^2 + 2}{\gamma^{2/3}} \overline{P}_2^2; \tag{11.9}$$

where

$$M_{2L} = -\int_0^\infty drC(r)r^{2L}, \ L = 1, 2 \tag{11.10}$$

are radial integrals over the DCF for an unoriented nematic state, the OPs can be expressed in terms of SODF $f_0(\cos\beta_i)$ as

$$\overline{P}_{2L} = \langle P_{2L}(\cos\beta_i)\rangle = \int_0^\pi P_{2L}(\cos\beta_i) f_0(\cos\beta_i) \sin\beta_i d\beta_i, \tag{11.11}$$

and $P_{2L}(\cos\beta_i)$ $(L = 1, 2, 3)$ denote Legendre polynomials of even rank. In contrast to the OH approach, in the framework of the Nemtsov-Zakharov (NZ) approximation,[9,10] the DCF has been chosen in the form $C(r/\sigma)$, where σ is the GB orientational-dependent size parameter, given in Ref. [1]. The Frank ECs are then expressed by[10,13]

$$K_1/K = 1 + \lambda_2(5 - 9z_2), \tag{11.12}$$

$$K_2/K = 1 - \lambda_2(1 + 3z_2), \tag{11.13}$$

$$K_3/K = 1 - 4\lambda_2(1 - 3z_2), \tag{11.14}$$

where $z_2 = (\langle\cos^4\beta\rangle - \langle\cos^6\beta\rangle)/(\langle\cos^2\beta\rangle - \langle\cos^4\beta\rangle)$, and $2\lambda_2 = \frac{\omega}{(3-\omega)}$ is also a geometric factor of the molecule. In the NZ density-functional approach the quantity K is given as[10,13]

$$K = \frac{1}{3}(K_1 + K_2 + K_3) = \mathcal{B}\frac{3-\omega}{6\omega}\frac{10\overline{P}_2 - 24\overline{P}_4 + 14}{105}\overline{P}_2^2, \tag{11.15}$$

where $\mathcal{B}$ is a factor which has the dimension of force and may be written as[10,13]

$$\mathcal{B} = 3\pi b^2\rho^2\sigma_\perp^5\omega^3 k_B T \frac{1 + \omega^2/14}{(1-\omega^2)^2} M_4, \tag{11.16}$$

with $b = 4\pi\sigma_\perp^3\rho\omega^2\frac{1+3\omega^2/14}{(1-\omega^2)^2}$, a dimensionless factor, $\rho = 1/v = N/V$ is the particle number density for a system of N molecules occupying a volume V at the temperature T, and M_4 is defined in Eq. (11.10). The averaged functions $\langle\cos^n\beta\rangle$, even n, are linear combinations of orientational order parameters of appropriate ranks, and given in Ref. [6]. Thus according to

Eqs. (11.6)-(11.8), (11.12)-(11.14), (11.9)-(11.10), (11.15), and (11.16), one can calculate the absolute values of K_i ($i = 1, 2, 3$) provided that $\overline{P}_{2L}$ ($L = 1, 2, 3$), $f_0(\cos\beta)$, $C(r)$ are known. Whereas values of the OPs, for various LC are usually fairly easily found, an accurate determination of the DCF for a nematic state is still a formidable task. It should be noted that both above-mentioned approaches OH and NZ predict a result of the form $K \sim \mathcal{I}\overline{P}_2^2$ for the average elastic constant, where $\mathcal{I}$ contains a weighted integral over the direct correlation function.

11.2.2. *Order parameters and direct correlation function: statistical-mechanical treatment*

The DCF for such anisotropic system as the nematic phase can be calculated in the framework of two independent approaches: (i) the equilibrium statistical-mechanical theory discussed in Ref. [14], which takes into account intercell pair correlations,[5,15] and (ii) molecular dynamics simulation applied to the description of nematic LCs.[16,17,19,33] Within the equilibrium statistical-mechanical theory, one can consider a classical one-component fluid of density ρ at temperature T. Particles can be identified, to a certain approximation, with prolate ellipsoids of revolution with major semiaxis $\sigma_{\parallel}$ and minor semiaxis $\sigma_{\perp}$.[10,15] One can ideally subdivide the total volume V into N cells, each with volume $v = V/N$, for instance, in a simple cubic lattice. The treatment further assumes that each cell is occupied by one and only one particle; this seems reasonable for a condensed system. Particles are assumed to interact via a pairwise additive potential $\Phi(ij)$, so that the total interaction energy for the N particles is $U = \sum_{i<j} \Phi(ij)$. The quantity $\exp\left[-\frac{U}{k_B T}\right]$ is the canonical probability density. Upon integrating over the coordinates of the remaining particles, one can define single-particle probability density $F(i)$, two-particle probability densities $F(ij)$, and so on.[10,14,15] The infinite hierarchy is truncated here at the two-particle level. By the above partitioning of space, $F(i)$ is the probability density for a single particle confined inside a cell of volume v, and $F(ij)$ is a joint probability density for two particles confined in two different cells. The above-mentioned functions can also be written in terms of mean-force potentials (MFPs) $\Psi_{i,j}(i)$,[10,14,15] where the subscript before the comma corresponds to the MFPs dependence on the coordinates of particles, and the subscript after the comma corresponds to the average states, i.e.,

$$F(i) = \frac{\Psi_i(i)}{\int_i d(i)\Psi_i(i)}, \tag{11.17}$$

$$F(ij) = F(i)F(j)V(ij)\Psi_{i,j}^{-1}(i)\Psi_{j,i}^{-1}(j). \tag{11.18}$$

where

$$\Psi_j(j) = \prod_{i \neq j} \Psi_{j,i}(j), \int d(i) = \int_w d\mathbf{r}_i d\hat{\mathbf{e}}_i, i \equiv (\mathbf{r}_i, \hat{\mathbf{e}}_i), \tag{11.19}$$

$w = v \bigotimes \alpha$ denotes the volume associated with orientation α and position v, and $V(ij) = \exp\left[-\frac{\Phi(ij)}{k_B T}\right]$ is defined by the pair intermolecular potential $\Phi(ij)$. In Eq. (11.19), $\prod_{i \neq j}$ runs over all neighbors of cell i. In the following, one takes into account only pair correlations between cells. It should be pointed out that the last three factors in Eq. (11.18) reflect correlations between cells, which distinguish the present approach from mean field approximations. The singlet function $F(i)$ automatically satisfies the normalization condition $\int_i F(i)d(i) = 1$. Using the relation between singlet and binary functions[10,14,15] $F(i) = \int d(j)F(ij)$, one obtains a closed system of nonlinear integral equations (IEs) involving MFPs[10,14,15]

$$\Psi_{i,j}(i) = \int_j d(j)V(ij)\Psi_{j,i}^{-1}(j)F(j). \tag{11.20}$$

After solving Eq. (11.20), one can compute a number of correlators of a liquid crystalline system, expressed in terms of one- and two-particle functions

$$\overline{P}_{2L} = \langle P_{2L}(\cos\beta)\rangle = \int_i d(i)F(i)P_{2L}(\cos\beta), \tag{11.21}$$

$$\langle \hat{\mathbf{e}}_i \cdot \hat{\mathbf{e}}_j \rangle = \int_i d(i) \int_j d(j)F(ij)(\hat{\mathbf{e}}_i \cdot \hat{\mathbf{e}}_j), \tag{11.22}$$

as well as the Helmholtz free energy $f = \frac{F}{N} = -k_B T \int_i d(i)\Psi_i(i)$. Solving a five-dimensional problem (11.20) is a very complicated task, and there are no good general methods for systems of more than one nonlinear equation. The solution $\Psi_{i,j}(i)$ is invariably obtained by a numerical iterative procedure, using the formula

$$\Psi^{m[k+1]}(Q_i) = \left[\Psi^{m[k]}(Q_i)\hat{L}\Psi^{m[k]}(Q_i)\right]^{1/2}, \tag{11.23}$$

or

$$\Psi^{m[k+1]}(Q_i) = \frac{1}{2}\left[\Psi^{m[k]}(Q_i) + \hat{L}\Psi^{m[k]}(Q_i)\right], \tag{11.24}$$

for $m = 1, 2, ..., l$. Here $\hat{L}$, $\|\hat{L}\| \leq 1$ is the $5-$dimensional nonlinear integral operator defined by Eq. (11.23) in ω, Q_i is a $5-$dimensional vector in the same space, k is the iteration number, and l is the number of neighbors. The algorithms, Eqs. (11.23) and (11.24), could be implemented as follows: the initial approximation was to set $\Psi^{m[0]}(Q_i)$ to a constant; the integrals in the right-hand side of Eqs. (11.23) or (11.24) were calculated by Sobol's method,[34] using the Haar functions for calculating multidimensional integrals[35]

$$\int_0^1 dx_1 ... \int_0^1 dx_n f(x_1, ..., x_n) \approx \frac{1}{M} \sum_{\alpha=1}^{M} f(Q_\alpha), \qquad (11.25)$$

where the points Q_α were uniformly distributed (mod 1) in the $n-$dimensional unit cube. These points belonged to a LP_τ sequence,[34] and were obtained by means of an LP_τ-generator;[35] M was the number of points, and precision of the calculation was $\mathcal{R} \sim \mathcal{O}(M^{-1} \ln^n M)$ versus $\sim \mathcal{O}(M^{-1/2})$ for Monte-Carlo integration. The iterative procedure was repeated for various neighbors of the ith cell; in order to keep the treatment computationally feasible, truncation of the interaction potential at second-nearest neighbors was chosen.[4–6,15,36] The convergence criterion $\delta = |\Psi^{[k+1]}(Q_i) - \Psi^{[k]}(Q_i)|/|\Psi^{[k]}(Q_i)|$ for the above iterative procedures was set to 10^{-4}. The kernel $V(ij)$ for the integral equations in Eq. (11.20) is determined by the pair interaction potential, chosen to be the sum of a Gay-Berne and a dipole-dipole term $\Phi(ij) = \Phi_{\rm GB}(ij) + \Phi_{\rm dd}(ij)$. The GB term has the form[1]

$$\Phi_{\rm GB}(ij) = 4\epsilon_0 \epsilon \left(R^{-12} - R^{-6}\right), \qquad (11.26)$$

where $R = (r - \sigma + \sigma_\perp)/\sigma_\perp$, and $r = |\mathbf{r}_i - \mathbf{r}_j|$; it is defined by a Lennard-Jones 12−6 potential,[1] with range $\sigma(\hat{\mathbf{e}}, \hat{\mathbf{e}}_i, \hat{\mathbf{e}}_j)$ and strength $\epsilon(\hat{\mathbf{e}}, \hat{\mathbf{e}}_i, \hat{\mathbf{e}}_j)$ which depend on the relative orientations of the particles $\hat{\mathbf{e}}_i$ and $\hat{\mathbf{e}}_j$, the orientations of the particles with respect to the intermolecular vector between their center-mass positions $\hat{\mathbf{e}} = \mathbf{r}/|\mathbf{r}|$, the molecular anisotropy parameter γ, and another set of parameters, which can be used to adjust the ratio between end-to-end and side-by-side well depths, denoted by ϵ_l/ϵ_s, and the exponent parameters ν and μ, which appear in[1] $\epsilon = \epsilon_1^\nu(\hat{\mathbf{e}}_i, \hat{\mathbf{e}}_j)\epsilon_2^\mu(\hat{\mathbf{e}}, \hat{\mathbf{e}}_i, \hat{\mathbf{e}}_j)$. The dipolar term is defined by[4,5,36]

$$\Phi_{\rm dd}(ij) = \frac{\Delta}{r^3} \left[(\hat{\mathbf{e}}_i \cdot \hat{\mathbf{e}}_j) - 3\,(\hat{\mathbf{e}} \cdot \hat{\mathbf{e}}_i)\,(\hat{\mathbf{e}} \cdot \hat{\mathbf{e}}_j)\right], \qquad (11.27)$$

where Δ is the magnitude of the molecular dipole moment. In order to keep the treatment computationally feasible, interactions are restricted to second nearest neighbors. Two points should be noted: first, in the present

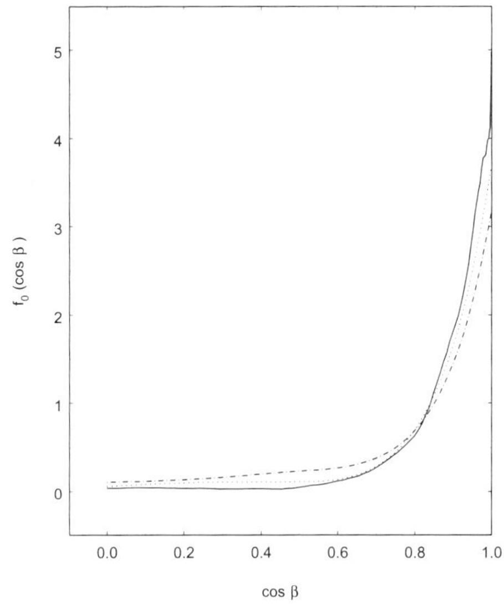

Fig. 11.1. Orientational distribution function at fixed density, $\rho = 0.33$, and temperature $\theta = 1.0$, for polar and nonpolar GB models; dashed line for $\mu_d = 2$, dashed-dotted line for $\mu_d = 0$,[5] respectively. The solid line corresponds to the MD simulation[5] of 5CB in the nematic phase, at T = 300 K and density 10^3 kg/m^3. Reproduced from figure 2 of Ref. [5] with permission of the APS.

calculation one has considered only a simple cubic structure, with six nearest neighbors and twelve next neighbors; this implies that one deals with a system of 18 nonlinear IEs in a five-dimensional space. In the case of polar LC, such as 5CB, the dipole moment ($\Delta \sim 5$D[36]) is directed from the polar head to the alkyl tail of the molecule. It has been shown that the dipolar GB pair potential provides a useful model for the interactions between real mesogenic molecules.[4–6,36] However, the variety of parameterizations that have been used in Refs. [1]–[3] yields very different potential forms. Whereas a value 3 for $\sigma_{\|}/\sigma_{\perp}$ is reasonable for 5CB molecules ($\sigma_{\|} \approx 1.8\ nm$, $\sigma_{\perp} \approx 0.6\ nm$), it is not known to what extent the values used for the other parameters are appropriate. It was chosen to adopt the potential for 5CB with parameters $\nu = 2$, $\mu = 0.98$, and $\epsilon_l/\epsilon_s = 0.245$.[4,6] It should also be pointed out that for the convergence of the iterative procedure the algorithm Eqs. (11.23)-(11.24) is very sensitive to the choice of both parameters ϵ_l/ϵ_s and μ, which vary the value of the well depth. The units of distance $\sigma_{\perp} = 0.6\ nm$ and energy $\epsilon_0 = 2.07 \times 10^{-21}$ J have been employed to give the reduced units for density $\rho = N\sigma_{\perp}^3/V$, temper-

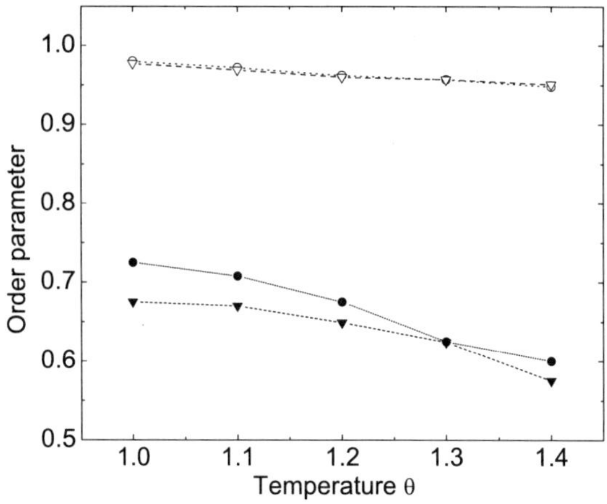

Fig. 11.2. Temperature dependence of the orientational order parameter $\overline{P}_2$ for both polar and nonpolar GB models, at constant density $\rho = 0.33$.[5] Open and closed symbols refer to polar and nonpolar systems, respectively. IE computations are shown using circles, the open triangles represent results from MC simulations,[3] and closed triangles correspond to MD[16] simulations. Reproduced from figure 1 of Ref. [5] with permission of the APS.

ature $\theta = k_B T/\epsilon_0$, and dipole moment $\mu_d = \Delta/(\epsilon_0 \sigma_\perp^3)^{1/2} \approx 2.5$.[4–6] The orientational order of LCs is traditionally quantified in terms of order parameters, however a complete description of the order is provided by the SODF $f_0(\cos\beta)$. At the temperature $T = 300$ K and density $\rho = 0.512$ for 5CB (corresponding to 10^3 kg/m^3), Figs. 11.1 and 11.2[5] show the SODF $f_0(\cos\beta) = \int d\mathbf{r}_i d\varphi F(\mathbf{r}_i, \hat{\mathbf{e}}_i)$ and the temperature dependence of OP $\overline{P}_2$ for both nonpolar ($\mu_d = 0$) and polar ($\mu_d = 2.0$) GB models, respectively. In both cases the density was fixed at $\rho = 0.33$ ($\sim 10^3$ kg/m^3). Figure 11.3[33] shows the temperature dependence of correlators $|\xi_\lambda|$ ($\lambda = 1, ..., 6$), for the nearest neighbors; the "vertical" ξ_λ ($\lambda = 1, 2$) are characterized by positive values, whereas the "horizontal" ξ_λ ($\lambda = 3, ..., 6$) are characterized by negative values. The average correlators for nn are defined by $\xi_\lambda = \frac{1}{2\lambda}\sum_l \langle \hat{\mathbf{e}}_0 \cdot \hat{\mathbf{e}}_\lambda \rangle$, where $\lambda = 1$ denotes the average of the two "vertical" neighbours nn of the central particle indicated by subscript 0, whereas the label $\lambda = 2$ corresponds to the average of four "horizontal" ones. The "vertical" correlators are characterized by a common positive value, while almost identical negative values were obtained for all four "horizontal" correlators. Figure 11.4[5] displays the temperature dependence of the average

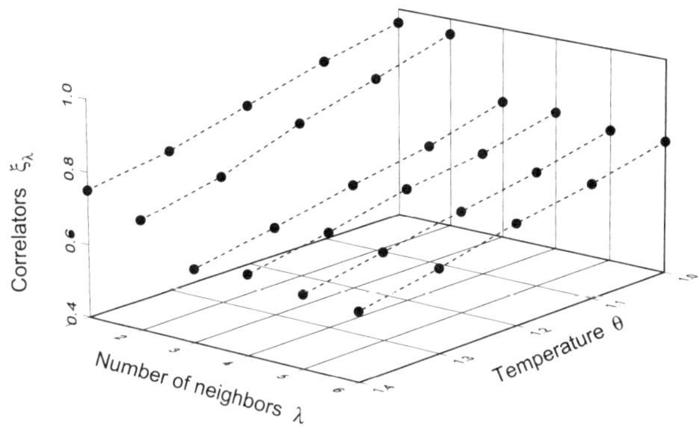

Fig. 11.3. Temperature dependence of the orientational correlators, $|\xi_\lambda|$ ($\lambda = 1, ..., 6$), for nearest neighbors; $\xi_\lambda > 0$ ($\lambda = 1, 2$) for "vertical" neighbors and $\xi_\lambda < 0$ ($\lambda = 3, ..., 6$) for the "horizontal" ones.[33] Reproduced from figure 3 of Ref. [33] with permission of the Pleiades Pub. Ltd.

correlators for nearest (nn) and next nearest (nnn) neighbors. Another average nn correlator $\xi = \frac{1}{3}\sum_\lambda \lambda\xi_\lambda$ shown in Fig. 11.4,[5] is therefore negative for all temperatures indicating an antiparallel orientation of the dipoles in neighbor cells. In analogy with nn, the correlations between nnn are defined $\eta_\lambda = \langle \hat{\mathbf{e}}_0 \cdot \hat{\mathbf{e}}_\lambda \rangle$, where $\lambda = 1, ..., 12$ labels the neighbors of the central particle. It has been found that these quantities are rather insensitive to λ, and therefore we report in Fig. 11.4 the temperature dependence of their average over the twelve neighbors, i.e., $\eta = \frac{1}{12}\sum_\lambda \eta_\lambda$. The average correlator η, is negative for all temperatures. Having obtained the singlet $F(i)$ and two binary $F_\lambda(ij)$ functions corresponding to the "vertical" nn cells ($\lambda = 1$), parallel to the director $\hat{\mathbf{n}}$, and to the "horizontal" nn cells ($\lambda = 2$), perpendicular to the director, respectively, one can determine two DCFs, using the classical Percus-Yevick closure approximation

$$C_\lambda(ij) = F_\lambda(ij)\left(1 - V^{-1}(ij)\right), \quad (\lambda = 1, 2). \tag{11.28}$$

After obtaining the DCF $C_\lambda(ij)$ and the set of $\overline{P}_{2L}$ ($L = 1, 2, 3$), it is possible to calculate ECs K_i. Figure 11.5[6] shows the temperature dependence of the K_i using NZ approach. Three different approximations have been combined for the calculation of the ECs:[6]

(i) the nonlinear IEs theory that requires access to the set of OPs and the binary correlation function,

(ii) the classical Percus-Yevick approximation for the DCF, and

(iii) NZ density-functional approach which implies that higher OPs (i.e.,

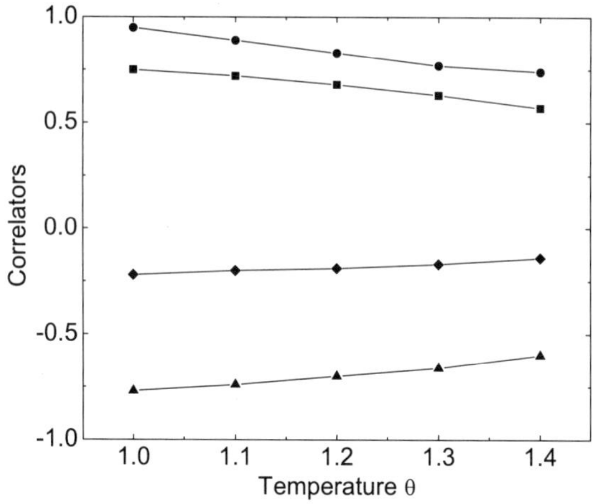

Fig. 11.4. Temperature dependence of the orientational correlators, ξ_λ, for nearest neighbors; circles correspond to the "vertical" ξ_1, whereas triangles correspond to the "horizontal" ξ_2,[5] respectively. The same dependence of the average orientational correlators ξ (squares) and η (rhombi), for the nearest and next nearest neighbors, respectively. Reproduced from figure 3 of Ref. [5] with permission of the APS.

$\overline{P}_6$) and higher moments (i.e.,M_4) of the DCF are used. Taking into account these approximations as well as the fact that the Percus-Yevick closure works well mostly for short-range intermolecular potentials, and also recalling the difficulty with measuring the absolute values of these coefficients, the qualitative agreement between these calculations and the experimental values obtained using the Freedericksz transition method[37] is heartening.

The statistical-mechanical approach used for calculating the ECs rests on models where the K_i $(i = 1, 2, 3)$ depend on the DCF and on set of OPs. The methods take into account both the intermolecular attraction and steric repulsion, but notice that it is limited to rigid particles, and thus unable to describe the effect of intramolecular degrees of freedom (molecular flexibility), usually present in most mesogens; for example, the structure of *cyanobiphenyls* is characterized by a rather rigid central core to which one or more flexible hydrocarbon chains are attached. Attempts have been carried out to construct a more general treatment allowing for molecular flexibility.

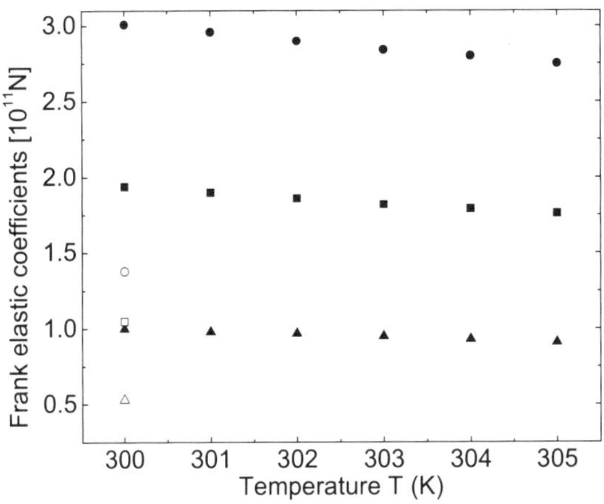

Fig. 11.5. The temperature dependence of the Frank ECs K_i ($i = 1, 2, 3$) for 5CB. Solid symbols represent points calculated using Eqs. (11.12),(11.13) and (11.14); circles (K_3), squares (K_1), and triangles (K_2),[6] respectively. Open circle (K_3), square (K_1), and triangle (K_2) are experimental data,[37] respectively. Reproduced from figure 3 of Ref. [6] with permission of the APS.

11.2.3. *Order parameters and direct correlation function: molecular dynamics simulation*

A useful route in studies of macroscopic properties of LCs is provided by computer simulation techniques.[16–19] Recently, a molecular dynamics (MD) simulation, based on realistic atom-atom interaction potential was used in order to facilitate an analysis of molecular structure in the nematic phase.[19] The orientational distribution and pair correlation functions were obtained from an investigation of a trajectory generated in the MD simulation. A detailed discussion of simulation procedure and force field parameters, as well as an extended analysis of results have been provided elsewhere.[19,38–40] Here we only summarize general aspects and give some properties of the simulated LC. The MD simulation was carried out on a system consisting of 110 5CB molecules in a rectangular cell, assuming a density of 10^3 kg/m^3. The total length of the simulation was 1.5 ns and a trajectory of 510 ps was used in the final analysis. The mesogens were modelled using the potential functions presented in Ref. [38], [39], which included intra- and intermolecular interactions. The temperature was kept at 300 K using an algorithm corresponding to the NVT ensemble. The instantaneous orientation of the director $\hat{\mathbf{n}}$ was determined from the Cartesian ordering matrix Q_{zz} (see, Eq. (4) Ref. [38]). Diagonalization of Q_{zz} gave three eigenvalues, and the eigenvector associated with the largest eigenvalue corresponded to the nematic director. Figures 11.1[5] and 11.6(a)[19] shows the SODF $f_0(\cos\beta_i)$ and the center-of-mass radial distribution function (RDF) $G(r)$, obtained from the MD trajectory. The OPs

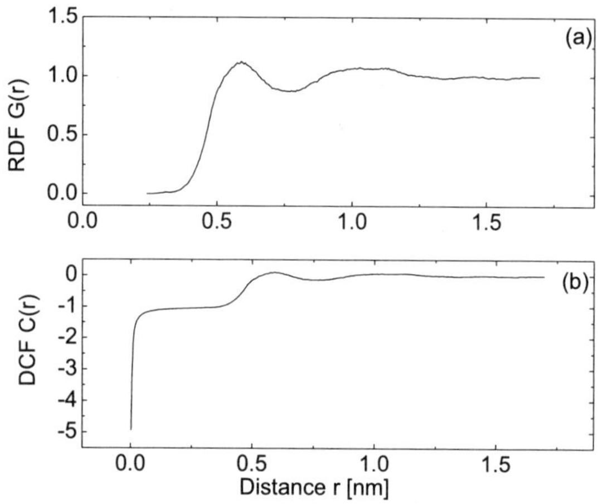

Fig. 11.6. (a) Center-of-mass RDF for 5CB, calculated from the MD trajectory.[19] (b) DCF for 5CB, obtained by solving the Ornstein-Zernike equation numerically.[19] Reproduced from figures 2 and 3 of Ref. [19] with permission of the EDP Sciences.

can be calculated by averaging appropriate functions over all molecules at every time step of the simulation. The following results are obtained for the lowest OPs:[19] $\overline{P}_2 = 0.504$, $\overline{P}_4 = 0.188$, $\overline{P}_6 = 0.06$, and the biaxiality parameter, $S_{xx} - S_{yy} = 0.03$. Here, $S_{\alpha\alpha}$, where $\alpha = x, y, z$ are the diagonal elements of the OP tensor. According to Eqs. (11.3)-(11.5), the bulk ECs K_i $(i = 1, 2, 3)$ of the nematic phase depend on the DCF of the corresponding reference state. In the case of Ref. [19] this reference state was an unoriented liquid crystal. It was achieved by averaging the DCF $C(r) = \int C(\mathbf{r}, \hat{\mathbf{e}}_i, \hat{\mathbf{e}}_j) d\Omega d\Omega_i d\Omega_j$ over all possible orientations of the director and representing the remaining part by a spherical series. Generally, $C(r)$ can be defined in terms of the RDF $G(r)$ through the Ornstein-Zernike

Table 11.1. Frank elastic constants K_i ($i = 1,2,3$, in pN) for 5CB at 300 K. Calculations were carried out using Eqs. (11.3)-(11.5) (MA), (11.6)-(11.8)(OH), and (11.12)-(11.14) (NZ),[19] experimental data (EXP) were determined using the Freedericksz transition technique.[37]

Model	OH	NZ	MA	EXP
$K_1 \times 10$	20.9	9.5	10.1	10.5
$K_2 \times 10$	8.3	5.1	5.2	5.4
$K_3 \times 10$	39.5	13.8	14.0	13.8

equation[17,19]

$$C(r) = G(r) - 1 - \frac{\rho}{2r} \int_0^{\infty} dt \left[tC(t) \int_{|r-t|}^{r+t} (G(s) - 1)\, s ds \right]. \qquad (11.29)$$

This equation can be solved by applying a numerical iterative procedure.[19] Figure 11.6(b)[19] shows the direct correlation function $C(r)$, obtained from Eq. (11.29). Having obtained from the MD trajectory the SODF $f_0(\cos \beta_i)$, radial $G(r)$, direct correlation $C(r)$ function, and a set of OPs $\overline{P}_{2L}$ ($L = 1, 2, 3$), it is possible to determine the elasticity coefficients K_i. This was performed using three different routes:[19] (i) the microscopic approach (MA) based on Eqs. (11.3)-(11.5), (ii) the OH model, Eqs. (11.6)-(11.8), and finally (iii) the NZ approximation, Eqs. (11.12)-(11.14). In all three cases the DCF defined by Eq. (11.29) was used. Note that the expression for the coefficients K_i in the framework of the MA approach contains seven scalar integrations;[19] these have been carried out numerically employing Simpson's rule. The ECs calculated using approaches (i)-(iii), together with the experimental values obtained via the Freedericksz transition technique are collected in Table 11.1.[19] Both the analytical MA theory, resting on the SODF, and the NZ theory, taking into account the higher-order parameters and moments of the DCF, show a much better agreement with experiment than the OH model.[19] However, in view of the crude approximation about the isotropic nature of the direct correlation function, such agreement must be considered as fortuitous.

11.2.4. *Flexoelectric effect*

The onset of a spontaneous polarization **P** in a LC composed of asymmetric polar molecules, in response to elastic deformations, is known as flexoelectric (FE) effect.[41] The flexoelectric effect has first been predicted almost 40 years ago but, although potentially very attractive for applications, it has yet to be fully exploited, partly due to difficulties in the experimental determination of the FE coefficients but also due to the lack of a precise understanding of their relation with the molecular structure. In the nematic phase, splay and bend deformations give rise to two independent FE coefficients (e_1,e_3), whose contribution to the induced polarization can be written as[41–43]

$$\mathbf{P} = e_1 \hat{\mathbf{n}} \cdot \nabla \hat{\mathbf{n}} - e_3 \left(\hat{\mathbf{n}} \times \nabla \times \hat{\mathbf{n}} \right). \tag{11.30}$$

In the case of nonpolar molecules the FE effect originates from a gradient of quadrupole moment density;[42] such a polarization does not need shape asymmetries nor dipole moments in mesogenic molecules. Flexoelectric polarization can influence electro-optical properties, defect formation and structural instability. Thus, different techniques have been suggested since the pioneering work of Prost and Marcerou[42] to observe both dipolar and quadrupolar mechanisms.[43,44] However, the accuracy of these measurements is influenced by a number of parameters such as dielectric and optical anisotropy, elastic moduli, etc.; problems also arise with the determination of the sign of the observed effects. In practice, these FE coefficients are not measured separately, but their combinations $e_+ = e_1 + e_3$, $e_- = e_1 - e_3$, or e_+/K_i (e_-/K_i) are, where $i = 2, 3$ and K_2 and K_3 are the bulk twist and bend elastic coefficients, respectively. Early experiments on some compounds, such as 4 − *octyl* − 4′ − *cyanobiphenyl* (8CB) ,[43] found that the e_-/K_2 ratio showed temperature-dependent values, and that $e_- \sim \overline{P}_2$. Later, the temperature dependence of e_-/K_2 for a number of *cyanobiphenyl* compounds (nCB) was reported.[44] For higher homologue nCB compounds ($n = 6, 7, 8$), e_-/K_2 went through maximum values as a function of temperature within their nematic phase, whereas e_-/K_2 of 5CB monotonically increased with increasing temperature. Moreover, recent experimental data obtained by pyroelectric (PE) technique for 5CB nematic cells have shown that the sign of e_+ is negative,[45] whereas the sign of e_+ (e_-) in an earlier study[44] is positive. It should be pointed out that the uncertainties on the measured values of e_-/K_2 are relatively large (40%).[44] Taking into account that e_- is mainly determined by the dipolar contributions,[46] which do not vary sufficiently in the case of *cyanobiphenyls* (dipole moments range from 4.5D to 5D, and directed from the cyano group to the alkyl tail of the molecule), the geometric factor of the molecules as well as interactions and correlations among molecules probably play a significant role to influence the magnitude and sign of the FE coefficients. Indeed, the density-functional formalism has shown that e_- is proportional to the molecular length and inversely proportional to the cube of the width (or diameter) of the molecule.[46] In view of the present experimental situations, a theoretical route to calculate individual e_1 and e_3 using the microscopic theory is very useful. In the framework of the density-functional approach for a system of N particles occupying a volume V at the temperature T, the FE coefficients are given by the formal expressions[47]

$$e_1 = \rho^2\Delta \int r_x C(\mathbf{r}, \hat{\mathbf{e}}_i, \hat{\mathbf{e}}_j) f_0'(\cos\beta_i) f_0(\cos\beta_j) e_{i,z} r^2 dr d\Omega d\Omega_i d\Omega_j, \tag{11.31}$$

$$e_3 = \rho^2\Delta \int r_x C(\mathbf{r}, \hat{\mathbf{e}}_i, \hat{\mathbf{e}}_j) f_0'(\cos\beta_i) f_0(\cos\beta_j) e_{i,x} r^2 dr d\Omega d\Omega_i d\Omega_j. \tag{11.32}$$

In order to make progress in the evaluation of the FE coefficients, a simplified approach was adopted for the DCF using the classical Perus-Yevick closure approximation.[36] Expressions for the FE coefficients e_1 and e_3 contain seven scalar integrations; these have been carried out numerically employing (i) the Haar function technique[35] and the nonlinear IEs theory[36] for the calculation of the DCF $C(ij)$, and (ii) both the SODF $f_0(\cos\theta_i)$ and DCF $C(r)$ obtained from the MD trajectory.[19,33] In the first case, temperature dependences of e_1 and e_3 are shown in Fig. 11.7.[36] The calculated $e_+ = -35.5$ pC/m at $T = 300$ K is slightly lower than the reported e_+[45] ($e_+ = -8.5 \div -13$ pC/m at T = 298 K) and disagrees with results for e_+ in Ref. [44]. The last calculations also show that both FE coefficients e_1 and e_3 are negative and differ, on the average, by a factor 3, and that the signs of e_+ and e_- are determined by the sign e_1. Thus, according to the weak temperature dependences of the coefficients e_1 and e_3, the measurable quantities e_+ are expected to show the same behavior with temperature, at least for all *cyanobiphenyls*. The sign e_1 is determined by an average of the projection of the unit vector $\hat{\mathbf{e}}$ on the $x-$axis with some "effective" distribution function. Since one has a congruent set of DCF for all *cyanobiphenyls*, one can expect, at least in the framework of that approach, that the coefficients

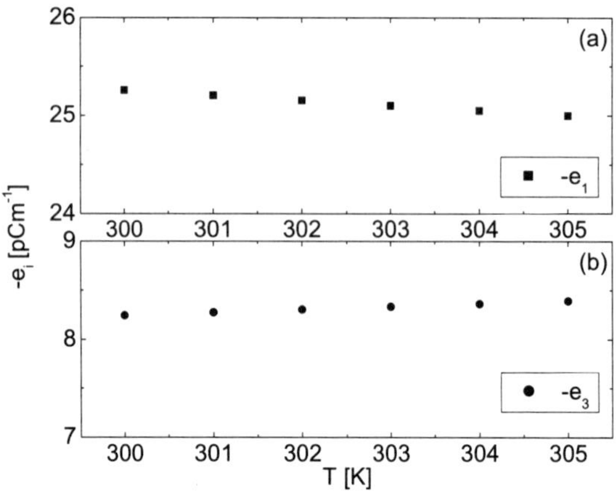

Fig. 11.7. Temperature dependence of e_1 (squares) and e_3 (circles) calculated at fixed density for 5CB, using Eqs. (11.31) and (11.32),[36] respectively. Reproduced from figure 1 of Ref. [36] with permission of the EDP Sciences.

Table 11.2. Flexoelectric coefficients e_1, e_3, and e_+ for 5CB at 300 K. Calculations were carried out using Eqs. (11.31)-(11.32), within the framework of statistical mechanics (SM)[36] and MD trajectory,[33] and the experimental data for e_+ obtained by the pyroelectric technique (PT),[45] respectively.

Model	$-e_1$(pC/m)	$-e_3$(pC/m)	$-e_+$(pC/m)
SM	25.3	8.2	33.2
MD	11.6	4.3	15.9
PT			13.0

e_1 (e_-) have the same sign for these compounds. The FE coefficients e_1 and e_3 calculated using the SODF $f_0(\cos\beta_i)$ and DCF $C(r)$ obtained from the MD trajectory[33] and within the framework of statistical mechanics are collected in Table 11.2.[36] The analytical theory rests on the microscopic approach with the DCF obtained from the MD trajectory, which, in turn, takes molecular flexibility into account, and produces a much better agreement with experiment[45] than the density-functional approach where the

DCF obtained from the nonlinear IE theory is employed. It should be noted that atomistic MD calculations, performed for another LC molecules $n-4-(trans-4-n-pentylcyclopexyl)benzonitrile$ (PCH5),[48] similar in structure to 5CB, found that the signs of both e_1 and e_3 are positive. For PCH5, experimental values for the FE coefficients have been determined with e_1 given as 5.3 pC/m and e_3 as 3.3 pC/m at 303 K, with quoted errors of about 40%.[44] Another measurements of the FE properties of 8CB and 8OCB shows that their effective FE coefficients $e^* = 2(e_1 + e_3)/(K_1 + K_3)$ are opposite in sign, and quite different in magnitude. For instance,[43] the magnitude of e^* in 8OCB (reported to be -0.44 $N^{1/2}$) is much greater than that in 8CB (0.09 $N^{1/2}$). The difference between FE properties of 8CB and 8OCB has been explained as follows:[43] adjacent 8CB molecules exhibit antiparallel ordering, owing to the dipolar interaction between the *cyanobiphenyl* moieties. The same antiparallel ordering is assumed to occur in 8OCB, with the microscopic cancellation both of the longitudinal dipoles conferred on the molecules by the presence of the nitrile unit and of the transverse one resulting from the presence of the oxygen atom. On the other hand, in a bent configuration of a pair of molecules, the transverse oxygen dipoles become correlated and do not completely cancel out; this mutual pair configuration is expected to be macroscopically populated.

This measurement confirms the prediction arising from the theoretical treatment[46] of FE coupling in nematic materials, which indicates that the dipolar contribution to the FE effect is dominated by transverse dipole moments. Notice that comparison with previous models[49–51] is possible only at qualitative level, due to the vast differences in the used models. New approaches are now needed to understand how the interplay of molecular structure and geometry results in the FE coefficients. Recently, by a direct method for measuring the FE coefficients of a bent-core nematic (BCN) mesogen, it was shown that the value of e_3 is about 3 orders of magnitude greater than in conventional calamatic (rod-shaped) nematics.[52] Based on this giant flexoelectric effect, BCN materials must be viewed as a highly promising candidate for a new technological breakthrough in the conversion of mechanical to electrical energy at the molecular scale.

We conclude by pointing out that the combination of theory and computer simulation provides a powerful tool for investigations not only elastic properties of nematic liquid crystals but also their spontaneous polarization in response to elastic deformations.

11.3. Viscous properties of a nematic liquid crystal

Dissipation processes in LCs are an extremely important property, whose underlying fundamental physics is far from being completely understood.[53,54] Despite the fact that certain qualitative advances have been made in the construction of a molecular theory of rheological properties of NLCs,[24–30] it is still too early to talk about the development of a theory which would make it possible to describe the rheological processes from first principles, based only on the form of the Hamiltonian. Yet, the rotational viscosity coefficients (RVCs), γ_i $(i = 1, 2)$ are parameters determining the dynamics of LCs, and a large amount of experimental data on LC viscosity has been accumulated,[21–23] and therefore the question of their theoretical interpretation inevitably arises.

11.3.1. *Statistical-mechanical formulas for the rotational viscosity*

The classical approach to LC viscosity, in the framework of the Ericksen-Leslie (EL) theory,[53,54] assumes that there is a time-dependent unit vector field $\hat{\mathbf{n}}(\mathbf{r}, t)$, representing the average alignment at each point in space-time. The general relation for the dynamic ST σ_{ij} of incompressible LC with velocity field $\mathbf{v}$ is given by the following expression $\sigma = \sigma^s + \sigma^a$,[53,54] with symmetric and antisymmetric parts

$$\sigma^s = \alpha_1 \hat{\mathbf{n}}\hat{\mathbf{n}} \colon M \hat{\mathbf{n}}\hat{\mathbf{n}} + \frac{1}{2}(\alpha_2 + \alpha_3)(\mathbf{N}\hat{\mathbf{n}} + \hat{\mathbf{n}}\mathbf{N}) + \alpha_4 M + \frac{1}{2}(\alpha_5 + \alpha_6)(M \cdot \hat{\mathbf{n}}\hat{\mathbf{n}} + \hat{\mathbf{n}}\hat{\mathbf{n}} \cdot M), \tag{11.33}$$

$$\begin{aligned} \sigma^a &= \frac{1}{2}(\alpha_2 - \alpha_3)(\hat{\mathbf{n}}\mathbf{N} - \mathbf{N}\hat{\mathbf{n}}) \\ &= \frac{1}{2}(\alpha_5 - \alpha_6)(\hat{\mathbf{n}}\hat{\mathbf{n}} \cdot M - M \cdot \hat{\mathbf{n}}\hat{\mathbf{n}}) \\ &= \frac{1}{2}\epsilon \cdot [(\gamma_1 \mathbf{N} + \gamma_2 \hat{\mathbf{n}} \cdot M) \times \hat{\mathbf{n}}], \end{aligned} \tag{11.34}$$

where ϵ denotes the Levi-Civita tensor, and $\mathbf{N} = \frac{d\hat{\mathbf{n}}}{dt} - W \cdot \hat{\mathbf{n}}$, $2M = \nabla \mathbf{v} + (\nabla \mathbf{v})^T$, $2W = \nabla \mathbf{v} - (\nabla \mathbf{v})^T$. Here M and W are symmetric and antisymmetric parts of the flow velocity gradient $v_{i,j} = \frac{\partial v_i}{\partial x_j}$, and $(\nabla \mathbf{v})^T$ denotes the transpose of the $(\nabla \mathbf{v})$ matrix. The Leslie coefficients α_1 to α_6 in Eqs. (11.33)-(11.34) satisfy the general Onsager-Parodi relation

$\alpha_2 + \alpha_3 = \alpha_6 - \alpha_5$, so that only five of the six coefficients α_i are independent. The coefficients $\gamma_1 = \alpha_3 - \alpha_2$ and $\gamma_2 = \alpha_3 + \alpha_2$ are the RVCs. The average of the symmetric part of ST σ^s can be calculated exactly just by comparison with its phenomenological form.[28] Calculations of γ_1 and γ_2, on the other hand, require an averaging of the antisymmetric part of the microscopic ST. This can be done using an appropriate nonequilibrium SODF f, that in case of a small velocity gradient can be written in the form $f = f_0(1+h)$, where f_0 is the local equilibrium SODF, which depends only on the polar angle β between the director and the molecular symmetry axis. In the case of small velocity gradient, the stationary correlation, h, takes the form[26] $h = g_0\hat{\mathbf{n}}\hat{\mathbf{n}}\colon M + g_1\hat{\mathbf{n}}\hat{\mathbf{m}}\colon W + g_2\hat{\mathbf{n}}\hat{\mathbf{m}}\colon M + g_3\hat{\mathbf{m}}\hat{\mathbf{m}}\colon M$, where $\hat{\mathbf{m}}$ is the unit vector perpendicular to the director, and g_i are scalar functions, only depending on the polar angle β. The averaging over the azimuthal angle, determined by the orientation of the vector $\hat{\mathbf{m}}$, can be performed and one can obtain the antisymmetric part of the tensor σ as[40]

$$\sigma^a = \frac{\rho}{4}\int_0^\pi f_0(\cos\beta)\frac{\partial U}{\partial\beta} \\ [g_2(\beta)\,(\hat{\mathbf{n}}\hat{\mathbf{n}}\colon M - M\colon \hat{\mathbf{n}}\hat{\mathbf{n}}) - g_1(\beta)\,(\hat{\mathbf{n}}\mathbf{N} - \mathbf{N}\hat{\mathbf{n}})]\sin\beta d\beta, \quad (11.35)$$

where U is the potential of the mean torque. Comparing Eq. (11.35) with Eq. (11.34), one can get expressions for the RVCs γ_i $(i = 1, 2)$,[26,28,30,40] which are determined by the functions $g_1(\beta)$ and $g_2(\beta)$, respectively,

$$\gamma_i = \frac{\rho J_0 \overline{P}_2}{2}\int_0^\pi f_0(\cos\beta)\sin 2\beta\widetilde{g}_i(\beta)\sin\beta d\beta, \quad (11.36)$$

where J_0 is the strength of the anisotropic ordering potential, and

$$\widetilde{g}_i(\beta) = \begin{cases} g_1(\beta), & \mathrm{i} = 1, \\ -g_2(\beta), & \mathrm{i} = 2. \end{cases}$$

The kinetic equation for the function $\widetilde{g}_i(\beta)$ can be derived using a Fokker-Planck-type relation together with the correlation function h[40]

$$\hat{L}\widetilde{g}_i(\beta) = G_i, \quad (11.37)$$

where $\hat{L}$ is the kinetic operator in terms of a polar coordinate

$$\hat{L} = \frac{\partial^2}{\partial\beta^2} + \left(\tan^{-1}\beta - \frac{1}{k_BT}\frac{\partial U}{\partial\beta}\right)\frac{\partial}{\partial\beta} - \sin^{-2}\beta, \quad (11.38)$$

$$G_i = \begin{cases} \frac{1}{D_\perp}\frac{1}{k_BT}\frac{\partial U}{\partial\beta}, & \mathrm{i} = 1, \\ -\frac{\omega}{D_\perp}\,(\frac{1}{k_BT}\frac{\partial U}{\partial\beta}\cos 2\beta + 3\sin 2\beta), & \mathrm{i} = 2, \end{cases}$$

and $D_\perp$ is the rotational diffusion coefficient (RDC), which corresponds to the molecular tumbling in the nematic. Equations (11.37) are linear, nonhomogeneous differential equations of second order. A general solution, $\widetilde{g}_i(\beta)$ for an equation of this type can be written as a sum $\widetilde{g}_i(\beta) = \widetilde{g}_{i,GS} + \widetilde{g}_{i,PS}$, where $\widetilde{g}_{i,GS}(\beta)$ is the general solution of the corresponding homogeneous problem, and $\widetilde{g}_{i,PS}(\beta)$ is a particular solution of the nonhomogeneous differential equation. The general solution depends on two constants that are determined by the boundary and normalization conditions: (i) the values of the functions $\widetilde{g}_i(\beta)$ should be symmetrical between the two extrema of the integration interval, and (ii) the integral $\int_0^\pi f_0(\cos\theta)h(\beta)\sin\beta d\beta$ should vanish to fulfill the normalization condition. It should be pointed out that the general solution of Eq. (11.37) for $g_2(\beta)$,[40] is $g_2(\beta) = g_{2,PS}(\beta) = \frac{\omega}{2D_\perp}\sin 2\beta$, because the general solution $g_{2,GS}(\beta)$ of the homogeneous differential equation (11.37) equals 0. It is apparent that conditions (i) and (ii) are fulfilled in this case. The general solution $g_1(\beta)$ for Eq. (11.37) can be obtained numerically, for example, using the second-order Runge-Kutta method.[55]

Thus, according to Eqs. (11.36)-(11.37) γ_i are found to be inversely proportional to the RDC $D_\perp$. A simple torque balance shows that if the director is aligned in the shear plane under flow influence, the nonzero component of the tensor $\overline{\sigma}^a \sim (\gamma_1 - \gamma_2\cos 2\theta_0)\, v_{i,j}$. This tensor vanishes at the preferred alignment angle $\theta_{eq} = \frac{1}{2}\arccos\lambda^{-1}$, where $\lambda = -\frac{\gamma_2}{\gamma_1}$. It is therefore apparent, that $|\lambda| \geq 1$ implies that the director might come to rest at a fixed angle relative to the stream lines, whereas for $|\lambda| < 1$ the director rotates continuously in the shear plane. Eqs. (11.36)-(11.37) enable us to calculate the coefficient λ,[40]

$$\lambda = \frac{\omega}{2D_\perp}\frac{\langle\sin 2\beta\rangle}{\langle g_2(\beta)\sin 2\beta\rangle}. \tag{11.39}$$

It should be pointed out that there exist several other approaches for calculating γ_1. According to Ref. [40], γ_1 was found to depend on temperature, density, and on certain polynomials of OPs via

$$\gamma_1 = \frac{k_B T\rho}{D_\perp}\xi^{\mathrm{NZ,F}}. \tag{11.40}$$

In particular, two models based on different physical approximations were considered: i) a microscopic approach, proposed by Nemtsov[25] and Zakharov[27] (NZ), and ii) a method based on a change of the Helmholtz free energy proposed by Fiałkowski[30] (F). The expressions for ξ in these two models become

$$\frac{\xi^{\mathrm{NZ}}}{\overline{P}_2^2} = \frac{9.54 + 2.77\overline{P}_2}{2.88 + \overline{P}_2 + 12.56\overline{P}_2^2 + 4.69\overline{P}_2^3 - 0.74\overline{P}_2^4},$$
$$\frac{\xi^{\mathrm{F}}}{\overline{P}_2^2} = \frac{70}{16\overline{P}_4 + 5\overline{P}_2 + 14}, \tag{11.41}$$

whereas the expression for γ_2, according to the Kuzuu-Doi (KD) method can be written as[56]

$$\gamma_2 = -\frac{k_B T \rho}{D_\perp}\overline{P}_2. \tag{11.42}$$

There are also a number of possible routes to γ_1 involving equilibrium Green-Kubo (GK) methods[57,58] and nonequilibrium molecular dynamics (NEMD) methods.[59,60] In the former case, the GK relation for the desired transport coefficient is evaluated by performing an equilibrium molecular dynamics (EMD) simulation. In the latter the system is coupled to a fictitious mechanical field. The field drives irreversible thermodynamic currents. The analytical expression for the field is chosen in such a way that the currents become the same as the ones induced by real Navier-Stokes forces. The transport coefficient in question is obtained as the ratio between current and field in the limit of zero field. NEMD methods are not particularly attractive for atomistic simulations, since the director must be constrained or reoriented, and these tasks are not simple at an atomistic level. However, EMD methods are more accessible.

Notice that the three microscopic theories (Eqs. (11.36), (11.40)-(11.41), and (11.42)) proposed for description of the rotational viscosity are based on different physical models. The analytical theory of the Eq. (11.36) is based on the SODF and contains complete information about orientational order, whereas the original NZ theory is based on including the effects of correlations of the stress tensor with the director and fluxes with OPs tensors, and contains only a few OPs. On the other hand, averaging in the framework of F approach is based on an explicit form of the nonequilibrium SODF, and also contains only averaged quantities. However, Eqs. (11.36), (11.40)-(11.41), and (11.42) allow us to calculate γ_i $(i = 1, 2)$ provided the temperature dependence of $f_0(\cos\beta)$, $\overline{P}_{2L}$ $(L = 1, 2)$, and $D_\perp$ are known. Whereas values of the OPs for various LC are usually fairly easily found, the determination of motional constants for rotational diffusion and orientational time correlation functions (TCFs) is still a formidable task. However, a number of useful routes for determining both RVCs and RDCs can be provided. The RVCs γ_i ($i = 1, 2$) for such anisotropic system as the nematic phase can be calculated in the framework of two independent approaches: (i) nonequilibrium statistical-mechanical theory,[25–28,40] which requires an averaging both of the symmetric and antisymmetric parts of ST, and (ii) molecular dynamics simulation, both EMD[57,58] and NEMD,[59,60] of nematic LCs.

The analytical determination of γ_i ($i = 1, 2$) requires access to $D_\perp$. In turn, the RDC, $D_\perp$, for nematic phase can be obtained in the framework of two independent techniques: (i) the MD simulation, based on realistic atom-atom interaction potential, and (ii) two of the most important spectroscopic techniques, NMR and dielectric spectroscopy (DS) techniques. Notice that there is one additional possible route to $D_\perp$, involving the theoretical modelling. That modelling is based upon random walk theory[65] together with some statistical-mechanical ideas to allow calculations of the RDC for nematic phase.

11.3.2. *Motional constants and orientational time correlation functions*

MD simulation, NMR and dielectric spectroscopy (DS) are three of the most important techniques for investigating microscopic dynamics in liquid crystals. Nuclear and dielectric relaxations can both give dynamic information over a wide frequency range, as realized by changing their characteristic measurement frequencies. In the case of NMR, for instance, spin-spin and spin-lattice relaxation times are used for characterizing slower (0.1 Hz to 100 KHz) and faster (10 MHz to 10^3 MHz) motions, respectively,[61] while a different instrumental apparatus allows the frequency range of 1 KHz $\leq \frac{\omega}{2\pi} \leq$ 10 GHz to be covered in DS.[62] Despite this, comparison between the results obtained by the two spectroscopic techniques is quite difficult, partly owing to their intrinsically different nature. In particular, NMR probes the system at a microscopic level, by measuring nuclear properties associated with different sites within the molecule, whereas DS monitors an overall macroscopic property of the system. Therefore, NMR and DS experimental data are determined by different sets of molecular motions. Moreover, NMR and DS usually deal with "rank-two" and "rank-one" tensorial properties, respectively. The perspective of quantitatively comparing quantities determined by the two techniques is, however, very attractive to reciprocally validate the reliability of the dynamic information. Models have been developed that allow the experimental relaxation times to be quantitatively analyzed to get individual diffusion coefficients for overall molecular motions (i.e., spinning and tumbling), internal reorientations of molecular fragments, and collective orientational order fluctuations. For instance, the rotational dynamics of molecules in anisotropic phase can be described by the small step rotational diffusion model.[63] This model is based on the assumption that molecular reorientations proceed through a random sequence of infinitesimal angular jumps.

In general, however, the rotational motion of an uniaxial molecule in a nematic phase is conveniently characterized by using orientational TCFs[63,64] $\Phi^L_{mn}(t) = \langle D^{L*}_{mn}(\Omega(0))D^L_{mn}(\Omega(t))\rangle$, where $D^L_{mn}(\Omega)$ is the Wigner rotational matrix element of rank L, and $\Omega = (\alpha, \beta, \gamma)$ is a set of time dependent Euler angles, which define orientation of the molecular axis system relative to the director frame. The index m is related to the director coordinate system whereas molecular properties are dictated by the index n. First rank ($L = 1$) TCFs are relevant for infrared and dielectric spectroscopies, while TCFs with $L = 2$ appear in the expressions for nuclear spin relaxation rates and Raman band shapes. The initial values of the $\langle D^{L*}_{mn}(\Omega(0))D^L_{mn}(\Omega(t))\rangle$ can be expressed in terms of the OPs.[64] Notice that the first rank TCFs at $t = 0$ depend on $\overline{P}_2$ only, whereas both $\overline{P}_2$ and $\overline{P}_4$ are required for the second rank correlation functions.

In general, the correlation functions may be written as infinite sums of exponentially decaying terms. Here, one employs a single exponential approximation[64] to give

$$\Phi^L_{mn}(t) = \Phi^L_{mn}(\infty) + \left[\Phi^L_{mn}(0) - \Phi^L_{mn}(\infty)\right] \exp\left(-\frac{t}{\tau^L_{mn}}\right). \qquad (11.43)$$

In order to determine molecular reorientations, one needs a model to express the correlation times τ^L_{mn} in terms of rotational diffusion constants.[64] Based on the short time expansion of the TCFs, expressions for the correlation times were proposed[64]

$$\tau^1_{00} = \left[D_\perp \frac{2 - 2\overline{P}_2}{1 - 2\overline{P}_2}\right]^{-1}, \qquad (11.44)$$

$$\tau^1_{10} = \left[D_\perp \frac{2 + \overline{P}_2}{1 - \overline{P}_2}\right]^{-1}, \qquad (11.45)$$

$$\tau^2_{00} = \left[6\ D_\perp \frac{7 + 5\ \overline{P}_2 - 12\ \overline{P}_4}{7 + 10\ \overline{P}_2 + 18\ \overline{P}_4 - 35\ \overline{P}_2^2}\right]^{-1}. \qquad (11.46)$$

Having obtained sets of both correlation times τ^L_{mn} and OPs $\overline{P}_{2L}$ $(L = 1,2)$, it is now possible to calculate the RDC $D_\perp$ and the RVC γ_i $(i = 1,2)$.

It should be noted that the several theoretical approaches for the description of rotational diffusion have been proposed as well.[26,39,65] In particular, for uniaxial molecules in a nematic phase, the model based on random walk theory has been applied to calculating RDC coefficient. The model is based on the assumption that molecular reorientation proceeds through a random sequence of large - amplitude jumps. As result one can write a simple expression for $D_\perp$ as[65,66]

$$D_\perp = \frac{\pi^3}{6}\left(\frac{k_B T}{2\pi I_\perp}\right)^{\frac{1}{2}} f_0(\cos\frac{\pi}{2}), \tag{11.47}$$

where $I_\perp$ is the molecular moment of inertia corresponding to the short axis of the molecule, and $f_0(\cos\frac{\pi}{2})$ is the value of SODF at $\theta = \frac{\pi}{2}$, which corresponds to the molecular orientation orthogonal to the director. Using the SODF $f_0(\cos\beta_i)$ obtained, for instance, by the atomistic MD simulation and plotted in Fig. 11.1,[5] the RDC $D_\perp$ for 5CB at $T = 300$ K, was estimated to be 8×10^8 s^{-1}, in agreement with the experimental values for ^{2}H NMR[61] $(3 \div 5)\times 10^8$ s^{-1}.

11.3.3. *Numerical results for RVC's and λ*

Having obtained SODF, $f_0(\cos\beta)$, RDC, $D_\perp$, and a set of OPs $\overline{P}_{2L}$ $(L = 1,2)$, based on realistic atom-atom interaction potentials,[38,39] it is possible to calculate the functions $g_1(\beta)$ of the linear, inhomogeneous differential equation (11.37) numerically, and to determine both RVCs γ_i and the coefficient $\lambda = -\frac{\gamma_2}{\gamma_1}$. The coefficients γ_1 for 5CB, derived in the framework of these two expressions (NZ, F) and the SODF-based approach (Eq. (11.40)) are given in Table 11.3.[40] Here the values of $D_\perp$ are calculated using Eqs. (11.44), (11.45) and (11.46). In order to determine τ^1_{00}, τ^1_{10}, and τ^2_{00} in these equations, the corresponding functions $\Phi^1_{00}(t)$, $\Phi^1_{10}(t)$, and $\Phi^2_{00}(t)$ (see, Fig. 11.8[40]) were used. For these TCFs the obtained values of correlation times were found to be $\tau^1_{00} = 14$ ns, $\tau^1_{10} = 2.7$ ns, and $\tau^2_{00} = 1.1$ ns. Using Eq. (11.45) with $\overline{P}_2 = 0.504$, the value $D_\perp = 0.7\times 10^8$ s^{-1} was obtained, in good agreement with experimental values for ^{2}H NMR $(0.5 \div 1.5)\times 10^8$ s^{-1}.[61] The experimental values of λ were determined using rotating magnetic field (EXP$_1$)[67] and the electro - optical (EXP$_2$)[68] methods, whereas the calculated ones were derived from Eqs. (11.36) and (11.37). Notice that the analytical theory which rests on

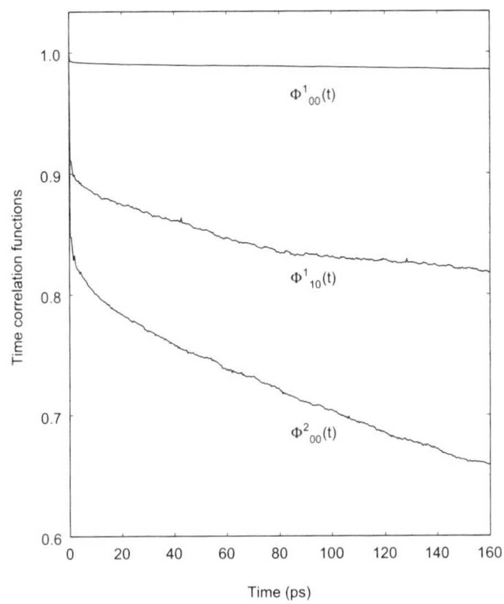

Fig. 11.8. Orientational time correlation functions for 5CB in the nematic phase, calculated from the MD trajectory.[40] Reproduced from figure 2 of Ref. [40] with permission of the EDP Sciences.

Table 11.3. Values of the RVCs γ_i $(i = 1, 2)$ (in poise) and λ, for 5CB nematic liquid crystal, at 300K. Calculations were carried out using Eq. (11.40) together with different functions[40] ξ_i: $\xi^{(NZ)}$, $\xi^{(F)}$; experimental data obtained were obtained using magnetic field (EXP_1[67]) and the electro-optical (EXP_2[68]) methods, respectively. SODF refers to using Eqs. (11.36)–(11.37).

Model	(SODF)	(NZ)	(F)	(EXP_1)	(EXP_2)
γ_1	0.45	0.18	0.22	0.65	0.74
$-\gamma_2$	0.49			0.72	
λ	1.09			1.10	

the SODF shows a better agreement with the experimental results than the two models presented in Eqs. (11.40)-(11.42). This is not surprising because the distribution functions contain complete information about orientational order, whereas only a few OPs are used in the other approaches.

Recently, the RVC γ_1 of *cyanobiphenyl* series (nCB) with $n = 5, 6, 7, 8$ in their nematic phase, have been studied by atomistic MD simulations,[69,70]

Table 11.4. $t = T/T_{\mathrm{NI}}$ is the dimensionless temperature.[69] The experimental values are those obtained in Ref. [23].

n	t	$D_\perp(10^8\ s^{-1})$	γ_1^{NZ}(mPa s)	γ_1^{F}(mPa s)	γ_1^{EXP}(mPa s)
5	0.983	0.9 ± 0.2	48.3	60.3	52.0
6	0.986	0.6 ± 0.1	59.6	68.4	65.0
7	0.984	0.7 ± 0.1	55.9	67.9	48.0
8	0.984	0.5 ± 0.1	69.1	83.8	68.0

T_{NI}(5CB) = 307.7 K; T_{NI}(6CB) = 302 K; T_{NI}(7CB) = 315 K; and T_{NI}(8CB) = 313.5 K, respectively. 1 mPa s = 10^{-2} poise.

using two models NZ and F. Both models (NZ and F) derive a prediction for γ_1 from the knowledge of the $D_\perp$ and OPs. The values of $D_\perp = \int_0^\infty \langle \omega_\perp(t)\omega_\perp(0)\rangle dt$ for $nCB's$, at various temperatures in the nematic phase were obtained from lengthy MD simulations.[69] Here $\omega_\perp$ is the perpendicular projection of the molecular angular velocity $\vec{\omega}$ onto the nCB para axis of the *biphenyl* moiety. These data are collected in Table 11.4.[69] The agreement with experiment shown by both theoretical methods (NZ and F) is satisfactory, but the NZ approach performs better, especially on higher homologues. The temperature dependences of the RVCs γ_i $(i = 1, 2)$ for $4 -$ *octyloxy* $- 4' -$ *cyanobiphenyl* (8OCB) are shown in Fig. 11.9.[71] The NZ, F and KD models have been employed to estimate the theoretical values. Using Eqs. (11.40), (11.41) and (11.42) together with the parameter $D_\perp$ and $\overline{P}_{2L}$ $(L = 1, 2)$ determined from NMR measurement,[71] the viscosity values have been obtained in close agreement with experiment.

As mentioned above, the time-domain dielectric spectroscopy enables us to study the complex permittivity of liquid crystals.[62] The complex dielectric constant parallel to the director allows us to calculate the relaxation time τ_{00}^1 for hindered rotation of molecules around their molecular short axes. Having obtained the relaxation time τ_{00}^1 and the data for $\overline{P}_2$, one can calculate from Eq. (11.44) the coefficient $D_\perp$, and then both γ_1 and γ_2. The temperature dependences of the RDC $D_\perp$, and RVCs γ_1 and γ_2, for 8CB in the temperature range corresponding to the nematic phase are shown in Fig. 11.10.[72] The behavior of a molecule in a simple shear flow depends strongly on whether the nematic is of the tumbling or flow-aligning type. According to the torque balance analysis, tumbling behavior occurs when $|\gamma_1| > |\gamma_2|$ (see Eq. (11.39)). For the flow-aligning nematic, the director in

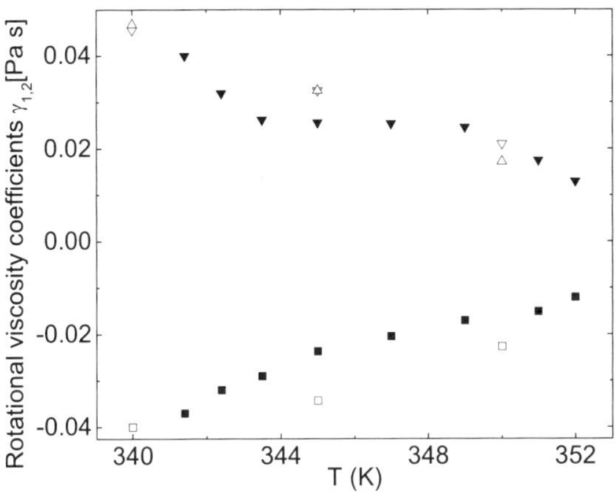

Fig. 11.9. Temperature dependence of the γ_i $(i = 1, 2)$ for 8OCB at fixed density,[71] calculated using Eq. (11.42) (γ_2, open squares), Eqs. (11.40) and (11.41) (F) (γ_1, open down triangles), Eqs. (11.40) and (11.41) (NZ) (γ_1, open up triangles), and measured values γ_1 (solid triangles) and γ_2 (solid squares),[23] respectively. Reproduced from figure 3 of Ref. [71] with permission of the APS.

the shearing flow is driven by viscous torques towards the flow-alignment angle $\theta_{\text{eq}} = \frac{1}{2}\cos^{-1}\lambda^{-1}$, where $\lambda = -\frac{\gamma_2}{\gamma_1}$ is the so-called tumbling parameter.[28,29] Because of the large effect λ has on the behavior of the nematic in the shearing flow, one would like to be able to determine its the value from the molecular theory,[40] and especially to predict whether it is greater than or less than unity. Recently, it has been shown that, in the limit of slow flows, an appropriate analytical expression for λ may be derived in terms of OPs $\overline{P}_2$ and $\overline{P}_4$ as[28,29]

$$\lambda = \omega\, \frac{14 + 5\overline{P}_2 + 16\overline{P}_4}{35\overline{P}_2}. \tag{11.48}$$

Assuming $\omega = 0.8$, λ was determined for two independent sets of OPs for 5CB in the nematic phase: i) $\overline{P}_2 = 0.504$, and $\overline{P}_4 = 0.188$, at temperature $T = 300$K, from the MD simulation,[40] and ii) $\overline{P}_2 = 0.59$, $\overline{P}_4 = 0.14$, at $T = 303$K from the experimental data, obtained using polarized laser Raman scattering.[28] It was found that $\lambda = 0.89$ using Eq. (11.48) in the simulated LC and $\lambda = 0.74$ from the experimental OPs. It should be noted that Eq. (11.48) predicts $|\lambda| < 1$ for both cases whereas both the complete calculation, based on Eq. (11.39), and experimental measurements give $|\lambda| > 1$ (Table 11.3[40]). These results are sensitive to the value of the

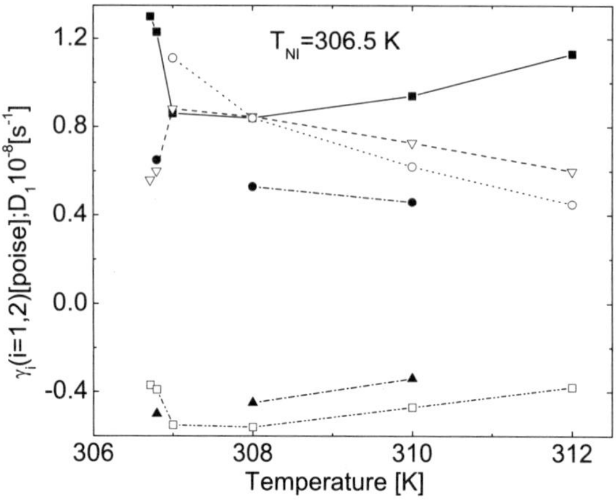

Fig. 11.10. Temperature dependence of $D_\perp$ and γ_i $(i = 1, 2)$[72] for 8CB at fixed density, calculated using Eq. (11.44) ($D_\perp = D_1$, solid squares), Eqs. (11.40)-(11.41)(NZ) (γ_1, open triangles), Eq. (11.42) (γ_2, open squares), and measured values of γ_1 (solid[73] and open circles[23]) and γ_2 (solid triangles), respectively. Reproduced from figure 2 of Ref. [72] with permission of the AIP.

parameter ω. However, the choice of $\omega = 0.8$, corresponds to a reasonable value of the length-breath ratio, $\gamma = 3$, for the 5CB molecule. Remembering that $|\lambda| \geq 1$ implies a fixed alignment angle, while $|\lambda| < 1$ corresponds to the rotation of the director, we can conclude that the approximate expression in Eq. (11.48) is not useful for a qualitative analysis of the alignment of the stress tensor, since it requires highly accurate values of ω. Calculations of the temperature dependence of λ, as well as comparisons of the values of the RVCs γ_1 and γ_2, both for 8CB and 8OCB, show that $\lambda(\text{8CB}) < 1$, but $\lambda(\text{8OCB}) > 1$. This means, in according with equation $\theta_{\text{eq}} = \frac{1}{2}\cos^{-1}\lambda^{-1}$, that 8CB molecules under Couette high shear flow exhibit a tumbling instability in the temperature interval where the nematic phase exists. At lower shear rates the director orientation is governed by the viscous, elastic and electric forces, both in the bulk and close to the surface.[72,74] On the other hand, for 8OCB a molecule that has an extra oxygen atom compared to 8CB, λ is always greater than unity and the director, at least in the high shear flow, aligns at the angle θ_{eq} to the flow direction. The small chemical difference between the molecules 8CB and 8OCB leads to different flow dynamics, whereas the phase diagrams for these LC compounds practically are the same, and characterized by the existence of a nematic-smectic A phase transition.

Two other compounds in the *cyanobiphenyl* series, i.e. 6CB and 7CB, were recently studied experimentally, by direct measurements of the flow-alignment angle θ_{eq}.[75] Both compounds were examined in their respective nematic temperature ranges, and it was found that 6CB exhibited a flow-aligning behaviour, whereas 7CB showed a transition from tumbling to flow-aligning somewhere between 305.5 and 306.4 K. These results show that, whereas both 5CB and 6CB flow-align over their entire nematic ranges, another *cyanobiphenyl* compound, 7CB, undergoes the transition from flow alignment to tumbling. Such a dramatic difference in the flow dynamics of 5CB or 6CB and 7CB is unexpected, since the thermodynamic phase diagrams for these LC compounds are practically the same. So much work is needed to generate new molecular theories which can provide more realistic picture of the shearing flow.

11.4. Conclusion

In this chapter, some recent progress made in the area of the elastic and flexoelectric coefficients and rotational viscosity has been discussed. The theoretical treatments for both of dynamic and static processes of flexible molecules in an anisotropic medium require a certain number of simplifying assumptions, which may only be justified by comparison between model predictions and experimental results. For instance, the statistical-mechanical approaches used for calculating elastic and flexoelectric coefficients take into account both intermolecular attraction and steric repulsion, but are unable to describe the effect due to molecular flexibility. Since the structure of mesogenic molecules is often characterized by a rather rigid central core to which one (or more) flexible hydrocarbon tails are attached, the ability to account for the chain flexibility in the final expressions for these material coefficients is an important step toward their better theoretical comprehension.

On the one hand, this limitation is counterbalanced by the simplicity of existing theoretical approaches, which can be used to calculate many physical parameters of nematic liquid crystals. On the other hand, the effect of chain flexibility can be accounted for by a combination of theoretical aspects of statistical mechanics and computer simulation techniques, in order to facilitate the analysis of molecular structure in the nematic phase. There is also an additional procedure which would allow us to further examine the validity of models. This entails carrying out a statistical-mechanical treatment of the coupling between macroscopic LC parameters and molecular structure, followed by comparisons of, for example, rotational viscosity and elastic coefficients with the experimental data. Thus, the combination of experimental techniques, such as NMR and dielectric spectroscopy, and theoretical approaches, such as statistical-mechanical theories, which take into account translational, orientational, and mixed correlations, or molecular dynamics simulation applied to the description of nematic LC provides a powerful tool for investigating both elastic and viscous properties of real nematics.

Acknowledgment

Discussions with Prof. S. Romano and Dr. A. A. Vakulenko are gratefully acknowledged.

References

1. J. G. Gay and B. J. Berne, *J. Chem. Phys.* **74**, 3316 (1981).
2. M. A. Bates and G. R. Luckhurst, *Computer Simulation of Liquid Crystal Phases Formed by Gay-Berne Mesogens*, in Liquid Crystals I, ed. by D. M. P. Mingos, (Springer, Berlin, 1999).
3. K. Satoh, S. Mita, and S. Kondo, *Chem. Phys. Lett.* **255**, 99 (1996).
4. A. V. Zakharov and S. Romano, *Phys. Rev. E.* **58**, 7428 (1998).
5. A. V. Zakharov, S. Romano, and A. Maliniak, *Phys. Rev. E.* **60**, R1142 (1999).
6. A. V. Zakharov and R. Dong, *Phys. Rev. E.* **64**, 031701 (2001).
7. D. K. Yang, S. T. Wu, *Fundamentals of Liquid Crystal Devices*, John Wiley, New York, 2006.
8. A. Poniewierski, J. Stecki, *Mol. Phys.* **38**, 1931 (1979).
9. V. B. Nemtsov, *Theor. Appl. Mech. (Minsk)* **13**, 16 (1987).
10. A. V. Zakharov, *Physica A.* **175**, 327 (1991).
11. M. A. Osipov and S. Hess, *Mol. Phys.* **78**, 1191 (1993).
12. P. I. C. Teixeira, V. M. Pergamenshchik, and T. Sluckin, *Mol. Phys.* **80**, 1339 (1993).
13. A. V. Zakharov, *Phys. Lett. A.* **187**, 331 (1994).
14. L. A. Rott, *Statistical-Mechanical Theory of Molecular Systems*, Nauka, Moscow, 1978.
15. A. V. Zakharov, *Phys. Rev. E.* **51**, 5880 (1995).
16. J. Stelzer, L. Longa, and H. R. Trebin, *J. Chem. Phys.* **103**, 3098 (1995).
17. J. Stelzer, M. Bates, L. Longa, and G. R. Luckhurst, *J. Chem. Phys.* **107**, 7483 (1997).
18. N. H. Phuong, G. Germano, and F. Schmid, *J. Chem. Phys.* **115**, 7227 (2001).
19. A. V. Zakharov and A. Maliniak, *Eur. Phys. J. E.* **4**, 85 (2001).

20. P. G. de Gennes and J. Prost, *The Physics of Liquid Crystals*, 2nd ed. Oxford University Press, Oxford, 1995.
21. Ch. Gahwiller, *Phys. Lett. A.* **36**, 311 (1971).
22. Orsay, L. C. G. *Mol. Cryst. Liq. Cryst.* **13**, 187 (1971).
23. A. G. Chmielewski, *Mol. Cryst. Liq. Cryst.* **132**, 319 (1986).
24. S. Hess, *Z. Naturforsch. A.* **30**, 1224 (1975).
25. V. B. Nemtsov, *Theor. Mat. Fiz.* **25**, 188 (1975)[*Theor. Math. Phys.* **25**, 1019 (1975)].
26. M. A. Osipov and E. M. Terentjev, *Phys. Lett. A.* **134**, 301 (1989).
27. A. V. Zakharov, *Phys. Lett. A.* **193**, 471 (1994).
28. M. Kroger and S. Sellers, *J. Chem. Phys.* **103**, 807 (1995).
29. L. A. Archer and R. G. Larson, *J. Chem. Phys.* **103**, 3108 (1995).
30. M. Fialkowski, *Phys. Rev. E.* **58**, 1955 (1998).
31. M. Kroger and P. Ilg, *J. Chem. Phys.* **127**, 034903 (2007).
32. H. Yokoyama, *Handbook of Liquid Crystal Research*, eds. by P. J. Collings and J. P. Patel, (Oxford University Press, Oxford, 1997), Chapt.6.
33. A. V. Zakharov and A. A. Vakulenko, *Crystallography Report.* **48**, 686 (2003).
34. I. M. Sobol, *SIAM. J. Number. Anal.* **16**, 790 (1979).
35. N. M. Korobov, *Dokl. Akad. Nauk. SSSR* **115**, 1062 (1957).
36. A. V. Zakharov and R. Dong, *Eur. Phys. J. E.* **6**, 3 (2001).
37. P. P. Karat and N. V. Madhusudana, *Mol. Cryst. Liq. Cryst.* **40**, 239 (1977).
38. A. V. Komolkin, A. Laaksonen, and A. Maliniak, *J. Chem. Phys.* **101**, 4103 (1994).
39. A. V. Zakharov, A. V. Komolkin, and A. Maliniak, *Phys. Rev. E.* **59**, 6802 (1999).
40. A. V. Zakharov and A. Maliniak, *Eur. Phys. J. E.* **4**, 435 (2001).
41. R. B. Meyer, *Phys. Rev. Lett.* **22**, 918 (1969).
42. J. Prost and J. P. Marcerou, *J. Phys. (Paris)* **38**, 315 (1977).
43. I. Dozov, P. Martinot-Lagarde, and G. Durand, *J. Phys. Lett. (Paris)* **44**, L-818 (1983).
44. P. R. M. Murthy, V. A. Raghunathan, and N. V. Madhusudana, *Liq. Cryst.* **14**, 483 (1993).
45. L. M. Blinov, M. I. Barnik, M. Ozaki, N. M. Shtykov, and K. Yoshino, *Phys. Rev. E.* **62**, 8091 (2001).
46. M. A. Osipov, *Sov. Phys. JETP* **58**, 1167 (1983).
47. A. M. Somoza and P. Tarazona, *Mol. Phys.* **72**, 911 (1991).
48. D. L. Cheung, S. J. Clark, and M. Wilson, *J. Chem. Phys.* **121**, 9131 (2004).
49. J. Steltzer, R. Berardi, and C. Zannoni, *Chem. Phys. Lett.* **299**, 9 (1999).
50. J. L. Billeter and R. A. Pelcovits, *Liq. Cryst.* **27**, 1151 (2000).
51. A. Ferrarini, *Phys. Rev. E.* **64**, 021710 (2001).
52. J. Harden, B. Mbanga, N. Eber, K. Fodor-Ksorba, S. Sprunt, J. T. Gleeson, and A. Jakli, *Phys. Rev. Lett.* **97**, 157802 (2006).
53. J. L. Ericksen, *Arch. Ration. Mech. Anal.* **4**, 231 (1960).
54. F. M. Leslie, *Arch. Ration. Mech. Anal.* **28**, 265 (1968).
55. I. S. Berezin and N. P. Zhidkov, *Computing Methods*, 4th ed. Pergamon Press, Oxford, 1965.

56. N. Kuzuu and M. Doi, *J. Phys. Soc. Jpn.* **52**, 3486 (1983).
57. M. P. Allen, *Phys. Rev. Lett.* **65**, 2881 (1990); S. Hess, D. Frenkel, and M. P. Allen, *Mol. Phys.* **74**, 765 (1991).
58. A. M. Smondyrev, G. B. Loriot, and R. A. Pelcovits, *Phys. Rev. Lett.* **75**, 2340 (1995).
59. D. J. Evans and G. P. Morriss, *Statistical Mechanics of Nonequilibrium Liquids*, Academic, London, 1990.
60. S. Sarman, *J. Chem. Phys.* **103**, 393 (1995).
61. R. Y. Dong, *Nuclear Magnetic Resonance of Liquid Crystals*, 2nd edition, Springer- Verlag, New York, 1997.
62. J. Thoen and T. K. Bose, *Handbook of Low and High Dielectric Constant Materials and Their Applications*, ed. by H. S. Nalwa, Academic Press, New York, 1999; Chapt.11.
63. P. L. Nordio and U. Serge, *The Molecular Physics of Liquid Crystals*, eds. by G. R. Luckhurst and G. W. Gray, Academic Press, London, 1979; Chapt.18.
64. I. Dozov, N. Kirov and M. P. Fontana, *J. Chem. Phys.* **80**, 2585 (1984).
65. E. T. Brook-Levinson and A. V. Zakharov, *Euro. Phys. Lett.* **22**, 439 (1993).
66. A. V. Zakharov, D. Taguchi, and M. Iwamoto, *Chem. Phys. Lett.* **458**, 143 (2008).
67. H. Kneppe, F. Schneider, and N. K. Sharma, *J. Chem. Phys.* **77**, 3203 (1982).
68. S. T. Wu and C. S. Wu, *Phys. Rev. A.* **42**, 2219 (1990).
69. M. Cifelli, L. De Gaetani, G. Prampolini, and A. Tani, *J. Phys. Chem. B.* **112**, 9777 (2008).
70. M. Ilk Capar and E. Cebe, *Chem. Phys. Lett.* **407**, 454 (2005); M. Ilk Capar and E. Cebe, *Phys. Rev. E.* **73**, 061711 (2005).
71. A. V. Zakharov and R. Y. Dong, *Phys. Rev. E.* **63**, 011704 (2001).
72. A. V. Zakharov, A. A. Vakulenko, and J. Thoen, *J. Chem. Phys.* **118**, 4253 (2003).
73. G. R. Luckhurst, B. A. Timimi, T. Miyamoto, and A. Sugimura, *Mol. Cryst. Liq. Cryst.* **394**, 77 (2003).
74. A. V. Zakharov and R. Y. Dong, *Phys. Rev. E.* **65**, 052701 (2002).
75. D. J. Ternet, R. G. Larson, and L. G. Leal, *Rheol. Acta* **38**, 183 (1999).

Chapter 12

Carbon-13 NMR Studies of Thermotropic Liquid Crystals

Ronald Y. Dong

Department of Physics and Astronomy,
University of British Columbia,
6224 Agricultural Road, Vancouver, BC, Canada, V6T 1Z1
email: rondong@phas.ubc.ca

This chapter reports recent studies by ^{13}C NMR spectroscopy of liquid crystals (LC) formed by bent-core and rod-shaped mesogenic molecules. The method to obtain molecular ordering and structure in uniaxial and biaxial mesophases from ^{13}C chemical shifts ($< \delta >$) and C-H dipolar splittings is considered. In particular, the observed differences in $< \delta >$'s for various carbons in the mesophase from isotropic chemical shifts depend on the order parameters as well as the carbon chemical shift anisotropy (CSA) tensors of the observed carbons. Moreover, the assignment of carbon peaks in the NMR spectrum is often a challenge, particularly in aligned mesophases. Incorrect ^{13}C CSA principal values and peak assignments can easily lead to erroneous order parameters. Various one- and two-dimensional solid-state NMR techniques can be employed to assist the task of ^{13}C peak assignments as well as to obtain CSA powder patterns for the determination of principal components of their CSA tensors. Density functional theory provides, abeit in vacuo, a reasonable means to determine CSA tensors and their principal axes in a chosen molecular frame. Both approaches to obtain ^{13}C CSA tensors are compared for a rodlike mesogen. Molecular order in the chiral smectic C* phase of several rodlike smectogens is also reviewed.

12.1. Introduction

NMR spectroscopy has been shown to be a powerful technique for studying both the structure and dynamics of liquid crystals (LC).[1–3] Nowaday liquid crystals[4–6] refer to a class of organic materials that form intermediate phases between the solid and isotropic (I) phases and possess different degrees of anisotropic structural organization. Their constituent molecules exhibit certain shape anisotropy (e.g., rod-like, disk-like, banana-like). Classical calamitics, discotics, metallomesogens, and bent-core mesogens are thermotropic in nature. Some common mesomorphic phases are nematic (N), smectic A (SmA), columnar, chiral smectic C (SmC*), and banana (B) phases. For instance, uniaxial nematics are optically uniaxial and apolar, i.e., the director $\vec{n}$ and -$\vec{n}$ are equivalent. Biaxial nematics can also occur in which lath-like molecules are stacked in such a way that molecules are also aligned along a minor director.[7,8] In addition to the orientational ordering, positional order may also occur in layered structures shown by different smectic phases[9] and lamellar (L) phases. The other class of LC is known as lyotropics, which are formed by dissolving amphiphilic molecules into a selective solvent (and co-solvent) to form a two or more components system and can exhibit mesophases like N, L, micellar, hexagonal, and/or cubic phases by varying the temperature, or the solute concentration, as well as both of these. Many fundamental advances in understanding LC were made before the mid 1960s. These included the model of long stiff rods by Onsager[10] which showed the role of repulsive forces, the Maier-Saupe statistical model[11] which pointed to the role of dispersion forces, and the model by Gelbart and Cotter[12] which united these two approaches in the Van der Waals theories. But the real push in LC research came only after the first application of these materials in electro-optical technology.[13] The structures of various mesophases are described in standard texts.[4–6] This chapter describes the recent advance in ^{13}C NMR study of LC, mainly based on works carried out in our laboratory.

12.2. Spin Hamiltonian in LC

Brownian motions of molecules average out all anisotropic spin interactions in ordinary liquids and their NMR spectra are governed by the spin Hamiltonian due to the Zeeman interaction, the isotropic chemical shift (δ^i_{iso}) and the isotropic indirect spin-spin coupling (J^{iso}_{ij})

$$H = -\hbar B \sum_i \gamma_i (1 - \delta^i_{iso}) I_{iZ} + \sum_{i<j} J^{iso}_{ij} \, \mathbf{I}_i \cdot \mathbf{I}_j \qquad (12.1)$$

where the external magnetic field B is along the laboratory Z axis. The above Hamiltonian is inappropriate in LC due to the presence of a potential of mean torque, arising from the shape anisotropy of the molecules. Three main types of spin interactions are of interest: dipole-dipole (D), nuclear electric quadrupole and chemical shift anisotropy (CSA). Nuclear

Table 12.1. Spin operators $T_{l,m}$ for dipolar (D) and chemical shift (CS) interactions.

λ	$T_{0,0}$	$T_{2,0}$	$T_{2,\pm1}$	$T_{2,\pm2}$
D	$-\frac{1}{\sqrt{3}}\mathbf{I}\cdot\mathbf{S}$	$\sqrt{\frac{1}{6}}\left(3I_ZS_Z-\mathbf{I}\cdot\mathbf{S}\right)$	$\mp\frac{1}{2}\left(I_ZS_\pm+I_\pm S_Z\right)$	$\frac{1}{2}I_\pm S_\pm$
CS	$-I_ZB/\sqrt{3}$	$\sqrt{2/3}I_ZB$	$\mp I_ZB/2$	0

quadrupolar interactions only exist for spin $I > 1/2$. The indirect spin-spin coupling J is usually small in comparison to the direct D interaction, but may be important in some studies (e.g., solute). For ^{13}C NMR, only the chemical shift (CS) and D interactions will normally be required. These spin interactions may be formally written in terms of irreducible spherical tensors of rank l:

$$H_\lambda = C_\lambda \sum_l \sum_{m=-l}^{l} (-1)^m T_{l,m}^\lambda R_{l,-m}^\lambda \tag{12.2}$$

where the rank $l = 0$, 1 and 2 and the scalar factor C_λ is an appropriate constant for the spin interaction labeled by λ ($C_D = -\mu_0\gamma_I\gamma_S\hbar^2/4\pi$, and $C_{CS} = \gamma_I\hbar$). For λ =D and CS, the spin operators $T_{l,m}$ with order m in the laboratory L (X, Y, Z) frame are listed in Table 12.1. The spin $\mathbf{I} \neq \mathbf{S}$, and $\mathbf{I} = \mathbf{S}$ are for the heteronuclear and homonuclear dipolar coupling, respectively. The coupling tensor $R_{l,m}$ in the L frame is time dependent due to motions of spin-bearing molecules. It can be expressed in terms of the rotational transformation of the irreducible components $\rho_{l,n}$ in the principal axis system (PAS) frame to the L frame by

$$R_{l,m} = \sum_n D_{m,n}^{l*}\left(\Omega_{PL}\right)\rho_{l,n} \tag{12.3}$$

where $D_{m,n}^l\left(\Omega_{PL}\right)$ denotes the Wigner rotation matrices, and Ω_{PL} denotes the Euler angles for transforming between the L frame and the PAS frame. For the D interaction $\rho_{0,0} = 0$, while for the CSA tensor, $\rho_{0,0} = -\left(\delta_{11}+\delta_{22}+\delta_{33}\right)/\sqrt{2} = -\left(\mathrm{Tr}\ \underset{\sim}{\delta}\right)/\sqrt{2}$. For $l = 2$, one has $\rho_{2,0} = \sqrt{3/2}\Delta_\lambda$, $\rho_{2,\pm1} = 0$ and $\rho_{2,\pm2} = \frac{1}{2}\Delta_\lambda\eta_\lambda$ where Δ_λ and η_λ can be obtained from V_{ii}, the principal values of the coupling tensor $\underset{\sim}{V}$. For CS, $\Delta_{CS} = \delta_{33} - \frac{1}{3}\ \mathrm{Tr}\ \underset{\sim}{\delta}$ and $\eta_{CS} = \left(\delta_{22}-\delta_{11}\right)/\Delta_{CS}$, where δ_{ii} are the CSA tensor elements, while for D, $\eta_D = 0$ and $\Delta_D = \left\langle 1/r_{IS}^3\right\rangle$, where r_{IS} is the internuclear distance and the brackets denote a vibrational average.

In the high-field limit, the dominant term in the Hamiltonian $H = H_Z + H_D + H_J$ is the Zeeman (H_Z) contribution, and the off diagonal matrix elements of $H_D + H_J$ which couple different $I_Z = \sum_i I_{iZ}$ can be neglected due to the perturbution theory. This leads to certain simplifications in the above tensorial spin interactions, yielding in the L frame[15]

$$H = -\hbar B \sum_i \gamma_i (1 - \delta^i_{ZZ}) I_{iZ} + \sum_{i<j} J^{iso}_{ij} \, \mathbf{I}_i \cdot \mathbf{I}_j + \sum_{i<j} (2D^{aniso}_{ij} + J^{aniso}_{ij}) [I_{iZ} I_{jZ} - \frac{1}{4} \{ I_{i+} I_{j-} + I_{i-} I_{j+} \}]) \quad (12.4)$$

where

$$J^{iso}_{ij} = \frac{1}{3} \mathrm{Tr} \, \underset{\sim}{J} \quad (12.5)$$

$$J^{aniso}_{ij} = J_{ij,ZZ} - J^{iso}_{ij} \quad (12.6)$$

$$D^{aniso}_{ij} = -\frac{\mu_0 \gamma_i \gamma_j \hbar^2}{4\pi < r^3_{ij} >} P_2(\cos \theta_{ij}) \quad (12.7)$$

and θ_{ij} is the polar angle of the internuclear vector between the i-j spin pair in the (X, Y, Z) frame.

12.2.1. *Molecular orientational order and motion*

In uniaxial mesophases, the orientational order of rigid mesogenic molecules is specified by a set of order parameters $S_{\mu\nu}$, which forms a symmetric, traceless orientational order matrix $\underset{\sim}{S}$, defined by Saupe,[14] as

$$S^{ZZ}_{\mu\nu} = \frac{1}{2} \langle 3 \cos \theta_{Z,\mu} \cos \theta_{Z,\nu} - \delta_{\mu\nu} \rangle \equiv S_{\mu\nu} \quad (12.8)$$

where the Z axis is along $\vec{n}$, μ, ν = x, y or z axis of a molecule-fixed (M) frame, $\theta_{Z,\mu}$ is the angle between $\vec{n}$ and the molecular μ axis, and the ensemble average, denoted by brackets, is carried out by averaging all molecules over their tumbling motions within a domain. In a magnetic or electric field of sufficient field strength, different domains of nematic and some smectic LC can align macroscopically to give a single domain sample. Since $S_{\mu\nu} = S_{\nu\mu}$ and $\sum_\mu S_{\mu\mu} = 0$, there are five independent order parameters:

$$\begin{aligned} S_{zz} &= <3\cos^2\theta - 1> /2 \equiv S \\ S_{xx} - S_{yy} &= 3 < \sin^2\theta \cos 2\psi > /2 \equiv D \\ S_{xy} &= 3 < \sin^2\theta \sin 2\psi > /4 \\ S_{xz} &= 3 < \sin\theta \cos\theta \cos\psi > /2 \\ S_{yz} &= 3 < \sin\theta \cos\theta \sin\psi > /2 \end{aligned} \tag{12.9}$$

where the polar angles (θ,ψ) specify the Z axis (the director) in the molecular M (x,y,z) frame. The number of non-zero, independent elements $S_{\mu\nu}$ depends on the point group symmetry of the molecule. For molecules with C_1 and C_i symmetry, all 5 order parameters are non-zero. On the other hand, only one nematic order parameter S is required for rigid ellipsoids of axial symmetry (C_3 and higher symmetry). Two order parameters S and D exist for molecules with C_{2v}, D_2 and D_{2h} symmetry, while S, D and S_{xy} are non-zero for molecules with C_2, C_{2h} and C_s symmetry. For non-rigid molecules, one needs to consider all possible conformers available to the molecule and applies statistical models.[1,15] To describe molecular order in a more general term, one needs to use an order supermatrix $S^{ii}_{\nu\mu}$, where $i = (X, Y, Z)$ refers to the principal axis of the L frame, and $\nu, \mu = (x, y.z)$ denote axes of the M frame. In the simple case like biaxial nematics, four common order parameters are required:[16]

$$\begin{aligned} S &= S^{ZZ}_{zz} \\ D &= S^{ZZ}_{xx} - S^{ZZ}_{yy} \\ P &= S^{XX}_{zz} - S^{YY}_{zz} \\ C &= (S^{XX}_{xx} - S^{XX}_{yy}) - (S^{YY}_{xx} - S^{YY}_{yy}). \end{aligned}$$

In the uniaxial nematic (N_U) phase S and D are nonzero whereas P and C are zero (for cylindrical rod, D is also identical to zero), while in the biaxial nematic phase (N_B) all four order parameters are nonzero.

The partial average of any second rank tensor element A_{ij} gives a component along the $\vec{n}$ ($\| Z$ axis) in uniaxial mesophases as

$$< A_{ZZ} >= A_0 + \frac{2}{3} \sum_{\mu,\nu} S_{\mu\nu} A_{\mu\nu} \tag{12.10}$$

where $A_0 = \frac{1}{3}\mathrm{Tr}\ \underset{\sim}{A}$ and is independent of the chosen axis system. While A_0 vanishes for the D interaction, $A_0 = \delta_{iso}$ and $A_0 = J^{iso}_{ij}$ for the CS and indirect spin-spin interaction, respectively. Eq. (12.10) is a result of tensor transformation from the L frame to the M frame. To express the molecular properties in the spin Hamiltonian using $S_{\mu\nu}$, one can write for the CS interaction

$$\delta^i_{ZZ} = \delta^i_{iso} + \delta^i_{aniso} \tag{12.11}$$

$$\delta^i_{aniso} = \frac{2}{3}\sum_{\mu,\nu} S_{\mu\nu}\delta_{\mu\nu}. \tag{12.12}$$

Similarly, one obtains for the indirect spin-spin interaction

$$J^{iso}_{ij} = \frac{1}{3}\left(J_{ij,xx} + J_{ij,yy} + J_{ij,zz}\right) \tag{12.13}$$

$$J^{aniso}_{ij} = \frac{2}{3}\sum_{\mu,\nu} S_{\mu\nu}J_{ij,\mu\nu} \tag{12.14}$$

and for the D interaction

$$D^{aniso}_{ij} = \frac{2}{3}\sum_{\mu,\nu} S_{\mu\nu}D_{ij,\mu\nu}. \tag{12.15}$$

Eqs. (12.11) - (12.15) can be used in Eq. (12.4) to give the average Hamiltonian in the L frame in terms of the order parameter tensor for LC. The NMR spectrum arising from the average Hamiltonian can provide a means to determine the orientational ordering of molecules in uniaxial LC.

The effect of tilting the director away from the magnetic field by an angle β (see Fig. 12.1) is to modify Eq. (12.10) via an additional coordinate transformation (i.e. $(X, Y, Z) \rightarrow$director frame $\rightarrow(x, y, z)$), giving[17]

$$< A_{ZZ} >= A_0 + [\frac{1}{2}(3\cos^2\beta - 1)]\frac{2}{3}\sum_{\mu,\nu} S_{\mu\nu}A_{\mu\nu} \tag{12.16}$$

This equation indicates that the average anisotropic part of A along the

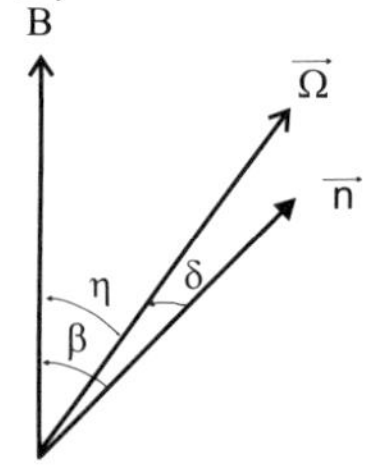

Fig. 12.1. Schematic diagram of the orientations of the director $\vec{n}$ and the spinning axis of a rotating sample $\vec{\Omega}$ in an external magnetic $\vec{B}$ field.

magnetic field is now scaled by a reduction factor $P_2(\cos\beta)$. As a consequence of the director tilt, the total Hamiltonian becomes

$$H = -\hbar B \sum_i \gamma_i I_{iZ} \left(1 - \delta^i_{iso} - P_2(\cos\beta)\delta^i_{aniso}\right) + \sum_{i<j} J^{iso}_{ij}\, \mathbf{I}_j \cdot \mathbf{I}_j$$
$$+ \sum_{i<j} P_2(\cos\beta)\left(2D^{aniso}_{ij} + J^{aniso}_{ij}\right)\left[I_{iZ}I_{jZ} - \frac{1}{4}\{I_{i+}I_{j-} + I_{i-}I_{j+}\}\right]. \tag{12.17}$$

In applying Eq. (12.17), the following conditions are necessary: 1. rotational motions of molecules are fast on the NMR time scale, 2. the director dynamics are governed by the magnetic torque and the mechanical motion of the sample if any, and 3. the B field is sufficient to quantize the nuclear spin. Under mechanical spinning of the sample about a spinning axis $\vec{\Omega}$ which is tilted at an angle η w.r.t. B (Fig. 12.1), the behavior of $\vec{n}$ is governed by the anisotropic diamagnetic susceptibility $\Delta\chi$, the magnetic and viscous torques, and the spinning speed ω_r of the sample. This is because the potential energy for the director alignment depends on $P_2(\cos\beta)$ and $\Delta\chi$, viz., $U = -\frac{1}{3}\Delta\chi B^2 P_2(\cos\beta)$. When the sample with $\Delta\chi > 0$ is spun at $\omega_r > \omega_c$ ($\omega_c = \Delta\chi B^2/\gamma_1$ is a characteristic frequency and γ_1 is the Leslie rotational twist viscosity coefficient), the director has no time to orient parallel to B, but would orient to minimize the average potential energy (or $\cos^2\beta$) over one cycle of the rotation. Suppose the director has polar angles δ and φ in the principal frame of the bulk liquid crystal, defined by the spinning axis $\vec{\Omega}$ (see Fig. 12.1). It can be shown that

$$\cos\beta = \cos\eta\cos\delta + \sin\eta\sin\delta\cos(\omega_r t + \varphi)$$

and

$$P_2(\cos\beta) = \left(\frac{3\cos^2\eta - 1}{2}\right)\left(\frac{3\cos^2\delta - 1}{2}\right). \tag{12.18}$$

For uniaxial nematics with $\Delta\chi > 0$ and being spun at $\omega_r > \omega_c$, the director tends to align along $\vec{\Omega}$ when $\eta < \theta_m$ (magic angle θ_m is 54.7°), or when $\eta > \theta_m$, the directors are distributed in a plane perpendicular to $\vec{\Omega}$ ($\delta = 90°$). If $\Delta\chi < 0$, the above conditions for η would reverse. When doing off-magic-angle-spinning (OMAS)[17] in the case of $\delta = 0$, substituting Eq. (12.18) into Eq. (12.17) gives all anisotropic interactions (e.g., D^{aniso}_{ij}) being scaled down by a factor $(3\cos^2\eta - 1)/2$. When $\eta = \theta_m$, the LC directors have no prefered orientation, and the sample behaves like a powder. In the limit of $\eta = \theta_m$, only the isotropic interactions δ_{iso} and J^{iso}_{ij}

survive in Eq. (12.17). Thus, the ^{13}C NMR spectrum obtained using magic-angle-spinning (MAS) from a mesophase or powder sample shows only the isotropic chemical shifts. In a particular case of the chemical shift difference $\delta^i - \delta^j >> D_{ij}$ in Eq. (12.17), the dipolar interaction can now be truncated to contain only the $I_{iZ}I_{jZ}$ term, resulting in the so-called first-order dipolar coupling,[18] and the corresponding NMR spectrum is called a first-order spectrum.

12.3. ^{13}C NMR Techniques

To achieve good spectral resolution of ^{13}C peaks in solids, mechanical rotation of the sample is needed to reduce anisotropic spin interactions even though the ^{13}C chemical shifts cover a relatively large range (> 200 ppm). If the sample is spun about an axis oriented at η w.r.t. the B field with a spinning rate ω_r considerably larger than the coupling constants Δ_λ, the orientational dependent nuclear spin interactions are scaled down by a factor of $\frac{1}{2}(3\cos^2\eta - 1)$. This fact is used in the magic-angle-spinning (MAS) technique ($\eta = \theta_m$).[19,20] Soon after the introduction of MAS, cross-polarization (CP)[21] under the Hartmann-Hahn (H-H) condition was found to enhance signals of rare spins. Thus CP and MAS have significantly increased the sensitivity of ^{13}C NMR spectroscopy. CP/MAS together with multiple-pulse proton decoupling can now achieve highly resolved ^{13}C spectra in solids. Under MAS, cross-polarization polarization inversion (CPPI)[22,23] experiment creates a 1D ^{13}C spectrum with null CH signals, negative CH_2 signals, and positive signals from quaternary carbons. This provides a means to assign various protonated carbons. The idea behind the coherent averaging of the anisotropic spin interactions involves either manipulating the spins in the spin space or spatially averaging them through mechanical rotation of the sample.

12.3.1. *Spatial averaging*

Mechanical spinning of LC materials can average the anisotropic spin interactions beyond what static samples can provide. For OMAS, $\vec{n}$ can be aligned either parallel or perpendicular to the rotor axis $\vec{\Omega}$. When the director alignment is parallel to the $\vec{\Omega}$ axis ($\delta = 0$), OMAS has been used to scale the dipolar coupling, and the CSA[17] by the factor $(3\cos^2\eta - 1)/2$. For the case of ^{13}C NMR, the overlap of ^{13}C peaks in the proton-coupled ^{13}C spectra is still too extensive to give useful information. To alleviate this difficulty, OMAS in combination with two-dimensional (2D) separated local field (SLF) spectroscopy[24,25] has been used to measure ^{13}C-^{1}H dipolar couplings in LC.[17]

12.3.2. *Spin averaging*

Heteronuclear ^{13}C-$\{^{1}H\}$ proton decoupling

Heteronuclear decoupling is used to remove spin interactions (i.e., dipolar and J couplings) between the abundant (^{1}H) and rare (^{13}C) spins. Since high radiofrequency ($r.f.$) power is often required to suppress relatively large ^{13}C-^{1}H dipolar couplings in a non-spinning LC sample, the continuous wave (CW) method can cause severe heating and create undesirable temperature gradients inside the LC sample. To avoid $r.f.$ heating, low power decoupling by modulating the $r.f.$ irradiation as in liquids (ca. 20 kHz or less) and a long recycling time ($\geq$10 s) with a short acquisition time are typical for high-resolution ^{13}C NMR studies of LC. LC samples are usually not spun at the magic angle in order to preserve the director alignment along the B field. Decoupling schemes must, therefore, address a non-zero average static Hamiltonian. In the past, many special decoupling sequences have been applied to LC, including SPINAL-64 (Small Phase INcremental ALternation) which evolves from TPPM:[26]

$$\text{Spinal-64} = Q\overline{Q}\overline{Q}Q\overline{Q}QQ\overline{Q}$$

where[27]

$$Q = P(10^\circ)P(-10^\circ)P(15^\circ)P(-15^\circ)P(20^\circ)P(-20^\circ)P(15^\circ)P(-15^\circ)$$

and pulse $P(\epsilon)$ denotes a pulse width of 165° and phase of ϵ; $\overline{Q}$ is Q with all its ϵ replaced by -ϵ. Thus, SPINAL-64 has a nominal rotation angle cycle of 64π. Recently, the decoupling scheme SDROOPY-1[28] was found to be very robust w.r.t. the proton $r.f.$ power setting, and sufficiently broadband in the proton transmitter offset, thereby giving good decoupling for both the aromatic and aliphatic protons.

Homonuclear decoupling

Homonuclear decoupling is used to remove dipolar couplings among the abundant spins such as ^{1}H. Both MAS and cyclic multiple-pulse sequences have been used. These approaches involve rotation, the former in the real space and the latter in the spin space. The similarity among them is particularly obvious in the case of dipolar interaction. The earliest method was demonstrated by off-resonance "magic angle CW" irradiation, proposed by Lee and Goldburg (LG).[29] In the LG method, a strong *r.f.* B_1 field is applied off-resonance at $\omega_0 \pm \Delta\omega$, satisfying the condition $\gamma B_1/\Delta\omega = \sqrt{2}$, so that in the rotating frame the spins would precess about the effective field at the θ_m angle w.r.t. the B field. Consequently, the zeroth order homonuclear dipolar coupling term in the effective spin Hamiltonian vanishes. The WAHUHA multiple pulse sequence[30] has become the building block of many subsequent decoupling sequences.[31,32] It is noted that BLEW-48[31] is windowless, since the r.f. phases of $\pi/2$ pulses are switched, and has a theoretical scaling factor of 0.424. The LG method has been improved to give the frequency-switched Lee-Goldburg (FSLG-2) decoupling, in which the CW *r.f.* irradiation frequency is switched between the two LG conditions $\omega_0 \pm \gamma B_1/\sqrt{2}$ for each 2π rotation of the proton magnetization about the effective field.[33,34] The scaling factor for FSLG-2 is 0.577. Another variant of the LG method is the phase-modulated Lee-Goldburg (PMLG) decoupling.[35] Both FSLG-2 and PMLG suppress the zero- and first-order dipolar terms in the average spin Hamiltonian.

12.3.3. *Two-dimensional techniques*

Various 2D ^{13}C experiments suitable to assist ^{13}C peak assignments and to access proton-carbon correlations will be briefly surveyed. As mentioned above, SLF experiment has been used to determine the C-H dipolar couplings in LC. ^1H-^{13}C 'HETCOR' experiment has been done[36] to observe long-range proton-carbon correlations. 2D ^{13}C MAS experiments can be useful when a LC sample[37] fails to align in an external B field. Under MAS, a solid powder LC sample can be used to study the CSA powder patterns of all the resolved carbon sites in a 2D Separation of Undistorted Powder patterns by Effortless Recoupling (SUPER) experiment.[38]

12.3.3.1. *HETCOR*

2D heteronuclear C-H correlation experiment has been routinely used in solution NMR to asign carbon peaks. However, strong dipolar couplings among protons in solids have limited its application. Homonuclear dipolar couplings are partially averaged by motions making the application of HETCOR in LC possible even without MAS. During the evolution t_1 period, FSLG-2 is used for homonuclear decoupling, while a carbon π pulse is applied at $t_1/2$ to effectively remove the C-H dipolar evolution, but leave the evolution of proton CS unaffected. LG-CP is then used to transfer the proton polarization to carbons for signal detection under heteronuclear decoupling.[39] The contact time can be varied to distinguish direct and long-range correlations between the protons and carbons.

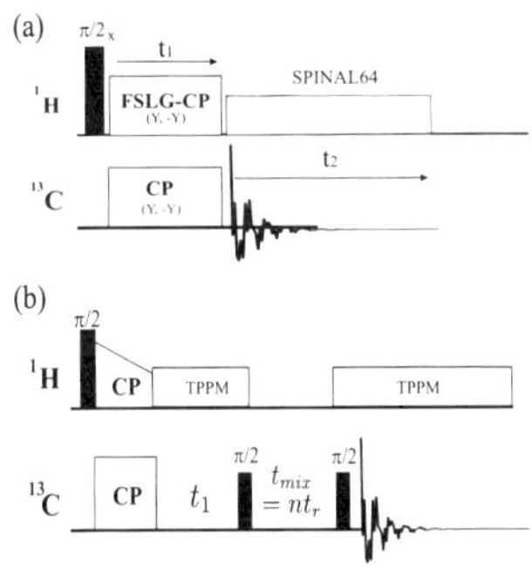

Fig. 12.2. (a) 2D SLF pulse sequence, which can be implemented with or eithout MAS, (b) 2D MAS exchange pulse sequence. t_r is the rotation period.

12.3.3.2. *SLF experiment*

^{13}C SLF spectroscopy utilizes the dispersion of carbon CS to allow the measurement of ^{13}C-^{1}H dipolar couplings, which can provide local ordering information for LC. In this experiment (see Fig. 12.2(a)), LG is applied during the t_1 period to remove the ^{1}H-^{1}H dipolar couplings while scaling down the ^{13}C-^{1}H dipolar couplings by a factor r, and broadband ^{13}C-$\{^{1}$H$\}$ decoupling is used during t_2 acquisition. Using double Fourier Transform (FT), the spectrum in the f_2 dimension shows a normal proton-decoupled ^{13}C spectrum, while the slices in the f_1 dimension show 2^m lines for different ^{13}C atoms, where m denotes the number of directly coupled protons. The splitting in each slice is given by[17]

$$\Delta\nu = r\left[(3\cos^2\eta - 1)D_{ij}^{aniso} + J_{ij}\right] \tag{12.19}$$

where $\eta = 0$ in static samples, $J_{ij} = J_{ij}^{iso} + J_{ij}^{aniso}$ can be obtained from the isotropic spectrum as $J_{ij}^{aniso} \simeq 0$ and the value of r is determined by the applied homonuclear decoupling scheme. For instance, FSLG homonuclear decoupling scales the dipolar coupling by $r = \sin\theta_m$. Thus, the diploar coupling constant D_{ij}^{aniso} for each carbon in a molecular segment can be readily determined. Beside the dipolar oscillations, there exists also a non-oscillatory part, which yields a zero frequency peak in the f_1 dimension, but can be suppressed by PI. It was demonstrated[40] in MBBA that PI leads to a doubling in the amplitude of the dipolar oscillations, with a reduction in the zero frequency component. Note that in the SLF experiment each ^{13}C magnetization evolves under the local fields of all the neighbouring protons. Further improvement of the resolution has been achieved by proton-detected local field (PDLF) spectroscopy,[41,42] since PDLF spectra are governed by two-spin ^1H-^{13}C couplings, A later version of PDLF used ^{13}C signal for detection instead.[42–44] By combining with OMAS at $\eta = 48.2°$, the superiority of PDLF over SLF has been demonstrated.[44] 3D PDLF spectroscopy has been used to obtain long-range ^{13}C-^1H dipolar couplings.[43] 2D SLF experiment can also be done in combination with MAS to assign the carbon peaks. The LGCP spin dynamics in a rotating sample is similar to that for normal CP. If the spinning frequency ω_r is much bigger than the C-H dipolar coupling $\omega_d = (\mu_0/4\pi)\gamma_I\gamma_S h^2/r^3$, this dipolar coupling can also be extracted using LGCP buildup curve as done in a stationary sample. For an isolated I-S pair, FT the LGCP buildup curve results in a Pake-like spectrum with two singularities at $\omega = \pm\frac{1}{2}\omega_d \cos\theta_m$.[45] The Pake-like splittings can only provide the C-H bond distances, which can be used to discern various carbon sites.

12.3.3.3. *MAS exchange experiment*

2D exchange experiment has been utilized to characterize molecular details of slow or intermediate motional processes.[18,41] In particular, deuterium 2D exchange experiments have been carried out by Spiess and co-workers in polymers and LC.[20] Due to the low ^{13}C sensitivity in non-labeled powder samples, MAS must be utilized to acquire 2D exchange spectrum. The pulse sequence for recording ^{13}C exchange spectra is shown in Fig. 12.2(b). A ramp-CP is used under MAS to enhance the ^{13}C NMR signal. During the t_1 and t_2 periods, TPPM decoupling is necessary to create narrow spectral lines in the f_1 and f_2 dimension. Proton decoupling is not applied during mixing time t_{mix} owing to its large value (order of second). Under slow MAS condition, the sidebands for different ^{13}C sites may create some cross peaks even when no real exchange process occurs among them. To avoid these artifacts, one simple way is to synchronize t_{mix} with the sample rotation period t_r, i.e. $t_{mix} = nt_r$, together with the appropriate phase cycling.[46] However, this coherence-selection scheme prevents the acquisition of a pure-phase 2D spectrum. This can dramatically decrease the spectral resolution. Despite the sign of f_2 frequency is still discriminated by the STATES method,[47] a magnitude type 2D exhange spectrum is obtained by calculating $\sqrt{I_r^2 + I_i^2}$, where I_r and I_i are the intensities obtained from the cosine and sine modulated spectra, respectively. The resulting 'star' structure of each resonance peaks still limits the resolution in the magnitude type exchange spectrum. If the ^{13}C spectral resolution is high enough, one can still separate most of carbon peaks to see their exchange intensities.

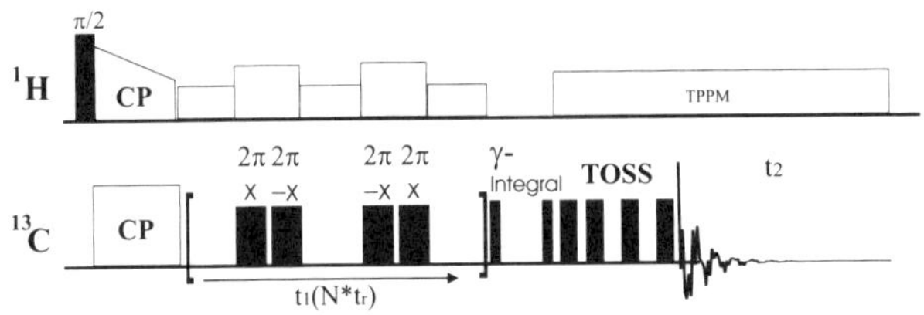

Fig. 12.3. The pulse sequence for SUPER experiment. The increment of t_1 is one rotation period t_r. The intervals of the z-filters during the γ-integral are governed by $t_z = kt_r/K$. $K = 4$ is used to make sidebands invisible.

12.3.3.4. *SUPER experiment*

As seen in Eq. (12.12), the anisotropic part of chemical shift of a particular carbon atom in a mesophase is governed by the order parameters S and its CSA tensor. If one wants to determine S from several carbon sites, their CSA tensors and orientations in a molecular (or fragment) frame must first be known. 2D SUPER experiment under MAS on a powder sample has employed a pulse sequence shown in Fig. 12.3.[38] After a ramp-CP, four ^{13}C 2π pulses are used in one t_r period to recouple CS interactions. To avoid the signal being dephased by strong heteronuclear couplings during the recouping 2π pulses, a strong CW proton decoupling field is applied. Acquisition is carried out under TPPM decoupling. The spinning sidebands are removed by the γ integral[48] and total suppression of spinning sidebands (TOSS),[49] since TOSS alone can not remove the spinning sidebands completely in 2D experiments.[20] By fitting some appropriate number of 2π pulses into one t_r, the CSA can be preserved under MAS but scaled by a factor. Detailed calculations in Ref. [38] give $t_a = 0.25t_r$, $t_b = 0.18t_r$ and a scaling factor of 0.155 for the anisotropic part of the CSA tensors. Moreover, the δ_{iso} values must be re-scaled by a factor of 0.667. The power level of 2π pulse is determined by the rotor's speed ω_r, i.e. $\omega_{rf} = 12.12\omega_r$. A SUPER spectrum gives CSA powder patterns in the f_1 dimension, and isotropic carbon chemical shifts in the f_2 dimension. Notice that the CSA powder patterns do show the correct principal values once δ_{iso}'s have been properly scaled and shifted to their correct isotropic values.[38]

12.4. Chemical Shifts in Uniaxial and Biaxial Phases

The time-averaged CS Hamiltonian in the high-field limit is given by Eq. (12.2) as:

$$\overline{H_{CS}} = \gamma\hbar B I_z \delta_{iso} + \sqrt{\frac{2}{3}}\gamma\hbar B I_z \langle R_{2,0}\rangle \tag{12.20}$$

where the bar over the Hamiltonian denotes a time average over internal bond rotations and overall motions of the molecule, $\langle R_{2,0}\rangle$ is the corresponding time-averaged second rank irreducible spherical tensor in the L frame. To describe internal motions of the fragment and overall motions of the molecule, the following coordinate transformations need to be considered, as illustrated in Fig. 12.4:

$$\begin{aligned}
\text{L frame } (\vec{B}) &\overset{(\phi_0,\theta_0,\psi_0)}{\Longrightarrow} \text{Director frame } (\vec{n})\\
\text{Director frame } (\vec{n}) &\overset{(\phi,\theta,\psi)}{\Longrightarrow} \text{M frame } (x,y,z)\\
\text{M frame } (x,y,z) &\overset{(\alpha_F,\beta_F,\gamma_F)}{\Longrightarrow} \text{Fragment frame } (x',y',z')\\
\text{Fragment frame } (x',y',z') &\overset{(0,\beta_P,\frac{\pi}{2})}{\Longrightarrow} \text{PAS frame } (\delta_{11},\delta_{22},\delta_{33}).
\end{aligned}$$

The transformation from the fragment frame (x',y',z') to PAS frame $(\delta_{11}, \delta_{22}, \delta_{33})$ involves $\beta_P = 60°, 120°$ for protonated carbons, and $\beta_P = 0°$ for quaternary carbons, while $(\alpha_F, \beta_F, \gamma_F)$ are Euler angles to transform between the M and fragment frames. Accordingly, $\langle R_{2,0}\rangle$ is given by[50]

$$\langle R_{2,0}\rangle = \sum_{qpn} D_{q0}(\phi_0,\theta_0,\psi_0)\left\langle D^*_{qp}(\phi,\theta,\psi)D^*_{pn}(\alpha_F,\beta_F,\gamma_F)\right\rangle R'_{2,n}$$

where the brackets denote the ensemble averaging, and the primed tensor $R'_{2,n}$ in the fragment frame (x',y',z') is given by

$$\begin{aligned}
R'_{2,0} &= \sqrt{\frac{2}{3}}[P_2(\cos\beta_P)(\delta_{33}-\delta_{22}) + \frac{1}{2}(\delta_{22}-\delta_{11})]\\
R'_{2,\pm1} &= \mp\frac{1}{2}\sin 2\beta_P(\delta_{33}-\delta_{22})\\
R'_{2,\pm2} &= \frac{1}{2}[(\sin^2\beta_P)\delta_{33} + (\cos^2\beta_P)\delta_{22} - \delta_{11}]
\end{aligned} \tag{12.21}$$

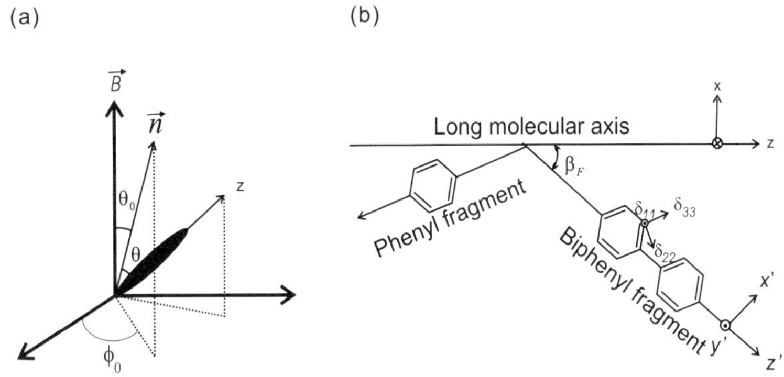

Fig. 12.4. (a) Schematic representation of the orientation of a molecule with respect to the director $\hat{\mathbf{n}}$ in the magnetic field ($\vec{B}$) which defines the laboratory frame, (b) A schematic of the PAS frame of a carbon CSA tensor (δ_{11},δ_{22},δ_{33}), a fragment coordinate system (x',y',z'), and a molecular frame (x,y,z) fixed on the molecule.

Now fast 180° ring flips in the phenyl or biphenyl groups have averaged $R'_{2,\pm1}$ to zero ($\beta_P \leftrightarrow -\beta_P$). To simplify the calculation, one may choose the fragment and molecular frames such that $\alpha_F = 90°$ and $\gamma_F = 0°$ to obtain $\langle R_{2,0}\rangle_q$ for $q = 0$:

$$\langle R_{2,0}\rangle_0 = P_2(\cos\theta_0) \times \left\{ \left[P_2(\cos\beta_F)S_{00} + \frac{3}{4}\sin(2\beta_F)S_{01} - \frac{3}{4}\sin^2\beta_F S_{02} \right] R'_{2,0} + \left[\sqrt{\frac{3}{2}}\sin^2\beta_F S_{00} - \sqrt{\frac{3}{8}}\sin(2\beta_F)S_{01} - \sqrt{\frac{3}{8}}(1+\cos^2\beta_F)S_{02} \right] R'_{2,2} \right\}$$

where molecular order parameters $S_{00} = S, S_{01} = \frac{4}{3}S_{yz}$, and $S_{02} = \frac{2}{3}D$. Similarly, by defining six phase biaxial order parameters:[51]

$$\begin{aligned} S_{10} &= \langle \sin 2\theta \sin\phi \rangle \\ S_{11} &= \langle \cos\theta\cos\phi\cos\psi - \cos 2\theta \sin\phi \sin\psi \rangle \\ S_{12} &= \left\langle \sin\theta\cos\phi\sin 2\psi + \tfrac{1}{2}\sin 2\theta \sin\phi\cos 2\psi \right\rangle \\ S_{20} &= \left\langle \sin^2\theta\cos(2\phi) \right\rangle = \tfrac{2}{3}P \\ S_{21} &= \left\langle \sin\theta\sin 2\phi\cos\psi + \tfrac{1}{2}\sin 2\theta\cos 2\phi\sin\psi \right\rangle \\ S_{22} &= \tfrac{1}{2}\left\langle (1+\cos^2\theta)\cos 2\phi\cos 2\psi - 2\cos\theta\sin 2\phi\sin 2\psi \right\rangle = \tfrac{1}{3}C \end{aligned} \tag{12.22}$$

one can get $\langle R_{2,0}\rangle_q$ for $q = \pm 1$ and ± 2. Finally the observed chemical shift of a particular carbon in a mesophase can therefore be calculated:

$$\begin{aligned}\langle\delta\rangle &= \delta_{iso} + \sqrt{\frac{2}{3}}\langle R_{2,0}\rangle \\ &= \delta_{iso} + a\left(\sqrt{\frac{2}{3}}P_2(\cos\beta_F)R'_{2,0} + (\sin^2\beta_F)R'_{2,2}\right) \\ &\quad + b\left(\sqrt{\frac{3}{8}}(\sin 2\beta_F)R'_{2,0} - \frac{1}{2}(\sin 2\beta_F)R'_{2,2}\right) \\ &\quad + c\left(\sqrt{\frac{3}{8}}(\sin^2\beta_F)R'_{2,0} + \frac{1}{2}(1+\cos^2\beta_F)R'_{2,2}\right) \qquad (12.23)\end{aligned}$$

where

$$\begin{aligned}a &= P_2(\cos\theta_0)S_{00} + \frac{3}{4}\sin 2\theta_0 \sin\phi_0 S_{10} + \frac{3}{4}\sin^2\theta_0\cos 2\phi_0 S_{20} \\ b &= P_2(\cos\theta_0)S_{01} + \sin 2\theta_0\sin\phi_0 S_{11} - \sin^2\theta_0\cos 2\phi_0 S_{21} \\ c &= -P_2(\cos\theta_0)S_{02} + \sin 2\theta_0\sin\phi_0 S_{12} - \sin^2\theta_0\cos 2\phi_0 S_{22} \qquad (12.24)\end{aligned}$$

are coefficients involving the nine generalized order parameters. Now, the phase biaxial order parameters appear always multiplied by either $\sin 2\theta_0$ or $\sin\theta_0$ in the coefficients a, b, c of Eq. (12.24). Thus, they become unobservable when θ_0=0, which is normally the case for an aligned ($\Delta\chi > 0$) sample with $\vec{n}$ lying along the external B field. As a consequence, the phase biaxiality is usually not needed in interpreting ^{13}C $< \delta >$'s in aligned mesophases. For a uniaxial phase, Eq. (12.23) reduces to:

$$\begin{aligned}\langle\delta\rangle &= \delta_{iso} + \sqrt{\frac{2}{3}}P_2(\cos\theta_0) \\ &\quad\times\left\{\left[P_2(\cos\beta_F)S + \frac{3}{4}\sin(2\beta_F)S_{01} - \frac{1}{2}\sin^2\beta_F D\right]R'_{2,0}\right. \\ &\quad+\left.\left[\sqrt{\frac{3}{2}}\sin^2\beta_F S - \sqrt{\frac{3}{8}}\sin(2\beta_F)S_{01} - \sqrt{\frac{1}{6}}(1+\cos^2\beta_F)D\right]R'_{2,2}\right\}. \qquad (12.25)\end{aligned}$$

Note that the observed chemical shift $< \delta >$ is still scaled by the second Legendre $P_2(\cos\theta_0)$ if the director is not along the B field. Since there are ample evidence of vanishing phase biaxiality in chiral SmC* phases from deuterium NMR studies, the above equation is also suitable for ^{13}C studies of SmC* phases. In this case, θ_0 corresponds to the tilt angle (θ') of the molecules. As the long molecular axis becomes hindered in the SmC* phase, S_{01} becomes important in interpreting the ^{13}C $< \delta >$'s. This implies that different segments of the molecule do not in general share a common principal axis system as found by optical[52] and x-ray measurements.[53] In a N_B phase, some experimental evidence of similar temperature behaviors of the molecular biaxial order D and phase biaxial order C exists.[54]

If molecules have fast rotations about their long axes, such as in uniaxial N and SmA phases, S_{01} is averaged to zero, and for the director aligned along the magnetic field ($\theta_0 = 0$), Eq. (12.25) then becomes

$$\begin{aligned}\langle\delta\rangle &= \delta_{iso} + \left(\sqrt{\frac{2}{3}}P_2(\cos\beta_F)R'_{2,0} + (\sin^2\beta_F)R'_{2,2}\right)S \\ &- \left(\sqrt{\frac{1}{6}}(\sin^2\beta_F)R'_{2,0} + \frac{1}{3}(1+\cos^2\beta_F)R'_{2,2}\right)D. \qquad (12.26)\end{aligned}$$

By setting β_F to 0, the segmental order parameters S' (or S'_{00}) and D' (or S'_{02}) can be obtained from

$$\begin{aligned}\langle\delta\rangle &= \delta_{iso} + \frac{2}{3}S'[P_2(\cos\beta_P)(\delta_{33}-\delta_{22}) + \frac{1}{2}(\delta_{22}-\delta_{11})] \\ &+ \frac{1}{3}D'[\delta_{11} - \delta_{22}\cos^2\beta_P - \delta_{33}\sin^2\beta_P]. \qquad (12.27)\end{aligned}$$

Thus we can get the local molecular order parameters S' and D' for the biphenyl fragment and the phenyl ring by fitting the observed ^{13}C δ's from these fragments, respectively. By comparing Eqs. (12.26) and (12.27), the relation[55]

$$S' \approx S\ P_2(\cos\beta_F) \qquad (12.28)$$

is found, which allows the determination of the β_F angle. The last two equations will be used in the next section.

12.5. Bent-core Mesogens

Chiral properties of tilted smectic phases have attracted much attention in recent years. The breaking of the mirror symmetry results from constituent molecules that either possess asymmetric carbon(s) (chiral center(s))[56] or are achiral.[57] Bent-core mesogens do not have cylindrical symmetry and require two unit vectors (the directors $\vec{n}$ and $\vec{m}$) for their description. While the $\vec{n}$ is along the long axis of the molecule, the $\vec{m}$ is normal to the average molecular plane. Each layer is polar because the efficient packing of the molecules produces an electric polarization pointing along the unit vector $\vec{b}$, that is normal to both $\vec{n}$ and $\vec{m}$ when the achiral molecule has a C_{2v} symmetry. When molecules tilt away from the layer normal $\vec{k}$, the tilt angle θ' is given by $\sin\theta' = \vec{n} \times \vec{k} = |\vec{\epsilon}|$. The $\vec{b}$ vector forms an angle ψ' with the $\vec{\epsilon}$ vector. All the tilted banana phases can be specified by the θ' and ψ' angles. The tilt direction of the molecular planes and the polar order $(\vec{b})$ axis provide two independent symmetry breaking elements.[58] Electro-optical switching in B_2 and B_5 phases are possible.[59] In this section, some bent-core mesogens listed in Fig. 12.5 are studied in either a N or SmA phase. The nPBBC homologue series (see Fig. 12.5(a)) has been investigated[60] initially by looking at the overall nematic order S, while ignoring molecular biaxiality D in interpreting the observed ^{13}C $< \delta >$'s in the N phase. More recently, the local order parameters for various fragments are found to be more useful in giving information on twist configurations.[61] The observed differences between ^{13}C $< \delta >$'s from the I phase and the

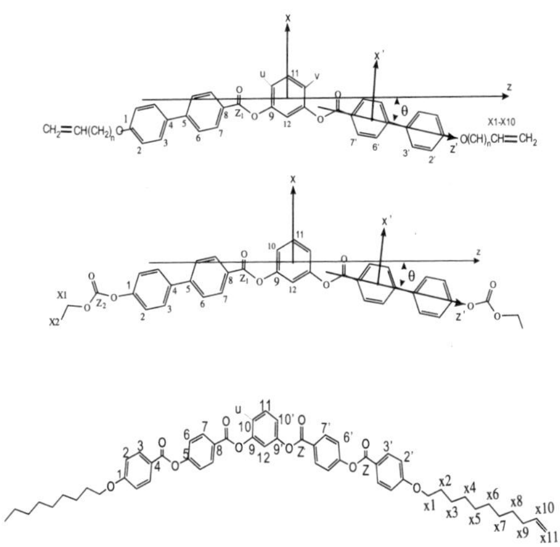

Fig. 12.5. Molecular structures of some bent-core LC (a)10DClPBBC (n=10, u=v=Cl), and 11ClPBBC (n=12, u=Cl, v=H), (b) PEBC, c) ClPbis11BB (u=Cl), and Pbis11BB (u=H). See text for full chemical names. Note β_F in Fig. 12.4 is relabeled here as θ.

corresponding ones in the mesophase depend critically on the correct peak assignments in these phases. ^{13}C assignments of high-resolution solution state spectra can be done based on ^{13}C DEPT, HETCOR and ^{1}H spectra. The δ^i_{iso} values for various carbon peaks of a particular compound can be predicted by commercial softwares, in vacuo density functional theory

Table 12.2. CSA tensors for some typical carbon sites in a solid sample of Pbis11BB.

Sites	1	3,7	4	5	6	11
δ_{11}	70±5	13±3	15±5	72±3	7±3	5±5
δ_{22}	177±3	151±3	136±3	139±3	143±3	153±3
δ_{33}	247±3	232±3	208±5	253±3	214±3	231±5

(DFT) calculations, and/or those determined from a 2D SUPER spectrum. In the latter case, the CSA powder lineshapes may be used to identify different types of carbon environment. It is also important to have CSA tensors specific to the LC compound in question when analyzing its $< \delta >$ data. However, the orientation of these CSA tensors in a M frame can not be obtained from SUPER, but GIAO-DFT calculations can provide such information.[62] We have employed SUPER to measure CSA tensors in several bent-core mesogens in the past.[63–65] To illustrate this, a particular example is now given.

12.5.1. *Measurement of CSA tensors*

The principal values of CSA tensors for 1,3-Phenylene-bis 4-[4-(10-undecenyloxy-benzoyloxy] benzoate (Pbis11BB) have been measured[65] using the SUPER method. Figure 12.6 shows the a typical 2D SUPER spectrum of Pbis11BB at 293 K. It can be seen from Fig. 12.7 that CSA powder patterns of protonated carbons are more broad than those of quaternary carbons due to their residual C-H dipolar couplings. These power patterns are, however, in good agreement with typical patterns observed in other compounds,[38,66] and their isotropic shifts can also be used to confirm the ^{13}C peak assignments in the solution spectrum. The CSA tensors should be similar for carbon sites having similar neighbouring atoms, such as carbon 5 and 9. For those carbons exhibiting doublets (reasons given later), the CSA powder patterns are identical for the two peaks in the doublets within the experimental uncertainty (each principal component is accurate to within a few ppm). The simulated powder patterns are also shown in Fig. 12.7 for direct comparison with the observed ones, and their CSA principal values are listed in Table 12.2. When CSA tensors of ^{13}C peaks in the studied sample are unavailable, one may use those from the same carbon types of similar or related LC compounds, but caution is needed in the derived order parameters from fitting ^{13}C $< \delta >$ data.

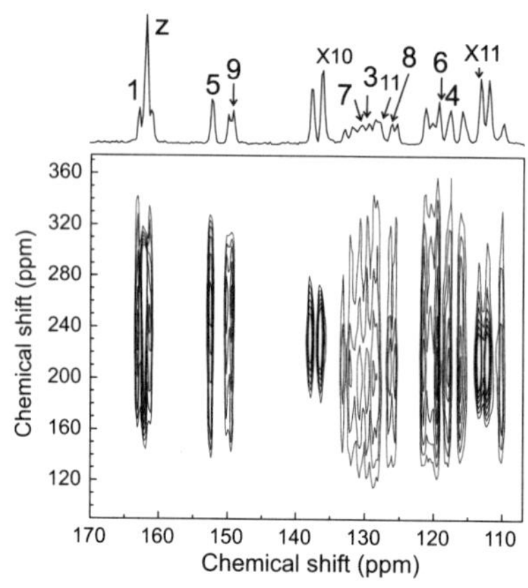

Fig. 12.6. 2D SUPER spectrum of Pbis11BB together with a 1D spectrum from a projection onto the f_2 dimension. Spinning rate of 6.5 kHz and 64 t_1 increment were used. The spectrum has been tilted in f_1 dimension. Only aromatic carbons are shown here. The labels for the ^{13}C slices are those shown in Fig. 12.5(c).

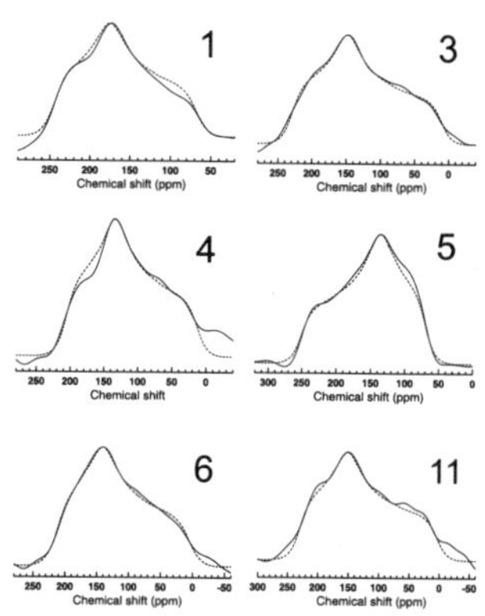

Fig. 12.7. Plots of simulated (dashed lines) and observed (solid lines) CSA powder patterns for some carbon sites obtained from Fig. 12.6. The chemical shifts have been shifted to the correct position based on their isotropic chemical shift values.

12.5.2. *10DClPBBC and 11ClPBBC*

Typical spectra in the I and N phases of 4,6-dichloro-1,3-phenylene bis[4'-(9-decenyloxy)-1,1'-biphenyl] carboxylate (10DClPBBC) are shown in Fig. 12.8(a) together with its phase sequence. The temperature dependent $< \delta >$ for the aromatic region and X9 and X10 in the N phase of 10DClPBBC[63] are shown in Fig. 12.9(a). At the I-N transition, ^{13}C

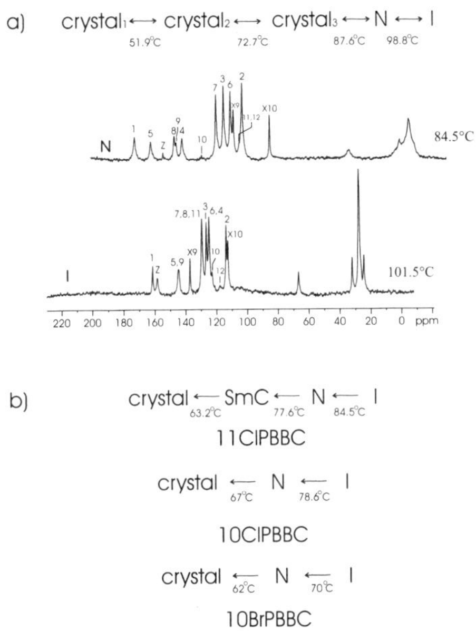

Fig. 12.8. (a) Typical ^{13}C NMR spectra taken in the I and N phases of 10DClPBBC, together with its phase sequence. The carbon labels are those indicated in Fig. 12.5, (b) Schematic of the phase sequence and transition temperatures in other members of nPBBC homologue series.

Table 12.3. CSA tensors for various carbon sites of 10DClPBBC (carbon labels same for both lateral wings) determined from SUPER with an uncertainty in each principal value of ca. $\pm 3ppm$.

Sites	6	7	9	10	12	z
δ_{11}	34.4	28.3	72.2	44.6	30	111
δ_{22}	140.7	145.7	145.2	109	149.3	133
δ_{33}	204.6	212.7	217.2	213.6	183	242

$<\delta>$'s of the aromatic carbons shift markedly downfield but those of the aliphatic carbons shift slightly upfield. Because aliphatic carbon peaks are often overlapped (as in this case except $X9$ and $X10$), the aromatic carbons are normally investigated in detail to extract both the orientational ordering of the bent-core and its 'average' molecular structure.

The ^{13}C $<\delta>$ in a N phase is related to its isotropic δ_{iso} value by Eq. (12.27). The CSA values measured by SUPER in 10DClPBBC listed

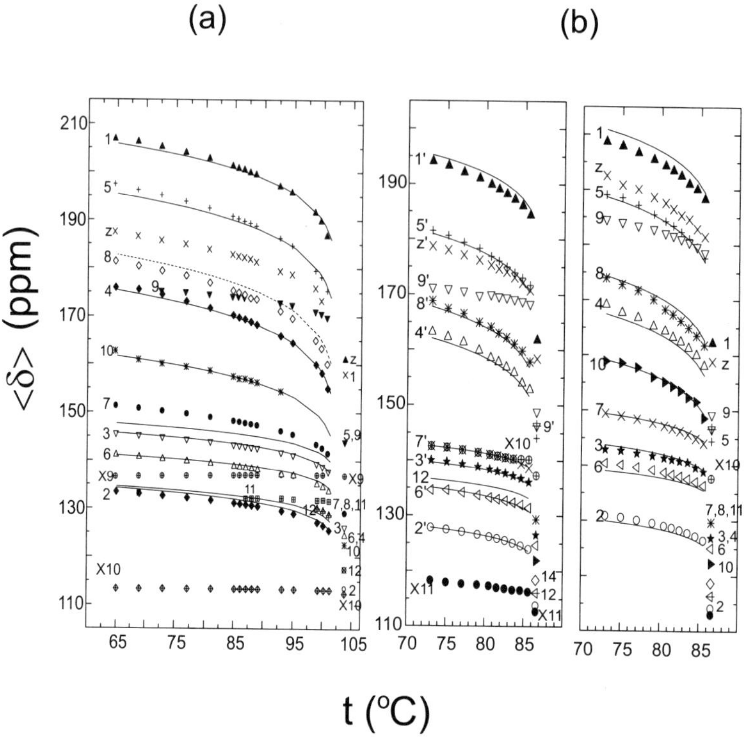

Fig. 12.9. Plot of chemical shifts for the aromatic carbons and aliphatic carbons X9 and X10 in the isotropic and nematic phases of 10DClPBBC (a) and of 11ClPBBC (b). The carbon labels are those indicated in Fig.12.5. Solid curves denote calculated chemical shifts after a global minimization.

in Table 12.3 and from 10ClPBBC listed[69] in Table 12.4 are used to fit the observed $< \delta >$'s in the N phase. Note the CSA tensors of 10ClPBBC reported in Ref. [69] have unfortunately missed a scaling of 0.667 for δ_{iso}, which has been corrected[61] in Table 12.4. Besides, this sample was found not to align by cooling in a B field of 9.4 T from the I phase, while the homologue 10BrPBBC could only be weakly aligned very close to the I-N transition. Now, the CSA tensor's PAS (1,2,3) is chosen by assuming that its 3 axis is along a C-H bond, or C-C bond (along the para axis), with the 1 axis normal to a ring plane. The assumption can be checked with those derived from DFT calculations.[62] In using Eq. (12.27), $\beta_P = 0$ for quaternary carbons (carbon 1, 4, 5, 8) and $\beta_P = 60°$ for protoned carbons (carbon 2, 3, 6, 7) on the biphenyl groups are used, while $\beta_P = 90°$ for carbon 12 and $\beta_P = 150°$ for carbon 10 on the central ring. All D' values

Table 12.4. CSA tensors for carbons obtained from SUPER in 10ClPBBC except C2[a].

Sites	1	3	4	5	7
δ_{11}	50.5	9.6	27.3	58.9	11.3
δ_{22}	161.2	146.8	127.4	129.9	154.5
δ_{33}	257.2	216.5	223.7	249.3	219.7
Sites	8	9	11	13	14
δ_{11}	8.6	44.6	29.3	44.6	30.7
δ_{22}	144.9	160.1	130.7	160.1	124.2
δ_{33}	229.0	261.8	201.7	261.8	189.6

a: Derived from Ref. [69] using $\delta_{ii}^{new} = (\delta_{ii} - \delta_{iso})/0.667 + \delta_{iso}$; site C2 (8.7, 135.5, 192.9) from Ref. [67].

were found to depend linearly on temperature. To globally analyze the $< \delta >$'s,[61] one can use the Haller equation for S', viz. $S'(T) = S_0(1 - T/T^*)^f$ where S_0 and f are empirical constants, and T^* ($T^* = T_c + 1\text{K}$) is the temperature slightly above T_c at which S' vanishes. Also a linear equation is used to describe the temperature dependence of D', i.e. $D' = \Delta(T - T_c)$ where Δ is a fitting constant. S_0, f and Δ were optimized to give the best calculated $< \delta >$'s for the fitted carbon sites shown as solid lines in Fig. 12.9(a). The local order S' values of the central ring and biphenyl groups obtained from the fittings are shown in Fig. 12.10(a). These results imply that the local z axis located on the central ring represents a more ordered axis, i.e. the long molecular axis that reflects the overall molecular ordering. The z' axis along the para axes in the biphenyl group is less ordered. Now one can relate the ordering of these local z axes using Eq. (12.28), where $\beta_F (= \theta)$ is the angle between the para axes of the biphenyl group and the more ordered z axis of the center ring. Using this equation, the θ angle is found to vary with temperature, increasing from 26° to 29° upon decreasing temperature. Hence, the bend angle Θ subtended by the two lateral arms in 10DClPBBC is also temperature dependent as shown in Fig. 12.11, range between 128° to 122° at the low temperature end. It is noted that among all the phenyl rings, the D' for the inner ring in the biphenyl group and the center ring have similar magnitude, but their signs are opposite. Thus these two phenyl ring planes oriented differently with respect to $\vec{n}$, implying a definite twist conformation in the molecular core. Also a slight twist of the phenyl ring planes in the biphenyl fragment can be inferred from Fig. 12.10(a). 2D SLF experiment was performed in 10DClPBBC at 78°C to observe the dipolar oscillations due to C-H dipolar couplings.[63] The observed dipolar splittings were used to get the coupling constants D_{ij}^{aniso}, from which the local order parameters could also be calculated. The derived S' is found to be consistent with the one found above from fitting the ^{13}C $< \delta >$ data.

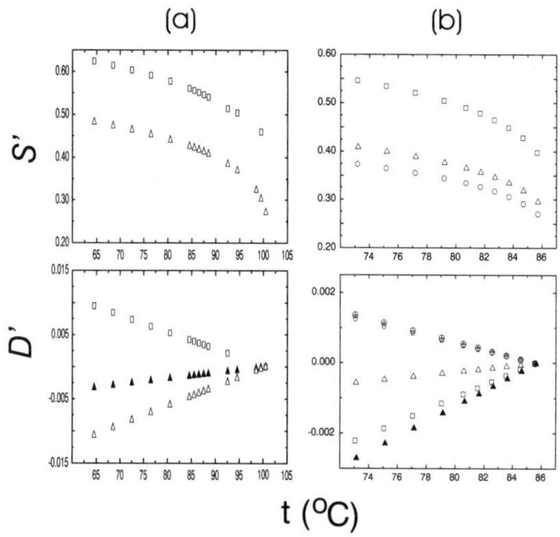

Fig. 12.10. (a) Plot of the local order parameter S' and biaxial order parameter D' versus temperature for the biphenyl group (triangles) and central ring (squares) (in the biphenyl group, D' for the ring (triangles) closer to the center ring and for the outer ring (full triangles)) in the N phase of 10DClPBBC, (b) Plot of S' for the biphenyl group (triangles for 'unprimed' fragment, circles for 'primed' fragment) and central ring (squares) and D' vs temperature, in the N phase of 11ClPBBC (same symbols as for S' with solid and open symbols denoting phenyl ring close to the center ring, and for the outer ring in the biphenyl group, respectively).

The same approach is applied to analyze the $< \delta >$ data observed in the N phase of 4,6-chloro-1,3-phenylene bis[4'-(10-undecenyloxy)-1,1'-biphenyl] carboxylate (11ClPBBC)[60] shown in Fig. 12.9(b). The I-N transition temperature is shown in Fig. 12.8(b). It is noted that 11ClPBBC has a mono-substituted chlorine on the center ring, making the two biphenyl fragments inequivalent so as to generate two sub-spectra of aromatic carbon peaks. The purpose is to see the effects, if any, of the di-substituted and mono-

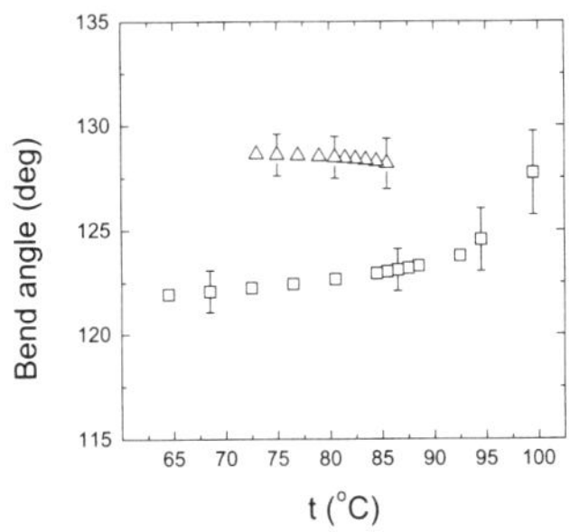

Fig. 12.11. Plot of the bend angle Θ in the N phase of 10DClPBBC (squares) and 11ClPBBC (triangles).

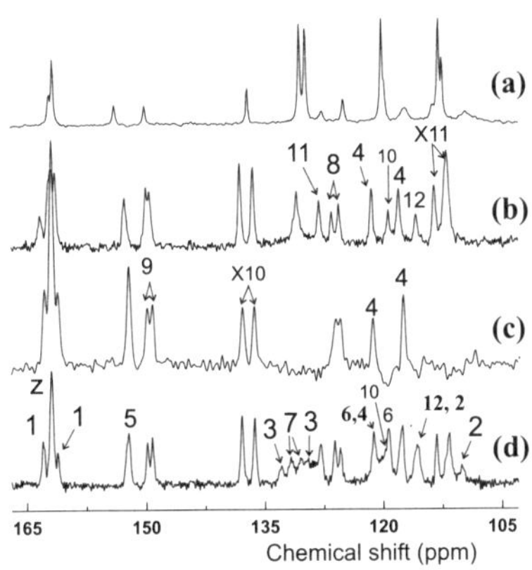

Fig. 12.12. ^{13}C NMR spectrum of Pbis11BB in the isotropic phase (a), CP-MAS ^{13}C NMR spectrum at 367.8K (b) and 310.4K (d) (MAS speed 5kHz was used for both cases), (c) CPPI ^{13}C NMR spectrum recorded at 367.8K with MAS.

substituted PBBC on the overall molecular ordering and packing in their N phases. The CSA tensors listed in Tables 12.3 and 12.4 are again used here to fit the $< \delta >$'s from carbons 10, 12 of the center ring, and carbons 1-8 of the 'unprimed' and 'primed' biphenyl fragments to yield their calculated $< \delta >$'s shown as solid curves in Fig. 12.9(b). The fits are very good. The derived S' and D' values of center ring and two biphenyl fragments are plotted versus temperature in Fig. 12.10(b). Again the z axis of the center ring is the most ordered axis. Furthermore, the two para axes of the biphenyl fragments are ordered more or less the same as if one has a dichloro-substituted compound. The θ angles for the two biphenyl lateral wings are very close, with decreasing temperature from 24.4° to 24°, and from 27.5° to 27.3° for the umprimed and primed lateral wings. The bend angle Θ is ca. 128.5° and relatively insensitive to temperature (see Fig. 12.11). This angle is larger than the Θ of 10DClPBBC. It is very interesting to examine the behaviors of local molecular biaxial order parameters shown in Fig. 12.10. There is a sign change in D' of the center ring between this compound and 10DClPBBC, implying different ordering of its x' and y' axes w.r.t. $\vec{n}$ for the two cases. For 11ClPBBC, D' for the two biphenyl fragments differ in sign. This is likely a result of the mono-substitution in the center wing, causing the two lateral wings to be distinct from each other. The phenyl ring of the unprimed wing closer to the center ring appears to behave with D' similar to that of the center ring.

In summary, a twisted conformation within the bent-core can be inferred from the local molecular biaxiality D' and the details in the orientations of ring planes depend on the substitution at the center ring. Moreover, the effect of substitution on the center ring and the chain length have a profound effect on the molecular alignment of the bent-core molecules in the mesophase. In this connection, the second molecule phenylenebis4-[(ethoxycarbonyl)oxy]-1-1-biphenyl-4-carboxylate (PEBC) in Fig. 12.5 has the same bent-core structure but with shorter and different side chains as PBBC. It shows an interdigitated SmA (SmA_{int}) phase[64] by cooling from above the I-SmA_{int} phase transition at 174.1°C. The $< \delta >$ of various aromatic carbons are almost temperature insensitive. The CSA tensors for carbons 1, 3, 7, 5, 9 were determined by SUPER. Based on these limited CSA tensors, the $< \delta >$'s of these carbons were analysed as above. Assuming the center ring having a vanishing D', its S' is 0.9. The biphenyl fragments have $S' = 0.5$. The Θ between the two lateral wings has been estimated as 112.6°.

12.5.3. *Pbis11BB*

A twisted bent-core molecular shape has been inferred by ^{13}C CP-MAS spectra in ester bent-core molecules which can form B_2 phases.[70,71] The bent-core mesogen 1,3-Phenylene-bis-4-[4-(10-undecenyloxy)-benzoyloxy] benzoate (Pbis11BB) (see Fig. 12.5) has a B_2 phase instead of a N phase upon cooling from the isotropic melt. This particular phase does not align in a 9.4 T magnet, thus preventing us in studying its orientational ordering. However, it has been studied under MAS to give some important insight on the molecular structure and dynamics in the B_2 and crystalline phases.[37] Fig. 12.12 shows some typical MAS spectra collected under dif-

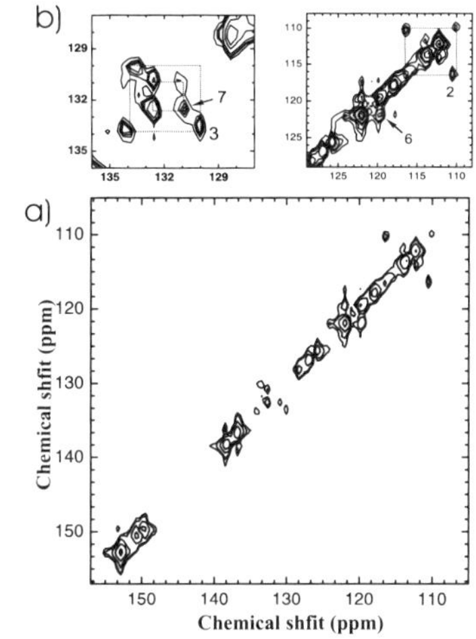

Fig. 12.13. (a) Contour plot the ^{13}C 2D exchange spectrum recorded at 293 K and MAS speed of 8kHz. Mixing time is 0.1s, and the ramp-CP contact time is $2ms$, recycle delay $5s$ and 256 t_1 increment were used in this experiment. 56kHz proton decoupling was used in t_1 and t_2 intervals, (b) Expanded regions near the carbon 3,7 and carbon 2, respectively.

ferent conditions. CPPI has been used to distinguish between the protonated and quaternary carbons (see Fig. 12.12(c)). It is also clear that some carbon peaks (either protonated or quaternary) in the I phase turn into doublets in the B_2 phase (see Fig. 12.12(b), and d). It is important to realize that the doubling in the carbon peaks is not a result of the two lateral wings being inequivalent, as in the case of mono-substitution at the center ring (see ClPbis11BB below). To understand the occurrence of these doublets, ^{13}C MAS-exchange spectrum at 293 K was collected as depicted in Fig. 12.13. There are four pairs of cross-peaks, each for the carbons 2, 3, 6 and 7. The doublet (eg. 2, 2') is due to a slight difference in the chemical environment of the ortho carbons in the phenyl ring when its motion is frozen in the solid state. The exchange intensities in the cross peaks imply slow ring flips (estimated < 100 Hz) at 293 K. Now the significant chemical shift differences (>3 ppm) between 3 and 3' as well as 6 and 6' suggest that the two lateral rings must be in a *trans-trans* configuration (though their ring planes may not be coplanar). The rather small difference between the chemical shifts of 7 and 7' would imply the phenyl ring(s) close to the center prefers a *cis-trans* configuration with the center ring.[37] This provides a twisted conformation in the 5-ring bent-core in the solid state.

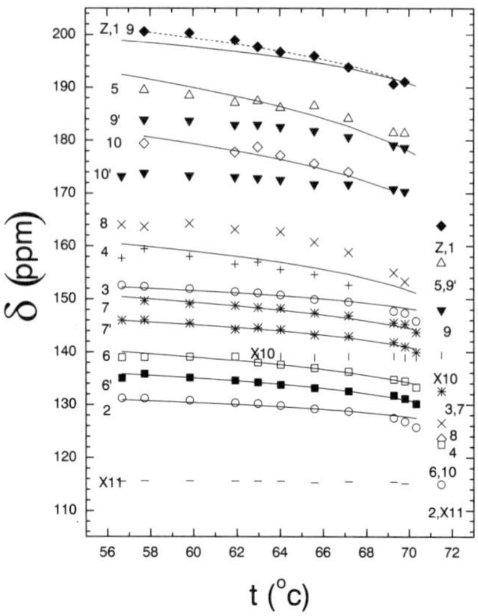

Fig. 12.14. Plots of $<\delta>$'s for the aromatic carbons and aliphatic carbons X10 and X11 in the I and N phases of ClPbis11BB. The carbon labels are those indicated in Fig.12.5. Solid curves denote calculated δ_i with the dashed line for carbon 9.

12.5.4. *ClPbis11BB*

The bent-core molecule 4-chloro-1,3-Phenylene-bis-4-[4-(10-undecenyloxy)-benzoyloxy] benzoate (ClPbis11BB) was the first one studied in our laboratory using ^{13}C NMR and this study has pre-dated our determination of CSA tensors by SUPER. In particular, the lack of CSA tensors for carbons of the center ring in the literature has hampered an in-depth analysis of the observed $<\delta>$'s in the N phase of ClPbis11BB shown in Fig. 12.14.[72] Since some CSA tensors have now been determined in Pbis11BB (see Table 12.2), this allows a re-analysis of the $<\delta>$ data in the manner described above. We have used carbons 9 and 10 in the center ring to investigate the most ordered axis of this molecule using the CSA principal values in Table 12.3. The ^{13}C spectra show a difference between the two phenyl rings closer to the center ring, giving doublings of carbon peaks (6, 6') and (7, 7'). As far as our spectra can tell, the outer rings of the two lateral wings are equivalent (i.e. no doublings were seen for carbon 2 or 3). The $<\delta>$'s of carbons 1 to 7 as well as 6' and 7' are analyzed globally using the CSA tensors in Table 12.2, except for carbon 2 ($\delta_{11} = 23, \delta_{22} = 134, \delta_{33} = 195$ ppm found[62] for a rod-like mesogen ZLL 7/*). The calculated $<\delta>$'s are also plotted in Fig. 12.14. The fits are quite satisfactory and the resulting local order parameters versus temperature are shown in Fig. 12.15. To give S', S_0 and f in the Haller equation are (1.065, 0.217) for the center ring, and for the 'primed' and 'unprimed' phenyl rings next to the center ring (0.838, 0.215) and (0.935, 0.336), respectively, while for the most outer ring (0.635, 0.179). Again the z axis of the center ring is most ordered, followed by z' axis of the inner ring of the 'primed' lateral wing. The θ angle between these two para axes is 22.9° and the temperature dependences of their S' are in total agreement with those found from deuterium NMR in deuterated isotopomers.[72] The para axis of the phenyl ring close to the chlorine atom is the least order axis, giving a $\theta = 33.3°$. The Θ angle subtended by the two lateral wings is estimated as 123°. From the D' in Fig. 12.15, one observes that the ring planes of the center ring and the two 'unprimed' rings are similar in orientations but differ from that of the 'primed' ring. To end this section on bent-core mesogens, it seems that the notion of an 'average' twisted conformation for the aromatic core due to different orientations of the phenyl ring planes is a prevalent picture for this class of LC.

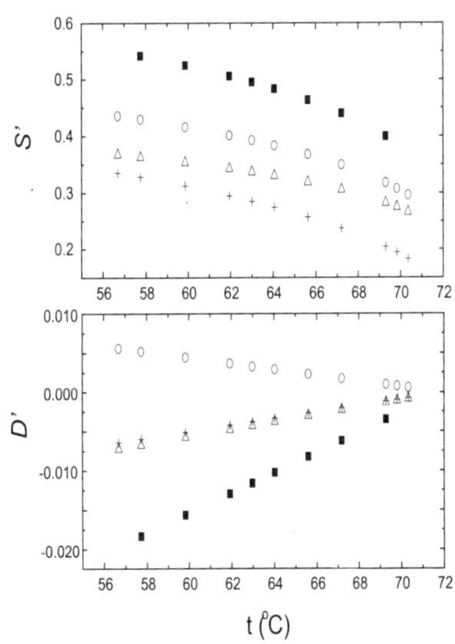

Fig. 12.15. (a) Plot of the local order parameters S' vs. temperature for the center ring (■), outer ring (△) and inner ring (+) of the 'unprimed' lateral wing, and the inner ring (○) of the primed lateral wing in the N phase of ClPbis11BB. (b) Plot of the local biaxial order parameter D' vs temperature using same symbols as in (a).

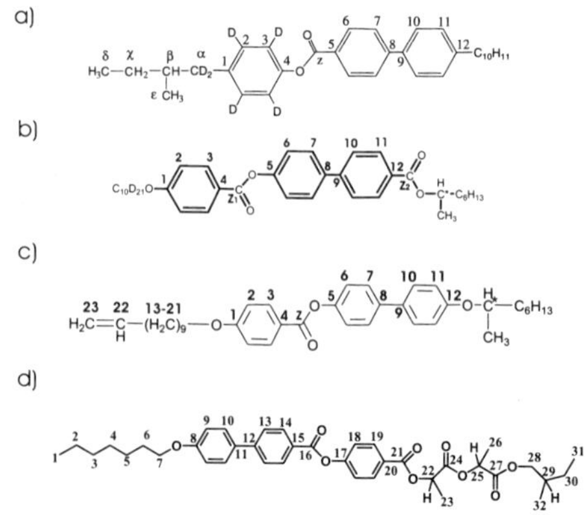

Fig. 12.16. Molecular formulae with carbon labels for (a) (S)[4-(2-methylbutyl)phenyl]-4'-octylbiphenyl carboxylate 8BEF5, (b) 1-methylhepty 4'-(4-n-decyloxybenzoyloxy) biphenyl-4-carboxylate 10B1M7, (c) 4-[4'-(1-methyl heptyloxy)] biphenyl 4-(10-undecenyloxy) benzoate 11EB1M7, and (d) (S)-2-methylbutyl-[4'-(4"-heptyloxyphenyl)-benzoyl-4-oxy-(S)-2-((S)-2')-benzoy-propionyl]-propionate ZLL 7/*.

12.6. Calamitic Mesogens

^{13}C studies of molecular ordering are surveyed in the SmA and SmC* phases of several chiral rodlike LC. A list of LC mesogens (10B1M7, 8BEF5, 11EB1M7, ZLL 7/*) is given in Fig. 12.16. The last mesogen in this list shows a rich and peculiar mesomorphic sequence including a re-entrant ferroelectric SmC^*_{re} phase. As an example, Fig. 12.17 shows some typical spectra from an aligned 8BEF5 sample in the I, SmA and SmC* phases together with peak assignments.

12.6.1. *Order parameters of 8BEF5*

The ^{13}C $< \delta >$ data of the biphenyl fragment in 8BEF5 were used[50] to derive the local order parameters (S', D') using Eq. (12.27) and CSA tensors from similar LC and model compounds found in the literature. The phenyl carbons could not be easily detected through CP due to deuteration of the phenyl ring in 8BEF5. Thus, the local order parameters of the phenyl ring are those from the deuteron data.[55] The S' of the two fragments are plotted in Fig. 12.18(a). As seen in this figure, the transition from SmA to SmC* phase is clearly delineated. Furthermore, the S' values of the phenyl ring are higher than those of the biphenyl fragment. The gradual decreasing S' for the phenyl and biphenyl fragments in the SmC* phase,

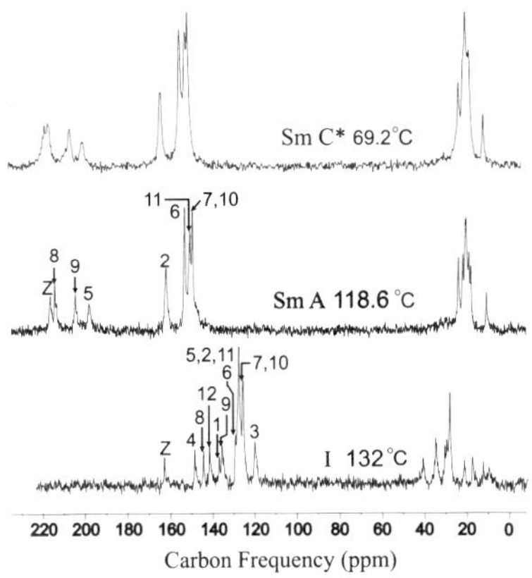

Fig. 12.17. Typical ^{13}C NMR spectra taken in the isotropic, smectic A and smectic C* phases of 8BEF5.

however, show different temperature behaviors. Thus, the behaviors of S' in the SmC* phase with decreasing temperature can not be just due to varying the tilt angle. A proper treatment of molecular order parameters in the SmC* phase is required and will be considered below. From the two

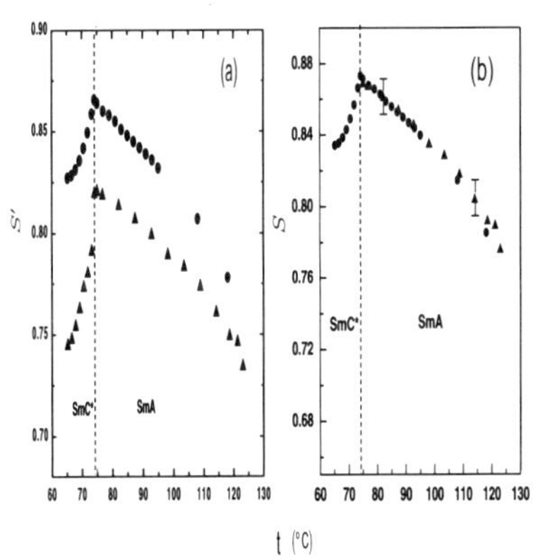

Fig. 12.18. (a) Plot of local order parameter S' for phenyl (•) and biphenyl (▲) fragments in the SmA and SmC* phases of 8BEF5, respectively, (b) Plot of the molecular order parameter S derived from phenyl (•) and biphenyl (▲) fragments as a function of temperature for 8BEF5.

local S' values of phenyl and biphenyl fragments at each temperature, one can calculate the angle between their para axes. A value of 15° from similar compounds in the literature[55] is, however, assumed for the angle between

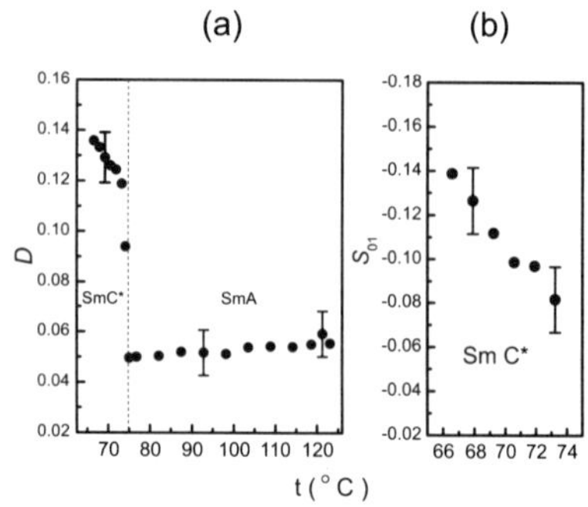

Fig. 12.19. (a) Plot of the molecular order parameter D as a function of temperature in SmA and SmC* phases of 8BEF5, (b) Plot of the molecular order parameter S_{01} in SmC* phase of 8BEF5.

the para axes in 8BEF5. Based on this information, one can then determine the location of the long molecular z axis for the core in terms of β_F^{biph} and β_F^{ph}. The average β_F^{ph} in 8BEF5 is found from Eq. (12.28) as 5.2°. The phenyl and biphenyl fragments are normally coplanar. Now the core molecular z axis is known, one can discuss the orientational ordering of the molecular core in terms of S, D and S_{01}, since the latter order parameter may be non-zero in SmC* phases. Using a β_F^{biph} of 9.8° for the biphenyl fragment and Eq. (12.25) to fit the $< \delta >$ for carbons 5 to 11 in the SmA phase, we have derived the molecular S (Fig. 12.18(b)) and D (Fig. 12.19(a)) values with S_{01} set to zero. For direct comparison with the deuteron data,[55] the phenyl S' is scaled to give the core S. It is noted that the molecular biaxial order D is relatively insensitive to temperature in this phase. Since phase biaxiality in the SmC* phase has not been observed, we have again used Eq. (12.25) to determine D and S_{01} while fixing the S values to the extrapolated values from those given in the SmA phase. The tilt angles are those found from the deuteron work.[55] The derived D and S_{01} in the SmC* phase, shown in Fig. 12.19, are now sensitive to temperature. In comparison with other mesogens, the present S_{01} value is slightly smaller at the same reduced temperature. The non-zero S_{01} values indicate that rotation of the molecular core in the SmC* phase is not free and increasingly becomes partially frozen out[73] upon decreasing temperature.

12.6.2. *Order parameters of 10B1M7 and 11EB1M7*

The above methodology has also been applied to analyze the ^{13}C $< \delta >$ of the phenyl and biphenyl carbons of 10B1M7 to first obtain the local order parameter S' of the para axes of these two fragments as shown in Fig. 12.20(a). Again the local order of the phenyl para axis is the more ordered axis, which forms an angle of 14.4° with the para axes of the biphenyl fragment. In this figure, it is again noted that the S' for the phenyl and biphenyl fragments show different temperature behaviors in the SmC* phase. The average β_F^{ph} is found to be about 0.1°, implying that the long molecular z axis is closely lined up to the para axis of the phenyl ring. The tilt angle in the SmC* phase has been determined from the molecular order parameter S which is taken to be identical to S' of the phenyl ring. Given the S and tilt angle, the order parameters D and S_{01} in the SmC* phase are derived from fitting the $< \delta >$'s according to Eq. (12.25) and plotted in Fig. 12.20(b). As noted in this figure, S_{01} becomes non-zero upon entering SmC* phase, and rises to a small plateau of ca. -0.23. This indicates that the orientation of PAS of the order matrix in the SmA phase has been altered in the SmC* phase. The negative S_{01} is simply a result of the present choice of molecular x and y axes. The 11EB1M7 has also

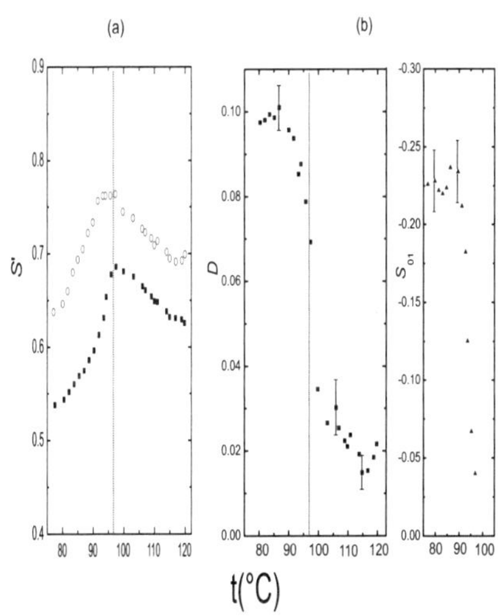

Fig. 12.20. Plot of local order parameter S' for phenyl (•) and biphenyl (▲) fragments and of the molecular order parameters D (a) and S_{01} (b) as a function of temperature in SmA and SmC* phases of 10B1M7.

been studied by ^{13}C NMR, but mainly in its SmA phase.[74] Because of the similarity in its molecular structure with 10B1M7, most of the CSA tensors used for 10B1M7 are also used here. By fitting the $< \delta >$'s from carbons 1-12, using $\beta_F^{biph} = 11.64°$ and $\beta_F^{ph} = 0.06°$, the molecular order parame-

Table 12.5. Comparison of CSA tensors for various carbon sites of ZLL7/* obtained by SUPER and DFT[a].

Sites	C8	C9	C10	C11	C12
δ_{11}	99.2(100.0)	23(22.9)	27.6(25.7)	38.8(36.9)	39.7(40.4)
δ_{22}	156.7(158.2)	134(130.1)	135.6(139.9)	166.8(171.0)	147.7(151.2)
δ_{33}	222.7(220.3)	195(199.2)	223.6(221.9)	255.8(253.4)	239.7(235.3)
$\beta_P(deg)$	0.0(0.0)	60.0(59.5)	120.0(119.6)	180.0(180.0)	0.0(0.0)
Sites	C13	C14	C15	C16	C17
δ_{11}	45(37.1)	45.2(42.5)	18(18.8)	122.7(122.0)	77.0(72.0)
δ_{22}	132(134.3)	157.2(160.5)	127(129.2)	128.7(125.3)	154.0(152.8)
δ_{33}	195(200.6)	221.2(220.0)	236(233.1)	234.7(237.2)	237.0(238.2)
$\beta_P(deg)$	60.0(60.9)	120.0(120.8)	180.0(180.0)	120.0(124.5)	0.0(0.0)
Sites	C18	C19	C20	C21	
δ_{11}	17.6(17.3)	45.2(45.8)	20.5(25.4)	128.0(127.6)	
δ_{22}	140.6(140.1)	157.2(157.9)	142.5(140.1)	135.0(136.5)	
δ_{33}	206.6(206.0)	221.2(218.1)	228.5(225.8)	234.5(232.4)	
$\beta_P(deg)$	60.0(59.2)	120.0(119.5)	180.0(180.0)	120.0(116.2)	

[a]The numbers in parentheses are results from DFT. Note that axes 1 and 2 are switched in Ref. [62].

ters of the core S and D have been derived. In addition, SLF spectra at several temperatures have also been measured to obtain some C-H dipolar splittings from the biphenyl fragment.[74]

12.6.3. *CSA tensors and molecular order in ZLL 7/**

The smectogen ZLL 7/* has three chiral centers and shows a rich polymorphic sequence of I-SmA-SmC*-SmC^*_A-Sm^*_{re}-Sm^*_{Hex}-Crystal. SUPER experiment was carried out[62] to obtain CSA powder patterns which were simulated to yield principal values for some carbon sites as listed in Table 12.5. DFT-GIAO calculations have been performed on methyl-[4'-(4"-ethyloxyphenyl)-benzoyl-4-oxy]-benzoate which serves as a model compound of ZLL 7/*. The CSA tensors were obtained by referring to the absolute shielding of TMS (185.97 ppm) which was calculated at the same level of theory as the ZLL 7/*. The corresponding principal values for the carbon sites obtained from DFT methods are also listed in Table 12.5 for direct comparison. The agreement between the in vacuo DFT and SUPER values is surprisingly good. One advantage of DFT calculations of CSA tensors is its ability to locate their PAS systems in the M frame. Such information is given in the same table for comparing with the 'regular' geometry. In analyzing the $^{13}C < \delta >$'s to obtain the molecular order parameters S and D, Eq. (12.25) has been simplified for all mesophases by setting S_{01} equal to zero as a first approximation. The main purpose of this study is to compare the derived S values based on different senerios, i.e. 1. 'regular' geometry and CSA tensors from SUPER, 2. DFT geometry and CSA tensors from SUPER, 3. 'regular' geometry and DFT CSA tensors, 4. DFT geometry and DFT CSA tensors. It was found that the choice of the orientations of CSA tensors w.r.t. the M frame is not particularly critical. On the contrary, despite the excellent agreement between principal values of CSA tensors of the two approaches the derived S do show detectable differences. Note that the angle between the para axes of the two aromatic fragments is varied during the fitting of $< \delta >$'s, yielding a value between 4.7° and 9.9° depending on temperature and the different fitting cases. This angle is 8.5° from DFT calculations. In summary, this work demonstrates the sensitivity of fitting ^{13} C $< \delta >$'s on the choice of CSA tensors, both their magnitudes and orientations, in the determination of molecular ordering in mesophases. It also serves to validate the SUPER method in obtaining reliable CSA powder patterns in LC.

12.7. Conclusion

This chapter has summarized our recent ^{13}C studies of both bent-core mesogens and chiral smectogens by providing the chemical shift theory in the anisotropic phases and its application to analyze the ^{13}C chemical shift data. The ability to derive order parameters from unlabelled compounds is attractive, since deuteration of new LC materials can often be time consuming and costly. However, the analyses of $^{13}C < \delta >$ data is often a nontrivial matter, and the correct assignment of carbon peaks in ^{13}C spectra from the mesophase is paramount to the success of the present approach. It is noted that DFT calculations can provide quite accurate CSA tensors and its use in LC is just beginning. Study of phase biaxiality in LC by ^{13}C remains a challenge to NMR spectroscopists. Bent-core mesogens still provide many challenging problems such as phase biaxiality in biaxial nematic[75] and various B phases. Indeed, a recent combined DFT-^{13}C solid-state NMR approach has demonstrated a fruitful way to investigate conformational states and their energies in the 5-ring bent-core of a biaxial nematogen A131. This important account is the subject of a separate chapter in this volume, where the computational details will be addressed.

Acknowledgment

The financial support of Natural Sciences and Engineering Council of Canada is gratefully acknowledged. The author thanks Drs. J. Xu and J. Zhang for their works on some of the materials given here.

References

1. R. Y. Dong, *Nuclear Magnetic Resonance of Liquid Crystals*, Springer-Verlag, New York 1994.
2. R. Y. Dong, *Prog. Nucl. Magn. Reson. Spectrosc.*, **41**, 115 (2002).
3. R. Y. Dong, *Annu. Rep. on NMR Spectrosc.* **53**, 67 (2004).
4. P. J. Collings and M. Hird, *Introduction to Liquid Crystals*, Taylor & Francis, London, 1997.
5. P. G. de Gennes, *The Physics of Liquid Crystals*, Clarendon, Oxford 1975.
6. S. Chandrasekhar, *Liquid Crystals*, Cambridge Univ. Press, Cmabridge, 1992.
7. M. J. Freisen, *phys. Rev. Lett.* **24**, 1041 (1970).
8. J. P. Straley, *Phys. Rev. A* **10**, 1881 (1974).
9. W. L. McMillan, *Phys. Rev. A* **4**, 1238 (1971); ibid **6**, 936 (1972).
10. L. Onsager, *Ann. NY Acad. Sci.* **51**, 627 (1949).
11. W. Maier and A. Saupe, *Z. Naturforsch.* **13a**, 564 (1958).
12. M. A. Cotter, *J. Chem. Phys.* **66**, 1098 (1977); W. M. Gelbart and A. Gelbart, *Mol. Phys.* **23**, 1387 (1977).
13. G. H. Heilmeier and W. Helfrich, *Appl. Phys. Lett* **13**, 91 (1968).
14. A. Saupe, *Z. Naturforsch.*, **19a**, 161 (1964).
15. E. E. Burnell and C. de Lange (Eds.), *NMR of Orientationally Ordered Liquids*, Kluwer Academic, Dordrecht, 2003.
16. D. A. Dunmur, and K. Toriyama, in *Physical Properties of Liquid Crystals*, edited by D. Demus, J. Goodby, G. W. Gray, H. W. Spiess, and V. Vill, Wiley-VHC, Weinheim, 1999.
17. J. Courtieu, J. P. Bayle, and B. M. Fung, *Prog. Nucl. Magn. Reson. Spectrosc.*, **26**, 141 (1994).
18. R. R. Ernst, G. Bodenhausen, and A. Wokaun, *Principles of Nuclear Magnetic Resonance in One and Two Dimensions*, Clarendon Press, Oxford, 1987.
19. E. R. Andrew and R. G. Eades, *Proc. R. Soc. London*, **A216**, 398 (1953).
20. K. Schmidt-Rohr, and H. W. Spiess, *Multidimensional Solid-state NMR and Polymers*, Acad. Press, 1994.
21. R. R. Hartmann and E. L. Hahn, *Phys. Rev.*, **128**, 2042 (1962).
22. X. L. Wu, and K. W. Zilm, *J. Magn. Reson. A*, **102**, 205 (1993).
23. X. L. Wu, S. T. Burns, and K. W. Zilm, *J. Magn. Reson. A*, **110**, 29 (1994).
24. J. S. Waugh, *Proc. Natl. Acad. Sci. USA*, **73**, 1394 (1976).
25. R. K. Hester, J. L. Ackerman, B. L. Neff, and J. S. Waugh, *Phys. Rev. Lett.*, **36**, 1081 (1976).

26. A. E. Bennett, C. M. Rienstra, M. Auger, K. V. Lakshmi, and R. G. Griffin, *J. Chem. Phys.*, **103**, 6951 (1995).
27. B. M. Fung, A. K. Khitrin, and K. Ermolaev, *J. Magn. Reson.*, **142**, 97 (2000).
28. G. De Paëpe, D. Sakellariou, P. Hodgkinson, S. Hediger, and L. Emsley, *Chem. Phys. Lett.*, 2003, **368**, 511.
29. M. Lee and W. Goldburg, *Phys. Rev.*, **140**, 1261 (1965).
30. J. S. Waugh, L. M. Huber, and U. Haeberlen, *Phys. Rev. Lett.*, **20**, 180 (1968).
31. D. P. Burum, N. Linder, and R. R. Ernst, *J. Magn. Reson.*, **44**, 173 (1981).
32. K. Takegoshi and C. A. McDowell, *Chem. Phys. Lett.*, **116**, 100 (1985).
33. M. Mehring and J. S. Waugh, *Phys. Rev. B*, **5**, 3459 (1972).
34. A. Bielecki, A. C. Kolbert, and M. H. Levitt, *Chem. Phys. Lett.*, **155**, 341 (1989).
35. E. Vinogradov, P. K. Madhu, and S. Vega, *Chem. Phys. Lett.*, **314**, 443 (1999).
36. N. Sinha, R. Bhattacharya, and K. V. Ramanathan, *ENC Conf.*, PJ426, (2003).
37. J. Xu, R. Y. Dong, V. Domenici, K. Fodor-Csorba, and C. A. Veracini, *J. Phys. Chem. B*, **110**, 9434 (2006).
38. S. F. Liu, J. D. Mao, and K. Schmidt-Rohr, *J. Magn. Reson.*, **155**, 15 (2002).
39. C. S. Nagaraja and K. V. Ramannathan, *Liq. Cryst.*, **26**, 17 (1999).
40. N. Sinha and K. V. Ramanathan, *Chem. Phys. Lett.*, **332**, 125 (2000).
41. K. Schmidt-Rohr, D. Nanz, L. Emsley, and A. Pines, *J. Phys. Chem.*, **98**, 6668 (1994).
42. S. Caldarelli, M. Hong, L. Emsley, and A. Pines, *J. Phys. Chem.*, **100**, 18696 (1996).
43. S. A. Caldarelli, A. Lesage, and L. Emsley, *J. Am. Chem. Soc.*, **118**, 1224 (1996).
44. B. M. Fung, K. Ermolaev, and Y. Yu, *J. Magn. Reson.*, **138**, 28 (1999).
45. B. J. van Rossum, C. P. de Groot, C. P. Ladizhansky, S. Vega, and H. J. de Groot, *J. Am. Chem. Soc.*, **122**, 3465 (2000).
46. A. F. de Jong, A. P. M. Kentgens, and W. S. Veeman, *Chem. Phys. Lett.*, **109**, 337 (1984).
47. D. J. States, R. A. Haberkorn, D. J. Ruben, *J. Magn. Reson.*, **48**, 286 (1982).
48. E. R. de Azevedo, W.-G. Hu, T. J. Bonagamba, K. Schmidt-Rohr, *J. Chem. Phys.*, **112**, 8988 (2000).
49. W. T. Dixon, *J. Chem. Phys.*, **77**, 1800 (1982)
50. R. Y. Dong, J. Xu, J. Zhang, and C. A. Veracini, *Phys. Rev. E*, **72**, 061701 (2005).
51. D. W. Allender, J. W. Doane, *Phys. Rev. A*, **17**, 1177 (1978).
52. B. Hodder, J. R. Sambles, S. Jenkins, R. M. Richardson, *Phys. Rev. Lett.*, **85**, 3181 (1999).
53. E. N. Keller, E. Nachaliel, D. Davidov, *Phys. Rev. A*, **34**, 4363 (1986).
54. K. Merkel, A. Kocot, J. K. Vij, R. Korlacki, G. H. Mehl, and T. Meyer, *Phys. Rev. Lett.*, **93**, 237801 (2004).

55. D. Catalano, M. Cavazza, L. Chiezzi, M. Geppi, C. A. Veracini, *Liq. Cryst.*, **27**, 621 (2000).
56. R. B. Meyer, L. Liebert, L. Strzelecki, and P. Keller, *J. Phys. (France) Lett.*, **36**, L69 (1975).
57. T. Niori, F. Sekine, J. Watanabe, T. Furukawa, and H. Takezoe, *J. Mater. Chem.*, **6**, 1231 (1996).
58. J. Xu, R. L. Selinger, J. V. Selinger, and R. Shashidhar, *J. Chem. Phys.*, **115**, 4333 (2001).
59. A. Eremin, I. Wirth, S. Diele, G. Pelzl, H. Kresse, H. Nadasi, K. Fodor-Csorba, E. Gacs-Baitz, and W. Weissflog, *Liq. Cryst.*, **29**, 775 (2002).
60. R. Y. Dong, J. Xu, G. Benyei, and K. Fodor-Csorba, *Phys. Rev. E*, **70**, 011708 (2004).
61. R. Y. Dong, *J. Phys. Chem. B*, **113**, 1933 (2009).
62. R. Y. Dong, M. Geppi, A. Marini, V. Hamplova, M. Kaspar, C. A. Veracini, and J. Zhang, *J. Phys. Chem. B*, **111**, 9787 (2007).
63. J. Xu, K. Fodor-Csorba, and R. Y. Dong, *J. Phys. Chem. A*, **109**, 1998 (2005).
64. K. Fodor-Csorba, A. Jakli, A. Vajda, S. K. Prasad, D. S. Shankar Rao, R. Y. Dong, J. Xu, and G. Galli, *ChemPhysChem*, **7**, 2184 (2006).
65. J. Xu, Ph.D. Thesis, University of Manitoba, 2006.
66. D.E. Wemmer, A. Pines, and D. D. Whitehurst, *Phil. Trans. R. Soc. Lond. A*, **300**, 12 (1981).
67. G. Zheng, J. Hu, X. Zhang, L. Shen, C. Ye, and G. A. Webb, *J. Mol. Struct.*, **428**, 283 (1998).
68. T. Nakai, H. Fujimori, D. Kuwahara, and S. Miyajima, *J. Phys. Chem. B*, **103**, 417 (1999).
69. R. Y. Dong, J. Zhang, and K. Fodor-Csorba, *Chem. Phys. Lett.*, **417**, 475 (2006). The CSA tensors reported herein are unfortunately incorrect because of a missing scaling factor.
70. T. Sekine, T. Niori, M. Sone, J. Watanabe, S.-W. Choi, Y. Takanishi, and H. Takezoe, *Jpn. J. Appl. Phys.*, **36**, 6455 (1997).
71. H. Kurosu, M. Kawasaki, M. Hirose, M. Yamada, S. Kang, J. Thisayukta, M. Sone, H. Takezoe, and J. Watanabe, *J. Phys. Chem. A*, **108**, 4674 (2004).
72. R. Y. Dong, K. Fodor-Csorba, J. Xu, V. Domenici, G. Prampolini, and C. A. Veracini, *J. Phys. Chem. B*, **108**, 7694 (2004).
73. D. J. Photinos, P. Bos, and J. W. Doane, *Phys. Rev. A*, **20**, 2203 (1979).
74. J. Zhang, V. Domenici, and R. Y. Dong, *Chem. Phys. Lett.*, **441**, 237 (2007).
75. R. Y. Dong, S. Kumar, V. Prasad, and J. Zhang, *Chem. Phys. Lett.*, **448**, 54 (2007). Note that CSA tensors reported herein are unfortunately incorrect due to missing the scaling factor in their isotropic shifts.

Chapter 13

A Combined DFT and Carbon-13 NMR Study of a Biaxial Bent-core Mesogen

Alberto Marini[1], Veena Prasad[2] and Ronald Y. Dong[3]

[1]*Dipartimento di Chimica e Chimica Industriale, via Risorgimento 35, Università di Pisa, Italy c/o Scuola Normale Superiore, Piazza dei Cavalieri 7, 56126 Pisa, Italy*

[2]*Center for Liquid Crystal Research, P.O. Box 1329, Jalahalli, Bangalore 560 013, India*

[3]*Department of Physics and Astronomy, University of British Columbia, Vancouver, BC, Canada V6T 1Z1*

email: rondong@phas.ubc.ca

This chapter provides a detailed conformational study of a bent-core molecule (A131) by means of density functional theory (DFT). The most probable conformational states for its bent-shaped core are found from the Potential Energy Surface (PES) of a 5-ring model of A131 mesogen. They belong to two distinct structural groups, namely the V-shaped and the hockey stick-shaped form. The chemical shielding tensors (CSTs) of the aromatic carbons, for the four prevalent conformers, have been calculated using GIAO-DFT approach. The derived CSTs are then related to chemical shift anisotropy (CSA) tensors, and compared with those measured by means of 2D NMR SUPER technique. The verified CSA tensors for A131 are used to assign the aromatic carbon peaks, and to interpret the observed ^{13}C chemical shifts from its uniaxial and biaxial nematic phases. Reliable structural and local orientational order parameters are obtained. The new ordering information are discussed in light of the conformational states found by in vacuo-DFT calculations. The local order parameters S and D of the aromatic rings are useful, particularly with the aid of DFT results, to postulate the preferred conformers of A131 in the two nematic phases.

13.1. Introduction

Bent-core Liquid Crystals (LC)[1] have received much interest in recent years because they may form a new class of "banana" B phases, and/or a biaxial nematic (N_B) phase. The latter has long been sought for in thermotropic LC systems since its prediction by Freiser.[2,3] A novel property of this class of bent-core LC (BLC) is the formation of tilted layer structures in various B_1–B_7 phases. Though these V-shaped molecules are achiral, due to the effective packing and polarity they can show chirality and become switchable electrooptically.[4] Thus BLC are attractive for their potential technical relevance. The V-shaped molecules usually contain a five aromatic rings system in the "rigid" core with flexible terminal chains, thereby exhibiting the necessary molecular biaxiality required to form biaxial nematics. Bent-core biaxial nematics have recently been studied by many experimental techniques like NMR,[5,6] X-rays[7,8] and Raman scattering.[9] One important structural factor of V-shaped molecules is the bend angle Θ subtended by the two lateral wings next to the central ring. The phase diagrams which predict the presence of N_B phase and the N_U-N_B phase transition seem to depend on the magnitude of the bend angle.[3,10] A Θ angle approaching 110° is most favored to give a biaxial nematic phase.

In the simple case of a biaxial nematic, four common order parameters appear in an order supermatrix S^{ii}_{lm}, where $i = (X, Y, Z)$ refers to the principal axis of the laboratory (L) frame, and $l, m = (x, y, z)$ denote an axis of the molecular (M) frame attached on the molecule. These are S, D, P and C.[11] As mentioned in the previous chapter, the order parameters S and D are nonzero whereas P and C are zero in the uniaxial nematic (N_U) phase, while in the N_B phase all four order parameters are nonzero. In addition, there exist more orientational order parameters depending on the symmetry of the phase and of the LC forming molecules. Incidentally, similar temperature behaviors for D and C have been observed in the biaxial nematic phase of tetrapode molecules.[12]

Nuclear Magnetic Resonance (NMR) spectroscopy has been widely used to characterize orientational order parameters in uniaxial and biaxial mesophases of LC.[13–15] It has recently been pointed out that local order supermatrices of phenyl rings (z axis is along para (p) axes) may be better than the principal order parameters of the V-shaped molecule to study the onset of phase biaxiality, i.e., S^{XX}_{pp} differs from S^{YY}_{pp} in computer simulation.[16] For this reason both ^{2}H and ^{13}C NMR can be considered an indispensable tool in studying biaxial nematic phases.[5,6] Indeed, ^{13}C NMR study of local order parameters of various fragments in a bent-core molecule can shed light on its average molecular structure, which provides a means to estimate the bend angle Θ in V-shaped molecules.[17,18] In the past few years, ^{13}C NMR chemical shifts $\langle\delta\rangle$'s have been reported in the mesophases of BLC as well as a two-dimensional SUPER technique[19] has been employed to determine the chemical shift anisotropy (CSA) tensors of various carbons.[6,17,18,20,21] Note that the derived order parameters and the bend angle based on the observed $\langle\delta\rangle$'s relative to their corresponding isotropic shifts may be problematic unless the principal components and the orientation of each CSA tensor are known. Density functional theory (DFT) has recently been found useful to calculate chemical shielding tensors (CSTs), which can be compared with those obtained by SUPER method,[22] once properly rescaled to a chosen external reference (such as TMS for ^{13}C nuclei). Furthermore, the orientation of CSA tensor in the molecular frame can be directly obtained in DFT calculations. Such information is lacking in SUPER method, as it can only provide the CSA powder patterns. The V-shaped molecule 2-methyl-3-[4-(4-octyl-benzoyloxy)-benzylidene]-amino-benzoic acid 4-(4-dodecyl-phenylazo)-phenyl ester (A131) provides a good model compound to study its known N_U-N_B phase transition. The DFT calculations of principal components of CSTs are implemented to verify those determined by SUPER. Furthermore, the conformational landscape for the molecular core of A131 is also investigated during the above process, yielding valuable information on the populations and energies of four prevalent conformers. The present chapter is organized as follows. Section 13.2 describes theorectical background necessary for the discussion of DFT and NMR results. The experimental and computational details are given in Sec. 13.3, while Sec. 13.4 provides an account of conformational analyses and the derived DFT CSA tensors. Section 13.5 contains the ^{13}C NMR data treatment and discussion of the orientational order. The last section contains some concluding remarks.

13.2. Theory

Here an overview of the pertinent DFT methods is given, while the general treatment of NMR chemical shifts in mesophases has been given in Chap. 12. This section starts with a description of quantum mechanical (QM) calculation of nuclear shieldings in vacuo, followed by how they are related to the NMR CSA tensors. The treatment of chemical shifts for a rigid fragment of C_s symmetry then follows.

13.2.1. *QM Calculation of nuclear shieldings in vacuo*

The nuclear magnetic shielding tensor $\boldsymbol{\sigma}^X$ of a nucleus X can be expressed, by theoretical definition,[23] as mixed second derivatives of the total electronic energy of the system with respect to the magnetic field **B** and the nuclear magnetic moment μ^X:

$$\sigma_{ij}^X = \frac{\partial^2 E}{\partial B_i \partial \mu_j^X} \tag{13.1}$$

where B_i and μ_j^X are the Cartesian components of the **B** field and of the nuclear magnetic moment, respectively. It is well-known that the presence of the magnetic field introduces the problem of defining the origin of the corresponding vector potential. Since $\boldsymbol{\sigma}$ is a molecular property, it must be invariant with respect to changes of the gauge origin. The problem of gauge variance is accomplished by using Gauge Invariant Atomic Orbitals (GIAO) (in Gaussian 03),[29] first introduced by Ditchfield within the coupled Hartree-Fock scheme.[24] A complete and detailed treatment of direct implementation of the GIAO and the Continuous Set of Gauge Transformations methods for calculating nuclear magnetic shielding tensors at both the levels of Hartree-Fock (HF) and DFT have been presented and discussed in Ref. 25. Here we only focus our attention on the relevant equations and essential tools that one has to use when calculating chemical shielding properties by quantum mechanical methods. For a solute described at a level of the HF or DFT, with an expansion of the molecular orbitals over the finite basis $\{\chi_\mu, \chi_\epsilon\}$, the components of the nuclear magnetic shielding tensor $\boldsymbol{\sigma}^X$ are obtained as:

$$\sigma_{ij}^X = tr[\mathbf{P}\mathbf{h}^{B_i,\mu_j^X} + \mathbf{P}^{B_i}\mathbf{h}^{\mu_j^X}] \tag{13.2}$$

where $\mathbf{P}^{B_i}$ is the derivative of the density matrix **P** with respect to the **B** field. Matrices $\mathbf{h}^{\mu_j^X}$ and $\mathbf{h}^{B_i,\mu_j^X}$ contain the derivatives of the standard one-electron Hamiltonian with respect to the nuclear magnetic moment and derivatives of the same operator with respect to the **B** field and the nuclear magnetic moment, respectively.

13.2.2. *Chemical Shielding: definition and commonly used conventions*

The chemical shielding tensor (CST) is characterized by a real three-by-three Cartesian matrix, that can be expressed in the M frame (x, y, z)

according to:

$$\boldsymbol{\sigma} = \begin{bmatrix} \sigma_{xx} & \sigma_{xy} & \sigma_{xz} \\ \sigma_{yx} & \sigma_{yy} & \sigma_{yz} \\ \sigma_{zx} & \sigma_{zy} & \sigma_{zz} \end{bmatrix}. \tag{13.3}$$

It can be decomposed into a scalar term, three antisymmetric pseudo-vector components, and five components corresponding to a symmetric tensor. Only the scalar and the five symmetric tensor elements can be observed in normal NMR spectra of solids. On the other hand, Eq. (13.3) gives an asymmetric second rank tensor (in our case for each ^{13}C nucleus) that contains up to nine unique components. For this reason a mathematical transformation described in Eq. (13.4) is required in order to separate out the antisymmetric contribution:[26]

$$\boldsymbol{\sigma} = \boldsymbol{\sigma}^{sym} + \boldsymbol{\sigma}^{anti} \tag{13.4}$$

where

$$\boldsymbol{\sigma}^{sym} = \frac{\boldsymbol{\sigma} + \boldsymbol{\sigma}^T}{2}; \qquad \boldsymbol{\sigma}^{anti} = \frac{\boldsymbol{\sigma} - \boldsymbol{\sigma}^T}{2} \tag{13.5}$$

with $\boldsymbol{\sigma}^T$ being the corresponding transposed matrix. In this way only the symmetric part of the raw calculated tensor $\boldsymbol{\sigma}^{sym}$ is retained, and can be diagonalized in its specific principal axis system (PAS) to produce its principal components (eigenvalues) and its principal directions (eigenvectors). The chemical shielding tensor is now expressed in terms of three principal components defined in descending order (such as $\sigma_{11} \geq \sigma_{22} \geq \sigma_{33}$). The chemical shielding is not directly measured by NMR experiments, but rather, the variation in resonance frequencies of different nuclei provides the measurement of their relative shieldings called the "chemical shifts". These experimentally measured chemical shifts, $\boldsymbol{\delta}$, obey the same tensor properties of the CST. The calculated shielding, σ_{ij}, and experimental shift, δ_{ij}, elements have a one-to-one correspondence given by the following relation:

$$\sigma_{ij} = K_{ij}\sigma_{ref} - \delta_{ij} \tag{13.6}$$

where the offset value, σ_{ref}, corresponds to the isotropic shielding of a reference compound and is only added to diagonal elements as expressed in Eq. (13.6) due to the Kronecker delta, K_{ij}. The isotropic shielding of the reference compound is either estimated from a gas-phase measurement or assumed from calculations (as detailed in the Sec. 13.3.2). In order to compare the calculated CSA data reported in this chapter with those reported in literature, we have decided to denote the principal elements of the symmetric part of the chemical shift tensor by $\delta_{33} \geq \delta_{22} \geq \delta_{11}$. This convention* states that δ_{33} corresponds to the direction of the least shielded, σ_{33}, with the highest frequency (highest shift), while δ_{11} corresponds to the direction of highest shielding, σ_{11}, with the lowest frequency (lowest shift).

13.2.3. *Chemical shift anisotropy*

This has been treated in Sec. 12.4 of the previous Chapter and the given formulae can be used here. However, another relevant aspect remains when dealing with the observed chemical shifts for a rigid fragment like the central ring of A131. Here the ring cannot flip or rotate like the remaining lateral phenyl rings. Consequently, it has a reduced (C_s) symmetry and the Saupe order matrix in the chosen molecular (x, y, z) frame (see Fig. 13.1(a)) is not diagonal. In other words, there are three nonzero order parameters S_{zz}, $S_{xx} - S_{yy}$ and S_{xz}. The last order parameter, however, is usually vanishingly small. Even so, this local order tensor (in the (x, y, z) frame) can be fully determined from the five ^{13}C chemical shifts $\langle\delta\rangle$'s obtained for the central ring of A131 using the following equation:

$$\langle\delta\rangle = \delta_{iso} + \frac{2}{3}[\Delta\delta S_{zz} + \frac{1}{2}(\delta_{xx} - \delta_{yy})(S_{xx} - S_{yy}) + 2\delta_{xy}S_{xy} + 2\delta_{xz}S_{xz} + 2\delta_{yz}S_{yz}] \tag{13.7}$$

where $\Delta\delta$, the anisotropy of δ with respect to the chosen z axis, is given by

$$\Delta\delta = \delta_{zz} - \frac{1}{2}[\delta_{xx} + \delta_{yy}] \tag{13.8}$$

Now the tensor δ of each carbon can be related to its PAS values by using

$$\delta_{ij} = \sum_{\eta} \cos\theta_{\eta i} \cos\theta_{\eta j} \delta_{\eta\eta} \tag{13.9}$$

where $\theta_{\eta i}$ is the angle between the PAS η axis and the i molecular axis. It is again emphasized that the use of ^{13}C chemical shifts to determine molecular or local order in LC requires the precise knowledge of CSA tenors and their PAS's orientations.

*In the literature, the convention that uses the reverse order for δ_{ii} elements is also encountered, especially when treating solid state systems. Moreover, in liquid crystals field, it is common to substitute δ_{33} with δ_{zz}, δ_{22} with δ_{xx} and, consequently, δ_{11} with δ_{yy}.

13.3. Experimental

The A131 sample was that used in Ref. [8]. Its transition temperatures (K) are: I (449.5) N_u (422) N_b (391.5) SmC (377.3) SmX (366.4) SmY (355.8) Cr and its molecular structure is shown in Fig. 13.1(a). All the information concerning the ^{13}C NMR solution and solid state measurements have been detailed and discussed in Ref. [6].

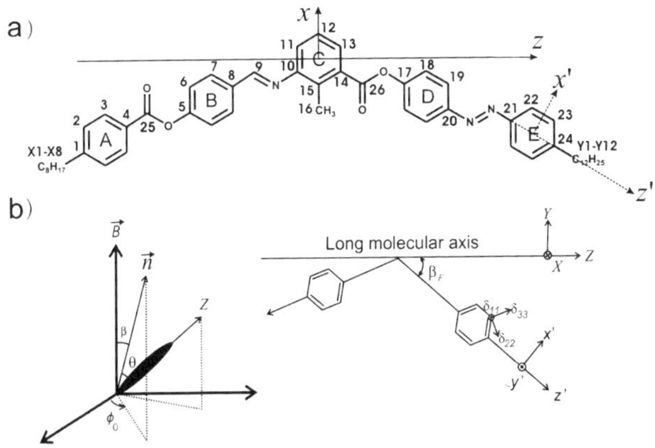

Fig. 13.1. a) Molecular structure of A131 with carbon site labels, (b) Schematic of the orientation of a molecule with respect to the director $\vec{n}$ in the magnetic field ($\vec{B}$) and of the axes systems used here.

13.3.1. *Computational details*

All the calculations have been performed using Gaussian 03 software,[29] where all the molecular models were built by means of GaussView 3.0. To shed light on the experimental NMR work of A131 mesogen,[6] DFT calculations have been performed to obtain the most populated conformational states within its five-ring molecular core. Moreover, the CST obtained from in vacuo GIAO-DFT calculations on each relevant conformer, found in the calculated Potential Energy Surface (PES), have been compared with the corresponding experimental values, measured by the 2D SUPER method.

13.3.1.1. *PES calculations*

In order to accurately study complex systems like A131, its complete structure has been reduced by substituting the long aliphatic chains with methyl units, namely the 2-methyl-3-[4-(4-methyl-benzoyloxy)-benzylidene]-amino-benzoic acid 4-(4-methyl-phenylazo)-phenyl ester (**ABCDE**) model as shown in Fig. 13.2. As far as the conformational analysis of the core is concerned, this approximation is valid, if the torsional barriers of the aromatic core rings are not influenced by the aliphatic chain length.

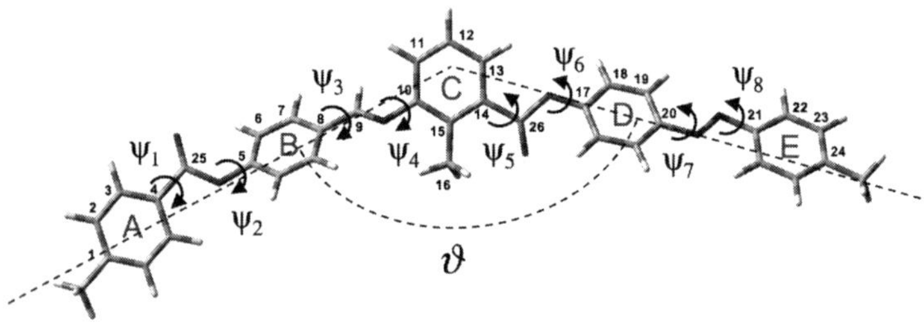

Fig. 13.2. Molecular structure of the **ABCDE** 5-ring **5R** model of A131 mesogen.

Moreover this assumption is specially true when investigating local properties such as chemical shielding interaction, because of its nature, i.e. it is closely linked to the environment of the nuclei.[23] In this way it is possible to investigate quite large molecules with an accurate description of their electronic structures offered by a high level QM approach. Preliminary tests on a full A131 model have shown that the torsional energies are almost independent from the length of the aliphatic chains. Thus, a 5-rings (**5R**) molecular model shown in Fig. 13.2 has been adopted. This model system has further been decomposed in its principal 3-rings molecular fragment (**3R**) model, namely

- the 3-(4-(4-methylbenzoyloxy)benzylideneamino)-2-methylbenzoic acid (**ABC**) model
- the 4-((3-(2-(4-(4-methyl)-phenylazaphenyl)acetyl)-2-methylphenylimino)methyl)phenyl formate (**BCD**) model
- the (4-(4-methyl)-phenylazaphenyl)-2-methyl-3-(methyleneamino) benzoate (**CDE**) model

reported in Fig. 13.3. On these **3R** models, three characteristic Potential Energy Surfaces (PES) have been obtained by scanning the following dihedrals angles:

(1) in **ABC** system, the C_{25}-O-C_5-C_6 (ψ_2) and C_9-N-C_{10}-C_{11} (ψ_4) dihedrals were scanned at every 30° intervals with respect to each other,

in the range of $-180 \div 180°$, to obtain a 12 x 12 matrix of energetic points;

(2) in **BCD** system, the C_9-N-C_{10}-C_{11} (ψ_4) and C_{13}-C_{14}-C_{26}-O (ψ_5) dihedrals were scanned at every 15° intervals with respect to each other, in the range of $-180 \div 180°$, to obtain a 24 x 24 matrix of energetic points;

(3) in **CDE** system, the C_{13}-C_{14}-C_{26}-O (ψ_5) and C_{19}-C_{20}-N-N (ψ_7) dihedrals were scanned at every 30° intervals with respect to each other, in the range of $0 \div 180°$, to give a 6 x 6 matrix of energetic points.

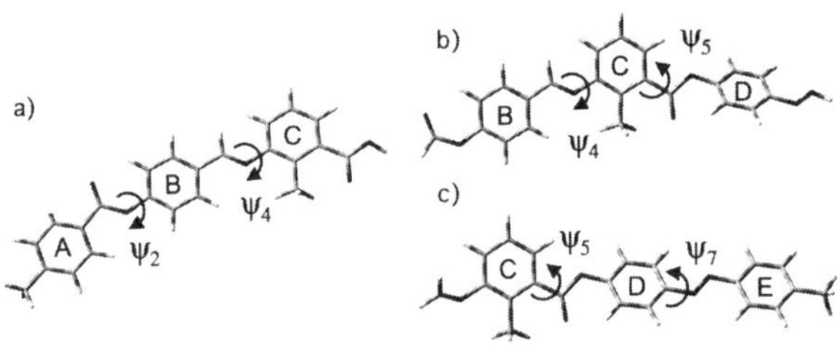

Fig. 13.3. Molecular structures of (a) **ABC**, (b) **BCD** and (c) **CDE** 3-ring models of A131 mesogen.

The three **3R** models could be considered as the simplest ones, but sufficient to study the conformational behaviors of the "wings" and of the central part separately, of this real banana mesogen. Relevant information on torsional barriers and on conformational states in each **3R** model have been exploited to build up reliable conformational structures for the **5R** model. It has been noticed that the conformations of **BCD** model determine the orientations of the lateral wings and the bend angle of the entire molecule, because the ψ_4 and ψ_5 dihedrals are expected to play a major role in determining the prevalent conformational states in A131 molecule. In fact, reliable **5R** model conformers have been built up in the basins of attraction found in the PES(**BCD**). Finally, the geometries of these conformers have been optimized without any constraints, until obtaining fully relaxed structures for the **5R** model, on which chemical shielding calculations have been performed. For computational and molecular complexity reasons, it has been decided to select the dihedrals which can reasonably influence the conformations of A131 molecule, once given a reliable thermal energy at the disposal of the mesogen. In order to choose the proper angles and to obtain reliable results, several conformational studies on banana shaped molecules have been considered.[30–34] Moreover, as the calculated shielding tensors are very sensitive to the size of the basis set and to the geometry employed, it has been decided to retain the polarized 6-31+G(d) basis set throughout our different geometry optimization at B3LYP[35] level of theory. Geometries obtained in this way are at least as good, and in several case (such as phenyl-benzoate) better than those obtained at the MP2/6-31+G(d) level, which shows greater sensitivity on the basis set.[31]

13.3.2. *Chemical shielding tensor calculations*

A complete and unambiguous interpretation of the experimental NMR data may rely upon ab initio calculations, which yield (i) an optimized geometry of the molecule, (ii) the chemical shielding principal components, (iii) the orientation of CST in the molecular coordinate system, and (iv) possibly, the dependence of CST components on important structural parameters, such as bond distance, angles and dihedrals. Recent achievements in ab initio methods and computer hardware and software have made it possible to accurately evaluate the chemical shieldings of many molecular systems.[23] Since the theory at the HF level neglects the instantaneous interaction between electrons (electron correlation), treating each one in an average or mean field of the others, it inevitably leads to serious consequences in the shielding calculations. However, by using HF methods with a sufficient large basis set, ^{13}C shielding tensors obtained for molecules where electron correlation effects are negligible such as most hydrocarbons,[36] have found good agreement with those experimentally observed. However, for aromatic compounds, where instantaneous electronic interaction contributions are more significant, it is extremely important to include electron correlation[37] contributions in order to obtain accurate prediction of CST. In general, there are basically several theoretical approaches to go beyond the HF theory:[38] Möller-Plesset (MP) or many body perturbation theory (MBPT), configuration interaction (CI), and density functional theory (DFT). In last decades, DFT with a suitable choice of both functional and basis sets, has been shown to be successful in predicting various molecular properties for systems in the ground state and at the equilibrium geometry. This approach, which has a computational cost of the same order as HF (considerably less than traditional correlation techniques), often gives a quality of calculations comparable to or even better than those of MP2.[39] In contrast to the HF based methods, it is impossible to grade these functionals at the theoy level.

The computational study of the NMR parameters has been performed principally using DFT methods, following the recommendations of Cheeseman et al.[25] for chemical shielding calculations of large molecules. The modified Perdew-Wang[40] exchange-correlation functional, called MPW1PW91[41] is particularly accurate in predicting the chemical shielding property, and is therefore used here for the chemical shielding prediction. Since nuclear chemical shielding is a very small effect, rather accurate wave functions are necessary for its calculations. As mentioned above, most calculations of the CST have been performed using the 6-311+G(d,p) basis set (one polarization function and one diffuse function), which seems to be the best compromise between accuracy and CPU time-intensive basis set. Moreover, from the variety of theories available to compute CSTs, we have decided to adopt the GIAO[24] for its numerous advantages.[44]

Finally, because quantum chemistry determines "absolute" displacement or shielding (σ scale) on the one hand, while on the other hand, experimental NMR shifts are referenced to some standard (δ chemical shift scale), a suitable scale for referencing can be built by computing the chemical shielding σ^{ref} of the reference. In particular, the chemical shift tensors can be obtained by referring the absolute chemical shielding tensors obtained by DFT to the absolute shielding of TMS (185.97 ppm), which is calculated at the same level of theory as for the **5R** model. It is noticed that the value of δ_{iso} for TMS obtained by GIAO-DFT calculations is very close to that (185.4 ppm) obtained after correction of vibrational averaging, bulk susceptibility, temperature, and relating to a secondary standard.[26] This result confirms the appropriateness of the basis set and functional used in our GIAO-DFT calculations. Indeed, the method adopted in this work has been previously tested, and found to be generally satisfactory for ^{13}C NMR shielding calculations of several different relatively large size organic compounds.[42,43] In particular, the integrated computational and Solid State NMR (SS-NMR) approach followed in this work has recently been successfully applied to study magnetic and orientational order properties of a ferroelectric liquid crystal, namely the ZLL 7/*, when combining with 2D SUPER technique for the measurement of ^{13}C CSA tensors.[22]

Data Analysis The chemical shielding tensors, obtained by Gaussian 03 software, mostly have been analyzed using a home-made program written in Mathematica 5.0 for Mac Os X (copyright 1988-2003, Wolfram Research, Inc.). This program allows us to manipulate a calculated CST for the evaluation of its eigenvalues and eigenvectors. Statistical analyses of

CST parameters have been done with commonly used software, such as Kaleidagraph 4.03 and/or Mathematica 5.0. The analyses of experimental chemical shift anisotropies are performed by using a home-made Fortran program, which exploits nonlinear least-squares methods to give molecular orientational order parameters based on the knowledge of CSA tensors.

13.4. Conformational Analysis

A conformational analysis of A131 is carried out based on breaking down the **5R** model into three **3R** models before the full **5R** model, and then the derived CSA tensors for the four prevalent conformers of the core are compared and contrasted with those determined by SUPER.

13.4.1. *ABC (3R) model*

The **AB** fragment of the **ABC** model is very close to a structural analogous of the phenyl-benzoate, studied in details by Wrzalik et al. in Ref. [31]. In particular, in their work, it was shown that the benzoate (Ph-COO) group in the phenyl-benzoate can be considered planar, due to strong conjugation effects between the aromatic core and the carbonylic group. This fact implies that the molecular flexibility is reduced to the torsional rotation around the C_a-O (and C_a-N) bond (where C_a stands for an aromatic carbon), leading us to investigate the torsional energetic barriers around ψ_2 and ψ_4 (see Fig. 13.3).

Figure 13.4 shows the torsional profiles of PES(**ABC**), corresponding to the energy variations with ψ_2 angle, together with a structural analogous of the **ABC** model calculated at DFT level by Cacelli et al. in Ref. 32. In particular the **AB** part is very similar to the central core of the **5R** model investigated by Cacelli et al.,[32] with the exception that in that case the central ring has a meta substitution for the lateral chains, while in our case the **A** and **C** rings are attached to the ring **B**, in para position with respect to each other. The comparison between the two PES profiles is very good, because they essentially give the same four conformers (with $\psi_2 \approx \pm 30$ and $\pm 150°$) in terms of relative energies as well as molecular structures. The energy barrier found is less than 1 Kcal/mol, thus basically all the conformational states from -180 to 180° are populated at room temperature, resulting in the fact the **AB** part of A131 is quite flexible. Instead, the energetic barrier concerning the rotation around ψ_4 is quite high, about 7 Kcal/mol, and only two conformational states are populated

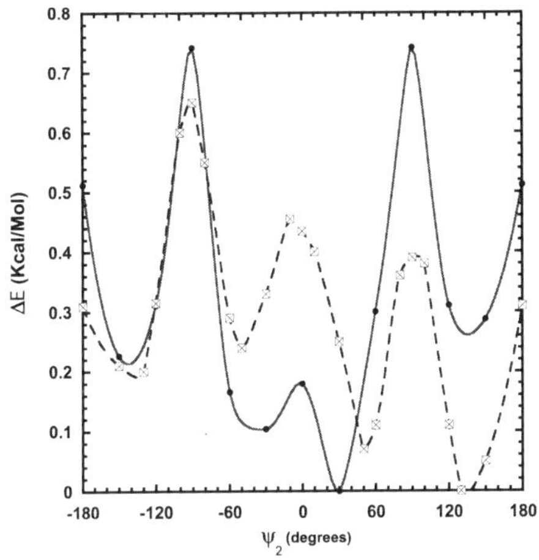

Fig. 13.4. Torsional energy profiles of the PES(**ABC**), as a function of ψ_2 angle, where ψ_4 is at $-60°$ (solid line) and of a structural analogous of the **ABC** model (dashed line), calculated with DFT by Cacelli et al. in Ref. [32].

($\psi_4 = \pm 60°$), due to the steric repulsion between the methyl (CH_3) unit and the C-H bond of the imine HC=N- unit. The rotations around ψ_1 and ψ_3 are essentially free with very low (less than 1 Kcal/mol) energetic barriers. In conclusion, four (eight symmetry related) different conformers can be distinguished in PES(**ABC**), having structural parameter values in well agreement with those reported by Wrzalik et al.[31] and by Cacelli et al.[32] The results are reported in Table 13.1.

13.4.2. *BCD (3R) model*

The torsional PES(**BCD**) profiles at two different values of the dihedral ψ_5, are reported as a function of dihedral ψ_4 in Fig. 13.5. These profiles clearly indicate that the dihedral ψ_4 can basically assume two different conformations ($\psi_4 = \pm 45°$), which always represent a *trans* state of C9 with respect to C16, as observed above for **ABC** model,† due to the steric repulsion mentioned above. It has already been noted that the conformations of

†It is noted that in the case of **ABC** model, which is similar to **BCD** model, we have obtained two conformational states characterized by $\psi_4 = \pm 60°$, which is close to $\pm 45°$ in **BCD** model. This is principally due to the fact that the two models are different but more due to the coarse sampling performed every 30°, with respect to that performed every 15° in the **BCD** model.

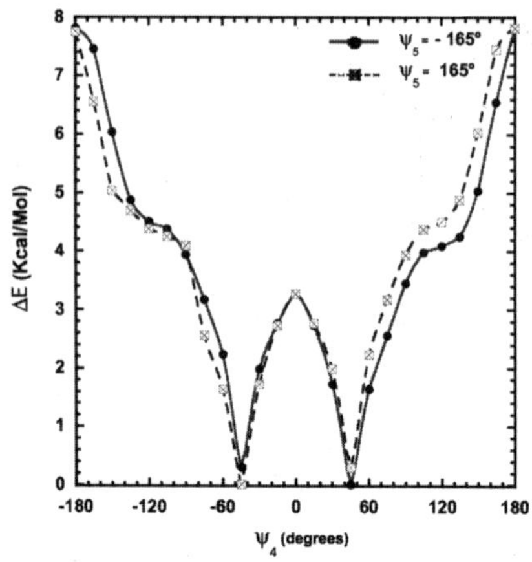

Fig. 13.5. Torsional PES(**BCD**) profiles as a function of dihedral ψ_4, at two different values of the dihedral $\psi_5 = \pm 165°$.

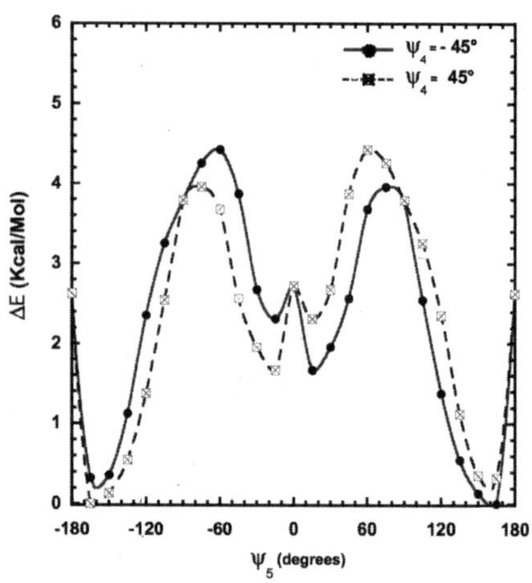

Fig. 13.6. Torsional PES(**BCD**) profiles in function of dihedral ψ_5, at two different values of the dihedral $\psi_4 = \pm 45°$.

BCD model determine the orientations of the lateral wings and the bend angle of the full molecule, because the ψ_4 and ψ_5 dihedrals are expected to play a major role in governing the main conformational states of A131 molecule. In Fig. 13.6, the torsional PES(**BCD**) profiles are reported as a function of dihedral ψ_5 for two different values of the dihedral $\psi_4 = \pm 45°$.

Table 13.1. Relative energies (in Kcal/mol) and populations (in %) for $\mathbf{ABC}_i$, $\mathbf{BCD}_i$ and $\mathbf{CDE}_i$ conformers.

Conformation	ψ_2 (°)	ψ_4 (°)	Δ E(Kcal/mol)	*P* (%)
$\mathbf{ABC}_1$	30	60	0.102	13.0
$\mathbf{ABC}_2$	-30	60	0.000	16.0
$\mathbf{ABC}_3$	150	60	0.212	11.0
$\mathbf{ABC}_4$	-150	60	0.263	10.0
$\mathbf{ABC}_5$	30	-60	0.000	16.0
$\mathbf{ABC}_6$	-30	-60	0.102	13.0
$\mathbf{ABC}_7$	150	-60	0.263	10.0
$\mathbf{ABC}_8$	-150	-60	0.212	11.0
Conformation	ψ_4 (°)	ψ_5 (°)	Δ E(Kcal/mol)	*P* (%)
$\mathbf{BCD}_1$	45	-165	0.000	30.0
$\mathbf{BCD}_2$	45	-15	1.665	1.9
$\mathbf{BCD}_3$	45	15	2.315	0.6
$\mathbf{BCD}_4$	45	165	0.325	17.5
$\mathbf{BCD}_5$	-45	-165	0.325	17.5
$\mathbf{BCD}_6$	-45	-15	2.315	0.6
$\mathbf{BCD}_7$	-45	15	1.665	1.9
$\mathbf{BCD}_8$	-45	165	0.000	30.0
Conformation	ψ_5 (°)	ψ_7 (°)	Δ E(Kcal/mol)	*P* (%)
$\mathbf{CDE}_1$	30	0	0.685	6.0
$\mathbf{CDE}_2$	-30	0	0.685	6.0
$\mathbf{CDE}_3$	150	0	0	19.0
$\mathbf{CDE}_4$	-150	0	0	19.0
$\mathbf{CDE}_5$	30	180	0.685	6.0
$\mathbf{CDE}_6$	-30	180	0.685	6.0
$\mathbf{CDE}_7$	150	180	0	19.0
$\mathbf{CDE}_8$	-150	180	0	19.0

These profiles clearly indicate that, in the two minima found for ψ_4, four different conformations ($\psi_5 = \pm 15°$ and $\pm 165°$), can be populated at room temperature. The results of this analysis are also given in Table 13.1, where it is clearly shown that the conformers characterized by $\psi_5 = \pm 165°$ have lower energy with respect to those having $\psi_5 = \pm 15°$, with corresponding higher population than that of the latter ones. The four conformational states obtained for ψ_5 dihedral roughly correspond to those found for ψ_2 in PES(**ABC**), because the **CD** fragment of **BCD** is very similar to the **AB** fragment of **ABC**, with the exception that in the case of **BCD** fragment the **C** ring has an additional methyl unit (C16) placed in a ortho position with respect to the C14-C26 bond.

The presence of this methyl unit introduces a symmetry-breaking point in the A131 central core, resulting in a direct selection of the most populated conformational states for ψ_4, as previously observed in the PES(**ABC**), and for ψ_5 investigated here. In fact, due to the presence of methyl unit (C16), the lowest energy states are the four ($\psi_4 = \pm 45°$ and $\psi_5 = \pm 165°$) symmetry related conformers: $\mathbf{BCD}_1$, $\mathbf{BCD}_4$, $\mathbf{BCD}_5$ and $\mathbf{BCD}_8$, where the position of C26=O unit is in *trans* with respect to the C13-C14 bond, and in synclinic position (on the same side) with respect to the central CH_3 (C16) group. While, $\mathbf{BCD}_2$, $\mathbf{BCD}_3$, $\mathbf{BCD}_6$ and $\mathbf{BCD}_7$ represent the higher energy states, characterized by $\psi_4 = \pm 45°$ and $\psi_5 = \pm 15°$, where the position of C26=O unit is in *cis* with respect to the C13-C14 bond and in anticlinic position (on the opposite side) with respect to the central CH_3 (C16) group, causing an increase of the total energy due to the steric repulsion between the methyl unit and the lateral ring D. It has to be underlined that, without the presence of the methyl unit (C16), the preferred orientation for the sp^2 carboxylic unit would be co-planar to the central aromatic **C** ring, due to the hyper-conjugation effect of these two groups, making the principal conformational state to have $\psi_5 = 0$, or 180°, as previously found for the phenyl-benzoate by Wrzalik et al.[31]

13.4.3. *CDE (3R) model*

The **CDE** fragment is a structural analogous of the very well known *trans*-azobenzene (TAB)[‡], often used in areas such as dyes, pH indicators, non-linear optical chromophores, photo-storage units and triggers for optical switching and actuators.[34] The conformations of this molecule have been recently studied by means of both DFT and MP2 computational methodologies.[34,45] In particular, Wang and coworkers found that for TAB: (i) the DFT-B3LYP/6-31+G* optimization calculations predict the planar conformation as the most stable one, and (ii) the MP2/6-31+G* geometry optimizations show two minima at ±19.5°, separated by a very low, only 0.15 kcal/mol, energy barrier at $\psi_7 = 0°$. It indicates that the energy variation

[‡]The inter-conversion barrier between *cis* and *trans* geometrical isomers is very high, close to 45 Kcal/mol, thus the latter configuration is the most prevalent one at room temperature. The conversion between these two forms can be induced by light (photoisomerization process), but the sample inside the magnet is not exposed to the light. Nevertheless, one can not a priori exclude the thermal isomerization process which is favored at high temperature, thus this fragment can in principle undergo a *cis-trans* isomerization, but it must be considered that the long aliphatic chain attached to **CDE** fragment diminishes its probability and favors the *trans* configuration.

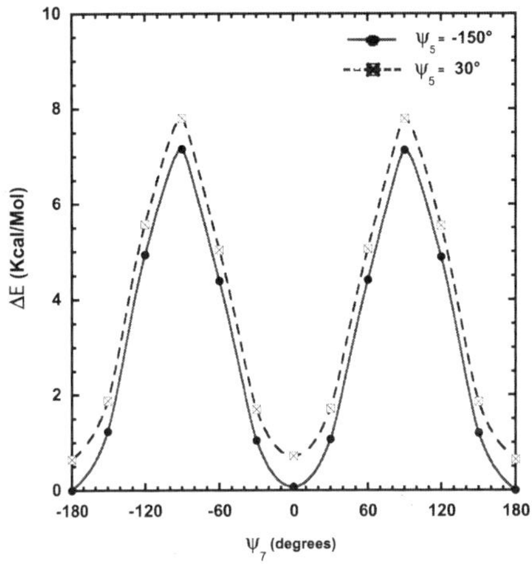

Fig. 13.7. Torsional profiles of the PES(**CDE**), corresponding to the ϕ_7 angle energy-variations, at two different values of $\psi_5 = 30°$ and $\psi_5 = -150°$.

along the NNCC dihedral angle (from -20° to 20°) is small and the energy barrier (at 0°) separating two equivalent non-planar stationary structures that can be overcome by thermal vibrations at the room temperature.[§]

Finally, compared with the experimental results, DFT gives the bond lengths and angles of TAB with higher accuracy than those obtained by MP2, whereas the latter can give more accurate NNCC dihedral angles.[34] Nevertheless, as a whole, the structural parameters optimized by these two methods show acceptable consistency. In Fig. 13.7, the torsional profiles of the PES(**CDE**), corresponding to the energy variations with ψ_7 angle, are reported at two different values of ψ_5, calculated at DFT level. Even in this case, as previously observed for **BCD** model, the rotation around ψ_5 introduces four different conformers (with $\psi_5 = \pm 30°$ and $\pm 150°$), separated by barriers of 7 Kcal/mol, while the rotation around ψ_6 is considerably free (energy barrier less than 1 Kcal/mol). The energetic barrier concerning the rotation around ψ_7 (or even around ψ_8, not scanned in this work, because of symmetry reasons), is quite high as well, about 8 Kcal/mol, leading to only two populated conformational states ($\psi_7 = 0°$ or $\psi_7 = 180°$). Thus, four (eight symmetry related) different conformers can be distinguished in

[§]This finding can be explained by assuming the existence of two competitive conjugate interactions. The conjugation between the π-orbit of azo group and phenyl rings (π-π conjugation) favors the planar structure, and the conjugation between the lone-pair electrons in nitrogen atoms and the π-orbit of phenyl rings (n-π conjugation) leads to the non-planar conformation.

PES(**CDE**), and the results are also summarized in Table 13.1. The values for structural parameters found for $\mathbf{CDE}_i$ conformers in this study agree very well with those reported by Wang et al.[34] for TAB system.

13.4.4. *ABCDE 5R model*

The PES(**BCD**) shows the highest rotational barriers found for all the **3R** models investigated here, and represents the central part of the aromatic semi-rigid core of A131. Thus it can better describe the orientations and the conformations of the lateral wings. Four basins of attraction labeled by $\Omega(\mathbf{\Gamma}_i)$ for the $\mathbf{\Gamma}_i$ conformers in the PES(**BCD**) have also been found in the **ABCDE** (**5R**) model. In these basins, the **5R** model of A131 has been built, by exploiting all the information gained in the conformational study performed on the **3R** models. The four wide basins, characterized by quite low energy regions (less than 2.5 Kcal/mol), separated by high energy barriers (from 6 to 12 Kcal/mol), contain the eight symmetry related conformers, whose central core structures are schematically represented in Fig. 13.8.

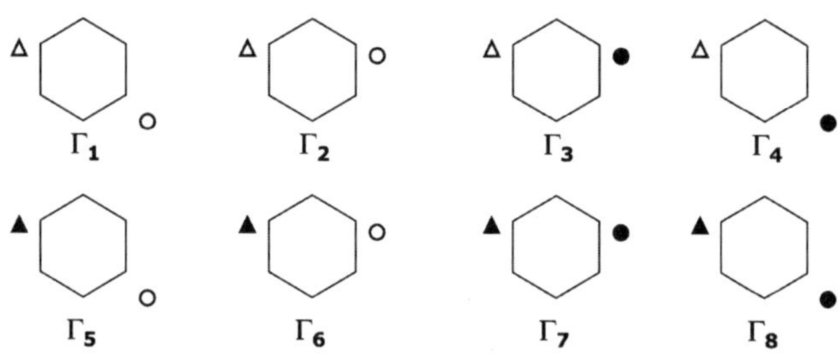

Fig. 13.8. Schematic representation of the eight symmetry related conformations discussed in the text, found for the **ABCDE** model of A131 mesogen. The hexagon stands for the central ring **C**, while the triangles and the circles stand for the iminic (>C=N-) and carboxylic (COO) sp^2 carbons, respectively. Moreover, the open (△ or ○) and filled (▲ or ●) symbols represent above and below positions of the two sp^2 carbons with respect to the aromatic plane of ring **C**, respectively.

The obtained $\mathbf{\Gamma}_i$ conformers were subsequently optimized without restraints, at the same level of theory used for the **3R** ones. These fully relaxed and optimized geometries lead to structural parameters very similar to those found for the minimum-energy conformers of **BCD** model, confirming that (i) the last configuration of **5R** model is good for interpreting the conformational behavior of the full molecule, and (ii) conformations

Table 13.2. Structural parameters: dihedral and bend angles (in degrees), relative energies (in Kcal/mol) and populations (in %) for $\mathbf{\Gamma}_i$ conformers of the **ABCDE** 5-rings model.

Conformation	ψ_2 (°)	ψ_4 (°)	ψ_5 (°)	ψ_7 (°)	ϑ (°)	Δ E(Kcal/mol)	P (%)
$\mathbf{\Gamma}_1$	43	44	-165	0	143	0.000	28.5
$\mathbf{\Gamma}_2$	42	45	-18	0	115	1.076	4.7
$\mathbf{\Gamma}_3$	42	45	17	0	116	1.413	2.65
$\mathbf{\Gamma}_4$	43	44	166	0	146	0.417	14.15
$\mathbf{\Gamma}_5$	43	-44	-166	0	146	0.417	14.15
$\mathbf{\Gamma}_6$	42	-45	-17	0	116	1.413	2.65
$\mathbf{\Gamma}_7$	42	-45	18	0	115	1.076	4.7
$\mathbf{\Gamma}_8$	43	-44	165	0	143	0.000	28.5

of the **A** and **E** rings have only small influences on the main conformational states of the central core. The results (on energy and population) of this analysis are summarized in Table 13.2, where only the dominant (symmetry related to ψ_4 and ψ_5) conformers of **5R** model are reported.

As far as the ψ_2 and ψ_7 dihedrals are concerned, for each $\mathbf{\Gamma_i}$ found there are at least 8 other (four principal states for ψ_2 x two principal states for ψ_7) populated conformations, not reported here because of symmetry reasons. The structural parameters found for ψ_2 and ψ_7 in the $\mathbf{\Gamma_i}$ conformers are in agreement with those found for the most populated conformers of PES(**ABC**) and PES(**CDE**), in particular the values of $\psi_2 \approx 45°$ and $\psi_7 = 0°$ represent conformations of the local lowest energy for these **3R** models. The final full-optimized conformers $\mathbf{\Gamma_1}$, $\mathbf{\Gamma_2}$, $\mathbf{\Gamma_3}$ and $\mathbf{\Gamma_4}$ have been adopted to calculate the chemical shielding tensors at GIAO-DFT level. The molecular structures of $\mathbf{\Gamma_1}$ and of $\mathbf{\Gamma_3}$ conformers are shown in Fig. 13.9. It has to be noticed that in vacuo, $\mathbf{\Gamma_1}$ and $\mathbf{\Gamma_4}$ (as well as the symmetry related $\mathbf{\Gamma_5}$ and $\mathbf{\Gamma_8}$) conformers are more predominant compared to the others.

13.4.4.1. *2D SUPER vs DFT chemical shielding anisotropy tensors*

The principal components of the CSA tensors for some carbon sites have been measured by SUPER.[6] In the preliminary work, the CSA tensors were unfortunately reported without taking into account the necessary scaling of the isotropic chemical shifts[19] in the SUPER experiment. These are now corrected and reported in Tables 13.3 and 13.4. The CSA tensors found by in vacuo-DFT calculations for all the four $\mathbf{\Gamma_i}$ conformers are also listed in these tables for direct comparison.

Table 13.3. Experimental (corrected 2D SUPER values from Ref. [6] unless specified otherwise), and calculated (GIAO-DFT) chemical shift tensors for $\mathbf{\Gamma}_i$ conformers of aromatic carbon sites of A131.

		δ^{Iso}	δ_{33}	δ_{22}	δ_{11}			δ^{Iso}	δ_{33}	δ_{22}	δ_{11}			δ^{Iso}	δ_{33}	δ_{22}	δ_{11}
	exp	149.4	–	–	–		exp	122.4	211.1	145.9	11.7		exp	123.6	203.5	145.0	25.1
	Γ_1	145.7	249.4	180.7	7.0		Γ_1	120.5	211	136.9	13.6		Γ_1	121.9	216.1	131.9	17.7
C1	Γ_2	146.2	249.6	181.2	7.9	**C6**	Γ_2	120.3	210.9	136.7	13.4	**C11**	Γ_2	121.5	215.4	131.7	17.5
	Γ_3	145.8	249.5	181.0	7.0		Γ_3	121.3	210.5	139.8	13.6		Γ_3	121.3	214.8	131.3	17.7
	Γ_4	146.2	249.6	181.2	7.7		Γ_4	121.2	210.7	139.7	13.1		Γ_4	121.4	215.6	130.7	18.0
	exp	129.1	224.9	137.9	28.5		exp	130.5	213.6	170.9	17.2		exp	126.5	–	–	–
	Γ_1	128.5	232.7	135.6	17.3		Γ_1	131.1	233.8	150.7	8.8		Γ_1	125	232.9	136.0	6.2
C2	Γ_2	128.6	232.5	135.5	17.8	**C7**	Γ_2	130.7	232.8	150.6	8.6	**C12**	Γ_2	125.2	233.3	135.4	7.0
	Γ_3	128.3	232.4	135.1	17.3		Γ_3	130.8	233.0	150.5	8.8		Γ_3	125.4	232.8	137.3	6.2
	Γ_4	128.4	232.4	135.5	17.5		Γ_4	130.4	232.7	150.4	7.9		Γ_4	125.5	232.6	136.9	7.1
	exp	130.5	213.6	170.9	17.2		exp	129.1	219.3	147.4	14.6		exp	129.1	224.9	137.9	28.5
	Γ_1	132.7	237.6	154.2	6.2		Γ_1	132.4	219.0	167.7	10.4		Γ_1	131.2	235.4	148.9	9.1
C3	Γ_2	132.0	237.8	153.5	4.8	**C8**†	Γ_2	132.7	219.7	167.4	11.1	**C13**	Γ_2	132.1	236.1	150.9	9.4
	Γ_3	132.2	237.0	153.4	6.2		Γ_3	132.8	219.9	168.1	10.4		Γ_3	130.8	239.3	143.9	9.1
	Γ_4	131.5	237.3	153.1	4.2		Γ_4	132.8	219.9	167.4	11.1		Γ_4	129.8	239.6	144.6	5.2
	exp	126.5	217.8	135.3	30.4		exp	159.0	245	163	70		exp	134.5	214.6	160.6	30.2
	Γ_1	126.3	216.4	140.6	22.0		Γ_1	158.7	237.5	161.8	76.7		Γ_1	128.7	216.6	140.3	29.1
C4	Γ_2	126.5	217.1	141.0	21.4	**C9**%	Γ_2	159.3	237.9	162.4	77.5	**C14**	Γ_2	127.4	216.0	138.2	27.9
	Γ_3	126.3	216.2	140.7	22.0		Γ_3	158.9	238.2	161.9	76.7		Γ_3	130.4	218.4	143.6	29.1
	Γ_4	126.7	217.1	141.0	21.9		Γ_4	159.1	238.1	162.3	77.0		Γ_4	132.1	220.1	143.0	33.3
	exp	154.1	243.5	150.5	66.5		exp	152.3	241	173	35.5		exp	134.5	–	–	–
	Γ_1	153.9	249.4	141.5	70.7		Γ_1	153.3	235.7	178.6	45.4		Γ_1	142.4	222.2	190.0	15.0
C5	Γ_2	154.8	250.0	142.6	71.7	**C10**#	Γ_2	153.4	235.7	178.5	45.9	**C15**	Γ_2	143.3	223.4	190.6	16.1
	Γ_3	154.0	249.7	141.6	70.7		Γ_3	152.9	235.6	177.8	45.4		Γ_3	137.7	214.2	184.0	15.0
	Γ_4	154.7	250.0	142.5	71.6		Γ_4	153.3	236.4	177.3	46.1		Γ_4	137.6	212.9	184.1	15.8

† Ref. [46]; % Ref. [47]; # Ref. [48].

Table 13.4. Experimental (corrected 2D SUPER values from Ref. [6] unless specified otherwise), and calculated (GIAO-DFT) chemical shift tensors for $\mathbf{\Gamma}_i$ conformers of the remaining aromatic carbon sites of A131.

		δ^{Iso}	δ_{33}	δ_{22}	δ_{11}			δ^{Iso}	δ_{33}	δ_{22}	δ_{11}
	exp	152.3	244.6	149.0	71.0		exp	123.6	203.5	145.0	25.1
	Γ_1	155.1	251.7	140.8	72.7		Γ_1	124.1	219.0	146.3	7.1
C17	Γ_2	155.4	252.9	139.7	73.7	**C22**	Γ_2	124.1	219	146.2	7.2
	Γ_3	154.1	251.4	138.1	72.7		Γ_3	124.2	218.8	146.7	7.1
	Γ_4	155.1	251.7	138.5	74.9		Γ_4	124.1	218.8	145.8	7.6
	exp	122.4	211.1	145.9	11.7		exp	130.5	191.4	159.9	47.4
	Γ_1	121.3	211.4	138.4	14.0		Γ_1	129.2	232.6	133.9	21.0
C18	Γ_2	121.8	212.2	137.9	15.3	**C23**	Γ_2	131.9	235.5	137.5	22.8
	Γ_3	121.1	211.4	137.9	14.0		Γ_3	129.4	233.0	134.1	21.0
	Γ_4	121.4	210.9	138.5	15.0		Γ_4	129.1	232.8	133.7	20.8
	exp	124.5	210.2	168.2	-2.7		exp	146.3	–	–	–
	Γ_1	124.5	218.2	148.2	7.0		Γ_1	142.8	247.6	172.3	8.3
C19	Γ_2	123.3	217.3	147.0	5.6	**C24**	Γ_2	142.8	247.5	171.6	9.2
	Γ_3	124.4	217.7	148.7	7.0		Γ_3	142.8	247.7	172.4	8.3
	Γ_4	124.4	218.3	147.7	7.1		Γ_4	142.7	247.6	172.2	8.3
	exp	151.2	236.4	158.4	57.9		exp	164.9	267.8	117.1	106.6
	Γ_1	150.2	232.1	174.7	43.8		Γ_1	163.7	263.3	123.5	104.2
C20	Γ_2	150.3	232.4	174.2	44.5	**C25**	Γ_2	164.4	264.3	123.7	105.3
	Γ_3	151.0	232.8	176.3	43.8		Γ_3	163.5	264.8	121.5	104.2
	Γ_4	149.6	232.3	173.2	43.2		Γ_4	164.9	264.4	124.8	105.3
	exp	151.2	239.4	164.5	43.0		exp	164.9	270.0	112.6	102.1
	Γ_1	151.1	234.1	176.9	42.4		Γ_1	164.0	265.5	117.1	109.3
C21	Γ_2	150.7	234.0	177.1	41.2	**C26**	Γ_2	163.5	264.4	117.5	108.5
	Γ_3	151.6	234.2	178.2	42.4		Γ_3	167.3	270.7	121.8	109.3
	Γ_4	151.2	233.6	176.5	43.6		Γ_4	166.4	269.3	124.0	105.9

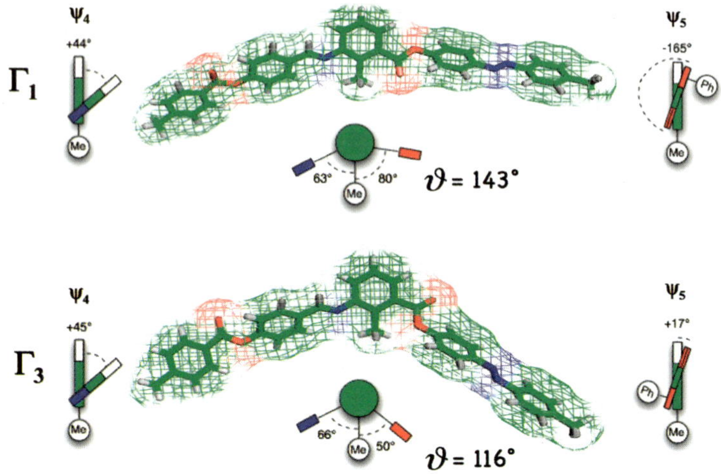

Fig. 13.9. Molecular structures and principal geometrical parameters of $\mathbf{\Gamma_1}$ and $\mathbf{\Gamma_3}$ conformers of A131 molecule. Moreover, the lateral side views are shown on the left and right sides of the molecular models.

Figure 13.10 shows the observed dependences of theoretical chemical shift parameters δ_{ii} (isotropic part and components) on the four relevant conformations $\mathbf{\Gamma_1}$, $\mathbf{\Gamma_2}$, $\mathbf{\Gamma_3}$ and $\mathbf{\Gamma_4}$ of A131. The Root Mean Squared Deviation evaluated among the $\mathbf{\Gamma_i}$ conformers is ca. ± 1 ppm for all the conformations, showing that there are no significant dependences of the CSA tensors on the minimum-energy conformational states. In particular, the differences are mainly observed for the carbons (C11, C12, C13, C14, C15, C17 and C26) belonging to the central ring **C** and close to it. These findings could be reasonably ascribed to the fact that the chemical shielding interaction is a local property, mostly dependent on the nature of chemical bonds with which the nucleus is involved, and only slightly influenced by nuclei placed at greater distances. Nevertheless, when the ψ_4 and ψ_5 dihedrals are scanned among the different conformers, the local electron densities of the chemical bonds involved in changes in rotations, indirectly influencing the local electron density of the central ring **C**, while almost not affecting the lateral rings. All this considered, the values obtained for the principal components of the calculated $\boldsymbol{\delta}$ tensors, in each $\mathbf{\Gamma_i}$ conformer, are in fairly good agreement with those reported in the literature for similar compounds.[22,49] A comparison of the DFT CST tensor's orientation of the 4 prevalent conformers with the nominal orientation geometry for various

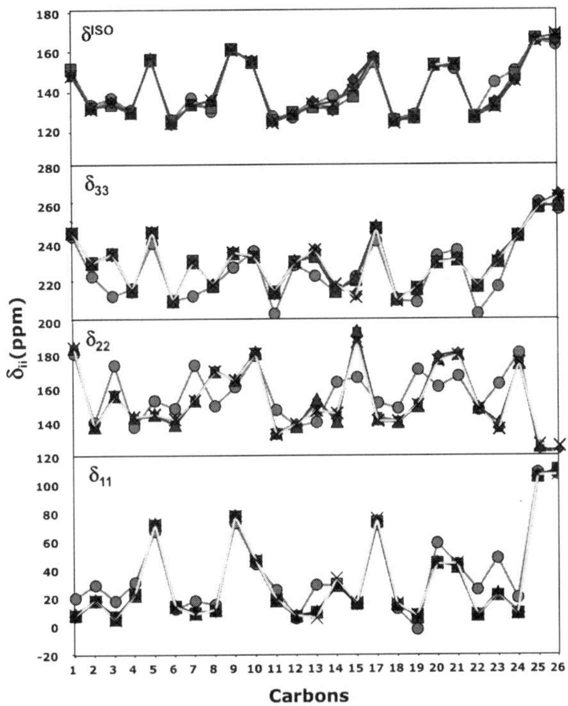

Fig. 13.10. Observed dependences of experimental (measured in $CDCl_3$ (asterisks) and in Solid State by SUPER (circles) at 25°C) and theoretical (calculated for the four relevant conformations $\boldsymbol{\Gamma_1}$ (squares), $\boldsymbol{\Gamma_2}$ (trangles), $\boldsymbol{\Gamma_3}$ (diamonds) and $\boldsymbol{\Gamma_4}$ (crosses) of A131) chemical shift parameters δ_{ii} (isotropic part and components).

carbon sites is summarized in Table 13.5. Note that the orientation of each carbon CST is rather insensitive to which particular conformer the A131 molecule has adopted.

Since the topology of each $\boldsymbol{\delta}$ tensor in the molecule can be studied by examining the relative magnitude of the CSA components in each $\boldsymbol{\delta}$ tensor, different topological situations experienced by different ^{13}C nuclei belonging to the same functional group (i.e. tertiary and quaternary aromatic carbons, carbonyl carbons) can be distinguished. By comparing the values reported in the table of Ref. [47] and in Tables 13.3 and 13.4, it is possible to assign each ^{13}C resonance in the ^{13}C NMR static spectrum of A131 and to map the electron density of the molecule, because the CST are sensible to this local property. In particular, the carbonyl carbons C25 and C26 have CSA tensors characterized by $\delta_{22} \approx \delta_{11} \ll \delta_{33}$[46] and show the highest

Table 13.5. β_P angles (degrees) between the z-axis of the chemical shift principal axes frame and the *para*-axis of the relevant aromatic fragments (see Fig. 13.1(a)) among the different cases: "regular" geometry and $\mathbf{\Gamma}_i$ conformers of **R5** model.

Carbons	reg. geom.	$\mathbf{\Gamma_1}$	$\mathbf{\Gamma_2}$	$\mathbf{\Gamma_3}$	$\mathbf{\Gamma_4}$	Carbons	reg. geom.	$\mathbf{\Gamma_1}$	$\mathbf{\Gamma_2}$	$\mathbf{\Gamma_3}$	$\mathbf{\Gamma_4}$
C1	180.0	180.0	180.0	180.0	180.0	**C14**	0.0	0.0	0.0	0.0	0.0
C2	120.0	119.6	119.8	119.5	119.3	**C15**	-60.0	-59.8	-59.6	-60.4	-60.7
C3	60.0	59.2	59.4	59.3	59.6	**C17**	180.0	180.0	180.0	180.0	180.0
C4	0.0	0.0	0.0	0.0	0.0	**C18**	120.0	119.8	119.6	120.5	120.3
C5	180.0	180.0	180.0	180.0	180.0	**C19**	60.0	59.6	59.2	60.3	60.4
C6	120.0	120.3	120.8	120.5	120.6	**C20**	0.0	0.0	0.0	0.0	0.0
C7	60.0	60.2	60.4	60.5	60.3	**C21**	180.0	180.0	180.0	180.0	180.0
C8	0.0	0.0	0.0	0.0	0.0	**C22**	120.0	120.2	120.1	119.4	119.8
C9	30.0	34.25	34.71	33.67	33.89	**C23**	60.0	60.4	60.3	60.1	60.6
C10	-120.0	-118.8	-118.7	-120.2	-120.4	**C24**	0.0	0.0	0.0	0.0	0.0
C11	180.0	180.0	180.0	180.0	180.0	**C25**	-30.0	-33.7	-33.4	-34.2	-34.5
C12	120.0	119.3	119.2	119.8	119.5	**C26**	30.0	34.3	34.7	32.6	32.9
C13	60.0	60.3	60.5	59.2	59.4	—	—	—	—	—	—

isotropic shift in the molecule. The carbon C9 of the imine group (>C=N-) shows a high isotropic shift and out-of-plane δ_{11} component,[47] similar to quaternary carbons C5 and C17, that are connected to an electron-donating group (C-O), and is easily distinguishable from the other quaternary aromatic carbons.[46] The quaternary carbons C10, C20 and C21, connected to an electron-donating group (=N-), are similar to C5 and C17, but they have lower electron density along the bond (where δ_{33} lies) and higher along the direction perpendicular to it (where δ_{22} is measured). Thus, δ_{33}(=N-) $< \delta_{33}$(C-O) and δ_{22}(=N-) $> \delta_{22}$(C-O), making them easily distinguishable from each other.[46] On the contrary, the quaternary carbons C4, C8, and C14, connected to an electron-withdrawing group (-COO or C=N-), exhibit very small δ_{11} components.[47]

The protonated aromatic carbons C2, C3, C6, C7, C11, C12, C13, C18, C19, C22 and C23 show δ_{33} and δ_{11} values which are smaller than those found for quaternary carbons connected to an electron-donating groups, such as C5, C9 and C17 (see Fig. 13.10). When comparing experimental (measured by 2D SUPER technique) and calculated (by ab initio methodologies) CSA tensors, the goodness of this direct comparison depends, on the one hand, the complexity of the investigated system, the resolution gained in the 2D spectra and the accuracy of the simulations needed to extract the CSA principal values from the 1D slices, while on the other hand, it is strongly affected by the accuracy in the QM description (basis set, and functional in the case of DFT, solvent effect if included) of the molecule under study.

In this work, the comparison between experimental and theoretical CSA components and isotropic shift, shown in Fig. 13.10, is quite satisfactory. Note that the accuracy of CSA tensors evaluated from 2D SUPER powder patterns is ca. ± 3 ppm for each component. In the comparison, the Root Mean Squared Error (RMSE) are about 3, 6, 5 and 4 ppm for δ^{iso}, δ_{11}, δ_{22} and δ_{33}, respectively, if we exclude carbons C3, C7, C11, C22 and C23,[¶] which show fairly noisy SUPER powder slices reported in Ref. [6]. If the comparison is extended to all the aromatic carbons of A131, the evaluated RMSE's for δ^{iso}, δ_{11}, δ_{22} and δ_{33} are 3, 10, 14 and 13, respectively, indicating that the experimental CSAs evaluated for carbons C3, C7, C11, C22 and C23 must be taken with some caution regarding their observed chemical shift anisotropies. Moreover, the relative values of CSA components

[¶]These carbons show very noisy 1D powder pattern slices, partially due to the overcrowding ^{13}C signals (derived from the complexity of the system under study), but mostly due to the small quantity of the available A131 sample.

found for these carbons have the largest deviations from those commonly found in carbons having the same topology and chemical nature.[47]

Nevertheless, one should consider that these results originate from two different electronic and structural situations: the 2D SUPER experiments being carried out in the solid phase, while the DFT calculations being performed in vacuo, without taking into account the effect of molecular packing in the mesophase. In general, the effect of the medium plays an important role in condensed phases; either structural modifications (*indirect effect*) or, for a given structure, electron density distribution modifications (*direct effect*) can induce significant changes in both the strength (principal components) and orientation of the chemical shielding tensor. Nonetheless, these effects have not been taken into account in this DFT work, because carbon-13 is only slightly affected by these effects. But this could probably represent an additional source of deviation between the calculated and experimental CSA tensors. All this considered, the assumption on the vacuo medium for carbon nuclei is likely satisfactory because of its topology and chemical nature. In fact, in organic molecules carbon atoms are always involved in the formation of inner framework of the core structure, so that they are less influenced by the solvent than other nuclei, such as ^{1}H and ^{19}F.[50] Moreover the 2D SUPER - ab initio combined approach strategy used in this work was found to be very reliable and to give satisfactory results in a similar study of a ferroelectric mesogen.[22] In conclusion, both experimental and theoretical values of the principal components and of isotropic shift for each $\boldsymbol{\delta}$ tensor are in agreement with those reported in the literature both for generic aromatic systems[47,51,52] and for similar compounds.[22,49]

Considering A131 molecule and its symmetry properties, it is obvious that CST of ^{13}C involved in the aromatic rings possesses a principal axis (y-axis, where δ_{11} is placed) oriented perpendicularly to the ring plane, with the two other axes (x and z) being in the plane (see Fig. 13.1). As far as the orientation of ^{13}C CSTs is concerned, the orientation of each tensor is only given by the calculated β_P angle between the CH (or CC) direction and the principal z axis of each fragment. All the tensor orientations (see Table 13.5), calculated for the different conformers, show only small variations (less than two degrees) from the "regular" orientation, often assumed and used in solid state NMR experiments. The "regular" geometry assumes that the transformation from the fragment frame (x',y',z') to PAS frame (δ_{33}, δ_{22}, δ_{11}) involves $\beta_P = 60°, 120°$ for protonated carbons, and $\beta_P = 0°$ for quaternary carbons in a phenyl ring (e.g. ring **A** in Fig. 13.1). As the deviations of the β_P angles from their nominal values for different CSA tensors are normally very small, and as shown before the use of DFT geometry does not have much of an impact on the determination of local order information, the following analyses of ^{13}C $\langle\delta\rangle$ are done using the standard geometry given in Table 13.5.

13.5. ^{13}C NMR Data Treatment

The ^{13}C NMR chemical shifts[6] in the N_U and N_B phases are analyzed with some ^{13}C peaks reassigned for correctness in order to get the proper local order parameters for various aromatic rings. The local ordering information can hopefully provide some insights on the bent-shaped molecular core in the two nematic phases of A131.

13.5.1. *Orientational Order Parameters*

The observed chemical shifts in the isotropic, N_U, N_B and SmC phases are reproduced in Fig. 13.11. To fit the observed $\langle\delta\rangle$'s in the N_U and N_B phase, the CSA values listed for A131 in Tables 13.3 and 13.4 are tested for all possible sets of CSA tensors. No substantial differences were found when calculating order parameters using "regular" or calculated geometry for the orientations of the shielding tensors, as we have previously found in the case of ZLL 7/* mesogen.[22] For this reason, we have decided to not further investigate the dependence of ordering from the CST orientations among the different conformers of A131. The ^{13}C $\langle\delta\rangle$ in mesophases is related to its isotropic δ_{iso} value by Eq. (12.27) in Chap. 12 for all aromatic carbon sites except those on the central ring. For the latter carbons, Eq. (13.7) is used instead. It was found that S_{xz} for the central ring is vanishingly small (ca. 0.1) for all the studied temperatures in the two nematic phases. When the derived Saupe matrix for the central ring is diagonalized to give the local S and D, the principal axes of its PAS can also be located. A rotation of the y axis of the $(x, y.z)$ molecular frame is needed for such a diagonalization at each temperature. The new principal z axis only shifts its orientation towards the C12-C26 bond by less than two degrees on average. This small change in the direction of the z axis is a result of the small S_{xz} value (Incidentally, the derived S and D for ring C using the Eq. (12.27) in this case produce the same results).

The fits are quite similar for cases in which DFT tensors of a particular conformer were used. The case for the $\mathbf{\Gamma_3}$ conformer is highlighted here as an example (Note that calculated chemical shift tensors and their tensor

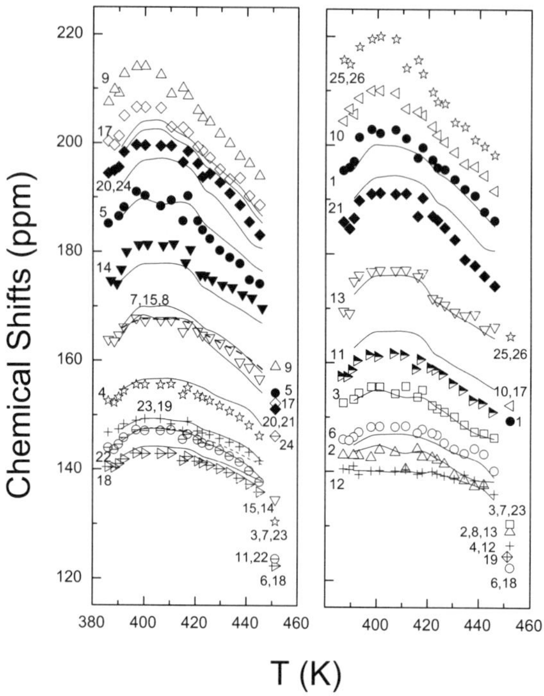

Fig. 13.11. Plots of chemical shifts for the aromatic carbons and linkage carbons of A131 in the isotropic and nematic phases. The carbon labels are those indicated in Fig. 13.1. Solid curves denote 'best' calculated chemical shifts after a global minimization (see text: *case 3*).

orientations have also been used to derive local orientational order in the aromatic rings of A131). It is believed that conformer $\mathbf{\Gamma_1}$ is supposed to be likely closer to the real molecular shape of A131 in the N_U phase, while $\mathbf{\Gamma_3}$ has a more bent structure, most likely to be suitable in the N_B phase. In simplifying our discussion, we first present two different scenarios in order to see the effects of these CSA tensors from $\mathbf{\Gamma_3}$ in vacuo (*case 1*), and from a solid (*case 2*) sample (note that when SUPER CSA is unavailable for C1, C12, C15 and C24, the corresponding CSA of $\mathbf{\Gamma_3}$ were used, while for C8, the CSA of Ref. [46] was used). The goodness of fits to all 13 chosen temperatures in the mesophases is measured by the Sum Squared Deviation (SSD). The SSD evaluated for *case 1* by fitting 21 carbon sites (C1-C8, C10-C15, and C17-C24) is 1258 (this corresponds to a RMSD value of about 2.2 ppm for each carbons on the average at each temperature), and the corresponding derived local order parameters S and D are plotted as a function of temperature in the left panel of Fig. 13.12. It is noted that the D's for all rings (**A-E**) are negative in the N_B phase, while D of ring **A** seems to go slightly positive as the temperature is increased closer to the isotropic phase. The *case 2* scenario uses mostly the available CSA tensors (see above) obtained experimentally by SUPER in fitting the same set of ^{13}C $\langle\delta\rangle$'s, and the fits are indeed improved somewhat to give a lower SSD of 1136 (which corresponds to a RMSD value of ca. 2.0 ppm). The improvement noticably comes from fitting of several carbons (e.g. C14). Hence it would be interesting to compare the S and D parameters of this scenario (*case 2*), which are summarized in the right panel of Fig. 13.12, with those of the former (*case 1*).

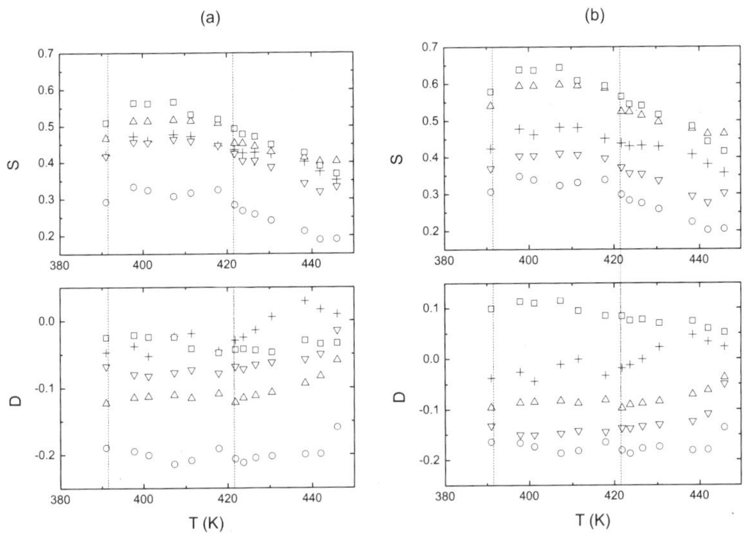

Fig. 13.12. Plots of local order parameters S and biaxial order parameters D of rings **A** (+), **B** (◯), **C** (△), **D** (□), and **E** (▽) for left panel: using entirely $\mathbf{\Gamma_3}$ CSA tensors (see text: *case 1*) and for right panel using a combination of SUPER and 4 CSA tensors calculated for $\mathbf{\Gamma_3}$ conformer (see text: *case 2*).

All S values are now slightly elevated in *case 2*, except that of ring **E**. The local S of ring **D** remains the largest among all aromatic rings as noted previously[6] for almost the entire temperature range. What is more interesting is to examine the behaviors of D in the N_B phase, particularly for ring **D**. This is because the D parameter reflects the relative degree of ordering of the two minor fragment axes x and y with respect to the major director. Note that ring **B** has the largest absolute D value, i.e. largest local biaxiality (or S_{yy} tends to zero) and is relatively insensitive to the temperature. This is accompanied by a smallest S for ring **B** among all the rings. The observation for ring **B** can be explained by taking into account two distinct observations: (i) the energy barriers for rotation around the dihedral angles ψ_2 and ψ_3 are very low, as found in the DFT calculations; and (ii) the orientation of para axis of ring **B** is highly tilted (it forms an angle of about 60°, close to the magic angle, 54.7°, see Fig. 13.9), because the rotation of ring **B** around ψ_4 is restricted, due to the presence of the methyl (C_{16}) unit in ring **C**. While all D's seem to level off in the N_B phase, the D of ring **D** is now positive (ca. 0.1), quite different from the negative value (ca. -0.03) in the *case 1*. Physically, this implies the ring **D**'s normal (y axis) to its ring plane is twisted in the opposite direction with respect to those of other rings in the N_B phase. Furthermore, D of ring **A** appears to go through zero at the N_U-N_B phase transition. Perhaps these two rings (**A** and **D**) are involved in providing a symmetry breaking at the uniaxial to biaxial phase transformation (see geometry in Fig. 13.9).

It is clear that the derived local order parameters S and D are sensitive to the specific CSA tensors used. To understand this, it would appear that the in vacuo CSA tensors may not be reflecting the slight perturbation due to the existence of packing or medium effects, even if this effect should be considered very small as found in Ref. [22]. On the other hand, as pointed out above some of the CSA tensors derived from noisy powder patterns may indeed not be too reliable. To check this possibility, we have replaced those of C7, C19, C22 and C23 by their corresponding $\boldsymbol{\Gamma_3}$ CSA values (this corresponds to *case 3*) for fitting the same set of ^{13}C $\langle\delta\rangle$'s. Indeed, the SSD of the new fits yields the best value of 1016 (which corresponds to a RMSD of 1.9 ppm), an improvement of ca. 20% when comparing with the SSD of *case 1*. The theoretical $\langle\delta\rangle$'s, determined here for *case 3*, are also plotted in Fig. 13.11. As seen in this figure, the overall fitting of all the aromatic carbon sites C1-C8, C10-C15, and C17-C24 is quite satisfactory, though systematic deviations do exist for some carbon sites (e.g. C14). One additional source for the discrepancy could be due to the use of nominal geometry for the orientations of all CSA tensors in their respective fragment frames, but this has been proven to have negligible variations in the values of order parameters (less than 0.5%), as previously found in Ref. [22] for the ferroelectric mesogen ZLL 7/*. The derived "best" S and D parameters for *case 3* are presented in Fig. 13.13.

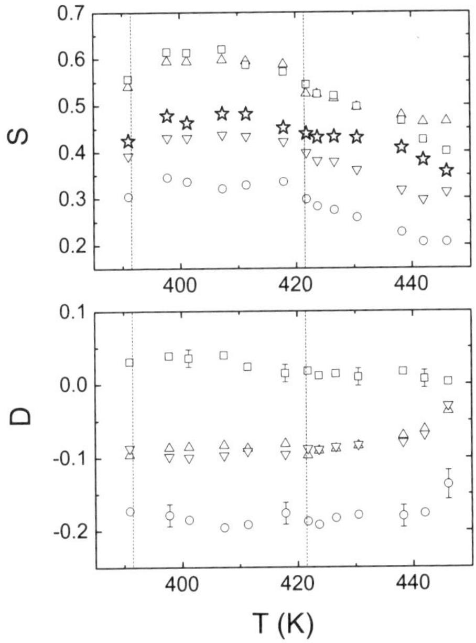

Fig. 13.13. Plots of local order parameters S and biaxial order parameters D of rings **A** ($\star$), **B** ($\bigcirc$), **C** ($\triangle$), **D** ($\square$), and **E** ($\triangledown$) using SUPER CSA tensors and 8 CSA tensors calculated for $\mathbf{\Gamma_3}$ conformer (see text: *case 3*).

The essential features for S and D, however, remain with ring **D** still having positive local D values in the N_B phase (nearly zero in the N_U phase), but slightly lower compared to *case*2. In principle, the best SUPER CSA values from the studied molecule should be used to gain reliable order information. In practice, this may require the augmentation by in vacuo DFT calculations for carbons sites with low signal sensitivity and for those having unavoidable spectral crowding. Furthermore, the present study has provided a reasonable means to obtain sound local S and D parameters in mesophases by means of ^{13}C NMR spectroscopy. Finally, the reliability and powerfulness of the combined ab initio – SS-NMR spectroscopy approach have been demonstrated.

To further shed light on the local orientational behaviors of various rings in A131 bent-core, we have decided to plot $-S_{xx}/S$, a normalized local order (ρ) of the x axis, as a function of temperature in Fig. 13.14. Note that ρ has two extreme limits: (i) when S_{yy} tends to zero, ρ has a limiting value of one corresponding to maximal ordering of the x axis with respect to the director; (ii) when $D = 0$, then $S_{xx} = S_{yy} = -S/2$ and ρ

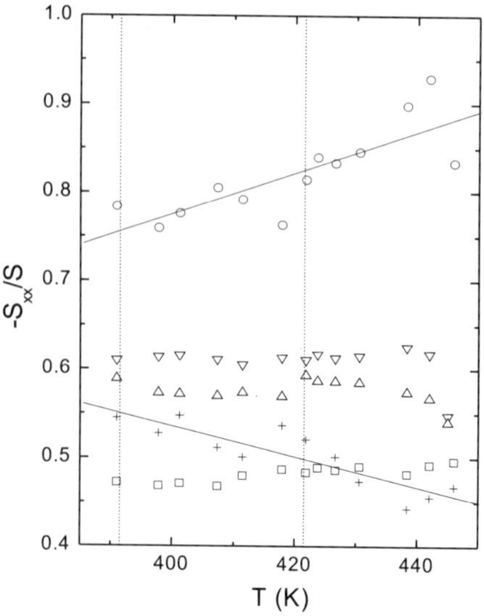

Fig. 13.14. Plots of 'normalized' local order parameters $-S_{xx}/S$ of rings **A** (+), **B** (○), **C** (△), **D** (□), and **E** (▽). Lines are drawn to aid the eyes.

approaches a limiting value of 0.5. Now both ρ of ring **A** and **B** show strong and opposite temperature dependences. Furthermore, the $|S_{yy}|$ is smallest for ring **B** as previously mentioned, making S_{xx} a larger negative value as demonstrated in this figure. The opposite temperature behaviors seen in ρ of ring **A** and **B** seem to indicate changes in the relative populations of the dominant conformers at different temperaures.

In summary, the ring **D** on one of the lateral wings (the longer one) is placed in the most rigid part of the A131 molecule (in fact the rotational barriers around ψ_5 and ψ_7 are very high, see Figs. 13.6 and 13.7), and it has been found to have the highest local order parameter S values, but rather small D values as noted above. The other lateral wing (the shorter one), where both ring **A** and **B** are placed, must play an important role in making the bent-core to be more like a banana-shaped (as observed in the case of $\mathbf{\Gamma_1}$ conformer), or more like a hockey stick-like conformation (as observed in the case of $\mathbf{\Gamma_3}$ conformer), when the temperature decreases from the N_U phase to the N_B phase. It is amazing to notice that the D's of these two rings change relative to each other with respect to the temperature (Fig. 13.13) in such a subtle manner to switch between the two possible conformers. Hence, it must be pointed out that the role of the flexible end chains can not be underplayed in the overall configuration of the V-shaped molecule in general. It is also very interesting to note that ring **B** also has small D values like those of ring **D**, i.e. close to zero in the N_U phase, but slightly negative in the N_B phase. Thus both rings **B** and **D** may arguably be sensing a phase transition at ca. 422 K, in particular in view of the measured biaxial order parameters by Raman scattering in a similar bent-core mesogen.[9]

13.6. Conclusion

The present study has demonstrated the potentiality and reliability of a combined DFT-solid state NMR approach to investigate conformations, their energy landscapes, shielding tensor properties and local ordering of the aromatic rings in the bent-core region of the biaxial nematic A131. DFT calculations of such a molecular structure have been carried out in steps of increasing complexity in terms of **3R** models and **5R** model to give the required potential energy surfaces, where the characteristic minimum-energy structures have been captured. Four prevalent conformations (and four symmetry related ones) showing two different bent-core shapes have been revealed by means of in vacuo DFT calculations. One group of conformers has a bend angle close to 145° and gives more like a 'banana'-shaped conformer. The other group has a smaller bend angle of about 115° and resembles more like a hockey stick-shaped structure.

Experimental (SUPER) and calculated (DFT) chemical shielding tensors have been employed to give reliable local orientational order parameters for the five distinct aromatic rings of the A131 molecule. The information obtained from the conformational analyses (in terms of structural parameters and local rotational barriers) have been combined with those derived from orientational order properties (specially in terms of relative orientations of the different rings), in order to shed further light on the possible shapes adopted by A131 in its N_U and N_B phases. When adding the flexible chains to the aromatic bent-core and viewing along their "long" axes, the former appears like a uniaxial rod (see Fig. 13.15(a)), while the latter more like a biaxial object (see Fig. 13.15(b)). These molecular shapes are inferred from subtle changes in the orientations of ring **A**, **B**, and **D** in these two nematic phases derived from a careful analysis of their ^{13}C chemical shifts $\langle\delta\rangle$'s. Specifically, one makes use of the temperature behaviors of the fragment biaxiality D's and the normalized local order parameters ρ $(-S_{xx}/S)$ of these rings.

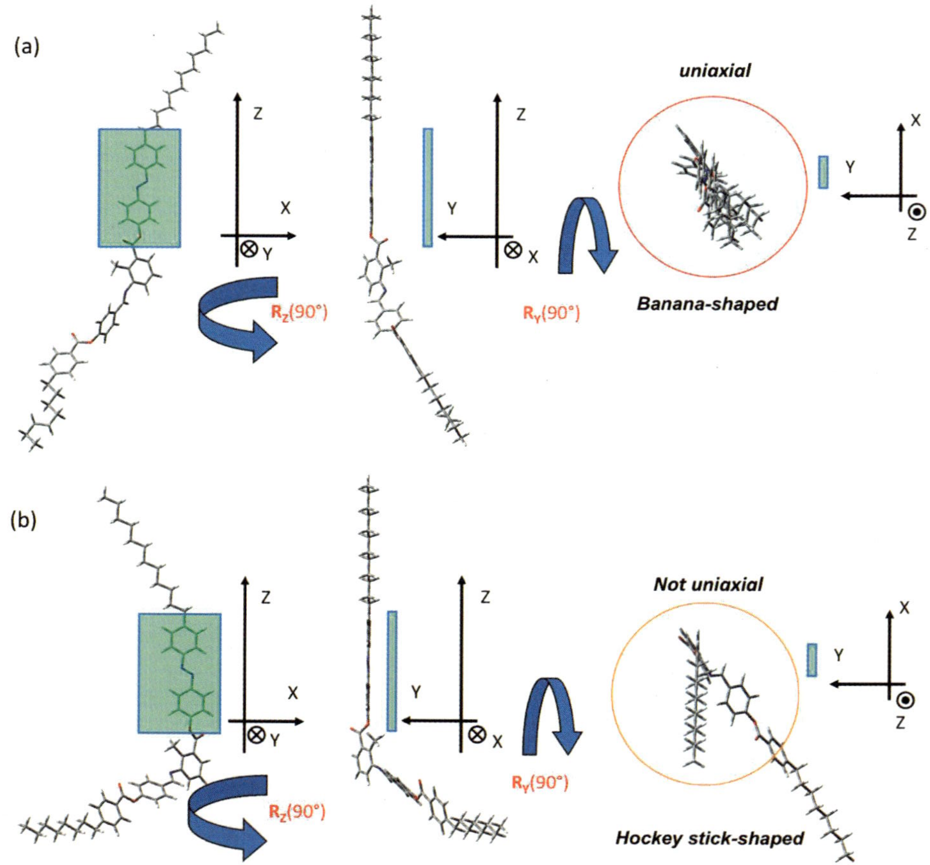

Fig. 13.15. Schematics of A131 molecule (a) uniaxial 'banana-shaped' conformation, (b) non-uniaxial 'hockey stick'-shaped conformation.

We conclude by postulating that the conformational changes among the aromatic rings of A131 do provide the required symmetry breaking at the uniaxial and biaxial phase transition as the sample is cooled from the high temperature N_U phase into the N_B phase. The slight drop in S's, when entering in the SmC phase, is a signature of a tilted layer structure. In this chapter we have also tested the sensitiveness of the derived local order properties on the nature of CSA tensors used in the analyses. Further 2D SUPER measurements could be performed in the different phases of A131 at several temperatures to obtain the experimental phase-dependent CSTs, as frequently done in the literature for discotic mesogens. On the other hand, the medium effects on shielding tensors may also need some further theoretical considerations, especially in the case when the dielectric tensor strength would be strongly affected by the different structures of the mesophases. Finally, further studies on other biaxial nematogens are needed to further test the postulates on conformational changes within the core region of V-shaped molecules in the presence of a N_U-N_B transition.

Acknowledgment

The financial support of Natural Sciences and Engineering Council of Canada is acknowledged. A.M. is particularly grateful to Scuola Normale Superiore of Pisa for financial support (Young Researcher 2007 Project Grant).

References

1. T. Niori, F. Sekine, J. Watanabe, T. Furukawa, and H. Takezoe, J. Mater. Chem. **6**, 1231 (1996).
2. M. J. Freiser, Phys. Rev. Lett. **24**, 1041 (1970).
3. G. R. Luckhurst, Thin Solid Films **393**, 40 (2001).
4. A. Eremin, I. Wirth, S. Diele, G. Pelzl, H. Schmalfuss, H. Kresse, H. Nadasi, K. Fodor-Csorba, and E. W. Gacs-Baitz, Liq. Cryst. **29**, 775 (2002).
5. L. A. Madsen, T. J. Dingemans, M. Nakata, and E. T. Samulski, Phys. Rev. Lett. **92**, 145505 (2004).
6. R. Y. Dong, S. Kumar, V. Prasad, and J. Zhang, Chem. Phys. Lett. **448**, 54 (2007). This preliminary report unfortunately misses the scaling of 0.667 in the chemical shift anisotropy tensors.
7. B. R. Acharya, A. Primak, and S. Kumar, Phys. Rev. Lett. **92**, 145506 (2004).
8. V. Prasad, S.-W. Kang, K. A. Suresh, L. Joshi, Q. Wang, and S. Kumar, J. Am. Chem. Soc. **127**, 17224 (2005).
9. C. D. Southern, P. D. Brimicombe, S. D. Siemianowski, S. Jaradat, N.

Roberts, V. Gortz, J. W. Goodby, and H. F. Gleeson, E. P. L. **82**, 56001 (2008).

10. P. I. C. Teixeira, A. J. Masters, B. M. Mulder, Mol. Cryst. Liq. Cryst. **323**, 167 (1998).
11. D. A. Dunmur, and K. Toriyama, in *Physical Properties of Liquid Crystals*, edited by D. Demus, J. Goodby, G. W. Gray, H. W. Spiess, and V. Vill, Wiley-VHC, Weinheim, 1999.
12. K. Merkel, A. Kocot, J. K. Vij, R. Korlacki, G. H. Mehl, and T. Meyer, Phys. Rev. Lett. **93**, 237801 (2004).
13. R. Y. Dong, *NMR of Liquid Crystals*, Springer-Verlag, New York, 1994.
14. E. E. Burnell, C. de Lange, eds., *NMR of Orientationally ordered Liquids*, Kluwer Academic, Dordrecht, 2003.
15. L. J. Yu, and A.Saupe, Phys. Rev. Lett. **45**, 1000 (1980).
16. M.A. Bates, Phys. Rev. E **74**, 061702 (2006).
17. R. Y. Dong, J. Xu, G. Benyei, K. Fodor-Csorba, Phys. Rev. E **70**, 011708 (2004).
18. J. Xu, K. Fodor-Csorba, R. Y. Dong, J. Phys. Chem. A **109**, 1998 (2005).
19. S. F. Liu, J. D. Mao, K. Schmidt-Rohr, J. Magn. Reson. **155**, 15 (2002).
20. R. Y. Dong, J. Zhang, and K. Fodor-Csorba, Chem. Phys. Lett. **417**, 475 (2006). The chemical shift anisotropy tensors reported herein unfortunately contain errors in scaling.
21. K. Fodor-Csorba, A. Jakli, A. Vajda, S. K. Prasad, D. S. S. Rao, R. Y. Dong, J. Xu, and G. Galli, ChemPhysChem **7**, 2184 (2006).
22. R. Y. Dong, M. Geppi, A. Marini, V. Hamplova, M. Kaspar, C. A. Veracini, and J. Zhang, J. Phys. Chem. B **111**, 9787 (2007).
23. M. Kaupp, M. Buhl, V. G. Malkin, *Calculation of NMR and EPR Parameters*, Wiley VCH, Weinheim, Germany, 2004.
24. R. Ditchfield, J. Chem. Phys. **56**, 5688 (1972).
25. J. R. Cheeseman, G. W. Trucks, T. A. Keith, M. J. Frisch, J. Chem. Phys. **104**, 5497 (1966).
26. J. Mason, *Multinuclear NMR*, Plenum Press, New York, 1987. Ch. 3.
27. R. Y. Dong, J. Xu, J. Zhang, and C. A. Veracini, Phys. Rev. E **72**, 061701 (2005).
28. D. W. Allender, J. W. Doane, Phys. Rev. A **17**, 1177 (1978).
29. Gaussian 03, Revision B.05, M. J. Frisch et al., Gaussian, Inc., Pittsburgh PA, 2003.
30. T. Imase, S. Kawauchi, J. Watanabe, J. Mol. Struc. **560**, 275 (2001).
31. R. Wrzalik, K. Merkel, A. Kocot, J. Mol. Model **9**, 248 (2003).
32. I. Cacelli, G. Prampolini, Chem. Phys. **314**, 283 (2005).
33. L. Calucci, C. Forte, K. Fodor-Csorba, B. Mennucci, S. Pizzanelli. J. Phys. Chem B **111**, 53 (2007).
34. L. Wang, X. Wang. Theochem **806**, 179 (2007).
35. (a) A. D. Becke, Phys. Rev. A **38**, 3098 (1988). (b) C. Lee, W. Yang, R. G. Parr, Phys. Rev. B **37**, 785 (1988). (c) P. J. Stephens, F. J. Devlin, C. H. Chabalowski, M. J. Frisch, J. Phys. Chem. **98**, 11623 (1994).

36. Y. He, D. H. Wu, L. F. Shen, B. W. Li, G. A. Webb, Magn. Reson. Chem. **33**, 710 (1995).
37. S. M. Cybulski, D. M. Bishop, Chem. Phys. Lett. **98**, 8057 (1993).
38. U. Fleischer, W. Kutzelnigg, C. V. Wullen, in *Encyclopedia of Computational Chemistry*, Eds. P. V. R. Schleyer, N. L. Allinger, T. Clark et al., Wiley, Chichester, 1998. pg. 1827.
39. K. B. Wiberg, J. Comput. Chem. **20**, 1299 (1999).
40. J. P. Perdew, In *Electronic Structure of Solids*, P. Ziesche, and H. Eschig, Eds. Akademie Verlag, Berlin, 1991. p. 11.
41. (a) C. Adamo, V. Barone, J. Chem. Phys. **108**, 664 (1998). (b) B. J. Lynch, Y. Zhao, D. G. Truhlar, J. Phys. Chem. A **107**, 1384 (2003).
42. D. Catalano, M. Geppi, A. Marini, C. A. Veracini, S. Urban, J. Czub, W. Kuczynski, R. Dabrowsky, J. Phys. Chem. C, **111**, 5286 (2007).
43. M. Geppi, A. Marini, C. A. Veracini, S. Urban, J. Czub, W. Kuczynski, R. Dabrowsky, J. Phys. Chem. B, **112**, 9663 (2008).
44. J. Gauss, J. Chem. Phys. **99**, 3629 (1993).
45. N. Kurita, T. Ikegami, Y. Ishikawa, Chem. Phys. Lett. **360**, 349 (2002).
46. G. Zheng, J. Hu, X. Zhang, L. Shen, C. Ye, and G.A. Webb, J. Mol. Struct. **428**, 283 (1998).
47. T.M. Duncan, *Principal Components of Chemical Shielding Tensors: a compilation*, 2nd edition, The Farragut Press, Madison, WI (1997).
48. P. Parhami, and B. M. Fung, J. Am. Chem. Soc. **107**, 7304 (1985).
49. T. Nakai, H. Fujimori, D. Kuwahara, S. Miyajima, J. Phys. Chem. B **103**, 417, (1999).
50. A. Marini, D. Catalano, B. Mennucci, manuscript in preparation.
51. A. C. Kolbert, R. G. Griffin, Chem. Phys. Lett. **166**, 87 (1990).
52. J. Z. Hu, A. M. Orendt, D. W. Alderman, R. J. Pugmire, C. Ye, D. M. Grant, Solid State Nucl. Magn. Reson. **3**, 181 (1998).

Index